The Role of PPARs in Disease

The Role of PPARs in Disease

Editors

Kay-Dietrich Wagner
Nicole Wagner

MDPI • Basel • Beijing • Wuhan • Barcelona • Belgrade • Manchester • Tokyo • Cluj • Tianjin

Editors
Kay-Dietrich Wagner
Institute of Biology Valrose
Université Côte d'Azur, CNRS,
INSERM
Nice
France

Nicole Wagner
Institute of Biology Valrose
Université Côte d'Azur, CNRS,
INSERM
Nice
France

Editorial Office
MDPI
St. Alban-Anlage 66
4052 Basel, Switzerland

This is a reprint of articles from the Special Issue published online in the open access journal *Cells* (ISSN 2073-4409) (available at: www.mdpi.com/journal/cells/special_issues/PPARs_Disease).

For citation purposes, cite each article independently as indicated on the article page online and as indicated below:

LastName, A.A.; LastName, B.B.; LastName, C.C. Article Title. *Journal Name* **Year**, *Volume Number*, Page Range.

ISBN 978-3-0365-3992-8 (Hbk)
ISBN 978-3-0365-3991-1 (PDF)

Contents

Editorial

The Role of PPARs in Disease

Nicole Wagner * and **Kay-Dietrich Wagner**

Université Côte d'Azur, CNRS, INSERM, iBV, 06107 Nice, France; kwagner@unice.fr
* Correspondence: nwagner@unice.fr; Tel.: +33-493-377665

Received: 23 October 2020; Accepted: 25 October 2020; Published: 28 October 2020

Abstract: Peroxisome proliferator-activated receptors (PPARs) are nuclear receptors that function as ligand-activated transcription factors. They exist in three isoforms: PPARα, PPARβ/δ, and PPARγ. For all PPARs, lipids are endogenous ligands, linking them directly to metabolism. PPARs form heterodimers with retinoic X receptors, and upon ligand binding, they modulate the gene expression of downstream target genes, depending on the presence of co-repressors or co-activators. This results in a complex, cell type-specific regulation of proliferation, differentiation, and cell survival. PPARs are linked to metabolic disorders and are interesting pharmaceutical targets. PPARα and PPARγ agonists are already in clinical use for the treatment of hyperlipidemia and type 2 diabetes, respectively. More recently, PPARβ/δ activation came into focus as an interesting novel approach for the treatment of metabolic syndrome and associated cardiovascular diseases; however, this has been limited due to the highly controversial function of PPARβ/δ in cancer. This Special Issue of *Cells* brings together the most recent advances in understanding the various aspects of the action of PPARs, and it provides new insights into our understanding of PPARs, implying also the latest therapeutic perspectives for the utility of PPAR modulation in different disease settings.

Keywords: peroxisome proliferator-activated receptors (PPARs); toxicology; pharmacology; ligands; vascular; proliferation; cellular metabolism; adipogenesis; hypertrophic obesity; insulin-resistance; lipidomics; inflammation; kidney; cancer; tumor angiogenesis; peroxisome proliferator-activated receptor-γ coactivator-1α; non-alcoholic fatty liver disease (NAFLD); non-alcoholic steatohepatitis (NASH); fibrosis; Alzheimer's disease (AD); addiction; alcohol; nicotine; opioids; psychostimulants; animal models; human studies

---

Peroxisome proliferator-activated receptors (PPARs) were identified around three decades ago and are ligand-activated transcription factors of the nuclear hormone receptor superfamily comprising the following three subtypes: PPARα, PPARγ, and PPARβ/δ. In the current nomenclature system, PPARs are grouped as nuclear receptor 1C subfamily (PPARα–NR1C1; PPARβ/δ–NR1C2; PPARγ–NR1C3) [1]. PPARα regulates energy homeostasis, the activation of PPARγ causes insulin sensitization and enhances glucose metabolism, and the activation of PPARβ/δ enhances fatty acid metabolism [2]. Thus, the PPAR family of nuclear receptors plays a major regulatory role in energy homeostasis and metabolic function. The present Special Issue of *Cells* critically analyzes the protective and detrimental effect of PPAR modulation in dyslipidemia, adipocyte differentiation, cancer, kidney disease, cardiovascular disease, neurodegenerative disorders, addiction, non-alcoholic fatty liver disease, and steatohepatitis, and it provides a systematic evaluation of PPAR-mediated toxicology and applied pharmacology.

The group of Valerio Costa describes in this Special Issue the generation of a model of human hypertrophic-like adipocytes, which is directly comparable to normal adipose cells and enables therefore evaluating the pathologic evolution toward a hypertrophic state as it is the case in hypertrophic obesity. Reduced neo-adipogenesis and dysfunctional lipid-overloaded adipocytes are hallmarks of hypertrophic obesity linked to insulin resistance. The authors performed a meticulous morphological and histological characterization of the immortalized hypertrophic-like adipocyte cell line. They further evidenced an unbalance of PPARγ isoforms in patients with hypertrophic obesity characterized by an increased relative amount of dominant negative and canonical transcripts (i.e., a higher PPARγΔ5/cPPARγ ratio). Their elegant model of hypertrophic-like adipocytes (HAs) mimics perfectly this increased PPARγΔ5/cPPARγ ratio as compared to mature adipocytes (MAs). Although cPPARγ expression was only slightly reduced in HAs (vs. MAs), PPARγ target genes (IRS2, ADIPOQ, FABP4, SLC2A4, PLIN1, PLIN2, and MRTFA) were highly reduced, explaining the impaired metabolic activity in hypertrophic adipose tissue. Only LPL was increased in HAs, which is in agreement with an increased expression also in subcutaneous adipose tissue from obese patients. The PPARγ-mediated induction of SLC2A4 (encoding the glucose transporter GLUT4) sets up an insulin sensitivity of adipose tissue, liver, and skeletal muscle. Costa's group further tested for a correlation between the PPARγΔ5/cPPARγ ratio and SLC2A4 expression in vivo using patient samples from subcutaneous adipose tissues of obese individuals. SLC2A4 negatively correlated with the PPARGγΔ5/cPPARγ ratio in these patients. In conclusion, it is likely that the disequilibrium of PPARγ isoforms in the adipose tissue—especially the high PPARγΔ5/cPPARγ ratio—contributes to the insulin resistance frequently observed in patients with obesity. The unbalanced ratio between dominant negative and canonical isoforms in the adipose tissue favors the transcriptional repression of metabolic genes and impairs neo-adipogenesis, which are two main features of hypertrophic obesity and are correlated with insulin resistance and the onset of type 2 diabetes [3].

To improve the understanding of PPAR signaling in diverse vascular tissues, David Bishop-Bailey's group investigated the in vivo generation of oxylipin PPAR ligands in different vascular tissues. Oxylipins are derived from the oxidation of polyunsaturated fatty acids (PUFAs), which act as important paracrine and autocrine signaling molecules. A subclass of oxylipins, the eicosanoids, has a broad range of physiological outcomes in inflammation, the immune response, cardiovascular homeostasis, and cell growth regulation. Using a targeted lipidomic analysis of ex vivo-generated oxylipins from porcine aorta, coronary artery, pulmonary artery, and perivascular adipose tissue, Bishop-Bailey's group determined that cyclooxygenase (COX)-derived prostanoids were the most predominant oxylipin from all tissues. Interestingly, the coronary artery produced significantly higher levels of oxylipins from cytochrome P450 (CYP450) pathways than the other tissues investigated. The CYP450 pathway is to provide anti-inflammatory oxylipins that prevent processes of inflammatory vascular disease progression. The Toll-like receptor 4 ligand lipopolysaccharide (LPS) induced prostanoide formation in all the vascular tissues tested. Treatment of primary coronary artery smooth muscle cells (pCASMCs) with LPS induced a high expression of pro-inflammatory genes, which could be prevented by soluble epoxide hydrolase (sEH) inhibitor TPPU administration, demonstrating that endogenous CYP-derived epoxy–oxylipin PPAR ligands were strongly anti-inflammatory [4]. In experimental animal models, PPAR ligands reduced aortic atherosclerosis. In human patients, PPARα and PPARγ agonists showed some clinical efficacy in reducing cardiovascular events; in contrast, the PPARγ ligand rosiglitazone seemed to increase cardiovascular events (reviewed in [5,6]). There is interest in developing selective modulators and pan/dual-PPAR agonists, which might have increased efficacy and reduced side effects. Bishop-Bailey's group presented work that suggests that endogenous oxylipin PPAR ligands are more likely to act as pan/dual or selective modulator-type agonists capable of reducing atherosclerosis. Importantly, the authors show that the CYP450 pathway in the coronary artery provides an anti-inflammatory tone, as the sEH inhibitor TPPU inhibited inflammatory mediators induced by TLR-4 activation in primary pCASMCs, thereby preventing processes of vascular disease progression. This important work describes for the first time the anti-inflammatory effects of sEH inhibitors in coronary tissue, supporting potential therapeutic cardioprotective actions of sEH inhibitors [4].

All PPARs are affecting angiogenesis. PPARα and PPARγ mediate mainly anti-angiogenic processes; in contrast, PPARβ/δ emerged as a pro-angiogenic factor (reviewed in [7]).

The role of PPARβ/δ in cancer is controversial. Our group specifically wanted to address the impact of vascular PPARβ/δ for cancer growth and progression. We first tested the effects of the specific PPARβ/δ modulation on the proliferation of Lewis lung carcinoma cells (LLC1) in vitro. The PPARβ/δ agonist GW0742 decreased tumor cell proliferation, whereas the antagonist GSK3787 increased LLC1 proliferation in culture. Next, we treated LLC1 tumor-bearing mice with GW0742 and astonishingly observed increased cancer growth as well as enhanced metastases formation. Tumor sample analyses revealed an increased vascularization upon treatment with the PPARβ/δ agonist [8]. To further decipher the relevance of vascular PPARβ/δ for tumor progression, we used a mouse model with an inducible conditional vascular-specific overexpression of PPARβ/δ [9]. The inducible vascular-specific overexpression of PPARβ/δ promoted cancer angiogenesis, growth, and spontaneous metastases formation in vivo. To examine the transcriptome of gene expression patterns in PPARβ/δ overexpressing tumor-derived endothelial cells, we sorted endothelial cells from the LLC1 tumors of controls and mice with vascular PPARβ/δ overexpression and performed RNA sequencing. RNA sequencing of tumor-sorted endothelial cells revealed a high number of upregulated pro-angiogenic genes in response to PPARβ/δ increase. Combining top ten network analysis with a search for PPAR responsive elements divulged the platelet-derived growth factor (PDGF)/platelet-derived growth factor receptor (PDGFR) pathway, tyrosinkinase KIT (c-Kit), and the VEGF/vascular endothelial growth factor receptor (VEGFR) pathway as mediators of the pro-angiogenic tumor-promoting effect of PPARβ/δ. To determine the relevance for human pathophysiology, we also investigated human tumor samples and confirmed a high expression of PPARβ/δ in the tumor vasculature. The treatment of human umbilical vein endothelial cells (HUVEC) with the PPARβ/δ agonist GW0742 increased proliferation, which was accompanied by an upregulation of PDGFRB, PDGFB, and c-KIT expression. In contrast, the antagonist GSK3787 decreased the expression of these genes in HUVECs. In conclusion, we showed that PPARβ/δ favors tumor angiogenesis. Independently from their action on different cancer cell types, the therapeutic use of PPARβ/δ agonists appears to be dangerous [8]. Our group also stressed this point in the review "PPAR Beta/Delta and the Hallmarks of Cancer" for this Special Issue of *Cells*. Hanahan and Weinberg defined the interplay of cancer cell proliferation, angiogenesis, resisting cell death, evading growth suppressors, activating invasion and metastasis, enabling replicative immortality, deregulating cellular metabolism, and avoiding immune destruction as the didactic concept of the "hallmarks of cancer" to determine cancer growth and progression [10]. We outline in this review the effects of PPARβ/δ on these hallmarks and their underlying molecular mechanisms. In conclusion, we postulate that the therapeutic activation of PPARβ/δ results in alterations of the hallmark capabilities, favoring a pro-tumorigenic profile [11]. Given the initially proposed therapeutic potential of PPARβ/δ agonists as "exercise mimetics" and treatment for metabolic syndrome [12,13], nowadays, extreme caution should be applied when considering PPARβ/δ agonists for therapeutic purposes [11].

In the excellent review about PPAR ligands as candidates for the treatment of non-alcoholic fatty liver disease, Anne Fougerat and colleagues describe in a very detailed manner the factors contributing to the development, progression, and complications of liver steatosis, non-alcoholic fatty liver disease (NAFLD) and non-alcoholic steatohepatitis (NASH). PPARs are strongly implicated in glucose and lipid metabolism, and inflammation. The deregulation of these pathways is the underlying factor for the development of NAFLD and NASH. Therefore, PPARs have become attractive targets in the treatment of this disease complex. The review provides extensive information about experimental and clinical studies concerning the use of first generation and novel PPAR agonists in the treatment of NAFLD. Finally, perspectives for the development of safe PPAR agonists with improved efficacy targeting this pathology are discussed [14].

The review from the group around Jean-Louis Géant focuses on the crosstalk between Sirtuin 1 (Sirt1) and PPARs in metabolic diseases and the inherited disorders of the one-carbon metabolism. They describe the anti-oxidant and anti-inflammatory roles of Sirt1 and PPARs in metabolic diseases as well as the interplay of Sirt1 and PPAR activators in this setting. Sirt1 modulates the acetylation status of peroxisome proliferator-activated receptor-γ coactivator-1α (PGC-1α) and PPARγ. This leads to a certain level of protection against metabolic syndrome, as fatty acid oxidation, mitochondrial biogenesis, and white to brown adipose tissue differentiation are enhanced. A decreased expression and activity of Sirt1 is a common hallmark in a high-fat diet and also methyl donor deficiency (MDD). Notably in some inherited disorders of intracellular metabolism of vitamin B12, decreased Sirt1 expression plays an important role. In their review, the authors present promising data concerning the therapeutic use of Sirt1 agonists in inherited disorders of vitamin B12 metabolism [15].

Joseph M. Chambers and Rebecca A. Wingert focus in this Special Issue on the role of peroxisome proliferator-activated receptor gamma co-activator 1 alpha (PGC-1α) in kidney development and disease. While PGC-1α is required for proper renal development in zebrafish, the situation is not well studied in other species. In mice, PGC-1α deficiency seems to be compensated by PGC-1β. However, in different kidney cancer subtypes, the direct role of PGC-1α is variable; given that PGC-1α functions as a stress sensor of glucose depletion and enhances energy production, it might in general fuel cancer cell proliferation and disease progression. In contrast, in acute kidney injury and chronic and polycystic kidney disease, PGC-1α seems to be beneficial [16].

Nathalie Pierrots et al. describe in their contribution to this Special Issue the potential role of PPARα in the therapy of Alzheimers Disease (AD). Metabolic dysfunction (dyslipidemia, glucose metabolism impairment, and insulin resistance) is one risk factor leading to alterations of amyloid precursor protein (APP) processing and brain amyloid-β (Aβ) deposition, which are the main features of AD. PPARs are all expressed in the brain (PPARβ/δ exclusively in neurons and PPARα and γ in neurons and astrocytes). Among PPARs, PPARα is of special interest for the therapy of AD, as it is the only PPAR described to have neuronal functions implied in memory processes. Treatment with PPARγ agonists improved cognitive behavior in animal models of AD, while results from human trials were less successful. PPARβ/δ agonists decreased brain neuroinflammation, neurodegeneration, amyloid burden, and improved cognitive function in several animal models, and in a small human clinical cohort study, promising results on cognitive functions were reported. Many studies reported the beneficial effects of PPARα agonists on cognitive behavior in preclinical AD models. In conclusion, PPAR modulation might be a beneficial approach in AD therapy [17].

Justin Matheson and Bernhard Le Foll focused on the therapeutic potential of PPAR agonists in drug addictions. They provide a detailed overview of the key studies providing behavioral evidence for a role of PPAR agonists in modulating addiction-related behaviors in animal models as well as of the results from clinical and human laboratory studies of PPAR agonists in drug-related outcomes. While PPARβ/δ agonists do not seem to be implicated in the modification of addiction-related behaviors, both PPARα and PPARγ agonists are effective in reducing both the positive and negative-reinforcing properties of various drugs. Certain discrepancies between the animal and human study outcomes are explained in this review, and the potency and selectivity of PPAR ligands and sex-related variability in PPAR physiology discussed. Mainly, PPARα agonists seem to provide promising results against alcohol and nicotine addiction [18].

Finally, Yue Xi from the group of Zhiying Huang reviews the latest findings on PPAR-mediated toxicology and applied pharmacology. Due to their huge ligand-binding domain, PPARs are capable of binding a broad variety of compounds: endogenous and synthetic ligands as well as xenobiotics. Pharmacological ligands include full agonists, partial agonists, neutral antagonists, and inverse agonists. The information from 18 clinical trials using PPAR agonists for the treatment of diabetes and/or dyslipidemia is summarized. Furthermore, the authors discuss the positive impact of PPAR ligands in applied pharmacology, but they also detail the various toxicities observed by PPAR modulation, referring especially to the cardiovascular system, the liver, and the gastrointestinal system. They further highlight the beneficial and detrimental effects of PPAR interference in reproduction, development, cancer, muscle pathologies, kidney disease, and central nervous system pathologies. The authors conclude that there is still a substantial need to improve the comprehension of PPAR function in pharmacology and toxicology and their underlying molecular mechanisms, as only a few studies address an integrated network of relationships of all these aspects [19].

Conclusions

Taken together, the Special Issue "The Role of PPARs in Disease" comprises the most recent studies that elucidate the physiological and pathological role of PPARs. The effects of PPAR modulation upon various pathological stimuli in cell/animal models of human diseases and in patients are discussed. The articles in this Special Issue further improve our understanding of the beneficial and detrimental consequences of PPAR modulation under physiological and pathological conditions, and they open new perspectives for the development of safer and more efficient PPAR-targeted therapies in the future.

Author Contributions: Conceptualization, N.W. and K.-D.W.; writing—original draft preparation, N.W. and K.-D.W.; writing—review and editing, N.W. and K.-D.W.; funding acquisition, N.W. and K.-D.W. Both authors have read and agreed to the published version of the manuscript.

Funding: This research was funded by "Fondation ARC pour la recherche sur le cancer", grant number n_PJA 20161204650 (N.W.), Gemluc (N.W.), and Plan Cancer INSERM, Fondation pour la Recherche Médicale (K.-D.W.).

Conflicts of Interest: The authors declare no conflict of interest.

References

1. Committee, N.R.N. A unified nomenclature system for the nuclear receptor superfamily. *Cell* **1999**, *97*, 161–163. [CrossRef]
2. Wagner, K.D.; Wagner, N. Peroxisome proliferator-activated receptor beta/delta (PPARbeta/delta) acts as regulator of metabolism linked to multiple cellular functions. *Pharmacol. Ther.* **2010**, *125*, 423–435. [CrossRef] [PubMed]
3. Aprile, M.; Cataldi, S.; Perfetto, C.; Ambrosio, M.R.; Italiani, P.; Tatè, R.; Blüher, M.; Ciccodicola, A.; Costa, V. In-Vitro-Generated Hypertrophic-Like Adipocytes Displaying. *Cells* **2020**, *9*, 1284. [CrossRef] [PubMed]
4. Edin, M.L.; Lih, F.B.; Hammock, B.D.; Thomson, S.; Zeldin, D.C.; Bishop-Bailey, D. Vascular Lipidomic Profiling of Potential Endogenous Fatty Acid PPAR Ligands Reveals the Coronary Artery as Major Producer of CYP450-Derived Epoxy Fatty Acids. *Cells* **2020**, *9*, 1096. [CrossRef]
5. Han, L.; Shen, W.J.; Bittner, S.; Kraemer, F.B.; Azhar, S. PPARs: Regulators of metabolism and as therapeutic targets in cardiovascular disease. Part I: PPAR-α. *Future Cardiol.* **2017**, *13*, 259–278. [CrossRef] [PubMed]
6. Han, L.; Shen, W.J.; Bittner, S.; Kraemer, F.B.; Azhar, S. PPARs: Regulators of metabolism and as therapeutic targets in cardiovascular disease. Part II: PPAR-β/δ and PPAR-γ. *Future Cardiol.* **2017**, *13*, 279–296. [CrossRef] [PubMed]
7. Wagner, N.; Wagner, K.D. PPARs and Angiogenesis-Implications in Pathology. *Int. J. Mol. Sci.* **2020**, *21*, 5723. [CrossRef]
8. Wagner, K.D.; Du, S.; Martin, L.; Leccia, N.; Michiels, J.F.; Wagner, N. Vascular PPARβ/δ Promotes Tumor Angiogenesis and Progression. *Cells* **2019**, *8*, 1623. [CrossRef] [PubMed]
9. Wagner, K.D.; Vukolic, A.; Baudouy, D.; Michiels, J.F.; Wagner, N. Inducible Conditional Vascular-Specific Overexpression of Peroxisome Proliferator-Activated Receptor Beta/Delta Leads to Rapid Cardiac Hypertrophy. *PPAR Res.* **2016**, *2016*, 7631085. [CrossRef] [PubMed]
10. Hanahan, D.; Weinberg, R.A. The hallmarks of cancer. *Cell* **2000**, *100*, 57–70. [CrossRef]
11. Wagner, N.; Wagner, K.D. PPAR Beta/Delta and the Hallmarks of Cancer. *Cells* **2020**, *9*, 1133. [CrossRef] [PubMed]
12. Fan, W.; Waizenegger, W.; Lin, C.S.; Sorrentino, V.; He, M.X.; Wall, C.E.; Li, H.; Liddle, C.; Yu, R.T.; Atkins, A.R.; et al. PPARδ Promotes Running Endurance by Preserving Glucose. *Cell Metab.* **2017**, *25*, 1186–1193.e1184. [CrossRef] [PubMed]
13. Wang, Y.X.; Lee, C.H.; Tiep, S.; Yu, R.T.; Ham, J.; Kang, H.; Evans, R.M. Peroxisome-proliferator-activated receptor delta activates fat metabolism to prevent obesity. *Cell* **2003**, *113*, 159–170. [CrossRef]
14. Fougerat, A.; Montagner, A.; Loiseau, N.; Guillou, H.; Wahli, W. Peroxisome Proliferator-Activated Receptors and Their Novel Ligands as Candidates for the Treatment of Non-Alcoholic Fatty Liver Disease. *Cells* **2020**, *9*, 1638. [CrossRef] [PubMed]
15. Kosgei, V.J.; Coelho, D.; Guéant-Rodriguez, R.M.; Guéant, J.L. Sirt1-PPARS Cross-Talk in Complex Metabolic Diseases and Inherited Disorders of the One Carbon Metabolism. *Cells* **2020**, *9*, 1882. [CrossRef] [PubMed]

16. Chambers, J.M.; Wingert, R.A. PGC-1α in Disease: Recent Renal Insights into a Versatile Metabolic Regulator. *Cells* **2020**, *9*, 2234. [CrossRef] [PubMed]
17. Sáez-Orellana, F.; Octave, J.N.; Pierrot, N. Alzheimer's Disease, a Lipid Story: Involvement of Peroxisome Proliferator-Activated Receptor α. *Cells* **2020**, *9*, 1215. [CrossRef] [PubMed]
18. Matheson, J.; Le Foll, B. Therapeutic Potential of Peroxisome Proliferator-Activated Receptor (PPAR) Agonists in Substance Use Disorders: A Synthesis of Preclinical and Human Evidence. *Cells* **2020**, *9*, 1196. [CrossRef] [PubMed]
19. Xi, Y.; Zhang, Y.; Zhu, S.; Luo, Y.; Xu, P.; Huang, Z. PPAR-Mediated Toxicology and Applied Pharmacology. *Cells* **2020**, *9*, 352. [CrossRef] [PubMed]

Publisher's Note: MDPI stays neutral with regard to jurisdictional claims in published maps and institutional affiliations.

Review

Peroxisome Proliferator-Activated Receptors and Their Novel Ligands as Candidates for the Treatment of Non-Alcoholic Fatty Liver Disease

Anne Fougerat [1],*, Alexandra Montagner [1,2,3], Nicolas Loiseau [1], Hervé Guillou [1] and Walter Wahli [1,4,5,*]

[1] Institut National de la Recherche Agronomique (INRAE), ToxAlim, UMR1331 Toulouse, France; alexandra.montagner@inserm.fr (A.M.); nicolas.loiseau@inrae.fr (N.L.); herve.guillou@inrae.fr (H.G.)
[2] Institut National de la Santé et de la Recherche Médicale (Inserm), Institute of Metabolic and Cardiovascular Diseases, UMR1048 Toulouse, France
[3] Institute of Metabolic and Cardiovascular Diseases, University of Toulouse, UMR1048 Toulouse, France
[4] Lee Kong Chian School of Medicine, Nanyang Technological University Singapore, Clinical Sciences Building, 11 Mandalay Road, Singapore 308232, Singapore
[5] Center for Integrative Genomics, Université de Lausanne, Le Génopode, CH-1015 Lausanne, Switzerland
*   Correspondence: Anne.Fougerat@inrae.fr (A.F.); Walter.Wahli@unil.ch (W.W.); Tel.: +33-582066376 (A.F.); +41-79-6016832 (W.W.)

Received: 29 May 2020; Accepted: 4 July 2020; Published: 8 July 2020

Abstract: Non-alcoholic fatty liver disease (NAFLD) is a major health issue worldwide, frequently associated with obesity and type 2 diabetes. Steatosis is the initial stage of the disease, which is characterized by lipid accumulation in hepatocytes, which can progress to non-alcoholic steatohepatitis (NASH) with inflammation and various levels of fibrosis that further increase the risk of developing cirrhosis and hepatocellular carcinoma. The pathogenesis of NAFLD is influenced by interactions between genetic and environmental factors and involves several biological processes in multiple organs. No effective therapy is currently available for the treatment of NAFLD. Peroxisome proliferator-activated receptors (PPARs) are nuclear receptors that regulate many functions that are disturbed in NAFLD, including glucose and lipid metabolism, as well as inflammation. Thus, they represent relevant clinical targets for NAFLD. In this review, we describe the determinants and mechanisms underlying the pathogenesis of NAFLD, its progression and complications, as well as the current therapeutic strategies that are employed. We also focus on the complementary and distinct roles of PPAR isotypes in many biological processes and on the effects of first-generation PPAR agonists. Finally, we review novel and safe PPAR agonists with improved efficacy and their potential use in the treatment of NAFLD.

Keywords: peroxisome proliferator-activated receptors (PPARs); synthetic agonists; non-alcoholic fatty liver disease (NAFLD); non-alcoholic steatohepatitis (NASH); fibrosis

---

1. Introduction

The aim of this review is to provide information for a better understanding of the factors that impact the development, progression and complications of liver steatosis, nonalcoholic fatty liver disease (NAFLD) and nonalcoholic steatohepatitis (NASH). We discuss the roles of the nuclear receptors peroxisome proliferator-activated receptors (PPARs) in the regulation of biological processes that are participating in NAFLD, which include energy metabolism, inflammation, and fibrosis. PPARs are ligand activated transcription factors; we present new agonists that are currently in clinical trials for their potential to treat NAFLD for which no effective therapy is available.

2. NAFLD

2.1. Epidemiology

NAFLD is the most common chronic hepatic disease. It comprises a spectrum of liver conditions that can eventually lead to cirrhosis and liver cancer. Hepatic steatosis in the absence of excessive alcohol consumption is the hallmark of NAFLD, which is characterized by abnormal accumulation of triglycerides (TGs) in hepatocytes, a condition called non-alcoholic fatty liver (NAFL). The condition is considered to be benign, but can evolve into NASH, which is accompanied by hepatocyte damage and inflammation, with or without fibrosis, resulting in an increased risk of progression to cirrhosis and hepatocellular carcinoma (HCC).

The worldwide prevalence of NAFLD is constantly increasing in parallel with the global obesity pandemic. The prevalence of NAFLD is currently estimated to be approximately 25% in the general population, with the highest rates reported in South America and the USA, and the lowest in Africa [1]. A rapid and massive increase in NAFLD prevalence has also been observed in China as a result of an increase in obesity due to urbanization and lifestyle changes [2]. The trends in NAFLD incidence were followed for 17 years in a US community, finding that the incidence of NAFLD increased 5-fold, and even more (7-fold) in young adults [3]. Importantly, due to the growing increase in childhood obesity and children presenting greater vulnerability to genetic and environmental factors, NAFLD is now affecting up to 20% of the general pediatric population [4,5]. NAFLD in non-obese patients, so-called lean NAFLD, is also increasing, particularly in Asian patients [6]. Lean NAFLD is not fully understood, but possible determinants may include genetic background, different fat distribution, high fructose intake, and altered gut microbiota. Both epidemiological and preclinical studies have shown that NAFLD is more common in men than in women before menopause [7]. However, the incidence of NAFLD increases in women after menopause, suggesting a protective role of estrogens [8]. Sex-specific NASH signatures were recently identified in human liver, suggesting that NASH is a sexually dimorphic disease [9].

2.2. Etiology

The etiology of NAFLD is complex and involves ethnic, genetic, metabolic, and environmental factors (Figure 1).

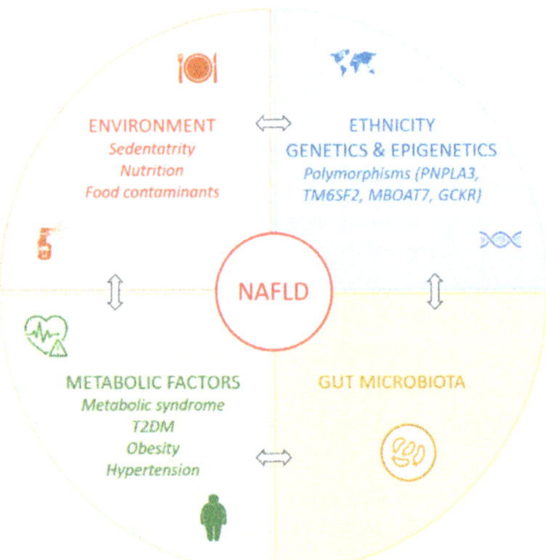

Figure 1. Non-alcoholic fatty liver disease (NAFLD) determinants. Multiple factors contribute to the development of NAFLD and its progression. Obesity and T2DM are closely associated with NAFLD

, and both drive the increasing prevalence of NAFLD. The genetic background also strongly influences disease development. In addition, the progression of NAFLD depends on complex interactions between genetic and environmental factors, especially dietary factors. More recently, the gut microbiota has emerged as an important determinant of NAFLD pathogenesis. Abbreviations: NAFLD, non-alcoholic fatty liver disease; T2DM, type 2 diabetes mellitus; PNPLA3, patatin-like phospholipase domain containing protein 3; TM6SF2, transmembrane 6 superfamily 2; MBOAT7, membrane bound O-acyltransferase domain-containing 7; GCKR, glucokinase regulator.

2.2.1. Ethnicity

Ethnic differences have been reported to be associated with the risk of NAFLD. For example, Hispanic individuals have a higher prevalence and severity of NAFLD [10]. Ethnic disparities are not yet well understood, but genetic and environmental factors are likely to influence the conditions associated with NAFLD, such as insulin resistance [11].

2.2.2. Genetic Factors

Genome-wide association studies have identified a number of genetic factors that influence NAFLD initiation and/or progression [12,13]. The most validated genes are involved in hepatic lipid metabolism and include *PNPLA3*, *TM6SF2*, *MBOAT7*, and *GCKR*. The most common and well-described polymorphism is in *PNPLA3*. Patatin-like phospholipase domain containing protein 3 (PNPLA3) is an enzyme highly expressed in the liver that hydrolyzes TGs in hepatocytes and retinyl esters in hepatic stellate cells (HSCs). The I198M variant (rs738409, isoleucine to methionine substitution at position 148) of PNPLA3 is strongly associated with the development and progression of NAFLD [14]. This variant has decreased hydrolase activity, resulting in an accumulation of TGs and retinyl esters in lipid droplets [15,16]. At the molecular level, PNPLA3 (I148M) accumulates on lipid droplets due to defective ubiquitylation, resulting in reduced proteasome degradation [17]. In preclinical studies, overexpression of mutant PNPLA3 (I148M) in mouse liver was shown to promote hepatic TG accumulation [18]. PNPLA3 (I148M) is present at high levels in Hispanics and may represent a major determinant of ethnicity-related differences in hepatic fat accumulation [19]. However, this variant increases the risk of severe hepatic fat accumulation, inflammation, fibrosis, and HCC in different ethnicities around the world [20]. A variant of the transmembrane 6 superfamily 2 (TM6SF2) protein has also been described as a major risk factor for NAFLD [21]. TM6SF2 is predominantly expressed in hepatocytes and enterocytes and localized in the endoplasmic reticulum and Golgi. The protein's exact function remains elusive, but it may be involved in very low-density lipoprotein (VLDL) formation in hepatocytes. The E167K (rs58542926, glutamate to lysine substitution at postion 167) loss of function variant of the protein causes higher liver fat content and fibrosis, but reduced secretion of VLDL and serum TGs [22–24]. This variant is also associated with reduced cardiovascular risk due to lower levels of circulating VLDL [25]. Results obtained from mouse studies, though often controversial, clearly indicate that the level of TM6SF2 protein is an important determinant of lipoprotein metabolism and NAFLD [26]. More recently, a polymorphism (rs641738) in the locus carrying the membrane bound O-acyltransferase domain-containing 7 (MBOAT7) gene has been associated with the risk and severity of NAFLD [27]. MBOAT7, also known as lysophosphatidylinositol acyltransferase (LPIAT1) is an enzyme involved in hepatic phospholipid remodeling by transferring polyunsaturated fatty acids to lysophospholipids. The variant rs641738 results in suppression of MBOAT7 at the messenger RNA and protein levels, altered phosphatidylinositol profiles, and was recently associated not only with steatosis development, but with more severe liver damage and advanced stages of fibrosis [28], as well as HCC in patients without cirrhosis [29]. In mice, downregulation of MBOAT7 leads to hepatic steatosis associated with obesity [30]. The rs1260326 polymorphism in the glucokinase regulator (GCKR) gene is a loss-of-function mutation that has also been linked to NAFLD development [31]. GCKR negatively regulates glucokinase in response to fructose-1-phosphate, modulating glucose uptake in the liver. The rs1260326 variant results in increased hepatic glucose uptake and malonyl-CoA concentration, providing more substrates for de novo lipogenesis [32]. This GCKR variant is also highly associated with fatty liver in obese youths [33]. Recent genetic and epidemiological studies have identified other polymorphisms associated with NAFLD progression in several genes involved in retinol metabolism (hydroxysteroid 17-beta dehydrogenase 3 (*HSD17B3*)), glycogen synthesis (protein phosphatase 1 regulatory subunit 3B (*PPP1R3B*)), bile acid homeostasis (beta-klotho (*KLB*)), oxidative stress (uncoupling protein 2 (*UCP2*), superoxide dismutase 2 (*SOD2*)), insulin signaling pathway (tribbles pseudokinase 1 (*TRIB1*)), and inflammation (suppressor of cytokine signaling 1 (*SOCS1*), interferon lambda 3 (*IFNL3*), MER proto-oncogene tyrosine kinase (*MERTK*)) [12,34–39]. In addition, epigenetic mechanisms, including post-translational histone modifications, DNA methylation, and micro-RNAs, are important in disease development [12]. A recent review presented a new prediction model that describes enriched genetic pathways in NAFLD, defined as the NAFLD-reactome [40]. Yet another layer of complexity has emerged with several genetic polymorphisms associated with both NAFLD and other liver diseases and metabolic disorders [41].

2.2.3. Metabolic Factors

In addition to ethnic and genetic factors, several metabolic and environmental factors contribute to NAFLD (Figure 1). Metabolic syndrome is defined as the presence of three of the five following conditions: high serum TGs, low serum high-density lipoprotein (HDL), elevated systemic blood pressure, hyperglycemia, and central obesity. Metabolic syndrome is recognized as a strong risk factor of NAFLD development and progression [42]. In a large cohort study including different ethnic groups, the prevalence of NAFLD increased in subjects with more metabolic syndrome criteria, reaching 98% when all five criteria were present [43]. NAFLD is not only associated with metabolic syndrome in general, but also with its individual conditions. Among NAFLD patients, the prevalence of metabolic syndrome is 42%, obesity 51%, type 2 diabetes mellitus (T2DM) 22%, dyslipidemia 69%, and hypertension 39% [1]. In patients with T2DM and normal circulating aminotransferase levels, the prevalence of NAFLD has been estimated to be 50% [44]. A recent meta-analysis including more than 35000 T2DM patients reported a pooled prevalence (24 studies) of NAFLD of 60% [45]. Based on histopathological assessment, T2DM patients also have a high risk of developing NASH and advanced fibrosis [46,47]. Obesity has been identified as an independent risk factor, with a 3.5-fold increased risk of developing NAFLD [48], and a linear relationship exists between body mass index (BMI) and NAFLD/NASH prevalence [49]. Several studies have highlighted the importance of fat distribution in showing that the amount of visceral fat is higher in NAFLD patients [50] and correlates with the severity of the disease [51], whereas large subcutaneous fat areas are associated with regression of NAFLD. These findings suggest that different types of body fat can increase or reduce the risk of NAFLD [52]. Several dyslipidemia phenotypes have been described in NAFLD patients [53], and are characterized by an increase in small dense low density lipoprotein (LDL) particles [54], higher postprandial lipemia after an oral fat meal [55], and HDL dysfunction [56]. Several studies have also reported that hypertension increases the risk of NAFLD [57] and the risk of NAFLD progression to fibrosis [58].

2.2.4. Environmental Factors

Environmental factors, especially dietary factors, also contribute to NAFLD development and progression [59,60]. The Western diet, which is particularly rich in added fructose, is associated with a greater risk of NAFLD, whereas the Mediterranean diet, which is high in polyunsaturated fatty acids (PUFAs), monounsaturated fatty acids (MUFAs), and fiber, has a beneficial effect on NAFLD [61,62]. Several nutrients impact the metabolic pathways leading to the lipid accumulation that characterizes the NAFLD initiation step, whereas others modulate key features in the pathogenesis of NASH, such as oxidative stress and inflammation.

NAFLD patients have been shown to have higher fructose intake due to sweetened beverage consumption [63], which is associated with the progression of fibrosis and inflammation [64]. Fructose consumption has dramatically increased in the last few decades, in parallel with the increased use of added sugars in the form of sucrose and high-fructose corn syrup in processed foods and beverages [65]. Mechanistically, fructose stimulates de novo lipogenesis, a central mechanism of hepatic lipid accumulation in NAFLD (see Section 2.3). Fructose metabolism rapidly induces precursors of lipogenesis, leads to ATP depletion, the suppression of mitochondrial fatty acid oxidation, and the production of carbohydrate metabolites, which activate the lipogenesis transcriptional program via the transcription factors carbohydrate-responsive element-binding protein (ChREBP) and sterol regulatory element binding protein 1c (SREBP1c) [65,66]. Recently, a novel pathway of lipogenesis activation by fructose, in which fructose is converted to acetate by the gut microbiota, was described. This pathway results in lipogenic pools of acetyl-CoA [67]. Fructose metabolism also leads to uric acid production, which has pro-oxidative and pro-inflammatory effects [68].

High-fat diets (HFDs) induce obesity and insulin resistance, which are strongly associated with NAFLD. A meta-analysis including 1400 NAFLD patients suggested that omega-3 PUFA supplementation has a beneficial effect on liver fat [69]. In contrast, no effect of omega-3 supplementation on NASH has been reported [70]. Omega-3 PUFAs, which are particularly abundant in fish oil, impact the activity of transcription factors, such as PPARα [71], liver X receptor (LXR) [72], ChREBP [73], SREBP1c [74], and peroxisome proliferator-activated receptor-gamma coactivator 1β (PGC1β) [75], which control the expression of genes involved in fatty acid homeostasis [76]. In a cohort of T2DM patients, a MUFA-rich diet induced a reduction in liver fat content [77], potentially through an increase in hepatic beta-oxidation [78]. In preclinical studies, dietary cholesterol has been shown to promote NASH and fibrosis, and contribute to HCC progression [79,80]. In human studies, high cholesterol levels have been associated mostly with cirrhosis and liver cancer [81].

High-protein diets have prevented hepatic lipid accumulation in animal studies [82–84]. In a small cohort of healthy men, a high-protein diet rich in glutamate increased plasma short-chain TG levels, which was interpreted as having resulted from increased de novo lipogenesis [85]. In contrast, alterations in plasma amino acid concentrations are clearly associated with the occurrence and severity of NAFLD [86], especially amino acids that are involved in glutathione synthesis, such as glycine, serine, and glutamate [87,88]. The current literature also suggests that branched chain amino acids (BCAAs) are increased in the plasma of NAFLD patients [89,90]. Interestingly, plasma BCAA levels correlate with NAFLD severity in a sex-dependent manner, increasing with disease severity in women, but decreasing in men [91]. Increased BCAA levels may be due to impaired BCAA catabolism by the gut microbiota [90]. Preclinical studies have proposed that BCAAs promote steatosis by increasing adipocyte lipolysis and decreasing the conversion of free fatty acids (FFAs) into TGs [92]. Micronutrients, such as vitamins, also play an important role in NAFLD. Plasma vitamin D levels are inversely associated with the severity of NAFLD [93]. In adult patients with NAFLD, vitamin E supplementation improves steatosis and hepatic inflammation, but has no effect on fibrosis [94]. Animal studies suggest that vitamin E ameliorates NAFLD/NASH by attenuating oxidative stress and inflammation [95].

Increasing epidemiological and experimental evidence suggests that exposure to some environmental contaminants could contribute to NAFLD progression [96–99]. Pesticides, insecticides, fungicides, and herbicides have demonstrated hepatotoxic effects by modulating lipid metabolism, inflammation, and oxidative stress [100].

2.2.5. Gut Microbiota

In recent years, gut microbiota and microbiota-derived compounds have emerged as important players in the pathogenesis of NAFLD in mice and humans [101,102]. Gut microbiota have been shown to cause NAFLD in animal studies. In humans, NAFLD severity is associated with gut dysbiosis, with an enrichment of Bacteroides in NASH patients compared to matched healthy individuals [103]. A recent review described the microbiome signature in human NAFLD according to the different stages of disease severity [104]. Suggested mechanisms by which the gut microbiota impact NAFLD and its progression include increased intestinal permeability [105], leading to the release of bacterial endotoxins (lipopolysaccharide (LPS)), and microbiota-derived factors (short-chain fatty acids), which may trigger inflammatory responses and affect hepatic metabolism via the modulation of metabolic gene expression [106]. As mentioned above, the gut microbiota converts fructose into acetate, which fuels hepatic lipogenesis [67]. Human NAFLD studies have some limiting factors that have not always been considered, such as possible confounding effects of obesity, insulin resistance, and T2DM on dysbiosis, as well as the variable demographic characteristics of the analyzed cohorts. Together with the use of different sequencing tools and NAFLD diagnostic methods, they may have been responsible for the discrepancy observed in microbiome signatures [104].

2.3. Pathophysiology

The pathogenesis of NAFLD and its complications are complex and not fully understood. As described previously, several factors acting in collaboration or synergy contribute to NAFLD development and its progression to NASH, leading to the multiple parallel hit hypothesis of NAFLD progression [107] (Figure 2).

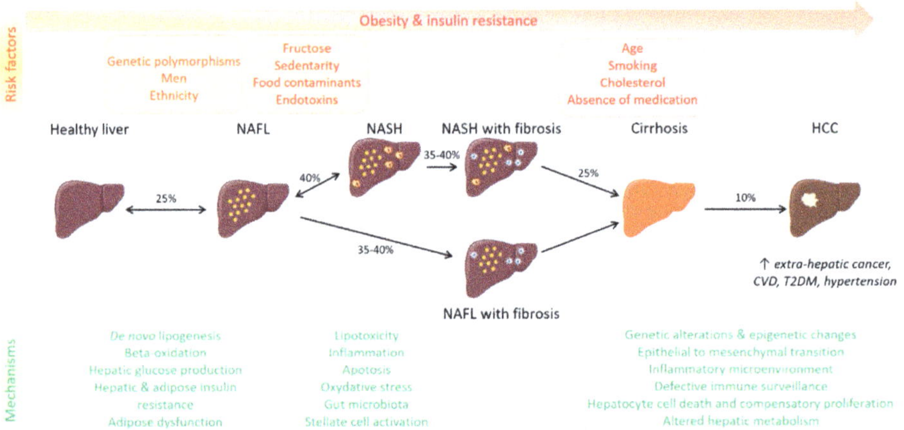

Figure 2. NAFLD progression. NAFLD is a progressive disease characterized by fat accumulation in hepatocytes, ranging from hepatic steatosis (NAFL) to non-alcoholic steatohepatitis (NASH), with additional inflammation with or without fibrosis. The latter is the strongest histological predictor of disease-related mortality. Though steatosis has previously been considered to be benign, some NAFL patients progress to NASH with or without fibrosis, whereas others develop fibrosis without having NASH. The pathogenesis of NAFLD is complex and involves several different pathways in multiple organs, including metabolic and inflammatory pathways. Abbreviations: NAFL, non-alcoholic fatty liver; NASH, non-alcoholic steatohepatitis; HCC, hepatocellular carcinoma; T2DM, type 2 diabetes mellitus; CVD, cardiovascular disease.

Hepatic steatosis is characterized by excessive accumulation of TGs in hepatocytes due to an imbalance between FFA influx and export, and/or catabolism. Increased FFAs within the livers of NAFLD patients originate primarily from adipose lipolysis (59%), followed by de novo lipogenesis (26%) and diet (15%) [108,109]. Both adipose lipolysis and de novo lipogenesis are normally regulated by insulin. However, NAFLD patients are usually insulin-resistant, and insulin is not able to suppress lipolysis, leading to increased circulating FFAs arriving to the liver. Adipose tissue contributes to NAFLD by modulating the lipid flux to the liver and via production of hormones and cytokines that impact hepatocyte physiology [110,111]. In the liver, insulin also fails to inhibit hepatic glucose production, but continues to stimulate lipid synthesis, leading to hyperglycemia, hyperlipidemia, and steatosis. This paradox of hepatic insulin resistance, which is also associated with obesity and T2DM, is still not fully understood. Current hypotheses to explain the selective hepatic insulin resistance involve extrahepatic pathways from peripheral metabolic organs [112], which underscores the multi-organ dimension of NAFLD pathogenesis. Insulin stimulates lipogenesis through the transcription factor SREBP1c, which regulates the expression of genes encoding enzymes involved in de novo lipogenesis [113,114]. The resulting hyperglycemia activates the glucose-responsive transcription factor ChREBP, which is also an important regulator of lipogenic gene expression [115,116]. Both SREBP1c and ChREBP are required for the maximal postprandial enhancement of lipogenesis [117]. The lipogenesis product malonyl-CoA inhibits fatty acids from associating with carnitine by down-regulating the enzyme carnitine acyltransferase, which reduces their entry into mitochondria and their beta-oxidation, thereby contributing to the overall increase in hepatic lipids. Adipose insulin resistance also leads to adipose tissue defects, including decreased secretion of adiponectin, an adipokine that increases beta-oxidation and decreases de novo lipogenesis in the liver [118]. When both fatty acid catabolism and export via VLDL secretion are not sufficient to compensate for the hepatic lipid overload, toxic fatty acid derivatives are produced that promote steatosis progression to NASH [119]. NASH is characterized by fat deposition, inflammation, ballooned hepatocytes, hepatocyte apoptosis and necrosis, and a variable rate of fibrotic progression. In hepatocytes, candidate lipotoxic lipids include saturated fatty acids, lysophosphatidylcholine, ceramides, sphingolipids, and diacylglycerol [120]. Hepatic free cholesterol levels are also elevated in NASH patients and contribute to liver toxicity [121]. A specific lipid signature that discriminates between control, steatotic, and NASH patients has been established and highlights dysregulation in the long-chain fatty acid (LCFA) synthesis pathway in NASH, leading to accumulation of LCFA and a decrease in phospholipids [122]. In response to lipid-induced hepatocellular injury, inflammasomes become activated, and endoplasmic reticulum (ER) and oxidative stress increase, leading to pro-inflammatory cytokine production, lipid peroxidation, hepatocyte cell death (apoptosis and necrosis), and aggravated liver damage. Chronic hepatocyte injury induces the recruitment and Toll-like receptor (TLR)-dependent activation of inflammatory cells, mainly liver macrophages or Kupffer cells, which amplifies inflammation and apoptosis. Kupffer cells also produce activating factors (platelet-derived growth factor [PDGF] and transforming growth factor β [TGFβ]) for the activation of HSCs, which proliferate and secrete collagen, as well as other extracellular matrix proteins, leading to fibrosis [123].

2.4. Progression and Associated Diseases

NAFLD progression is still not clearly understood due, in part, to its heterogeneity. Data indicate that all NAFLD patients have a risk of developing progressive liver disease over time. However, fibrosis is currently the best histopathological predictor of hepatic complications and disease-related mortality [124,125], and stage 2–4 fibrosis is predictive of cirrhosis-related issues [126]. In general, NAFLD is a slowly progressive disease, and many patients will develop cirrhosis or liver-related mortality; among NAFL patients who are considered to suffer from a benign condition, approximately 25% may progress to liver fibrosis. Identifying these patients and providing effective treatment remains a challenge [124]. Other patients will develop NASH, and these patients are more prone to progress to advanced stages of the disease. Overall, it means that some patients will remain at a stable steatosis stage, some will progress to NASH with or without fibrosis, and others will develop fibrosis without NASH (Figure 2). Using paired biopsies, McPherson et al. reported that 44% of patients with NAFL developed NASH, but that fibrosis progression was not different between patients with NASH and patients with NAFL at baseline [127]. A meta-analysis of paired liver biopsy studies in patients with NAFLD confirmed that fibrosis progression does not differ between NAFL and NASH patients, with an overall 35–40% of patients developing fibrosis [58]. Compared to matched controls, NAFLD patients are at higher risk of HCC [128], and the incidence of HCC was higher in NAFLD patients with cirrhosis than in those without cirrhosis [129]. Collectively, these data confirm the heterogeneous nature of NAFLD and led to classifying patients as fast and slow progressors. Slow progressors may develop

NASH but have a low risk of fibrosis, whereas fast progressors rapidly progress from steatosis to advanced fibrosis [130].

Due to its culmination in cirrhosis and HCC, NAFLD is becoming the major cause of liver transplantation. In addition to liver-related complications, NAFLD is also highly associated with an increased risk of extra-hepatic cancer [131], as well as cardiovascular and metabolic diseases. As described above, T2DM, hypertension, and cardiovascular disease (CVD) are major risk factors for NAFLD, but the link between these cardiometabolic diseases and NAFLD is more complex than initially thought. Clinical and experimental evidence now suggests a bi-directional relationship and indicate that NAFLD may precede and promote T2DM, hypertension, and CVD, rather than being the result of these conditions [132]. The incidence of metabolic comorbidities, cardiovascular events, and mortality was studied in a cohort of NAFLD patients followed for 17 years [3]. Patients with NAFLD had more diabetes, hypertension, and hyperlipidemia, increased risk of cardiovascular events and mortality, and shorter life expectancy than patients without NAFLD.

Altogether, NAFLD is a complex, multi-factorial, metabolic disease, the development and progression of which are strongly influenced by ethnicity, genetic predisposition, and metabolic and environmental risk factors (Figure 1). In addition, interactions between all of these factors, especially gene-diet interaction, promote NAFLD development, which has boosted the emergence of nutrigenomics as a novel approach for the management of NAFLD patients [133]. The pathogenesis of NAFLD is complex and involves many hepatic mechanisms, such as defects in lipid and glucose metabolism and insulin resistance, and important cross-talk between the liver and other organs in the adipose-liver and gut-liver axes, including important roles of the microbiota (Figure 2). Moreover, in contrast to NAFL, which can easily be detected by ultrasound and plasma biochemistry, the diagnosis of NASH and fibrosis requires liver biopsy for precise staging, which remains a limitation for the diagnosis of advanced phases of the disease. Despite many drugs being in development, there is currently no U.S. Food and Drug Administration (FDA)-approved pharmacological therapy for NAFLD treatment.

3. Current Therapeutic Strategies for NAFLD

3.1. Lifestyle Modification and Bariatric Surgery

NAFLD is considered the hepatic expression of metabolic syndrome and is closely associated with morbidities, such as obesity and insulin-resistance. Thus, weight loss represents the primary effective strategy for NAFLD management. Weight loss can be achieved through different interventions, including lifestyle changes, pharmacotherapy, and surgical procedures, and improves NAFLD biomarkers, though it effect on liver fibrosis is not significant [134]. In the absence of an approved drug therapy for NAFLD/NASH, weight loss through lifestyle interventions (exercise, diet) remains the first-line treatment. Bariatric surgery, which can be performed using minimally invasive techniques, also represents an effective option.

3.1.1. Exercise

Aerobic exercise refers to physical exercise usually performed at light-to-moderate intensity over a relatively long period, during which increased breathing brings oxygen into the body to sustain aerobic metabolism. Eight weeks of aerobic exercise in different forms reduces hepatic fat independently from the dose and intensity of the exercise [135]. Liver fat content is also reduced in pre-diabetic patients with NAFLD who are subjected to Nordic walking for 8 months [136]. These results are supported by a recent meta-analysis that found that exercise training alone has a beneficial effect on liver fat content, even in the absence of significant weight loss [137].

High-intensity interval training (HIIT), which alternates short periods of intense exercise with less intense recovery periods, performed three times per week for 12 weeks has been reported to reduce liver fat by 27% in adult NAFLD patients compared to individuals on standard care, and to also improve cardiac function [138]. Eight-week HIIT also has a beneficial effect on intra-hepatic TGs in obese diabetic patients with NAFLD [139]. Twelve weeks of HIIT reduces inflammatory markers and improves hepatic stiffness in obese men with NAFLD, suggesting that HIIT may have beneficial effects in patients with NASH [140]. Collectively, these data show that HIIT regimens significantly reduce hepatic fat in NAFLD/NASH patients.

Resistance training is a form of physical activity that causes muscle contraction against an external resistance and improves strength and endurance. Resistance exercise for 8 weeks reduces hepatic lipids in NAFLD patients [141], and 3 months of resistance training reduces liver fat content in NAFLD patients, but without a significant change in weight [142]. Interestingly, combined aerobic and resistance training improves aerobic capacity and skeletal muscle strength, and may be the most effective exercise program for improving NAFLD [143].

Collectively, exercise in whatever form appears to reduce the liver fat content, even in the absence of weight loss. No significant difference has been found between aerobic or resistance training in the reduction of liver fat, whereas continuous training of moderate volume and moderate intensity seems to be more beneficial [144,145]. Interestingly, although combining an exercise program with dietary interventions augments the reduction in hepatic fat content, exercise only is also effective in reducing hepatic lipid content in NAFLD patients [136,146]. As most of the studies have been performed with diabetic and/or obese NAFLD patients, the beneficial effect of exercise still needs confirmation in large-scale prospective studies, as a recent meta-analysis showed that physical activity only slightly reduces liver fat content in non-diabetic NAFLD patients [147]. Interestingly, starting to exercise has been independently associated with NAFLD remission only in men, suggesting a sex-specific hepatic response to exercise [148].

The above studies are informative, but the mechanisms underlying the reduction in hepatic fat following exercise are poorly studied. Proposed mechanisms of action include changes in liver physiology, such as increased VLDL clearance and improved mitochondrial fatty acid oxidation, together with extra-hepatic effects, such as improved peripheral insulin sensitivity, decreased visceral fat, and improved cardiovascular function [149,150].

3.1.2. Dietary Interventions

Dietary modifications remain the most effective physiological intervention for losing weight. Therefore, several studies have analyzed the effects of different dietary patterns on NAFLD development and progression. Currently, the Mediterranean diet is recommended for the management of NAFLD [151]. The Mediterranean diet has been shown by proton magnetic resonance spectroscopy to reduces liver steatosis in obese NAFLD patients without changes in body weight [152]. Adherence to the Mediterranean diet reduces the severity of liver disease among NAFLD patients and is associated with lower insulin resistance [153]. These findings are supported by two recent systematic reviews, which reported a reduction in hepatic steatosis in patients with NAFLD following the Mediterranean diet [154,155].

Caloric restriction leading to weight loss has also been associated with improved metabolic parameters in patients with NAFLD. A 12-month hypocaloric diet improved NASH-related histological parameters (steatosis, inflammation, and ballooning) in a paired biopsy study. In addition, individuals with weight loss > 10% have better NASH resolution and present with a regression of fibrosis, reinforcing the importance of weight loss in NAFLD management [156].

Given the detrimental hepatic effects of carbohydrates, especially fructose as described above, very low-carbohydrate ketogenic diets have received attention for the management of NAFLD. However, though ketogenic diets have largely been analyzed in rodents, only a few studies have been performed in humans. A pilot study in obesity-associated fatty liver disease showed that patients on a 6-month ketogenic diet lost weight and presented with histological improvements in steatosis, inflammation, and fibrosis [157]. Recently, a short-term ketogenic diet was shown to decrease hepatic lipids in obese patients in only 6 days, despite increased plasma FFA levels. This effect is attributed to an increase in hepatic TG hydrolysis and the use of released fatty acids for ketogenesis [158]. Another recent study reported that 1 year on a carbohydrate-restricted diet reduces the risk of fatty liver and advanced fibrosis in obese diabetic patients [159]. Notably, these two studies included obese and/or diabetic patients with suspicion of NAFLD, but imaging- or biopsy-proven NAFLD was not documented.

High-protein foods for weight loss have received much attention in recent years, and have started to be tested in NAFLD patients, but still remain poorly studied. A 2-week isocaloric, low-carbohydrate diet with increased protein content promotes multiple metabolic benefits in obese NAFLD patients, including a reduction in hepatic lipids due to decreased de novo lipogenesis and increased beta-oxidation. Interestingly, these changes are associated with an alteration in the composition of the gut microbiota [160]. Ketone bodies produced in response to carbohydrate restriction can induce additional protective effects in NAFLD, such as anti-oxidant and anti-inflammatory effects [161]. Another recent study analyzed the effects of isocaloric diets rich in animal proteins or plant proteins for 6 weeks in diabetic patients with NAFLD and found that both high-protein diets reduce liver fat [162]. As several studies have highlighted the role of the gut microbiota in NAFLD pathogenesis, supplementation with probiotics has been tested in NAFLD patients. To date, clinical data from such studies are disputed, but most of them report loss of body weight, suggesting that probiotic supplementation can be used as a complementary approach for patients with NAFLD [163]. Furthermore, high intake of insoluble dietary fiber correlates with a lower prevalence of NAFLD [164], and high-fiber diets promote short-chain fatty acid producing microbiota with beneficial effects in T2DM patients [165]. Clinical data on such diets in NAFLD patients are currently lacking. Recently, 1-year administration of a symbiotic combination (one probiotic and one prebiotic) was shown to change the fecal microbiome but had no effect on liver fat or fibrosis compared to placebo in NAFLD patients [166].

Taken together, observations from dietary interventions show that the Mediterranean diet and caloric restriction are beneficial for patients with NAFLD. As mentioned above, the macronutrient composition of the diet also appears to be important; saturated fatty acids and simple sugars damage the liver, whereas MUFAs, PUFAs, and dietary fiber induce beneficial hepatic effects [167]. More recently, studies have underscored that meal timing and frequency may also be important [143]. Studies in rodents have suggested that intermittent fasting and restricted feeding can have beneficial effects on NAFLD, and the few human studies agree that regular meals combined with regular fasting periods may provide physiological benefits (inflammation, circadian rhythm, autophagy, stress resistance, and gut microbiota) [168]. Combined diet and exercise interventions may induce greater benefits, though the current data are controversial [150]. Moreover, though lifestyle interventions (diet and exercise) are effective in reducing intrahepatic lipids without changes in body weight, weight loss appears to be required for improvement in NASH and fibrosis. Interestingly, lessening of NAFLD was measured in 67% of non-obese patients following lifestyle intervention [169].

The susceptibility to developing NAFLD comprises inherited risk factors, as described earlier, such as I148M PNPLA3, E167K TM6SF2, P446L GCKR, and rs641738 in *MBOAT7*. As these variants are nutrient-sensing, nutritional genomics approaches can be utilized in the future as interventions that make use of beneficial nutrients suitable to the patients' genomes and avoid those that have unhealthy effects. This avenue remains to be explored, though several ongoing clinical trials are already testing nutrigenomic diets in NAFLD patients [170].

3.1.3. Bariatric Surgery

Bariatric surgery is another effective non-pharmacological weight-loss therapy, and is indicated for patients with a BMI > 35 and severe comorbidities, such as T2DM and hypertension. Several studies have reported resolution of steatosis, as well as NASH and fibrosis, in patients who have undergone weight-loss surgery [171,172]. According to a meta-analysis of 21 studies, bariatric surgery results in histological or biochemical improvement of steatosis, NASH, and fibrosis in 88%, 59%, and 30% of NAFLD patients, respectively [173]. Furthermore, patients with NAFLD who undergo bariatric surgery have a lower risk of progression to cirrhosis compared to matched controls without surgery [174]. Bariatric surgery has beneficial effects through both weight loss and effects on metabolic pathways involved in NAFLD, including improved glucose and lipid homeostasis and decreased inflammation [175]. In a prospective study evaluating fibrosis and NASH in severely obese patients, most of the patients had low levels of NAFLD 5 years after surgery, but fibrosis had slightly increased [176]. Overall, bariatric surgery is very effective for reducing weight, but its effect on fibrosis progression is not yet clear and requires further attention. In addition, complications associated with this invasive procedure, such as sepsis and hemorrhage, limit its application.

In conclusion, lifestyle interventions and bariatric surgery are effective in NAFLD, especially through the induction of weight loss. However, studies are still needed to clarify the long-term effect of these interventions.

3.2. Pharmacotherapy

For most patients, lifestyle interventions such as diet and exercise, although effective, are difficult to achieve, and even more difficult to maintain. Thus, the development of pharmacological treatments is necessary. Most of the current pharmacological interventions aim at decreasing metabolic risk factors, such as obesity, insulin resistance, dyslipidemia, and hypertension. A systemic review of 29 randomized controlled trials testing several anti-diabetic drugs in NAFLD patients with and without T2DM reported that all anti-hyperglycemic agents have beneficial effects, at least on serum liver enzymes [177]. Among these anti-diabetic agents, pioglitazone is recommended for NAFLD patients with T2DM [178]. Vitamin E, which has anti-oxidant activity, is another current strategy for NASH management in patients without T2DM.

3.2.1. Pioglitazone

Pioglitazone is a thiazolidinedione that improves insulin resistance and glucose and lipid metabolism in T2DM. The phase 3 Pioglitazone vs Vitamin E vs Placebo for Treatment of Non-Diabetic Patients With Nonalcoholic Steatohepatitis (PIVENS) trial examined the effect of pioglitazone and vitamin E in non-diabetic patients with biopsy-proven NASH after 96 weeks of treatment. Compared to placebo, pioglitazone was associated with reduced hepatic steatosis, inflammation, and ballooning, but it did not improve fibrosis [94,179]. Several other studies have reported that pioglitazone treatment leads to histological improvement of steatosis and inflammation in subjects with NASH from 6 months of treatment onwards [180,181]. Interestingly, a retrospective analysis of data collected from the PIVENS trial suggested a strong link between the histological features of NASH resolution and improved fibrosis in NASH [182]. Nevertheless, the benefit of pioglitazone on fibrosis remains to be clarified because of divergent results. Some studies have reported an improvement in fibrosis [181,183,184], whereas others have reported no change in fibrosis [94,180].

In a recent study, patients with biopsy-proven NASH and prediabetes or T2DM were given pioglitazone or placebo for 18 months. A reduction in intrahepatic TG content and NASH resolution was observed in both groups, whereas fibrosis was reduced only in the T2DM patients [185]. Interestingly, genetic factors could contribute to the variability in the response to pioglitazone in NASH patients [186]. Adverse effects of pioglitazone include body weight gain, fluid retention, bone loss, and heart failure [187]. Furthermore, prediabetic and diabetic NASH patients treated with pioglitazone for 3 years exhibit decreased bone mineral density at the level of the spine, which is already present after 18 months of treatment compared to placebo [188]. A systematic review and meta-analysis concluded the risk of bladder cancer may be increased by pioglitazone and, therefore, recommend that patients on high-dose and long-term pioglitazone treatment be examined regularly for manifestations of bladder cancer [189]. Another systemic review of observational studies of the association between pioglitazone use and bladder cancer concluded that further research needs be conducted to clarify the role of pioglitazone use in the incidence of this cancer [190].

Preclinical studies have greatly contributed to our understanding of the mechanisms underlying the beneficial effects of pioglitazone. Pioglitazone is a ligand of PPARγ, a member of the nuclear receptor superfamily that is highly expressed in adipose tissue and plays a key role in glucose regulation and lipid metabolism [191]. Hepatoprotective effects of pioglitazone include increased insulin sensitivity, adipose TG storage, and adiponectin production, as well as decreased pro-inflammatory cytokine production by adipose tissue and macrophages [192,193]. These effects lead to a reduction in fatty acid delivery to the liver and decreased inflammation. In a murine model of NASH (high fructose and high trans fat), pioglitazone improves the toxic lipid profile by increasing the hepatic mitochondrial oxidative capacity and changing whole body BCAA metabolism [194]. Pioglitazone reduces HFD-induced steatosis in mice by stimulating the hepatic expression of genes and proteins involved in lipolysis, beta-oxidation, and autophagy [195]. In adiponectin-deficient mice, the reduction of HFD-induced steatosis by pioglitazone is blunted, revealing a role of adiponectin in this process [196].

3.2.2. Vitamin E

Vitamin E, which is known for its anti-oxidant effects, is considered the first-line treatment in NAFLD patients without T2DM. The PIVENS trial showed that vitamin E improves NASH compared to placebo (43% vs. 19%) in NAFLD patients without diabetes. As for pioglitazone, there was no improvement of fibrosis after 96 weeks of treatment [94]. Resolution of NASH in this cohort correlated with increased HDL levels, decreased TG levels, and reduced lipoprotein-related CVD risk compared to patients without an improvement in NASH [197,198]. The effect of vitamin E on NASH resolution was confirmed in non-diabetic children with NASH in the Treatment of Nonalcoholic Fatty Liver Disease in Children (TONIC) trial despite no improvement in liver enzyme levels [199]. Interestingly, the vitamin E response in non-diabetic NASH patients has been linked to the genotype of haptoglobin (Hp), an anti-oxidant protein that prevents hemoglobin-mediated oxidative injury. Two alleles of Hp (Hp 1 and Hp 2) generate three distinct genotypes (Hp 1-1, Hp 2-1, and Hp 2-2). NASH patients carrying at least one Hp 2 allele respond better to vitamin E treatment in terms of steatohepatitis resolution, histological improvement, and NAFLD activity score (NAS) compared to those with the Hp 1-1 genotype [200]. In contrast, in diabetic patients with biopsy-proven NASH, vitamin E supplementation for 18 months did not significantly reduce the NAS compared to placebo, despite resolution of NASH in 42% of patients vs. 18% with placebo. In this study, the effects of a combination of vitamin E and pioglitazone on liver histology were also examined. Though no change in fibrosis was observed, steatosis, inflammation, and ballooning were reduced by the combination therapy [201]. However, whether the combination of vitamin E and pioglitazone is more beneficial than pioglitazone alone was not examined. Others have seen differences regarding the vitamin E response between diabetic and non-diabetic individuals. The serum vitamin E concentration is higher in diabetic NAFLD patients, and there is an inverse relationship between vitamin E levels and all-cause mortality only in NAFLD patients without diabetes [202]. Clinical use of vitamin E has been limited because its long-term treatment has been associated with prostate cancer [203] and hemorrhagic stroke [204]. Vitamin E may also increase the risk of overall mortality, though this remains controversial [205,206].

The mechanisms of vitamin E action have been investigated in several rodent models of NAFLD. Well known for its anti-oxidant activities, vitamin E contributes to the scavenging of reactive oxygen species (ROS) and reactive nitrogen species (RNS), increase in the anti-oxidative enzyme superoxide dismutase (SOD), and inhibition of lipid peroxidation [207]. Recently, vitamin E supplementation for 2 weeks in HFD-fed mice showed beneficial effects on lipid accumulation and glucose homeostasis through activation of the transcription factor nuclear factor erythroid-2-related factor 2 (Nrf2) and upregulation of carboxylesterase 1 (CES1) [208]. In addition, vitamin E reduces apoptosis and inflammation through regulation of M1/M2 macrophage polarization and inhibition of T-cell recruitment [95]. Moreover, vitamin E induces adiponectin expression via a PPARγ-dependent mechanism [209].

In summary, the two classic therapies vitamin E and pioglitazone have beneficial effects on steatosis and inflammation. Vitamin E does not improve liver fibrosis, which is the strongest indicator of mortality in NAFLD patients, and the effect of pioglitazone on fibrosis varies from study to study. Furthermore, adverse effects and uncertain long-term benefits associated with both pioglitazone and vitamin E have limited their clinical use in NAFLD.

3.2.3. Other Current and Emerging Medications

Several other known molecules have been investigated or are currently under investigation in clinical trials for their effectiveness in NASH patients. Most of these medications target metabolic comorbidities and have been approved for the treatment of other diseases closely associated with NAFLD, such as obesity, dyslipidemia, and T2DM. For example, orlistat is an intestinal lipase inhibitor indicated for the treatment of obesity; statins are inhibitors of the enzyme hydroxymethylglutaryl-coenzyme A (HMG-CoA) reductase that are used to treat dyslipidemia due to their lipid-lowering effect; glucagon-like peptide (GLP-1) receptor agonists and dipeptidyl peptidase-4 (DPP-4) inhibitors increase incretins and are approved for the treatment of diabetic patients. In addition, the GLP-1 receptor agonist, liraglutide, is being investigated in a phase 2 clinical trial in NASH patients, the Liraglutide Efficacy and Action in NASH (LEAN) study [210]. All of these medications are effective in reducing hepatic steatosis, but no changes in liver inflammation or fibrosis have been reported [211]. As NAFLD is characterized by a disturbance in lipid and glucose homeostasis, drugs targeting de novo lipogenesis and glucose metabolism, such as stearoyl-CoA desaturase 1 (SCD1) and acetyl CoA carboxylase (ACC) inhibitors, sodium-glucose cotransporter-2 (SGLT2) inhibitors, and fibroblast growth factor (FGF) analogues, are currently being tested in phase 2 or 3 clinical trials. Several late-stage clinical trials are also investigating the effects of agents that target the mechanisms involved in advanced stages of NAFLD, such as inflammation (C-C chemokine receptor CCR2/CCR5 antagonist cenicriviroc), apoptosis (caspase inhibitor emricasan, apoptosis signal-regulating kinase 1 ASK1 inhibitor selonsertib), and fibrosis (galectin-3 inhibitor belapectin). Given the multiple-hit pathogenesis of NAFLD, a multifactorial approach based on combination treatments simultaneously targeting several pathways (metabolic syndrome conditions, hepatic lipid accumulation, and NASH features) should be more effective than single drug therapy [211–213].

3.2.4. Drugs Targeting Nuclear Receptors

Hepatic metabolic pathways, the alteration of which characterizes the first step of NAFLD, are mainly regulated at the transcriptional level. Therefore, transcription factors, and nuclear receptors in particular, may represent therapeutic targets in NAFLD. Within the nuclear receptor superfamily, PPARs, farnesoid X receptor (FXR), constitutive androstane receptor (CAR), pregnane X receptor (PXR), LXR, and thyroid hormone receptor-β (THR-β) are key regulators of the gut-liver-adipose tissue axis and control the expression of genes involved in lipid and glucose metabolism, bile acid homeostasis, and inflammation, which are all features of NAFLD/NASH [214–216]. Obeticholic acid is an FXR agonist that improves the histological features of NASH in patients without cirrhosis [217] and is currently being investigated in a phase 3 clinical trial [218]. An 18-month interim analysis of this ongoing study reported improved fibrosis in NASH patients treated with obeticholic acid compared to placebo [218]. A selective THR-β agonist, resmetirom, has demonstrated a highly significant reduction in hepatic fat and decreased hepatic inflammation in NASH patients following a 36-week treatment [219].

The three PPAR isotypes play distinct roles in lipid metabolism, energy homeostasis, and inflammation, which make them attractive targets in NAFLD, and they are discussed in more detail in the next section.

4. PPARs as Promising Targets for the Treatment of NAFLD

4.1. Overview of PPARs

PPARs are ligand-activated transcription factors belonging to the nuclear receptor family. Three isotypes of PPARs have been identified that are encoded by different genes: PPARα, PPARβ/δ, and PPARγ. Globally, the PPARs are activated by different ligands, have different tissue distribution, and distinct biological functions, but there is some overlap in these features (Table 1) and the three PPAR isotypes have a conserved protein structure and similar mechanisms of action (Figure 3). In addition, they all regulate energy homeostasis through lipid and glucose metabolism, and inflammation via modulation of largely specific target gene transcription.

Table 1. Expression, ligands, and functions of Peroxisome proliferator-activated receptors (PPARs) related to NAFLD and therapeutic potential.

Isotypes	PPARα	PPARβ/δ	PPARγ
Main tissue expression	Liver Skeletal muscles Heart Kidney BAT Intestine	Skeletal & cardiac muscles Liver WAT BAT Macrophages	WAT BAT Macrophages
Main natural ligands	FA Eicosanoids Phospholipids	FA VLDL components	FA Arachidonic acid metabolites
Main synthetic single agonists	Fenofibrate Wy14643 Gemfibrazil Pemafibrate	GW501516 GW0742 Seladelpar	Pioglitazone Rosiglitazone
Biological functions related to NAFLD	TG hydrolysis FA catabolism Ketogenesis FGF21 production Glycerol metabolism Anti-Inflammatory	Muscle FA storage FA catabolism Lipoprotein metabolism Glucose utilization Anti-inflammatory	Adipogenesis Adipose FA storage Adipokine secretion Anti-inflammatory Anti-fibrotic
Potential therapeutic target for	Hypertriglyceridemia Atherogenic dyslipidemia NAFLD	Atherogenic dyslipidemia Insulin resistance Obesity T2DM NAFLD	Insulin resistance Obesity T2DM NAFLD

Abbreviations: BAT, brown adipose tissue; WAT, white adipose tissue; FA, fatty acid; VLDL, very low density lipoprotein; TG, triglyceride; FGF21, fibroblast growth factor 21; T2DM, type 2 diabetes mellitus; NAFLD, non-alcoholic fatty liver disease.

4.1.1. Structure, Tissue Expression, and Mode of Action

PPAR proteins contain four domains. The *N*-terminal A/B domain contains the ligand-independent transactivation function called activation function (AF)-1. The C domain is the DNA binding domain (DBD), which consists of two zinc-finger motifs that bind a specific DNA sequence called the peroxisome proliferator response element (PPRE), which is usually localized in gene promoters. The D domain is a flexible hinge region connecting the DBD and the ligand-binding domain (LBD). The C terminus domain contains the LBD and the ligand-dependent transactivation function AF-2, which includes the region for dimerization and interaction with regulatory proteins [220] (Figure 3A).

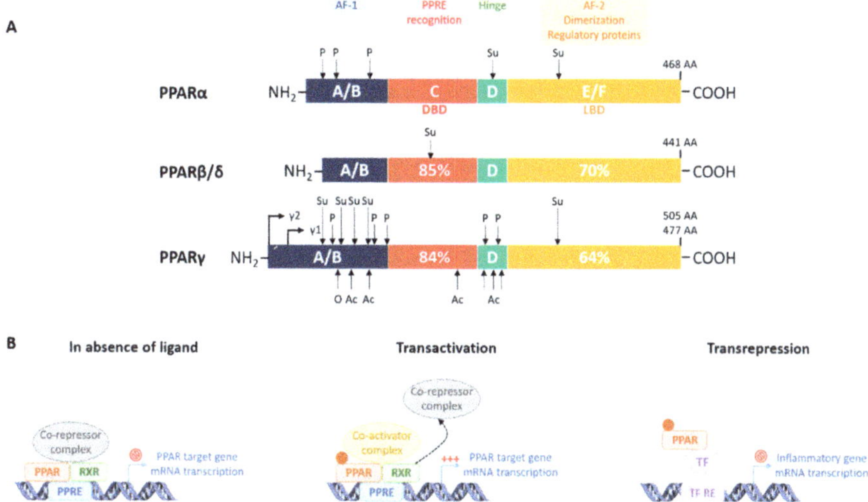

Figure 3. Protein structure and mechanisms of transcriptional regulation of PPARs. (**A**) Functional domains and posttranslational modifications of human PPARs. PPARs contain four distinct domains: a *N*-terminal A/B domain (ligand-independent AF-1), a C-domain (DNA-binding domain), a D- domain (hinge domain), and a *C*-terminal E/F domain (ligand-binding domain). Main functions of the 4 domains are listed. The number inside each domain corresponds to the percentage of amino acid sequence identity of human PPARβ/δ and γ relative to PPARα. The number of amino acids indicated at the COOH-terminus are for the human receptors. The locations of posttranslational modification sites are indicated by arrows. The 2 splice variants of PPARγ are indicated by γ1 and γ2. (**B**) PPAR mechanism of action. In the absence of ligand, PPAR-RXR heterodimers are bound to corepressor complexes and prevent gene transcription. Binding of an endogenous ligand or a synthetic agonist to the PPAR LBD triggers a conformational change, leading to corepressor complex dissociation and recruitment of coactivator complex. The activated PPAR/RXR heterodimer then binds to a specific DNA sequence in the promotor region of target genes (PPRE) and stimulates target gene transcription (transactivation). Through the binding to inflammatory transcription factors such as NF-κB and AP-1 (identified by TF), PPARs inhibit their binding to DNA and negatively regulate expression of proinflammatory genes (transrepression). Abbreviations: AF-1, activation function-1; AF-2, activation function-2; PPRE, peroxisome proliferator response element; DBD, DNA binding domain; LBD, ligand binding domain; AA, amino acid; P, phosphorylation; Su, SUMOylation; Ac, acetylation; O, *O*-GlcNacylation; RXR, 9-*cis* retinoic acid receptor; TF, transcription factor; TF RE, transcription factor response element; NH₂, protein N terminus; COOH, protein carboxyl terminus.

PPARα is highly expressed in oxidative tissues, such as the liver, skeletal muscle, brown adipose tissue (BAT), heart, and kidney. PPARβ/δ is most abundant in skeletal and cardiac muscles, adipose tissue, and skin, but also in inflammatory cells and liver cells, including hepatocytes, Kupffer cells, and HSCs. PPARγ is expressed predominantly in white and brown adipose tissue and macrophages [221] (Table 1).

PPARs have a large ligand-binding pocket, which contributes to their ability to bind various endogenous and synthetic ligands, as well as xenobiotics. The receptors are activated by endogenous ligands, including fatty acids and their derivatives, such as eicosanoids, which originate from dietary lipids, de novo lipogenesis, and adipose lipolysis [222–226] (Table 1). The development of several synthetic PPAR ligands, including molecules used in experimental research and pharmaceutical agents, has greatly contributed to the understanding of PPAR functions. Several studies reported that environmental pollutants also activate PPARs, supporting a role of PPARs in xenobiotic-induced toxicity in several organs [227–229].

All PPAR isotypes have a similar mechanism of action and function as heterodimers with the 9-cis retinoic acid receptor (RXR). In the absence of ligand, PPAR and its heterodimerization partner RXR are bound to corepressor complexes, leading to the repression of some target genes. Upon ligand binding to the LBD, a conformational change occurs, leading to corepressor dissociation and recruitment of coactivators. The activated PPAR/RXR heterodimers then bind to a DNA-specific sequence in the promoter of target genes (i.e., the PPRE) and stimulate transcription of the gene [226,230]. PPARs can also negatively regulate gene transcription via a PPRE-independent mechanism involving protein-protein interactions termed transrepression. In this process, PPARs bind other transcription factors, especially inflammatory transcription factors, inhibiting their binding to DNA and repressing their target gene transcription. Transrepression is the main mechanism involved in the anti-inflammatory effect of PPARs [231] (Figure 3B).

4.1.2. PPARs in Glucose and Lipid Metabolism

Through modulation of gene transcription, the three PPAR isotypes play distinct roles in lipid and glucose metabolism, which are key processes in NAFLD pathogenesis.

PPAR expression and activity are regulated at several levels, including gene and protein expression, as well as ligand availability, post-translational modifications, and cofactor recruitment, and by different factors, such as hormones, cytokines, and growth factors [220,226,232]. Interestingly, hepatic expression of PPARs fluctuates in a circadian manner that is linked to the nutritional status [233]. For example, hepatic PPARα peaks in the early night, which corresponds to the end of the day-time fasting period in nocturnal rodents [234,235], whereas PPARβ/δ is active during the dark/feeding period [236]. Accordingly, PPARα is mainly active in the fasted state [235,237,238]. In response to fasting, hepatocyte PPARα controls the expression of several genes involved in whole-body fatty acid homeostasis, allowing the liver to use fatty acids and provide energy-rich fuel for other organs. PPARα facilitates fatty acid uptake by the liver and mitochondrial transport by controlling the transcription of genes encoding fatty acid transport proteins (fatty acid transport protein-1 [FATP1], CD36, fatty acid binding protein-1 [L-FABP]) and carnitine palmitoyltransferases (CPT1A, CPT2). PPARα is the central regulator of hepatic fatty acid catabolism, it regulates gene transcription of rate-limiting enzymes required for microsomal (cytochrome P450 family 4 subfamily A [CYP4A]), peroxisomal (acyl-CoA oxidase 1 [ACOX], enoyl-CoA hydratase and 3-hydroxyacyl CoA dehydrogenase [EHHADH]), and mitochondrial beta-oxidation (acyl-CoA dehydrogenase medium chain [ACADM], acyl-CoA dehydrogenase long chain [ACADL], acyl-CoA dehydrogenase very long chain [ACADVL]) [220,235,239]. In addition, hepatic PPARα regulates the expression of ketogenic enzymes, such as 3-hydroxy-3-methylglutaryl-CoA synthase 2 (HMGCS2), leading to the production of ketone bodies, which are a vital alternative source of energy in the absence of glucose for several organs, including the brain and heart [237,240]. Ketone bodies also act as cell signaling mediators and modulate inflammation [241]. Furthermore, PPARα is required for the hepatic expression of murine and human fibroblast growth factor 21 (FGF21) [242–244], an hepatokine with systemic metabolic effects and hepatoprotective properties [245]. Hepatocyte PPARα is also essential for fasting-induced angiopoietin-like protein 4 (Angptl4; inhibitor of lipoprotein lipase) expression, whereas expression of the genes encoding growth differentiation factor 15 (Gdf15) and Igfbp1 is increased in the absence of PPARα in hepatocytes [246,247]. During fasting, PPARα also increases the transcription of genes involved in autophagy, leading to lipophagy, a mechanism involved in hepatic lipid catabolism [248]. Interestingly, there is reciprocal regulation of PPARα and the autophagy-lysosomal signal [249]. Lysosomal inhibition leads to downregulation of PPARα and its target genes, decreasing peroxisomal lipid oxidation and biogenesis [250]. The class 3 PI3K, Vps15, which plays a central role in autophagy, has been shown to control PPARα activation for lipid degradation and mitochondrial biogenesis [251]. In hepatocytes, PPARα activation promotes lipoprotein TG hydrolysis by increasing the enzyme activity of lipoprotein lipase (LPL) through a direct increase in its transcription, and decreases the expression of genes encoding lipoproteins, such as apolipoprotein C3 and apolipoprotein A4, which act as inhibitors of LPL activity [220,225]. Consequently, activation of mouse and human PPARα reduces plasma TG levels, indirectly leading to increased plasma HDL-cholesterol levels and decreased plasma LDL-cholesterol levels [252]. A few studies have reported that, in the fed state, PPARα regulates hepatic lipogenesis, mainly indirectly through transcriptional upregulation of SREBP1c [253] and increased proteolytic cleavage into its active form [254]. PPARα also modulates glucose metabolism by regulating the expression of genes involved in hepatic glycerol metabolism, promoting gluconeogenesis [255], which could explain the marked hypoglycemia in fasted PPARα-deficient mice [238]. In addition to lipid and glucose metabolism, PPARα also regulates amino acid metabolism in the liver through regulation of the expression of enzymes involved in the transamination and deamination of amino acids and urea synthesis, which correlates with a modulation of the plasma urea concentration [256]. From the above information, regulation of the hepatic activity of PPARα is expected to impact liver physiology, especially lipid metabolism. One example of such regulation is that of the NAD+-dependent protein deacetylase sirtuin 1 (SIRT 1), which increases the activity of PPARα primarily through the activation of peroxisome proliferator-activated receptor-gamma coactivator 1α (PGC-1α). Deletion or overexpression of SIRT1 in hepatocytes decreases or increases the expression of PPARα target genes. Accordingly, hepatocyte-specific SIRT1-knockout mice fed a HFD develop liver steatosis, inflammation, and ER stress [257]. Lipid oxidation in the skeletal muscle [258] and white adipose tissue [259] is also controlled by PPARα. Overexpression of PPARα in the heart resulting in high PPARα-dependent fatty acid oxidation contributes to diabetic cardiomyopathy through a mechanism involving the cardiac lipoprotein lipase as a source of PPARα ligand [260]. Interestingly, PPARα-dependent regulation of fatty acid oxidation in extrahepatic tissues plays an important role during fasting and can compensate, at least in part, for the absence of PPARα in hepatocyte-specific *Ppara*-null mice [261]. A role for adipose PPARα in the β-adrenergic regulation of lipolysis has been suggested [262]. Overexpression of PPARα in adipose tissue is associated with improvement in HFD-induced alterations in glucose metabolism, mostly through modulation of BCAA metabolism [263]. The role of PPARα in brown adipose tissue thermogenesis and white adipose tissue browning remains unclear, as some studies have suggested that PPARα is required to maintain body temperature [235,264] and for adipocyte browning [265], whereas other studies indicate that PPARα is dispensable for cold-induced adipose browning [266] and brown adipocyte function in vivo [267]. Redundant roles of PPARα and PPARγ in brown adipose tissue may account for these discrepancies [268]. A recent study identified hepatocyte B-cell lymphoma 6 protein (BCL6) as a negative regulator of the PPARα-dependent transcription program during fasting. BCL6 interacts with a high number of the same genes as PPARα and represses lipid catabolism in the fed state [269]. Intriguingly, though PPARα is required for the adaptive response to fasting, it is dispensable during intermittent fasting, a condition that ameliorates hepatic steatosis [270]. Finally, PPARα has demonstrated interesting functions in hepatic sexual dimorphism. Its SUMOylation in the female liver causes repression of genes involved in steroid metabolism and immunity, which safeguards female mice against estrogen-induced intrahepatic cholestasis, the most common liver disease during pregnancy [271].

PPARβ/δ is well-studied in skeletal muscles [272], where its expression is induced by exercise training and promotes mitochondrial biogenesis and glucose uptake by increasing PGC-1α [273]. PPARβ/δ also increases PGC-1α expression, even after exercise cessation, by preventing its degradation [274]. In addition, PPARβ/δ is required to maintain oxidative fibers in muscles via the transcription of PGC-1α [275]. Transgenic mice overexpressing PPARβ/δ in adipose tissue are protected from HFD-induced obesity and exhibit decreased adipose lipid accumulation through thermogenic gene regulation [276]. In the liver, PPARβ/δ regulates both lipid and glucose metabolism [230]. Its expression is highly reduced by fasting and rapidly restored by refeeding [277]. PPARβ/δ activation improves insulin sensitivity in diabetic mice, mostly by regulating genes related to hepatic fatty acid synthesis and the pentose phosphate pathway [278]. Accordingly, liver PPARβ/δ overexpression through adenovirus improves glucose tolerance and insulin sensitivity in mice fed a HFD. PPARβ/δ regulates glucose utilization by increasing the transcription of genes involved in lipogenesis, glucose utilization, and glycogen synthesis through direct and indirect mechanisms [279]. Such indirect mechanisms include upregulation of the lipogenic transcription factor SREBP-1c and co-activator PGC-1β [279]. Intriguingly, hepatic PPARβ/δ overexpression leads to decreased liver damage, suggesting that it may protect from lipotoxicity by regulating MUFA synthesis [279]. In contrast, another study showed that PPARβ/δ regulates SREBP-1 activity via induction of insulin-induced gene-1 (Insig-1), which inhibits the proteolytic cleavage of SREBP-1 into its mature form and consequently leads to reduced lipogenesis [280]. PPARβ/δ also regulates the expression of genes involved in lipoprotein metabolism (*APOA4, VLDLR*) [281], which is consistent with the reduced plasma TG levels observed after PPARβ/δ ligand treatment [282,283]. PPARβ/δ deficiency induces an increase in VLDL receptor (VLDLR) levels and hepatic steatosis through the activating transcription factor 4 (ATF4) ER stress pathway [284]. Interestingly, *Pparβ/δ* deletion in CD11b+ Kupffer cells leads to hepatic lipid accumulation in early life, during the suckling period [230]. Recently, intestinal PPARβ/δ was shown to participate in reducing obesity, insulin resistance, and dyslipidemia in mice fed a HFD, but the underlying mechanism is unknown [285]. Notably, outside the scope of this review article, several aspects of PPARβ/δ function are relevant to cancer growth [286].

PPARγ is mainly active in the fed state and controls fat storage in adipose tissue. It transcriptionally regulates the expression of genes involved in adipogenesis and adipose differentiation, and in lipid metabolism, including fatty acid uptake (fatty acid binding protein 4 [FABP4], CD36) and TG lipolysis (LPL) in adipose tissues. Consequently, adipose-specific deficiency of PPARγ induces a dramatic loss of adipose tissue and severe insulin resistance, leading to hepatic fat accumulation [287,288]. PPARγ enhances insulin sensitivity not only by reducing adipose fatty acid influx into the liver, but also by inducing adipokines, such as adiponectin and leptin [192,193], as well as FGF1 [289]. A recent study indicated that adipose PPARγ also regulates the plasma levels of BCAA, which may participate in the insulin-sensitizing effects [290]. Another mechanism contributing to increased insulin sensitivity upon PPARγ activation is the induction of FGF21 in adipose tissue, which acts in an autocrine manner to reciprocally regulate PPARγ activity by suppressing its SUMOylation [291]. A more recent study indicated that PPARγ is required to maintain brown adipose tissue thermogenesis [267]. PPARγ expression in the liver is low under ordinary physiological conditions but increases during the development of steatosis in rodents. Hepatocyte-specific deletion of PPARγ in diabetic mice improves steatosis through decreased expression of lipogenic genes (fatty acid synthase (FASN), ACC, SCD1), but aggravates systemic insulin resistance, likely by decreasing insulin sensitivity in adipose tissue [292]. PPARγ also promotes hepatic lipid accumulation by regulating the expression of lipid-droplet-binding protein FSP27 [293,294]. The activator protein-1 (AP-1) complex is an important regulator of hepatic PPARγ signaling, and distinct AP-1 dimers differentially regulate human and mouse PPARγ transcription in the liver and, thus, hepatic lipid content [295]. In addition to lipid droplet formation, PPARγ is also involved in TG synthesis, which may prevent peripheral lipotoxicity by storing FFAs as TGs [296]. PPARγ activation can also promote hepatic steatosis induced by genetic insults through the upregulation of glycolytic enzymes (pyruvate kinase M2 (PKM2), hexokinase 2 (HK2)) [297].

Cross-talk between the different PPAR isotypes have been reported but are relatively little documented so far. The three PPAR isotypes contain a highly conserved DNA-binding domain and bind the same response element (PPRE) in the regulatory regions of target genes. Furthermore, they present overlapping expression patterns in several organs. Therefore, cross-talks between PPARs is likely. In fact, an interplay between PPARα and PPARγ has been reported in BAT. A set of genes involved in BAT function is activated by both a PPARα agonist (fenofibrate) and a PPARγ agonist (rosiglitazone) in mice, which suggests a functional redundancy, which may explain why some findings suggest that PPARα is dispensable for thermogenesis while others clearly indicate a role of PPARα in BAT function. As an example of redundancy, the gene coding for lysosomal protease cathepsin Z, a regulator of BAT thermogenic function, is a shared PPARα and PPARγ target gene [268]. Compensation between PPARs has also been observed. In PPARα-deficient mice fed a HFD, in which PPARγ is overexpressed in the liver, characteristic PPARα targets involved in fatty acid oxidation are up-regulated, indicating that PPARγ can compensate for PPARα in gene regulation [222]. Similarly, a compensatory role of PPARβ/δ in the repression of hepatic Cyp7b1 in female mice has been shown in the absence of PPARα [271]. Collectively, these studies reveal cross-talk and compensatory mechanisms between PPAR isotypes, which may be important to consider when testing PPAR agonists.

Overall, all three PPAR isotypes regulate lipid and glucose metabolism by regulating both overlapping and distinct genes in multiple organs [298] (Figure 4). PPARα is the master regulator of hepatic lipid catabolism in response to fasting. PPARγ promotes insulin sensitivity by controlling adipose lipid storage and adipocyte differentiation, whereas its role in the liver remains unclear. PPARβ/δ promotes hepatic glucose utilization and fatty acid synthesis, as well as fat catabolism in muscles.

4.1.3. PPARs in Inflammation and HSC Activation

All PPARs play an important role in inflammation [299]. Evidence supports a role of PPARα in the control of hepatic inflammation [220]. One of the mechanisms by which PPARα exerts anti-inflammatory effects is through the down-regulation of acute phase genes and genes such as IL-1 receptor antagonist (IL-1Ra) and the nuclear factor kappa B subunit 1 (NF-κB) inhibitor IκB [220,300]. However, PPARα regulates inflammation mostly through a transrepression mechanism in which it binds to inflammatory transcription factors, such as NF-κB components (p65 and c-Jun), AP-1, and signal transducer and activator of transcription (STAT), thereby suppressing their transcriptional activity. An elegant study found that mice with a mutation in the DBD of PPARα, which limits its transcriptional activity to transrepression, are protected against liver inflammation through downregulation of pro-inflammatory genes and do not progress to liver fibrosis in dietary-induced NASH [301]. In addition, PPARα modulates the duration of inflammation by controlling the catabolism of its ligand leukotriene B4, a chemotactic agent involved in the inflammatory response [299,302]. Interestingly, hepatic PPARα contributes to the regulation of circulating monocytes during fasting through the modulation of bone marrow C-C motif chemokine ligand 2 (CCL2) production [303]. Few studies have examined the role of PPARα in Kupffer cells. A study of macrophage-specific PPARα-deficient mice compared to wild-type mice showed that Kupffer cell PPARα activation downregulates the expression of pro-inflammatory cytokines IL-15 and IL-18, which are mainly produced by M1 macrophages and Kupffer cells. This observation suggests that PPARα activation in these cells may prevent M1 polarization and mediate the anti-inflammatory effects of PPARα agonists [304]. Although anti-fibrogenic effects of PPARα activation have been reported in mouse models of liver fibrosis [301,305], the role of PPARα in HSCs is poorly defined. One study indicated that PPARα may inhibit TGFβ-stimulated HSC activation [306].

The role of PPARβ/δ in inflammation is less studied. PPARβ/δ is required for M2 macrophage activation in both adipose tissue and the liver. Bone marrow transfer experiments have shown that hematopoietic PPARβ/δ protects against HFD-induced insulin resistance, obesity, and fat accumulation in the liver. Moreover, PPARβ/δ appears to be necessary in Kupffer cells for oxidative phosphorylation, suggesting that these liver macrophages directly influence lipid homeostasis [307]. In addition, PPARβ/δ in CD11+ Kupffer cells has been suggested to prevent lipid accumulation in hepatocytes during the suckling stage in young mice [308]. PPARβ/δ is highly expressed in HSCs, but its role in these cells is not yet completely understood [230]. Studies in a mouse model of carbon tetrachloride (CCl4)-induced liver injury reported that PPARβ/δ activation stimulates HSC proliferation and promotes liver fibrosis [309], and contributes to HSC proliferation during acute and chronic hepatic inflammation in rats [310]. Another study in a mouse model of CCl4-induced liver fibrosis indicated that PPARβ/δ activation has anti-fibrotic effects [311]. The main difference between these studies is the use of different PPARβ/δ agonists.

PPARγ is expressed in macrophages, where it inhibits the expression of activated macrophage markers by reducing the activity of other transcription factors, including AP-1, STAT1, and NF-κB [312]. PPARγ activation induces monocyte differentiation towards M2 anti-inflammatory macrophages in vitro and in human blood [313]. In addition, PPARγ in macrophages is required for M2 macrophage activation and to protect mice against diet-induced obesity [314]. In line with these observations, PPARγ activation decreases HFD-induced M1 polarization through inhibition of the NF-κB pathway, reducing local inflammation and hepatic steatosis [315]. Interestingly, PPARγ promotes T regulatory cell accumulation in adipose tissue. Moreover, the PPARγ expressed by T regulatory cells is required for the insulin-sensitizing effect of PPARγ activation [316]. PPARγ expression and activation are reduced during HSC activation in vitro and in vivo [317]. In culture-activated HSCs, restoration of PPARγ levels using an adenoviral vector induces a phenotypic switch back to quiescence associated with inhibition of HSC activation markers [318]. Accordingly, ligand activation of PPARγ was shown to reduce HSC activation and proliferation, as well as collagen deposition, in a mouse model of CCl4-induced liver fibrosis [319]. The contribution of non-parenchymal cell PPARγ to the regulation of hepatic inflammation and fibrosis has been confirmed in the CCl4 model of liver injury [320].

To summarize, all PPARs play important roles in hepatic inflammation (Figure 4). PPARα negatively regulates pro-inflammatory genes, PPARβ/δ and PPARγ control macrophage M2 polarization. In addition, PPARγ has anti-fibrotic effects, but the roles of PPARα and PPARβ/δ in HSCs are not fully elucidated and require further study.

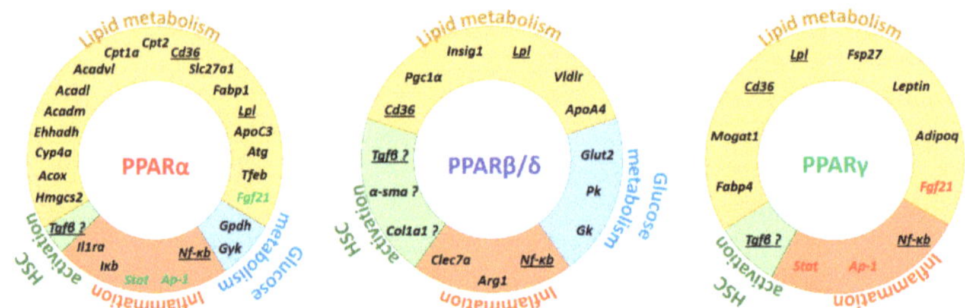

Figure 4. PPAR target genes and their implications in major functions associated with NAFLD pathogenesis. Key target genes of PPARα (**A**), PPARβ/δ (**B**), and PPARγ (**C**) and their association with four main biological processes driving NAFLD development and progression, i.e., lipid and glucose metabolism, inflammation, and hepatic stellate cell (HSC) activation. Genes in red, blue, and green are also regulated by PPARα, PPARβ/δ, and PPARγ, respectively. Genes whose expression is regulated by all three PPARs are underlined. A question mark (?) after a gene name indicates that PPAR may potentially regulate it. (**A**) PPARα promotes the expression of genes involved in fatty acid catabolism (Hmgcs2, Acox, Cyp4a, Ehhadh, Acad, Cpt1a, Cpt2, Cd36, Slc27a1, Fabp1) [235,237, 321,322], autophagy (Atg, Tfeb) [248,249], and glycerol metabolism (Gyk, Gpdh) [255], and regulates lipoprotein metabolism (Lpl, ApoC3) [220] and hepatic Fgf21 expression [242,243]. It also downregulates inflammatory genes and transcription factors (Nf-κb, Ap-1, Stat, Iκb, Il1ra) [220,281,300], and may downregulate Tgfβ expression [306]. (**B**) PPARβ/δ increases PGC-1α in muscles [273,275], Insig1 [280] and Cd36 [279] in the liver, and regulates the expression of genes involved in lipoprotein metabolism (ApoA4, Vldlr) [281,284] and glucose utilization [279]. PPARβ/δ regulates the expression of genes induced during alternative macrophage activation (Arg1, Clec7a) [307] and may also influence HSC activation (Tgfβ, α-Sma, Col1a1) [309,310]. (**C**) PPARγ controls the expression of genes involved in adipogenesis (Fabp4, Cd36, Lpl, Mogat1) [296,314] and genes encoding adipokines (Adipoq, leptin, Fgf21) [192,193,289,291]. It also promotes Fsp27 expression in the liver during steatosis [293]. PPARγ downregulates inflammatory transcription factors (Nf-κb, Ap-1, Stat) [312,315] and may also reduce expression of Tgfβ [318,319]. Abbreviations: Hmgcs2, 3-hydroxy-3-methylglutaryl-CoA synthase 2; Acox, peroxisomal acyl-coenzyme A oxidase 1; Cyp4a, cytochrome P450 family 4 subfamily A; Ehhadh,

enoyl-CoA hydratase and 3-hydroxyacyl CoA dehydrogenase; Acadm, acyl-CoA dehydrogenase medium chain; Acadl, acyl-CoA dehydrogenase long chain; Acadvl, acyl-CoA dehydrogenase very long chain; Cpt1a, carnitine palmitoyltransferase 1a; Cpt2, carnitine palmitoyltransferase 2; Slc27a1, solute carrier family 27 member 1; Fabp1, fatty acid binding protein 1; Fabp4, fatty acid binding protein 4; Lpl, lipoprotein lipase; ApoC3, apolipoprotein C3; ApoA4, apolipoprotein A4; Atg, autophagy-related genes; Tfeb, transcription factor EB; Fgf21, fibroblast growth factor 21; Pgc1α, peroxisome proliferator-activated receptor gamma coactivator 1-alpha; Fasn, fatty acid synthase; Acc, acetyl-CoA carboxylase; Scd1, stearoyl-CoA desaturase; Insig1, insulin-induced gene 1; Vldlr, very-low density lipoprotein receptor; Mogat1, monoacylglycerol O-acyltransferase 1; Fsp27, fat specific protein 27; Adipoq, adiponectin; Gpdh, glyceraldehyde-3-phosphate dehydrogenase; Gyk, glycerol kinase; Glut2, glucose transporter type 2; Pk, pyruvate kinase; Gk, glucokinase; Nf-κb, nuclear factor kappa B subunit 1; Ap-1, activator protein; Stat, signal transducer and activator of transcription; Iκb, Nf-κb inhibitor; Il1ra, IL-1 receptor antagonist; Arg1, arginase 1; Clec7a, C-type lectin domain containing 7A; Tgfβ, transforming growth factor beta; Col1a1, collagen type I alpha 1 chain; α-Sma, alpha-smooth muscle actin.

4.1.4. PPARs in NAFLD

Human studies indicate a link between PPAR functions and NAFLD pathogenesis. In a cohort of obese patients with NAFLD, the hepatic expression of PPARβ/δ and PPARγ remained unchanged during NAFLD progression, but the expression of PPARα and its target genes negatively correlated with the histological severity of steatosis and NASH both at baseline and after 1 year of follow-up. In addition, decreased liver PPARα expression is associated with increased insulin resistance and decreased adiponectin levels [323]. More recently, reduced PPARβ/δ expression and activity were observed in patients with severe hepatic steatosis [284].

Preclinical evidence indicates a role of PPARs in mouse models of NAFLD. Hepatic expression of PPARα and its target genes is increased in mice undergoing chronic high-fat feeding. Interestingly, an increase in PPARγ expression has been observed in PPARα-deficient mice fed a HFD [222]. Whole-body PPARα-deficient mice develop obesity, which is more pronounced and associated with higher fat deposition in females [324]. PPARα deficiency in mice fed a HFD promotes hepatic steatosis and inflammatory gene expression [325,326]. In a mouse model of steatohepatitis induced by methionine and choline deficiency diet (MCD), PPARα-null mice develop more severe steatosis and steatohepatitis is associated with increased lipid peroxidation compared to control mice [327]. The systemic deletion of PPARα also leads to more severe steatosis in response to a trans fatty acid-rich diet [328]. Liver-specific PPARα-deficiency has revealed the importance of hepatocyte PPARα in protecting the animals from HFD-induced NAFLD, including steatosis and hepatic inflammation [235]. Interestingly, PPARα-null mice and hepatocyte-specific PPARα-deficient mice do not present with increased glucose intolerance when fed a HFD [329]. In addition, hepatocyte-specific deletion of PPARα induces spontaneous steatosis in aging mice and aggravates MCD-induced liver damage [235]. Interestingly, hepatocyte-specific depletion of G protein pathway suppressor 2 (Gps2), a co-repressor of PPARα, protects mice from HFD-induced steatosis and improves MCD-induced fibrosis through PPARα activation. In humans, liver Gps2 expression positively correlates with NASH and fibrosis [330]. In response to HFD, whole-body and hepatic deficiencies in PPARα differentially alter the lipid profiles, suggesting that extrahepatic PPARα is involved in lipid metabolism and the adaptive response to HFD [329]. PPARα in extrahepatic tissues also contributes to the protection of fasting-induced hepatosteatosis [261].

PPARβ/δ-deficient mice exhibit impaired thermogenesis and increased HFD-induced obesity [276]. PPARβ/δ deletion also leads to fat deposition in the liver and exacerbated hepatic steatosis induced by ER stress, which is accompanied by an increase in hepatic VLDLR levels [284]. In response to CCl4-induced liver toxicity, PPARβ/δ-deficient mice present with increased hepatotoxicity, which is associated with an increase in NF-κB signaling [331].

Systemic deletion of PPARγ induces embryonic lethality due to placental defect [332,333]. Recently, mice with whole-body PPARγ deletion except in the placenta were obtained. These mice are completely lipodystrophic, which is consistent with PPARγ being required in mature white and brown adipocytes for their survival [287], develop T2DM, and get a fatty liver [334]. As mentioned above, specific deletion of PPARγ in adipose tissue also leads to hepatic steatosis [288]. Increased hepatic expression of PPARγ is also observed in mice fed a HFD [335]. Intriguingly, hepatocyte-specific deletion of PPARγ protects mice against HFD-induced steatosis and glucose intolerance, but has no effect on insulin sensitivity, hepatic inflammation, or obesity [335,336]. In contrast, PPARγ deficiency in non-parenchymal liver cells (Kupffer cells and HSCs) aggravates acute and chronic CCl4-induced liver damage, increasing inflammatory and fibrogenic responses, whereas the deletion of PPARγ in hepatocytes does not have this effect [320]. Finally, a role of hepatic PPARγ in tumorigenesis has been shown in a mouse model of liver cancer [337].

Collectively, all PPAR isotypes regulate not only many aspects of glucose and lipid metabolism, but also contribute to anti-inflammatory responses, and potentially to HSC function (Figure 4). In addition, caloric restriction, which has beneficial effects in NAFLD patients, reduces the expression of PPARα and its target genes involved in lipid oxidation in the duodenum. Interestingly, this change in duodenum gene expression influences the microbiota composition [338]. Reciprocally, the gut microbiota appears to influence hepatic PPAR activity [339]. Moreover, some beneficial effects of gut microbiota on NAFLD were recently suggested to involve PPARs [340]. Overall, PPARs modulate the transcription of both overlapping and distinct downstream target genes involved in many NAFLD-related functions in multiple organs, including lipid and glucose metabolism and inflammation (Figure 4). Therefore, PPARs represent relevant targets for NAFLD.

4.2. Available PPAR Agonists

Several experimental and clinical studies have reported the use of PPAR agonists in the treatment of NAFLD [341,342], which we review below.

4.2.1. PPARα Agonists

Fibrates are lipid-lowering agents used in clinical practice to treat hypertriglyceridemia and atherogenic dyslipidemia [343]. In rodent models of NAFLD, fibrates have demonstrated beneficial effects on hepatic steatosis, inflammation, and fibrosis. In MCD-induced mouse steatohepatitis, the PPARα agonist Wy14643 reduces hepatic TG levels and histological inflammation [327], as well as liver fibrosis in association with a decrease in HSC activation [305]. In this model, the beneficial effect of Wy14643 on MCD-induced liver damage is independent of its impact on fat accumulation in the liver, and due to the expression of genes involved in anti-inflammatory and anti-fibrogenic pathways [301]. PPARα activation by Wy14643 also decreases steatosis and inflammatory pathways in foz/foz mice, a genetic model of NASH, fed a HFD [344]. In the thioacetamide rat model of liver cirrhosis, Wy14643 and fenofibrate reverse histological liver fibrosis, in part by reducing the activity of the hepatic anti-oxidant enzyme catalase [345]. Fenofibrate also reduces CCl4-induced hepatic fibrosis in rats [346]. In a recent study, fenofibrate prevented liver damage induced by chronic intoxication of mice with 3,5-diethoxycarbonyl-1,4-dihydrocollidine (DDC), a model that induces key morphological features of NASH [347].

Conversely, fibrates only exhibit an effect on TG levels in humans. In obese patients with NAFLD, fenofibrate reduces plasma TGs by increasing VLDL-TG clearance from plasma, but does not change intrahepatic TG levels after 8 weeks of treatment [348]. Similarly, administration of fenofibrate for 48 weeks improves TG and glucose levels, but not liver histology in NAFLD patients [349]. Liver stiffness and biochemical markers of fibrosis (hyaluronic acid, TGF-β, and tumor necrosis factor-alpha (TNFα)) were decreased after 24 weeks of fenofibrate treatment, but no data on liver histology were given in this study [350]. In the Effects of Epanova Compared to Placebo and Compared to Fenofibrate on Liver Fat Content in Hypertriglyceridemic Overweight Subjects (EFFECT) I trial, 12 weeks of fenofibrate also reduced plasma TG levels, but increased liver fat content and liver volume in overweight or obese patients with NAFLD, suggesting a complex effect of fenofibrate on human hepatic lipid metabolism that requires further investigation [351]. Gemfibrozil has also demonstrated

PPARα-dependent hypolipidemic actions [352] and attenuated hepatic lipid accumulation in vitro [353]. However, in NAFLD patients, gemfibrozil has only shown beneficial effects on plasma levels of liver enzymes [354,355].

4.2.2. PPARβ/δ Agonists

Current PPARβ/δ agonists, including GW501516, GW0742, and MBX-8025 (Sedalpar), have mostly been tested in experimental models of NAFLD, and clinical studies are lacking. Though treatment of mice with GW501516 results in increased liver TG content after 4 weeks, long-term treatment (8 weeks) leads to reduced hepatic fat content. Interestingly, both PPARα and PPARβ/δ are required for the effect of GW501516 on hepatic lipid accumulation, as GW501516-dependent reduction in hepatic steatosis is abolished in PPARα-null mice. PPARβ/δ may modulate the levels of 1-palmitoyl-2-oleoyl-sn-glycero-3-phosphocholine (POPC), an endogenous activator of PPARα [356]. GW501516 treatment protects against HFD-induced obesity and insulin resistance, and reduces hepatic lipid accumulation by increasing muscle lipid oxidation [357]. GW501516 also increases the expression of hepatic VLDLR in mice fed a HFD [358]. Furthermore, GW501516 administration for 8 weeks decreases hepatic steatosis and insulin resistance in LDLR$^{-/-}$ mice fed a HFD via the increased expression of genes involved in hepatic fatty acid oxidation and decreased expression of hepatic fatty acid synthesis genes [359]. However, GW501516 does not improve liver injury induced by CCl4 [311]. One human study reported that administration of GW501516 to healthy individuals for 2 weeks reduced liver fat content and serum TG levels [282].

In a diabetic rat model, GW0742 decreased hepatic TGs, glucose intolerance, epididymal fat weight, and inflammatory cytokines [360]. Another study indicated that GW0742 reduces hepatic TGs, glucose intolerance, and insulin resistance in mice fed a HFD. These effects were associated with several changes in hepatic gene expression, including an increase in PPARα and beta-oxidation gene expression and decreased expression of PPARγ and lipogenic genes, as well as genes involved in inflammation and ER stress [361]. GW0742 also reduces CCl4-induced hepatotoxicity, which is associated with modulation of NF-κB signaling [362].

The more recent PPARβ/δ agonist seladelpar reduces glucose intolerance and hepatic TGs in the foz/foz mouse model of NASH when fed an atherogenic diet. Seladelpar also decreased the NAS by 50% and reversed NASH in all mice. In addition, seladelpar improved liver histology, with decreased hepatic apoptosis and fibrosis and a reduction in the number of macrophages around hepatocytes (crown-like structures) [363].

4.2.3. PPARγ Agonists

Thiazolidinediones (pioglitazone, rosiglitazone) are synthetic ligands of PPARγ that are clinically used as insulin sensitizers in the treatment of T2DM [193]. Though pioglitazone effectively improves hepatic steatosis in humans, preclinical data in rodents have been controversial, and the exact molecular mechanisms underlying the action of pioglitazone are not fully understood.

Several studies indicate that pioglitazone reduces HFD-induced steatosis in mice by increasing adiponectin production and the hepatic expression of genes involved in lipolysis, beta-oxidation, and autophagy [195,196]. In contrast, pioglitazone was shown to have no effect on liver histology in a rat dietary model of NASH (high fat, high cholesterol and cholate) [364]. A recent study showed that the effect of pioglitazone on NAFLD is influenced by CAR activity, as pioglitazone improves hepatic steatosis much better in CAR-deficient mice, suggesting an interaction between CAR and PPARγ [365]. Finally, another study indicates that pioglitazone promotes hepatic steatosis. In this study, the global expression profiles in the livers of mice fed a HFD and treated with pioglitazone reveal that pioglitazone upregulates the expression of genes involved in fatty acid uptake and de novo lipogenesis, and reduces the expression of inflammatory genes, leading to hepatic TG accumulation and improved insulin resistance [366].

As discussed above, several studies have reported that pioglitazone treatment is effective in NAFLD patients [94,180,181,183]. Pioglitazone improves the histological features of NAFLD, including steatosis and inflammation, whereas its effect on fibrosis is less clear. A recent meta-analysis reported that pioglitazone therapy is associated with an improvement in advanced fibrosis in NAFLD patients, even in non-diabetic patients. This meta-analysis also indicated that weight gain and limb edema is associated with pioglitazone treatment [367].

Another thiazolidinedione that has shown promising results in preclinical studies is rosiglitazone. In animal models, rosiglitazone protects against HFD-induced hepatic steatosis and reduces hepatic lipid content by increasing the expression of genes involved in beta-oxidation and decreasing the expression of lipogenic genes. These beneficial effects of rosiglitazone on lipid metabolism are accompanied by a decrease in hepatic M1 macrophages and modulation of the TLR4/NF-κB signaling pathways [368]. In the MCD model, rosiglitazone improves hepatic steatosis, inflammation, and fibrosis and reduces the expression of the HSC activator TGF-β [369]. In a model of liver cholestasis and fibrosis induced by bile duct ligation, rosiglitazone reduces fibrosis and hepatocyte apoptosis by inhibiting NF-κB-TNFα signaling in a PPARγ-dependent manner [370]. Interestingly, a recent study indicated that adipose PPARγ is dispensable for the whole-body insulin-sensitizing effect of rosiglitazone, suggesting the presence of PPARγ-independent targets of rosiglitazone in adipocytes, or that rosiglitazone activates PPARγ in other tissues [371]. In a small paired biopsy study, rosiglitazone treatment of NASH patients for 48 weeks results in improved hepatic steatosis, necroinflammation, and ballooning. In most patients, body weight increases during the treatment period, and the weight gain remains after a 6-month post-treatment follow-up [372]. The Fatty Liver Improvement With Rosiglitazone Therapy (FLIRT) trial assessed the effect of rosiglitazone in patients with biopsy-proven NASH. Treatment with rosiglitazone for 1 year increased adiponectin levels and reduced insulin resistance in most patients, and reduced hepatic steatosis in half of patients, but did not improve liver inflammation or fibrosis. The main adverse effect was weight gain in 40% of responders [373]. In a post hoc analysis of this cohort, patients treated with rosiglitazone presented with increased hepatic expression of PPARγ, which was associated with increased expression of several pro-inflammatory genes in the liver (monocyte chemoattractant protein-1 [MCP1], IL8, SOCS3), suggesting a potential long-term deleterious effect [374].

Although all PPAR agonists have had beneficial effects in preclinical models of NAFLD, their effectiveness in human pathology is limited. In NAFLD patients, PPARα activation only reduces plasma TG levels, whereas PPARγ agonists improve insulin sensitivity and steatosis, but do not seem to impact liver fibrosis. The efficacy of current PPARβ/δ agonists against NAFLD in humans is not known. Moreover, some PPAR agonists have adverse effects (weight gain and fluid retention following pioglitazone) or limited potency (fibrates) that limit their application.

Novel PPAR agonists, called selective PPAR modulators (SPPARMs), aim to maximize the beneficial effects and minimize the adverse effects of current agonists. Furthermore, given the multiple and distinct effects of PPARs in the liver and other organs, targeting two or three isotypes has emerged as a promising novel therapeutic strategy for treating NAFLD (Table 2).

Table 2. PPAR agonists currently in late-stage clinical trials (phase 2 and phase 3). Overview of new PPAR agonists: trivial name, chemical structure, and short description.

Compounds	Chemical Structure	Description
Pemafibrate		Selective and potent PPARα modulator Clinical application for dyslipidemia Safety profile in clinical studies Currently in phase 3: effect on CV events in T2DM
Elafibranor		High activity on PPARα, moderate activity on PPARβ/δ Safety profile in clinical studies Currently in phase 3: effect on fibrosis in NASH
Saroglitazar		High activity on PPARα, moderate activity on PPARγ Clinical application for dyslipidemia Safety profile in clinical studies Currently in phase 2: effect on NAFLD/NASH, phase 3: effect on fibrosis compared to pioglitazone and vitamin E
Lanifibranor		Moderate and balanced activity on PPARα, PPARβ/δ, PPARγ Currently in phase 2: effect on NAFLD/NASH

Abbreviations: CV, cardiovascular; NAFLD, non-alcoholic fatty liver disease; NASH, non-alcoholic steatohepatitis.

4.3. Novel PPAR Agonists

4.3.1. Pemafibrate

Pemafibrate (K-877) is a novel selective PPARα modulator that, compared to fenofibrate, exhibits high potency for human PPARα and enhanced PPARα selectivity and activity in vitro [375]. The crystal structure of the PPARα-pemafibrate complex showed that pemafibrate is highly flexible and can change its conformation following coactivator binding. In addition, several hydrophobic interactions between PPARα and pemafibrate may improve the binding affinity for PPARα [376]. The hepatic transcriptome of primary human hepatocytes and mice treated with pemafibrate indicates a PPARα-dependent increase in the expression of genes involved in lipid catabolism and ketogenesis. Interestingly, VLDLR and FGF21 are also induced by pemafibrate in humans and mice, and at a higher level than by fenofibrate [377,378]. Pemafibrate decreases plasma TG and total cholesterol levels in the LDLR$^{-/-}$ mouse model of atherosclerosis, which is associated with increased expression of PPARα and its target genes in both the liver and intestine [379]. In Western diet-fed ApoE2KI mice, pemafibrate also improves lipoprotein metabolism, resulting in a greater reduction in TG and increase in HDL-cholesterol levels compared to fenofibrate. Pemafibrate also decreases atherosclerotic lesions, lesion macrophage infiltration, and inflammatory markers [380]. In mice fed a HFD, pemafibrate reduces postprandial accumulation of TGs at the same level as fenofibrate, but at lower doses [381]. In addition, pemafibrate protects against HFD-induced obesity, glucose intolerance, and insulin resistance, and decreases the cell size in white adipose tissue and brown adipose tissue, but has no effect on hepatic TG accumulation. Pemafibrate-activated PPARα in the liver increases hepatic and plasma levels of FGF21, whereas in inguinal adipose tissue, pemafibrate increases the expression of genes involved in fatty acid oxidation and thermogenesis, and the mitochondrial marker elongation of very long chain fatty acids protein 3 (Elovl3) in brown adipose tissue [382]. One study examined the effects of pemafibrate in a mouse model of NASH induced by an amylin diet that exhibits the different stages of NASH, including steatosis, inflammation, hepatocyte ballooning, and fibrosis. Pemafibrate reduces hepatic TG levels, inflammation, and fibrosis, and increases expression of PPARα and its target genes involved in beta-oxidation. In addition, pemafibrate increases the expression of lipogenic genes. However, in contrast to previous reports, fenofibrate tested in parallel with pemafibrate was also effective in reducing fibrosis [383]. Recently, the therapeutic potential of pemafibrate was tested in a mouse model of a diabetes-based NASH-HCC model. In this model, combined chemical (one low dose of streptozotocin just after birth) and dietary (continuous HFD feeding) interventions leads to diabetes in 1 week and sequential liver damage from steatosis, NASH, and HCC but did not induce obesity and insulin resistance. Pemafibrate reduces macrophage recruitment and inflammation in the liver, potentially through the downregulation of endothelial adhesion molecules. Intriguingly, hepatic TG accumulation is not improved with pemafibrate in this model [384].

In clinical studies, pemafibrate has demonstrated safety and efficacy in patients with atherogenic dyslipidemia [385] and appears to be superior to fenofibrate to reduce plasma TG levels [386–388]. In a phase 3 clinical trial, treatment of Japanese T2DM patients with pemafibrate for 24 weeks reduced fasting serum TG levels by 45%. Interestingly, in this cohort, pemafibrate increased plasma FGF21 [389]. The ongoing Pemafibrate to Reduce Cardiovascular Outcomes by Reducing Triglycerides in Patients with Diabetes (PROMINENT) study is a placebo-controlled trial testing the effect of pemafibrate on cardiovascular events in T2DM patients with elevated TG and low HDL-cholesterol levels [390].

Due to the multiple and distinct effects of PPARs, dual or pan-PPAR agonists represent attractive approaches for targeting the multiple biological processes involved in the pathogenesis of NAFLD. Below, we review the most promising of them.

4.3.2. PPARα and β/δ Dual Agonist Elafibranor

The PPARα and PPARβ/δ dual agonist elafibranor (GFT-505) has preferential activity on human PPARα in vitro and additional but lower activity on human PPARβ/δ [391]. Several experimental studies indicate that elafibranor has beneficial effects on NAFLD/NASH and fibrosis in rodent models. Efficiency was first demonstrated in Western diet-fed human apolipoprotein E2 transgenic mice, in which elafibranor improves lipid profiles and reduces hepatic expression of pro-inflammatory and pro-fibrotic genes. Histological examination has demonstrated that elafibranor decreases steatosis, inflammation, and fibrosis. Similar results have been reported in ob/ob mice fed MCD. Using PPARα-deficient mice, this study demonstrates the importance of PPARα in the effects of elafibranor, but also reveals PPARα-independent mechanisms [392]. In other mouse models of diet-induced NASH, elafibranor induces weight loss and improves steatosis, as well as inflammation and fibrosis. Hepatic transcriptome analysis has revealed that elafibranor modulates the expression of genes involved in lipid metabolism, inflammation, fibrogenesis, HSC activation, and apoptosis [393,394]. Elafibranor has also shown efficiency in a rapid diet-induced NASH model with additional cyclodextrin in drinking water, which induces NASH in 3 weeks without obesity [395]. Elafibranor also prevents and reverses CCl4-induced liver fibrosis and inflammation in rats [392]. Interestingly, in alcoholic steatohepatitis, elafibranor reduces adipose tissue autophagy dysfunction, leading to hepatoprotective and anti-inflammatory effects in several organs, including the liver, intestines, and adipose tissue [396].

Few clinical studies have reported the impact of elafibranor in humans. Eight-week treatment with elafibranor reduces fasting plasma TG levels and improves both hepatic and whole-body insulin sensitivity in obese insulin-resistant patients. Elafibranor also improves liver enzyme levels, suggesting beneficial effects on liver functions [397]. A phase 2 study examined the efficacy of elafibranor treatment for 1 year in NASH patients. According to the updated definition of resolution for NASH, elafibranor resolves NASH without fibrosis worsening, but only in patients with severe disease (NAS > 4). Elafibranor is not efficient in patients with mild disease (NAS < 4) and fails to demonstrate a beneficial effect on fibrosis [398]. In these human studies, elafibranor had a safety profile with no specific adverse effects.

The combination of its insulin-sensitizing and hepatoprotective effects makes elafibranor a good candidate for treating NAFLD. However, no beneficial effects on fibrosis have been demonstrated. It is currently being tested in a phase 3 clinical trial in NASH patients with fibrosis (NAS > 4).

4.3.3. PPARα and γ Dual Agonist Saroglitazar

Glitazars are dual PPARα/γ agonists developed to combine the beneficial effects of PPARα on plasma TGs and lipoproteins and PPARγ on insulin resistance. Most of these agonists have been discontinued due to adverse effects, but saroglitazar (Lipaglyn) was clinically approved in India in 2013 to treat diabetic dyslipidemia [399]. In vitro, saroglitazar has predominant activity on PPARα and moderate activity on PPARγ, reducing the adverse effects associated with PPARγ activation by pioglitazone [400].

In a diet-induced mouse model of NASH (choline-deficient, L-amino acid-defined, HFD), saroglitazar leads to a greater reduction in NAS than the PPARα agonist fenofibrate and PPARγ agonist pioglitazone. Histological examination of the liver tissue demonstrated a strong reduction in steatosis and decreased hepatocyte ballooning and inflammation, but only a trend in reduced fibrosis. Saroglitazar reduces hepatic expression of pro-inflammatory and pro-fibrotic genes. In vitro, saroglitazar decreases lipid-mediated oxidative stress and HSC activation. In addition, saroglitazar reduces CCl4-induced liver fibrosis in rats [401] and regulates adipose tissue homeostasis in mice [402]. In HFD-fed mice, saroglitazar improves serum TG levels and insulin resistance and reduces body weight and white adipose tissue mass. Histological examination of the adipose tissue has shown that saroglitazar reduces adipocyte hypertrophy by increasing the expression of thermogenic genes. In addition, saroglitazar treatment increases M2 macrophages and decreases M1 macrophages in adipose tissue, indicating that saroglitazar promotes an anti-inflammatory environment in adipose tissue [402]. In a rapid rat model of NASH induced by high-fat emulsion and small doses of LPS, saroglitazar improved adipocyte dysfunction through increased plasma adiponectin. In the liver, saroglitazar induced a decrease in TLR4 signaling upon LPS administration, with reduced NF-κB, TLR4, and TGFβ, which suggests a role of saroglitazar in response to gut endotoxins [403]. Saroglitazar also reduces thioacetamide-induced liver fibrosis in rats and decreases leptin, TGF-β, and platelet-derived growth factor (PDGF-BB) in the liver [404].

Several clinical studies have indicated that saroglitazar treatment in patients with diabetic dyslipidemia results in improved lipid and glucose parameters, including a reduction in plasma TG levels and fasting plasma glucose [405,406], and improves whole body insulin sensitivity in these patients [407]. In a review summarizing 18 studies on the effect of saroglitazar in patients with diabetic dyslipidemia, saroglitazar treatment was associated with a reduction in ALT levels and fatty liver in NAFLD patients with diabetic dyslipidemia [408]. A phase 2 study is evaluating the safety and efficacy of saroglitazar in patients with NASH. The primary endpoint is to assess the changes in NAS with no worsening of fibrosis from baseline to week 24 of treatment. The 16-week efficacy of saroglitazar in reducing serum ALT in NAFLD patients is also being tested in a phase 2 trial. Furthermore, two phase 3 clinical trials are currently evaluating saroglitazar in NAFLD with an amelioration of the fibrosis score as the primary outcome. The first study is investigating the efficacy of saroglitazar compared to pioglitazone in NAFLD patients over a period of 24 weeks. The second study is comparing the effect of combined saroglitazar and vitamin E treatment vs. vitamin E alone vs. saroglitazar alone (NCT04193982). Based on all observations thus far, saroglitazar shows promise as a potential NASH drug.

4.3.4. Pan-PPAR Agonist Lanifibranor

Lanifibranor (IVA337) is a moderately potent and well-balanced modulator of the three PPAR isotypes and has a good safety profile. Compared to glitazones, lanifibranor has demonstrated differences in co-regulator recruitment [409]. Treatment of db/db mice with lanifibranor induces a dose-dependent decrease in circulating glucose and TG levels [409]. Lanifibranor has also been tested in mouse models of NASH [410,411]. In the MCD model, lanifibranor reduces steatosis and hepatic TG levels, as well as inflammation. In foz/foz mice fed a HFD, lanifibranor attenuates steatosis, inflammation, and hepatocyte ballooning. Hepatic gene expression analysis has shown increased expression of genes involved in fatty acid beta-oxidation and decreased expression of pro-inflammatory and pro-fibrotic genes. In addition, lanifibranor treatment improves metabolic parameters, such as glucose intolerance, and increases plasma adiponectin levels [410]. Beneficial effects of lanifibranor on NASH histology, including reduced fibrosis, were confirmed recently in a preclinical model of NASH and fibrosis (choline-deficient amino acid-defined HFD mouse model). Interestingly, decreased macrophage infiltration in the liver has been observed upon lanifibranor treatment, suggesting that Kupffer cells may be important targets of lanifibranor to improve NAFLD. Similar results of NASH histology were obtained in the Western diet model, together with a reduction in plasma TG levels [411]. Lanifibranor is also effective in reducing collagen deposition and increasing plasma adiponectin in mice with CCl4-induced liver fibrosis [409,410]. In vitro results have demonstrated that lanifibranor inhibits the proliferation and activation of HSCs, as well as the activation of hepatic macrophages [410,411]. The anti-inflammatory and anti-fibrotic effects of lanifibranor were also demonstrated in preclinical mouse models of skin and pulmonary fibrosis [412,413].

Lanifibranor is currently undergoing phase 2 clinical trials in NAFLD. The first study is evaluating the efficacy and the safety of two doses of lanifibranor for 24 weeks vs. placebo in adult NASH patients with liver steatosis and moderate to severe necroinflammation without cirrhosis. The second study is designed to study lanifibranor in patients with T2DM and NAFLD.

The therapeutic potential of the novel PPAR agonists discussed above appears to be well-established in experimental models. However, none of the current preclinical models of NASH reproduce all features of human NASH [414]. In addition, differences exist between humans and mice regarding the PPARs. For example, hepatic PPARα expression is higher in rodents than in humans, which may explain why PPARα activation has stronger beneficial effects in rodent NAFLD. In this respect, "humanized" preclinical strategies, for example using transgenic mice expressing human PPARs or mice with a humanized liver may represent a relevant strategy for the evaluation of PPAR agonists [342]. In addition, a better understanding of the molecular mechanisms, especially the transcriptional coregulator network, underlying PPAR-dependent transcription in different species, tissues and diseases may be needed for designing more-specific and more-potent PPAR ligands [415]. Finally, PPAR functions are also regulated at the level of posttranslational modifications that influence protein stability and localization, ligand binding, and co-factor interaction. Understanding the role of these posttranslational modifications and their association with diseases might help in the development of novel molecules that specifically inhibit or promote such modifications [232].

In a relatively near future, the results of phase 2 and phase 3 clinical trials will determine the therapeutic potential of these novel compounds in NAFLD. Given the role of PPARs in multiple pathways involved in NAFLD and the beneficial effects of each single isotype agonist, we consider combined activation of several PPARs as a promising approach for NAFLD treatment because of potential optimization of the benefits and reduction of the side effects (Figure 5).

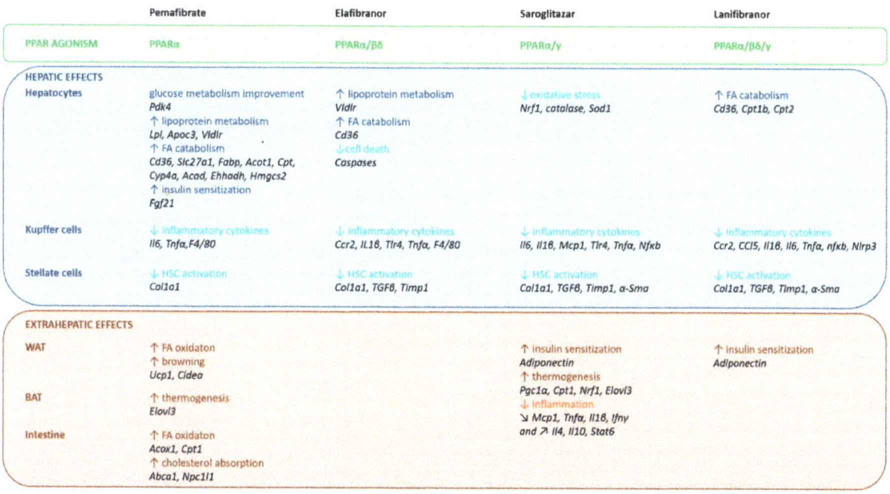

	Pemafibrate	Elafibranor	Saroglitazar	Lanifibranor
PPAR AGONISM	PPARα	PPARα/βδ	PPARα/γ	PPARα/βδ/γ
HEPATIC EFFECTS				
Hepatocytes	glucose metabolism improvement *Pdk4* ↑ lipoprotein metabolism *Lpl, Apoc3, Vldlr* ↑ FA catabolism *Cd36, Slc27a1, Fabp, Acot1, Cpt, Cyp4a, Acad, Ehhadh, Hmgcs2* ↑ insulin sensitization *Fgf21*	↑ lipoprotein metabolism *Vldlr* ↑ FA catabolism *Cd36* ↓ cell death *Caspases*	↓ oxidative stress *Nrf1, catalase, Sod1*	↑ FA catabolism *Cd36, Cpt1b, Cpt2*
Kupffer cells	↓ inflammatory cytokines *Il6, Tnfα, F4/80*	↓ inflammatory cytokines *Ccr2, IL1β, Tlr4, Tnfα, F4/80*	↓ inflammatory cytokines *Il6, Il1β, Mcp1, Tlr4, Tnfα, Nfκb*	↓ inflammatory cytokines *Ccr2, CCl5, Il1β, Il6, Tnfα, nfκb, Nlrp3*
Stellate cells	↓ HSC activation *Col1a1*	↓ HSC activation *Col1a1, TGFβ, Timp1*	↓ HSC activation *Col1a1, TGFβ, Timp1, α-Sma*	↓ HSC activation *Col1a1, TGFβ, Timp1, α-Sma*
EXTRAHEPATIC EFFECTS				
WAT	↑ FA oxidaton ↑ browning *Ucp1, Cidea*		↑ insulin sensitization *Adiponectin* ↑ thermogenesis *Pgc1α, Cpt1, Nrf1, Elovl3* ↓ inflammation ↘ *Mcp1, Tnfα, Il1β, Ifnγ* and ↗ *Il4, Il10, Stat6*	↑ insulin sensitization *Adiponectin*
BAT	↑ thermogenesis *Elovl3*			
Intestine	↑ FA oxidaton *Acox1, Cpt1* ↑ cholesterol absorption *Abca1, Npc1l1*			

Figure 5. Novel PPAR agonists with potential for NAFLD treatment. The main genes regulated by the PPARα agonist pemafibrate, the dual PPARα and β/δ agonist elafibranor, the dual PPARα and γ agonist saroglitazar, and the pan-PPAR agonist lanifibranor in preclinical models of NAFLD. Through the modulation of gene expression, these novel compounds regulate several hepatic and extrahepatic pathways, including lipid and glucose metabolism, inflammation, and hepatic stellate cell activation, which are all key processes involved in NAFLD. Abbreviations: BAT, brown adipose tissue; WAT, white adipose tissue; Pdk4, pyruvate dehydrogenase kinase 4; Lpl, lipoprotein lipase; ApoC3, apolipoprotein C3; Vldlr, very-low density lipoprotein receptor; Cpt, carnitine palmitoyltransferase; Slc27a1, solute carrier family 27 member 1; Fabp, fatty acid binding protein; Cyp4a, cytochrome P450 family 4 subfamily A; Ehhadh, enoyl-CoA hydratase and 3-hydroxyacyl CoA dehydrogenase; Acad, acyl-CoA dehydrogenase; Hmgcs2, 3-hydroxy-3-methylglutaryl-CoA synthase 2; Acot1, acyl-CoA thioesterase 1; Fgf21, fibroblast growth factor 21; Nrf1, nuclear respiratory factor 1; Sod1, superoxide dismutase 1; Il1β, interleukin-1 beta; Il6, interleukin-6; Tnfα, tumor necrosis factor-alpha; Ccr2, C-C motif chemokine receptor 2; Tlr4, Toll-like receptor 4; Mcp1, monocyte chemoattractant protein 1; Ccl5, C-C motif chemokine ligand 5; Nlrp3, NLR family pyrin domain containing 3; Nfκb, nuclear factor kappa B subunit 1; Tgfβ, transforming growth factor beta; Col1a1, collagen type I alpha 1 chain; Timp1, metalloproteinase inhibitor 1; α-Sma, alpha-smooth muscle actin; Ucp1, uncoupling protein 1; Cidea, cell death inducing DFFA like effector A; Elovl3, elongation of very long chain fatty acids protein 3; Acox, peroxisomal acyl-coenzyme A oxidase 1; Abca1, ATP binding cassette subfamily A member 1; Npc1l1, Niemann-Pick C1-like protein 1; Pgc1α, peroxisome proliferator-activated receptor gamma coactivator 1-alpha; Ifnγ, interferon gamma; Il4, interleukin-4; Il10, interleukin-10; Stat6, signal transducer and activator of transcription 6.

5. Concluding Remarks

The prevalence of NAFLD is dramatically increasing in developed countries, but no approved therapy is available. Most of the current pharmacological strategies target comorbidities, such as the manifestations of metabolic syndrome. Vitamin E and pioglitazone have beneficial effects on steatosis and inflammation, but can induce adverse effects in some patients. In addition, none of the currently used medications improve fibrosis, which is the strongest indicator of mortality in NAFLD. As highlighted in this review, the pathogenesis of NAFLD is multifactorial, which represents both a challenge and an opportunity for developing intervention strategies. As regulators of gene expression, the three PPARs impact, in some way, all currently known functions associated with NAFLD pathogenesis. The PPARs have emerged as crucial regulators of the whole organism and cellular metabolic functions. As links between lipid signaling and inflammation, they also fine-tune the crosstalk between metabolic processes and the innate immune system. All of these attributes make them relevant targets for treating NAFLD. Different approaches may be successful. One approach would be to selectively modify the pharmacological characteristics of agonists, as has been done with the PPARα selective modulator pemafibrate, to ameliorate the profile of beneficial effects with respect to issues associated with fibrate treatment. Developing molecules simultaneously targeting two or all three PPAR isotypes is another promising approach for NAFLD treatment that allows targeting of the multifaceted roles of PPARs. The dual agonists—elafibranor and saroglitazar—and the pan agonist lanifibranor have demonstrated many beneficial effects on liver histology with minimal adverse effects. Some of these novel agonists are currently in phase 3 clinical trials and appear promising for NASH treatment. As discussed herein, PPARs not only impact liver, but also other organs. In particular, they can have both positive and negative effects on heart physiology, pathology and injury [416,417]. Therefore, it is of importance that the new potential NASH drugs are evaluated for potential beneficial as well as deleterious effects on cardiac functions. In a post hoc analysis, elafibranor resolved NASH without fibrosis worsening and did not cause cardiac events [398]. Saroglitazar showed a potential to lower the cardiovascular risk in T2DM patients [418]. Pemafibrate is currently being tested for its effect on reducing cardiovascular events in diabetic patients with high TG levels in the PROMINENT study (NCT03071692) [390].

In parallel with the study of these novel promising agonists, it will be important to increase knowledge of the liver-specific functions of the PPAR isotypes, particularly in hepatocytes, Kupffer cells, and HSCs, and deepen our understanding of their roles in inflammation and fibrosis. Drugs that combine PPAR activation and other PPAR-independent pathways, which converge in ameliorating the manifestations of NAFLD, are also worth exploring. Interestingly, telmisartan, an angiotensin receptor blocker, is also a PPARα/γ dual agonist and worth exploring in the treatment of NAFLD. Inhibition of angiotensin converting enzyme (ACE) and angiotensin II type 1 receptor (AT1) improves liver fibrosis by keeping HSCs in a quiescent state through suppression of TGF-β [419]. These effects, combined with those known of PPARα/γ activation, deserve further investigation. Given the importance of circadian clock proteins in coordinating energy metabolism, the clock regulation of drug targets should also be taken into account in the development of pharmaceuticals for the treatment of NAFLD [420]. Despite recent remarkable and fast progress in the field, there are still many challenges imposed by the complex physiopathology underlying the development and progression of NAFLD, not least of all its heterogeneity, which is not fully understood. For example, why some patients will progress to advanced stages and others will not is not clear, and NAFLD in lean patients is also not completely understood.

Author Contributions: Conceptualization, A.F., W.W., H.G.; Writing—Original Draft Preparation, A.F., W.W., N.L.; Writing—Review & Editing, A.F., W.W., H.G., A.M., N.L.; Funding Acquisition, A.F., W.W., H.G., A.M., N.L. All authors have read and agreed to the published version of the manuscript.

Funding: A.F. is supported by the AgreenSkills+ fellowship program. W.W. was supported by a Start-Up Grant from the Lee Kong Chian School of Medicine, Nanyang Technological University Singapore. This work was also funded by ANR Hepadialogue and by grants from Région Occitanie (to A.F., W.W., H.G., A.M., N.L.).

Conflicts of Interest: The authors declare no conflicts of interest.

References

1. Younossi, Z.M.; Koenig, A.B.; Abdelatif, D.; Fazel, Y.; Henry, L.; Wymer, M. Global epidemiology of nonalcoholic fatty liver disease-Meta-analytic assessment of prevalence, incidence, and outcomes. *Hepatology* **2016**, *64*, 73–84. [CrossRef]
2. Zhou, F.; Zhou, J.; Wang, W.; Zhang, X.J.; Ji, Y.X.; Zhang, P.; She, Z.G.; Zhu, L.; Cai, J.; Li, H. Unexpected Rapid Increase in the Burden of NAFLD in China From 2008 to 2018: A Systematic Review and Meta-Analysis. *Hepatology* **2019**, *70*, 1119–1133. [CrossRef] [PubMed]
3. Allen, A.M.; Therneau, T.M.; Larson, J.J.; Coward, A.; Somers, V.K.; Kamath, P.S. Nonalcoholic fatty liver disease incidence and impact on metabolic burden and death: A 20 year-community study. *Hepatology* **2018**, *67*, 1726–1736. [CrossRef] [PubMed]

4.   Temple, J.L.; Cordero, P.; Li, J.; Nguyen, V.; Oben, J.A. A guide to non-alcoholic fatty liver disease in childhood and adolescence. *Int. J. Mol. Sci.* **2016**, *17*, 947. [CrossRef] [PubMed]

5.   Goldner, D.; Lavine, J.E. Nonalcoholic Fatty Liver Disease in Children: Unique Considerations and Challenges. *Gastroenterology* **2020**. [CrossRef]

6.   Younes, R.; Bugianesi, E. NASH in Lean Individuals. *Semin. Liver Dis.* **2019**, *39*, 86–95. [CrossRef]

7.   Lonardo, A.; Bellentani, S.; Argo, C.K.; Ballestri, S.; Byrne, C.D.; Caldwell, S.H.; Cortez-Pinto, H.; Grieco, A.; Machado, M.V.; Miele, L.; et al. Epidemiological modifiers of non-alcoholic fatty liver disease: Focus on high-risk groups. *Dig. Liver Dis.* **2015**, *47*, 997–1006. [CrossRef]

8.   Lonardo, A.; Nascimbeni, F.; Ballestri, S.; Fairweather, D.; Win, S.; Than, T.A.; Abdelmalek, M.F.; Suzuki, A. Sex Differences in NAFLD: State of the Art and Identification of Research Gaps. *Hepatology* **2019**, *70*, 1457–1469. [CrossRef]

9.   Vandel, J.; Dubois-Chevalier, J.; Gheeraert, C.; Derudas, B.; Raverdy, V.; Thuillier, D.; Van Gaal, L.; Francque, S.; Pattou, F.; Staels, B.; et al. Hepatic molecular signatures highlight the sexual dimorphism of Non-Alcoholic SteatoHepatitis (NASH). *Hepatology* **2020**. [CrossRef]

10.   Rich, N.E.; Oji, S.; Mufti, A.R.; Browning, J.D.; Parikh, N.D.; Odewole, M.; Mayo, H.; Singal, A.G. Racial and Ethnic Disparities in Nonalcoholic Fatty Liver Disease Prevalence, Severity, and Outcomes in the United States: A Systematic Review and Meta-analysis. *Clin. Gastroenterol. Hepatol.* **2018**, *16*, 198–210.e2. [CrossRef]

11.   Younossi, Z.; Anstee, Q.M.; Marietti, M.; Hardy, T.; Henry, L.; Eslam, M.; George, J.; Bugianesi, E. Global burden of NAFLD and NASH: Trends, predictions, risk factors and prevention. *Nat. Rev. Gastroenterol. Hepatol.* **2018**, *15*, 11–20. [CrossRef] [PubMed]

12.   Eslam, M.; Valenti, L.; Romeo, S. Genetics and epigenetics of NAFLD and NASH: Clinical impact. *J. Hepatol.* **2018**, *68*, 268–279. [CrossRef]

13.   Trépo, E.; Valenti, L. Update on NAFLD genetics: From new variants to the clinic. *J. Hepatol.* **2020**, *72*, 1196–1209. [CrossRef] [PubMed]

14.   Romeo, S.; Kozlitina, J.; Xing, C.; Pertsemlidis, A.; Cox, D.; Pennacchio, L.A.; Boerwinkle, E.; Cohen, J.C.; Hobbs, H.H. Genetic variation in PNPLA3 confers susceptibility to nonalcoholic fatty liver disease. *Nat. Genet.* **2008**, *40*, 1461–1465. [CrossRef] [PubMed]

15.   Pingitore, P.; Pirazzi, C.; Mancina, R.M.; Motta, B.M.; Indiveri, C.; Pujia, A.; Montalcini, T.; Hedfalk, K.; Romeo, S. Recombinant PNPLA3 protein shows triglyceride hydrolase activity and its I148M mutation results in loss of function. *Biochim. Biophys. Acta Mol. Cell Biol. Lipids* **2014**, *1841*, 574–580. [CrossRef]

16.   Pirazzi, C.; Valenti, L.; Motta, B.M.; Pingitore, P.; Hedfalk, K.; Mancina, R.M.; Burza, M.A.; Indiveri, C.; Ferro, Y.; Montalcini, T.; et al. PNPLA3 has retinyl-palmitate lipase activity in human hepatic stellate cells. *Hum. Mol. Genet.* **2014**, *23*, 4077–4085. [CrossRef] [PubMed]

17.   BasuRay, S.; Smagris, E.; Cohen, J.C.; Hobbs, H.H. The PNPLA3 variant associated with fatty liver disease (I148M) accumulates on lipid droplets by evading ubiquitylation. *Hepatology* **2017**, *66*, 1111–1124. [CrossRef] [PubMed]

18.   Li, J.Z.; Huang, Y.; Karaman, R.; Ivanova, P.T.; Brown, H.A.; Roddy, T.; Castro-Perez, J.; Cohen, J.C.; Hobbs, H.H. Chronic overexpression of PNPLA3 I148M in mouse liver causes hepatic steatosis. *J. Clin. Invest.* **2012**, *122*, 4130–4144. [CrossRef]

19.   Martínez, L.A.; Larrieta, E.; Kershenobich, D.; Torre, A. The Expression of PNPLA3 Polymorphism could be the Key for Severe Liver Disease in NAFLD in Hispanic Population. *Ann. Hepatol.* **2017**, *16*, 909–915. [CrossRef]

20.   Krawczyk, M.; Liebe, R.; Lammert, F. Toward Genetic Prediction of Nonalcoholic Fatty Liver Disease Trajectories: PNPLA3 and Beyond. *Gastroenterology* **2020**. [CrossRef]

21.   Kozlitina, J.; Smagris, E.; Stender, S.; Nordestgaard, B.G.; Zhou, H.H.; Tybjærg-Hansen, A.; Vogt, T.F.; Hobbs, H.H.; Cohen, J.C. Exome-wide association study identifies a TM6SF2 variant that confers susceptibility to nonalcoholic fatty liver disease. *Nat. Genet.* **2014**, *46*, 352–356. [CrossRef] [PubMed]

22.   Liu, Y.-L.; Reeves, H.L.; Burt, A.D.; Tiniakos, D.; Mcpherson, S.; Leathart, J.B.S.; Allison, M.E.D.; Alexander, G.J.; Piguet, A.-C.; Anty, R.; et al. ARTICLE TM6SF2 rs58542926 influences hepatic fibrosis progression in patients with non-alcoholic fatty liver disease. *Nat. Commun.* **2014**, *5*, 4309. [CrossRef] [PubMed]

23.   Kahali, B.; Liu, Y.L.; Daly, A.K.; Day, C.P.; Anstee, Q.M.; Speliotes, E.K. TM6SF2: Catch-22 in the fight against nonalcoholic fatty liver disease and cardiovascular disease? *Gastroenterology* **2015**, *148*, 679–684. [CrossRef] [PubMed]

24. Holmen, O.L.; Zhang, H.; Fan, Y.; Hovelson, D.H.; Schmidt, E.M.; Zhou, W.; Guo, Y.; Zhang, J.; Langhammer, A.; Løchen, M.L.; et al. Systematic evaluation of coding variation identifies a candidate causal variant in TM6SF2 influencing total cholesterol and myocardial infarction risk. *Nat. Genet.* **2014**, *46*, 345–351. [CrossRef] [PubMed]

25. Dongiovanni, P.; Petta, S.; Maglio, C.; Fracanzani, A.L.; Pipitone, R.; Mozzi, E.; Motta, B.M.; Kaminska, D.; Rametta, R.; Grimaudo, S.; et al. Transmembrane 6 superfamily member 2 gene variant disentangles nonalcoholic steatohepatitis from cardiovascular disease. *Hepatology* **2015**, *61*, 506–514. [CrossRef] [PubMed]

26. Ehrhardt, N.; Doche, M.E.; Chen, S.; Mao, H.Z.; Walsh, M.T.; Bedoya, C.; Guindi, M.; Xiong, W.; Ignatius Irudayam, J.; Iqbal, J.; et al. Hepatic Tm6sf2 overexpression affects cellular ApoB-trafficking, plasma lipid levels, hepatic steatosis and atherosclerosis. *Hum. Mol. Genet.* **2017**, *26*, 2719–2731. [CrossRef]

27. Mancina, R.M.; Dongiovanni, P.; Petta, S.; Pingitore, P.; Meroni, M.; Rametta, R.; Borén, J.; Montalcini, T.; Pujia, A.; Wiklund, O.; et al. The MBOAT7-TMC4 Variant rs641738 Increases Risk of Nonalcoholic Fatty Liver Disease in Individuals of European Descent. *Gastroenterology* **2016**, *150*, 1219–1230.e6. [CrossRef]

28. Luukkonen, P.K.; Zhou, Y.; Hyötyläinen, T.; Leivonen, M.; Arola, J.; Orho-Melander, M.; Orešič, M.; Yki-Järvinen, H. The MBOAT7 variant rs641738 alters hepatic phosphatidylinositols and increases severity of non-alcoholic fatty liver disease in humans. *J. Hepatol.* **2016**, *65*, 1263–1265. [CrossRef]

29. Donati, B.; Dongiovanni, P.; Romeo, S.; Meroni, M.; McCain, M.; Miele, L.; Petta, S.; Maier, S.; Rosso, C.; De Luca, L.; et al. MBOAT7 rs641738 variant and hepatocellular carcinoma in non-cirrhotic individuals. *Sci. Rep.* **2017**, *7*, 4492. [CrossRef]

30. Helsley, R.N.; Varadharajan, V.; Brown, A.L.; Gromovsky, A.D.; Schugar, R.C.; Ramachandiran, I.; Fung, K.; Kabbany, M.N.; Banerjee, R.; Neumann, C.; et al. Obesity-linked suppression of membrane-bound o-acyltransferase 7 (MBOAT7) drives non-alcoholic fatty liver disease. *Elife* **2019**, *8*, 1–69. [CrossRef]

31. Speliotes, E.K.; Yerges-Armstrong, L.M.; Wu, J.; Hernaez, R.; Kim, L.J.; Palmer, C.D.; Gudnason, V.; Eiriksdottir, G.; Garcia, M.E.; Launer, L.J.; et al. Genome-wide association analysis identifies variants associated with nonalcoholic fatty liver disease that have distinct effects on metabolic traits. *PLoS Genet.* **2011**, *7*. [CrossRef]

32. Beer, N.L.; Tribble, N.D.; McCulloch, L.J.; Roos, C.; Johnson, P.R.V.; Orho-Melander, M.; Gloyn, A.L. The P446L variant in GCKR associated with fasting plasma glucose and triglyceride levels exerts its effect through increased glucokinase activity in liver. *Hum. Mol. Genet.* **2009**, *18*, 4081–4088. [CrossRef] [PubMed]

33. Santoro, N.; Zhang, C.K.; Zhao, H.; Pakstis, A.J.; Kim, G.; Kursawe, R.; Dykas, D.J.; Bale, A.E.; Giannini, C.; Pierpont, B.; et al. Variant in the glucokinase regulatory protein (GCKR) gene is associated with fatty liver in obese children and adolescents. *Hepatology* **2012**, *55*, 781–789. [CrossRef] [PubMed]

34. Ma, Y.; Belyaeva, O.V.; Brown, P.M.; Fujita, K.; Valles, K.; Karki, S.; de Boer, Y.S.; Koh, C.; Chen, Y.; Du, X.; et al. 17-Beta Hydroxysteroid Dehydrogenase 13 Is a Hepatic Retinol Dehydrogenase Associated With Histological Features of Nonalcoholic Fatty Liver Disease. *Hepatology* **2019**, *69*, 1504–1519. [CrossRef] [PubMed]

35. Anstee, Q.M.; Darlay, R.; Cockell, S.; Meroni, M.; Govaere, O.; Tiniakos, D.; Burt, A.D.; Bedossa, P.; Palmer, J.; Liu, Y.-L.; et al. Genome-wide association study of non-alcoholic fatty liver and steatohepatitis in a histologically-characterised cohort. *J. Hepatol.* **2020**. [CrossRef]

36. Dongiovanni, P.; Meroni, M.; Mancina, R.M.; Baselli, G.; Rametta, R.; Pelusi, S.; Männistö, V.; Fracanzani, A.L.; Badiali, S.; Miele, L.; et al. Protein phosphatase 1 regulatory subunit 3B gene variation protects against hepatic fat accumulation and fibrosis in individuals at high risk of nonalcoholic fatty liver disease. *Hepatol. Commun.* **2018**, *2*, 666–675. [CrossRef]

37. Liu, Q.; Xue, F.; Meng, J.; Liu, S.S.; Chen, L.Z.; Gao, H.; Geng, N.; Jin, W.W.; Xin, Y.N.; Xuan, S.Y. TRIB1 rs17321515 and rs2954029 gene polymorphisms increase the risk of non-alcoholic fatty liver disease in Chinese Han population. *Lipids Health Dis.* **2019**, *18*. [CrossRef]

38. Dongiovanni, P.; Crudele, A.; Panera, N.; Romito, I.; Meroni, M.; De Stefanis, C.; Palma, A.; Comparcola, D.; Fracanzani, A.L.; Miele, L.; et al. β-Klotho gene variation is associated with liver damage in children with NAFLD. *J. Hepatol.* **2020**, *72*, 411–419. [CrossRef]

39. Kempinska-Podhorodecka, A.; Wunsch, E.; Milkiewicz, P.; Stachowska, E.; Milkiewicz, M. The Association between SOCS1−1656G>A Polymorphism, Insulin Resistance and Obesity in Nonalcoholic Fatty Liver Disease (NAFLD) Patients. *J. Clin. Med.* **2019**, *8*, 1912. [CrossRef]

40. Sookoian, S.; Pirola, C.J.; Valenti, L.; Davidson, N.O. Genetic pathways in nonalcoholic fatty liver disease: Insights from systems biology. *Hepatology.* **2020**. [CrossRef]

41. Eslam, M.; George, J. Genetic contributions to NAFLD: Leveraging shared genetics to uncover systems biology. *Nat. Rev. Gastroenterol. Hepatol.* **2020**, *17*, 40–52. [CrossRef] [PubMed]

42. Marchesini, G.; Bugianesi, E.; Forlani, G.; Cerrelli, F.; Lenzi, M.; Manini, R.; Natale, S.; Vanni, E.; Villanova, N.; Melchionda, N.; et al. Nonalcoholic fatty liver, steatohepatitis, and the metabolic syndrome. *Hepatology* **2003**, *37*, 917–923. [CrossRef] [PubMed]

43. Smits, M.M.; Ioannou, G.N.; Boyko, E.J.; Utzschneider, K.M. Non-alcoholic fatty liver disease as an independent manifestation of the metabolic syndrome: Results of a US national survey in three ethnic groups. *J. Gastroenterol. Hepatol.* **2013**, *28*, 664–670. [CrossRef] [PubMed]

44. Portillo-Sanchez, P.; Bril, F.; Maximos, M.; Lomonaco, R.; Biernacki, D.; Orsak, B.; Subbarayan, S.; Webb, A.; Hecht, J.; Cusi, K. High prevalence of nonalcoholic fatty liver disease in patients with type 2 diabetes mellitus and normal plasma aminotransferase levels. *J. Clin. Endocrinol. Metab.* **2015**, *100*, 2231–2238. [CrossRef] [PubMed]

45. Dai, W.; Ye, L.; Liu, A.; Wen, S.W.; Deng, J.; Wu, X.; Lai, Z. Prevalence of nonalcoholic fatty liver disease in patients with type 2 diabetes mellitus: A meta-analysis. *Medicine (USA)* **2017**, *96*, e8179. [CrossRef]

46. Leite, N.C.; Villela-Nogueira, C.A.; Pannain, V.L.N.; Bottino, A.C.; Rezende, G.F.M.; Cardoso, C.R.L.; Salles, G.F. Histopathological stages of nonalcoholic fatty liver disease in type 2 diabetes: Prevalences and correlated factors. *Liver Int.* **2011**, *31*, 700–706. [CrossRef]

47. Loomba, R.; Abraham, M.; Unalp, A.; Wilson, L.; Lavine, J.; Doo, E.; Bass, N.M. Association between diabetes, family history of diabetes, and risk of nonalcoholic steatohepatitis and fibrosis. *Hepatology* **2012**, *56*, 943–951. [CrossRef]

48. Li, L.; Liu, D.W.; Yan, H.Y.; Wang, Z.Y.; Zhao, S.H.; Wang, B. Obesity is an independent risk factor for non-alcoholic fatty liver disease: Evidence from a meta-analysis of 21 cohort studies. *Obes. Rev.* **2016**, *17*, 510–519. [CrossRef] [PubMed]

49. Katrina Loomis, A.; Kabadi, S.; Preiss, D.; Hyde, C.; Bonato, V.; Louis, M.S.; Desai, J.; Gill, J.M.R.; Welsh, P.; Waterworth, D.; et al. Body mass index and risk of nonalcoholic fatty liver disease: Two electronic health record prospective studies. *J. Clin. Endocrinol. Metab.* **2016**, *101*, 945–952. [CrossRef] [PubMed]

50. Yu, S.J.; Kim, W.; Kim, D.; Yoon, J.H.; Lee, K.; Kim, J.H.; Cho, E.J.; Lee, J.H.; Kim, H.Y.; Kim, Y.J.; et al. Visceral obesity predicts significant fibrosis in patients with nonalcoholic fatty liver disease. *Medicine (USA)* **2015**, *94*. [CrossRef] [PubMed]

51. van der Poorten, D.; Milner, K.L.; Hui, J.; Hodge, A.; Trenell, M.I.; Kench, J.G.; London, R.; Peduto, T.; Chisholm, D.J.; George, J. Visceral fat: A key mediator of steatohepatitis in metabolic liver disease. *Hepatology* **2008**, *48*, 449–457. [CrossRef] [PubMed]

52. Kim, D.; Chung, G.E.; Kwak, M.S.; Seo, H.B.; Kang, J.H.; Kim, W.; Kim, Y.J.; Yoon, J.H.; Lee, H.S.; Kim, C.Y. Body Fat Distribution and Risk of Incident and Regressed Nonalcoholic Fatty Liver Disease. *Clin. Gastroenterol. Hepatol.* **2016**, *14*, 132–138.e4. [CrossRef] [PubMed]

53. Du, T.; Sun, X.; Yuan, G.; Zhou, X.; Lu, H.; Lin, X.; Yu, X. Lipid phenotypes in patients with nonalcoholic fatty liver disease. *Metabolism* **2016**, *65*, 1391–1398. [CrossRef] [PubMed]

54. Sugino, I.; Kuboki, K.; Matsumoto, T.; Murakami, E.; Nishimura, C.; Yoshino, G. Influence of fatty liver on plasma small, dense LDL- cholesterol in subjects with and without metabolic syndrome. *J. Atheroscler. Thromb.* **2011**, *18*, 1–7. [CrossRef]

55. Musso, G.; Gambino, R.; De Michieli, F.; Cassader, M.; Rizzetto, M.; Durazzo, M.; Fagà, E.; Silli, B.; Pagano, G. Dietary habits and their relations to insulin resistance and postprandial lipemia in nonalcoholic steatohepatitis. *Hepatology* **2003**, *37*, 909–916. [CrossRef]

56. Katsiki, N.; Mikhailidis, D.P.; Mantzoros, C.S. Non-alcoholic fatty liver disease and dyslipidemia: An update. *Metabolism* **2016**, *65*, 1109–1123. [CrossRef]

57. Wu, S.J.; Zou, H.; Zhu, G.Q.; Wang, L.R.; Zhang, Q.; Shi, K.Q.; Han, J.B.; Huang, W.J.; Braddock, M.; Chen, Y.P.; et al. Increased levels of systolic blood pressure within the normal range are associated with significantly elevated risks of nonalcoholic fatty liver disease. *Medicine (USA)* **2015**, *94*. [CrossRef]

58. Singh, S.; Allen, A.M.; Wang, Z.; Prokop, L.J.; Murad, M.H.; Loomba, R. Fibrosis Progression in Nonalcoholic Fatty Liver vs Nonalcoholic Steatohepatitis: A Systematic Review and Meta-analysis of Paired-Biopsy Studies. *Clin. Gastroenterol. Hepatol.* **2015**, *13*, 643–654.e9. [CrossRef]

59. Barrera, F.; George, J. The role of diet and nutritional intervention for the management of patients with NAFLD. *Clin. Liver Dis.* **2014**, *18*, 91–112. [CrossRef]

60. Berná, G.; Romero-Gomez, M. The role of nutrition in non-alcoholic fatty liver disease: Pathophysiology and management. *Liver Int.* **2020**, *40*, 102–108. [CrossRef]

61. Torres, M.C.P.; Aghemo, A.; Lleo, A.; Bodini, G.; Furnari, M.; Marabotto, E.; Miele, L.; Giannini, E.G. Mediterranean diet and NAFLD: What we know and questions that still need to be answered. *Nutrients* **2019**, *11*, 2971. [CrossRef] [PubMed]

62. Marjot, T.; Moolla, A.; Cobbold, J.F.; Hodson, L.; Tomlinson, J.W. Nonalcoholic Fatty Liver Disease in Adults: Current Concepts in Etiology, Outcomes, and Management. *Endocr. Rev.* **2020**, *41*, 66–117. [CrossRef]

63. Ouyang, X.; Cirillo, P.; Sautin, Y.; McCall, S.; Bruchette, J.L.; Diehl, A.M.; Johnson, R.J.; Abdelmalek, M.F. Fructose consumption as a risk factor for non-alcoholic fatty liver disease. *J. Hepatol.* **2008**, *48*, 993–999. [CrossRef] [PubMed]

64. Abdelmalek, M.F.; Suzuki, A.; Guy, C.; Unalp-Arida, A.; Colvin, R.; Johnson, R.J.; Diehl, A.M. Nonalcoholic Steatohepatitis Clinical Research Network Increased fructose consumption is associated with fibrosis severity in patients with nonalcoholic fatty liver disease. *Hepatology* **2010**, *51*, 1961–1971. [CrossRef] [PubMed]

65. Softic, S.; Cohen, D.E.; Kahn, C.R. Role of Dietary Fructose and Hepatic De Novo Lipogenesis in Fatty Liver Disease. *Dig. Dis. Sci.* **2016**, *61*, 1282–1293. [CrossRef]

66. Hannou, S.A.; Haslam, D.E.; McKeown, N.M.; Herman, M.A. Fructose metabolism and metabolic disease. *J. Clin. Investig.* **2018**, *128*, 545–555. [CrossRef]

67. Zhao, S.; Jang, C.; Liu, J.; Uehara, K.; Gilbert, M.; Izzo, L.; Zeng, X.; Trefely, S.; Fernandez, S.; Carrer, A.; et al. Dietary fructose feeds hepatic lipogenesis via microbiota-derived acetate. *Nature* **2020**, *579*, 586–591. [CrossRef]

68. Jensen, T.; Abdelmalek, M.F.; Sullivan, S.; Nadeau, K.J.; Green, M.; Roncal, C.; Nakagawa, T.; Kuwabara, M.; Sato, Y.; Kang, D.H.; et al. Fructose and sugar: A major mediator of non-alcoholic fatty liver disease. *J. Hepatol.* **2018**, *68*, 1063–1075. [CrossRef]

69. Yan, J.H.; Guan, B.J.; Gao, H.Y.; Peng, X.E. Omega-3 polyunsaturated fatty acid supplementation and non-alcoholic fatty liver disease: A meta-analysis of randomized controlled trials. *Medicine (USA)* **2018**, *97*, e12271. [CrossRef]

70. Argo, C.K.; Patrie, J.T.; Lackner, C.; Henry, T.D.; De Lange, E.E.; Weltman, A.L.; Shah, N.L.; Al-Osaimi, A.M.; Pramoonjago, P.; Jayakumar, S.; et al. Effects of n-3 fish oil on metabolic and histological parameters in NASH: A double-blind, randomized, placebo-controlled trial. *J. Hepatol.* **2015**, *62*, 190–197. [CrossRef]

71. Martin, P.G.P.; Guillou, H.; Lasserre, F.; Déjean, S.; Lan, A.; Pascussi, J.M.; SanCristobal, M.; Legrand, P.; Besse, P.; Pineau, T. Novel aspects of PPARα-mediated regulation of lipid and xenobiotic metabolism revealed through a nutrigenomic study. *Hepatology* **2007**, *45*, 767–777. [CrossRef]

72. Ducheix, S.; Montagner, A.; Polizzi, A.; Lasserre, F.; Marmugi, A.; Bertrand-Michel, J.; Podechard, N.; Al Saati, T.; Chétiveaux, M.; Baron, S.; et al. Essential fatty acids deficiency promotes lipogenic gene expression and hepatic steatosis through the liver X receptor. *J. Hepatol.* **2013**, *58*, 984–992. [CrossRef] [PubMed]

73. Dentin, R.; Benhamed, F.; Pégorier, J.P.; Foufelle, F.; Viollet, B.; Vaulont, S.; Girard, J.; Postic, C. Polyunsaturated fatty acids suppress glycolytic and lipogenic genes through the inhibition of ChREBP nuclear protein translocation. *J. Clin. Invest.* **2005**, *115*, 2843–2854. [CrossRef] [PubMed]

74. Sekiya, M.; Yahagi, N.; Matsuzaka, T.; Najima, Y.; Nakakuki, M.; Nagai, R.; Ishibashi, S.; Osuga, J.I.; Yamada, N.; Shimano, H. Polyunsaturated Fatty Acids Ameliorate Hepatic Steatosis in Obese Mice by SREBP-1 Suppression. *Hepatology* **2003**, *38*, 1529–1539. [CrossRef] [PubMed]

75. Lin, J.; Yang, R.; Tarr, P.T.; Wu, P.H.; Handschin, C.; Li, S.; Yang, W.; Pei, L.; Uldry, M.; Tontonoz, P.; et al. Hyperlipidemic effects of dietary saturated fats mediated through PGC-1β coactivation of SREBP. *Cell* **2005**, *120*, 261–273. [CrossRef] [PubMed]

76. Desvergne, B.; Michalik, L.; Wahli, W. Transcriptional regulation of metabolism. *Physiol. Rev.* **2006**, *86*, 465–514. [CrossRef]

77. Bozzetto, L.; Prinster, A.; Annuzzi, G.; Costagliola, L.; Mangione, A.; Vitelli, A.; Mazzarella, R.; Longobardo, M.; Mancini, M.; Vigorito, C.; et al. Liver fat is reduced by an isoenergetic MUFA diet in a controlled randomized study in type 2 diabetic patients. *Diabetes Care* **2012**, *35*, 1429–1435. [CrossRef]

78. Bozzetto, L.; Costabile, G.; Luongo, D.; Naviglio, D.; Cicala, V.; Piantadosi, C.; Patti, L.; Cipriano, P.; Annuzzi, G.; Rivellese, A.A. Reduction in liver fat by dietary MUFA in type 2 diabetes is helped by enhanced hepatic fat oxidation. *Diabetologia* **2016**, *59*, 2697–2701. [CrossRef]

79. Liang, J.Q.; Teoh, N.; Xu, L.; Pok, S.; Li, X.; Chu, E.S.H.; Chiu, J.; Dong, L.; Arfianti, E.; Haigh, W.G.; et al. Dietary cholesterol promotes steatohepatitis related hepatocellular carcinoma through dysregulated metabolism and calcium signaling. *Nat. Commun.* **2018**, *9*, 4490. [CrossRef]

80. Wang, X.; Cai, B.; Yang, X.; Sonubi, O.O.; Zheng, Z.; Ramakrishnan, R.; Shi, H.; Valenti, L.; Pajvani, U.B.; Sandhu, J.; et al. Cholesterol Stabilizes TAZ in Hepatocytes to Promote Experimental Non-alcoholic Steatohepatitis. *Cell Metab.* **2020**, *31*, 969–986.e7. [CrossRef]

81. Ioannou, G.N.; Morrow, O.B.; Connole, M.L.; Ioannou, G.N.; Kuver, R.; Lee, S.P.; Teoh, N.C.; Farrell, G.C. Association between dietary nutrient composition and the incidence of cirrhosis or liver cancer in the United States population. *Hepatology* **2009**, *50*, 175–184. [CrossRef] [PubMed]

82. Garcia-Caraballo, S.C.; Comhair, T.M.; Verheyen, F.; Gaemers, I.; Schaap, F.G.; Houten, S.M.; Hakvoort, T.B.M.; Dejong, C.H.C.; Lamers, W.H.; Koehler, S.E. Prevention and reversal of hepatic steatosis with a high-protein diet in mice. *Biochim. Biophys. Acta Mol. Basis Dis.* **2013**, *1832*, 685–695. [CrossRef] [PubMed]

83. Garcia Caraballo, S.C.; Comhair, T.M.; Houten, S.M.; Dejong, C.H.C.; Lamers, W.H.; Koehler, S.E. High-protein diets prevent steatosis and induce hepatic accumulation of monomethyl branched-chain fatty acids. *J. Nutr. Biochem.* **2014**, *25*, 1263–1274. [CrossRef]

84. Schwarz, J.; Tomé, D.; Baars, A.; Hooiveld, G.J.E.J.; Müller, M. Dietary Protein Affects Gene Expression and Prevents Lipid Accumulation in the Liver in Mice. *PLoS ONE* **2012**, *7*. [CrossRef] [PubMed]

85. Charidemou, E.; Ashmore, T.; Li, X.; McNally, B.D.; West, J.A.; Liggi, S.; Harvey, M.; Orford, E.; Griffin, J.L. A randomized 3-way crossover study indicates that high-protein feeding induces de novo lipogenesis in healthy humans. *JCI Insight* **2019**, *4*. [CrossRef]

86. Gaggini, M.; Carli, F.; Rosso, C.; Buzzigoli, E.; Marietti, M.; Della Latta, V.; Ciociaro, D.; Abate, M.L.; Gambino, R.; Cassader, M.; et al. Altered amino acid concentrations in NAFLD: Impact of obesity and insulin resistance. *Hepatology* **2018**, *67*, 145–158. [CrossRef]

87. Mardinoglu, A.; Agren, R.; Kampf, C.; Asplund, A.; Uhlen, M.; Nielsen, J. Genome-scale metabolic modelling of hepatocytes reveals serine deficiency in patients with non-alcoholic fatty liver disease. *Nat. Commun.* **2014**, *5*. [CrossRef]

88. Mardinoglu, A.; Bjornson, E.; Zhang, C.; Klevstig, M.; Söderlund, S.; Ståhlman, M.; Adiels, M.; Hakkarainen, A.; Lundbom, N.; Kilicarslan, M.; et al. Personal model-assisted identification of NAD + and glutathione metabolism as intervention target in NAFLD. *Mol. Syst. Biol.* **2017**, *13*, 916. [CrossRef]

89. Lake, A.D.; Novak, P.; Shipkova, P.; Aranibar, N.; Robertson, D.G.; Reily, M.D.; Lehman-Mckeeman, L.D.; Vaillancourt, R.R.; Cherrington, N.J. Branched chain amino acid metabolism profiles in progressive human nonalcoholic fatty liver disease. *Amino Acids* **2015**, *47*, 603–615. [CrossRef]

90. Hoyles, L.; Fernández-Real, J.-M.; Federici, M.; Serino, M.; Abbott, J.; Charpentier, J.; Heymes, C.; Luque, J.L.; Anthony, E.; Barton, R.H.; et al. Molecular phenomics and metagenomics of hepatic steatosis in non-diabetic obese women. *Nat. Med.* **2018**, *24*, 1070–1080. [CrossRef]

91. Grzych, G.; Vonghia, L.; Bout, M.-A.; Weyler, J.; Verrijken, A.; Dirinck, E.; Joncquel, M.; Van Gaal, L.; Paumelle, R.; Francque, S.; et al. Plasma BCAA changes in Patients with NAFLD are Sex Dependent. *J. Clin. Endocrinol. Metab.* **2020**. [CrossRef] [PubMed]

92. Zhang, F.; Zhao, S.; Yan, W.; Xia, Y.; Chen, X.; Wang, W.; Zhang, J.; Gao, C.; Peng, C.; Yan, F.; et al. Branched Chain Amino Acids Cause Liver Injury in Obese/Diabetic Mice by Promoting Adipocyte Lipolysis and Inhibiting Hepatic Autophagy. *EBioMedicine* **2016**, *13*, 157–167. [CrossRef] [PubMed]

93. Liu, S.; Liu, Y.; Wan, B.; Zhang, H.; Wu, S.; Zhu, Z.; Lin, Y.; Wang, M.; Zhang, N.; Lin, S.; et al. Association between vitamin d status and non-alcoholic fatty liver disease: A population-based study. *J. Nutr. Sci. Vitaminol. (Tokyo)* **2019**, *65*, 303–308. [CrossRef] [PubMed]

94. Sanyal, A.J.; Chalasani, N.; Kowdley, K.V.; McCullough, A.; Diehl, A.M.; Bass, N.M.; Neuschwander-Tetri, B.A.; Lavine, J.E.; Tonascia, J.; Unalp, A.; et al. Pioglitazone, Vitamin E, or Placebo for Nonalcoholic Steatohepatitis. *N. Engl. J. Med.* **2010**, *362*, 1675–1685. [CrossRef]

95. Nagashimada, M.; Ota, T. Role of vitamin E in nonalcoholic fatty liver disease. *IUBMB Life* **2019**, *71*, 516–522. [CrossRef]

96. Zein, C.O.; Unalp, A.; Colvin, R.; Liu, Y.C.; McCullough, A.J. Smoking and severity of hepatic fibrosis in nonalcoholic fatty liver disease. *J. Hepatol.* **2011**, *54*, 753–759. [CrossRef]

97.	La Merrill, M.A.; Johnson, C.L.; Smith, M.T.; Kandula, N.R.; Macherone, A.; Pennell, K.D.; Kanaya, A.M. Exposure to Persistent Organic Pollutants (POPs) and Their Relationship to Hepatic Fat and Insulin Insensitivity among Asian Indian Immigrants in the United States. *Environ. Sci. Technol.* **2019**, *53*, 13906–13918. [CrossRef] [PubMed]

98.	al-Eryani, L.; Wahlang, B.; Falkner, K.C.; Guardiola, J.J.; Clair, H.B.; Prough, R.A.; Cave, M. Identification of Environmental Chemicals Associated with the Development of Toxicant-associated Fatty Liver Disease in Rodents. *Toxicol. Pathol.* **2015**, *43*, 482–497. [CrossRef] [PubMed]

99.	Marmugi, A.; Ducheix, S.; Lasserre, F.; Polizzi, A.; Paris, A.; Priymenko, N.; Bertrand-Michel, J.; Pineau, T.; Guillou, H.; Martin, P.G.P.; et al. Low doses of bisphenol a induce gene expression related to lipid synthesis and trigger triglyceride accumulation in adult mouse liver. *Hepatology* **2012**, *55*, 395–407. [CrossRef] [PubMed]

100.	Armstrong, L.E.; Guo, G.L. Understanding Environmental Contaminants' Direct Effects on Non-alcoholic Fatty Liver Disease Progression. *Curr. Environ. Health. Rep.* **2019**, *6*, 95–104. [CrossRef] [PubMed]

101.	Schwenger, K.J.; Clermont-Dejean, N.; Allard, J.P. The role of the gut microbiome in chronic liver disease: The clinical evidence revised. *JHEP Reports* **2019**, *1*, 214–226. [CrossRef] [PubMed]

102.	Leung, C.; Rivera, L.; Furness, J.B.; Angus, P.W. The role of the gut microbiota in NAFLD. *Nat. Rev. Gastroenterol. Hepatol.* **2016**, *13*, 412–425. [CrossRef]

103.	Boursier, J.; Mueller, O.; Barret, M.; Machado, M.; Fizanne, L.; Araujo-Perez, F.; Guy, C.D.; Seed, P.C.; Rawls, J.F.; David, L.A.; et al. The severity of nonalcoholic fatty liver disease is associated with gut dysbiosis and shift in the metabolic function of the gut microbiota. *Hepatology* **2016**, *63*, 764–775. [CrossRef]

104.	Aron-Wisnewsky, J.; Vigliotti, C.; Witjes, J.; Le, P.; Holleboom, A.G.; Verheij, J.; Nieuwdorp, M.; Clément, K. Gut microbiota and human NAFLD: Disentangling microbial signatures from metabolic disorders. *Nat. Rev. Gastroenterol. Hepatol.* **2020**, *17*, 279–297. [CrossRef]

105.	Luther, J.; Garber, J.J.; Khalili, H.; Dave, M.; Bale, S.S.; Jindal, R.; Motola, D.L.; Luther, S.; Bohr, S.; Jeoung, S.W.; et al. Hepatic Injury in Nonalcoholic Steatohepatitis Contributes to Altered Intestinal Permeability. *Cell. Mol. Gastroenterol. Hepatol.* **2015**, *1*, 222–232.e2. [CrossRef]

106.	Kolodziejczyk, A.A.; Zheng, D.; Shibolet, O.; Elinav, E. The role of the microbiome in NAFLD and NASH. *EMBO Mol. Med.* **2019**, *11*. [CrossRef] [PubMed]

107.	Tilg, H.; Moschen, A.R. Evolution of inflammation in nonalcoholic fatty liver disease: The multiple parallel hits hypothesis. *Hepatology* **2010**, *52*, 1836–1846. [CrossRef] [PubMed]

108.	Donnelly, K.L.; Smith, C.I.; Schwarzenberg, S.J.; Jessurun, J.; Boldt, M.D.; Parks, E.J. Sources of fatty acids stored in liver and secreted via lipoproteins in patients with nonalcoholic fatty liver disease. *J. Clin. Invest.* **2005**, *115*. [CrossRef]

109.	Lambert, J.E.; Ramos–Roman, M.A.; Browning, J.D.; Parks, E.J.; Ramos-Roman, M.A.; Browning, J.D.; Parks, E.J. Increased De Novo Lipogenesis Is a Distinct Characteristic of Individuals With Nonalcoholic Fatty Liver Disease. *Gastroenterology* **2014**, *146*, 726–735. [CrossRef]

110.	Anghel, S.I.; Wahli, W. Fat poetry: A kingdom for PPARγ. *Cell Res.* **2007**, *17*, 486–511. [CrossRef]

111.	Azzu, V.; Vacca, M.; Virtue, S.; Allison, M.; Vidal-Puig, A. Adipose Tissue-Liver Cross Talk in the Control oef Whole-Body Metabolism: Implications in Nonalcoholic Fatty Liver Disease. *Gastroenterology* **2020**. [CrossRef] [PubMed]

112.	Santoleri, D.; Titchenell, P.M. Resolving the Paradox of Hepatic Insulin Resistance. *CMGH* **2019**, *7*, 447–456. [CrossRef] [PubMed]

113.	Shimano, H.; Yahagi, N.; Amemiya-Kudo, M.; Hasty, A.H.; Osuga, J.I.; Tamura, Y.; Shionoiri, F.; Iizuka, Y.; Ohashi, K.; Harada, K.; et al. Sterol regulatory element-binding protein-1 as a key transcription factor for nutritional induction of lipogenic enzyme genes. *J. Biol. Chem.* **1999**, *274*, 35832–35839. [CrossRef] [PubMed]

114.	Horton, J.D.; Goldstein, J.L.; Brown, M.S. SREBPs: Activators of the complete program of cholesterol and fatty acid synthesis in the liver. *J. Clin. Invest.* **2002**, *109*, 1125–1131. [CrossRef]

115.	Iizuka, K.; Bruick, R.K.; Liang, G.; Horton, J.D.; Uyeda, K. Deficiency of carbohydrate response element-binding protein (ChREBP) reduces lipogenesis as well as glycolysis. *Proc. Natl. Acad. Sci. USA* **2004**, *101*, 7281–7286. [CrossRef]

116.	Dentin, R.; Benhamed, F.; Hainault, I.; Fauveau, V.; Foufelle, F.; Dyck, J.R.B.; Girard, J.; Postic, C. Liver-specific inhibition of ChREBP improves hepatic steatosis and insulin resistance in ob/ob mice. *Diabetes* **2006**, *55*, 2159–2170. [CrossRef]

117. Linden, A.G.; Li, S.; Choi, H.Y.; Fang, F.; Fukasawa, M.; Uyeda, K.; Hammer, R.E.; Horton, J.D.; Engelking, L.J.; Liang, G. Interplay between ChREBP and SREBP-1c coordinates postprandial glycolysis and lipogenesis in livers of mice. *J. Lipid Res.* **2018**, *59*, 475–487. [CrossRef]

118. Liu, Q.; Yuan, B.; Lo, K.A.; Patterson, H.C.; Sun, Y.; Lodish, H.F. Adiponectin regulates expression of hepatic genes critical for glucose and lipid metabolism. *Proc. Natl. Acad. Sci. USA* **2012**, *109*, 14568–14573. [CrossRef]

119. Koliaki, C.; Szendroedi, J.; Kaul, K.; Jelenik, T.; Nowotny, P.; Jankowiak, F.; Herder, C.; Carstensen, M.; Krausch, M.; Knoefel, W.T.; et al. Adaptation of Hepatic Mitochondrial Function in Humans with Non-Alcoholic Fatty Liver Is Lost in Steatohepatitis. *Cell Metab.* **2015**, *21*, 739–746. [CrossRef]

120. Farrell, G.C.; Haczeyni, F.; Chitturi, S. Pathogenesis of NASH: How metabolic complications of overnutrition favour lipotoxicity and pro-inflammatory fatty liver disease. In *Advances in Experimental Medicine and Biology*; Springer LLC: New York, NY, USA, 2018; Volume 1061, pp. 19–44.

121. Ioannou, G.N.; Landis, C.S.; Jin, G.; Haigh, W.G.; Farrell, G.C.; Kuver, R.; Lee, S.P.; Savard, C. Cholesterol Crystals in Hepatocyte Lipid Droplets Are Strongly Associated With Human Nonalcoholic Steatohepatitis. *Hepatol. Commun.* **2019**, *3*, 776–791. [CrossRef]

122. Chiappini, F.; Coilly, A.; Kadar, H.; Gual, P.; Tran, A.; Desterke, C.; Samuel, D.; Duclos-Vallée, J.C.; Touboul, D.; Bertrand-Michel, J.; et al. Metabolism dysregulation induces a specific lipid signature of nonalcoholic steatohepatitis in patients. *Sci. Rep.* **2017**, *7*. [CrossRef] [PubMed]

123. Friedman, S.L.; Neuschwander-Tetri, B.A.; Rinella, M.; Sanyal, A.J. Mechanisms of NAFLD development and therapeutic strategies. *Nat. Med.* **2018**, *24*, 908–922. [CrossRef] [PubMed]

124. Adams, L.A.; Ratziu, V. *Journal of Hepatology*; Elsevier, B.V.: Amsterdam, The Netherlands, 1 May 2015; pp. 1002–1004.

125. Taylor, R.S.; Taylor, R.J.; Bayliss, S.; Hagström, H.; Nasr, P.; Schattenberg, J.M.; Ishigami, M.; Toyoda, H.; Wai-Sun Wong, V.; Peleg, N.; et al. Association Between Fibrosis Stage and Outcomes of Patients With Nonalcoholic Fatty Liver Disease: A Systematic Review and Meta-Analysis. *Gastroenterology* **2020**, *158*, 1611–1625.e12. [CrossRef] [PubMed]

126. Angulo, P.; Kleiner, D.E.; Dam-Larsen, S.; Adams, L.A.; Bjornsson, E.S.; Charatcharoenwitthaya, P.; Mills, P.R.; Keach, J.C.; Lafferty, H.D.; Stahler, A.; et al. Liver fibrosis, but no other histologic features, is associated with long-term outcomes of patients with nonalcoholic fatty liver disease. *Gastroenterology* **2015**, *149*. [CrossRef]

127. McPherson, S.; Hardy, T.; Henderson, E.; Burt, A.D.; Day, C.P.; Anstee, Q.M. Evidence of NAFLD progression from steatosis to fibrosing-steatohepatitis using paired biopsies: Implications for prognosis and clinical management. *J. Hepatol.* **2015**, *62*, 1148–1155. [CrossRef]

128. Kanwal, F.; Kramer, J.R.; Mapakshi, S.; Natarajan, Y.; Chayanupatkul, M.; Richardson, P.A.; Li, L.; Desiderio, R.; Thrift, A.P.; Asch, S.M.; et al. Risk of Hepatocellular Cancer in Patients With Non-Alcoholic Fatty Liver Disease. *Gastroenterology* **2018**, *155*, 1828–1837.e2. [CrossRef]

129. Reig, M.; Gambato, M.; Man, N.K.; Roberts, J.P.; Victor, D.; Orci, L.A.; Toso, C. Should Patients with NAFLD/NASH Be Surveyed for HCC? *Transplantation* **2019**, *103*, 39–44. [CrossRef]

130. Haas, J.T.; Francque, S.; Staels, B. Pathophysiology and Mechanisms of Nonalcoholic Fatty Liver Disease. *Annu. Rev. Physiol.* **2016**, *78*, 181–205. [CrossRef]

131. Sanna, C.; Rosso, C.; Marietti, M.; Bugianesi, E. Non-alcoholic fatty liver disease and extra-hepatic cancers. *Int. J. Mol. Sci.* **2016**, *17*, 717. [CrossRef]

132. Lonardo, A.; Nascimbeni, F.; Mantovani, A.; Targher, G. Hypertension, diabetes, atherosclerosis and NASH: Cause or consequence? *J. Hepatol.* **2018**, *68*, 335–352. [CrossRef]

133. Dongiovanni, P.; Valenti, L. A nutrigenomic approach to non-alcoholic fatty liver disease. *Int. J. Mol. Sci.* **2017**, *18*, 1534. [CrossRef] [PubMed]

134. Koutoukidis, D.A.; Astbury, N.M.; Tudor, K.E.; Morris, E.; Henry, J.A.; Noreik, M.; Jebb, S.A.; Aveyard, P. Association of Weight Loss Interventions with Changes in Biomarkers of Nonalcoholic Fatty Liver Disease: A Systematic Review and Meta-analysis. *JAMA Intern. Med.* **2019**, *179*, 1262–1271.e12. [CrossRef] [PubMed]

135. Keating, S.E.; Hackett, D.A.; George, J.; Johnson, N.A. Exercise and non-alcoholic fatty liver disease: A systematic review and meta-analysis. *J. Hepatol.* **2012**, *57*, 157–166. [CrossRef] [PubMed]

136. Cheng, S.; Ge, J.; Zhao, C.; Le, S.; Yang, Y.; Ke, D.; Wu, N.; Tan, X.; Zhang, X.; Du, X.; et al. Effect of aerobic exercise and diet on liver fat in pre-diabetic patients with non-alcoholic-fatty-liver-disease: A randomized controlled trial. *Sci. Rep.* **2017**, *7*. [CrossRef]

137. Smart, N.A.; King, N.; McFarlane, J.R.; Graham, P.L.; Dieberg, G. Effect of exercise training on liver function in adults who are overweight or exhibit fatty liver disease: A systematic review and meta-analysis. *Br. J. Sports Med.* **2018**, *52*, 834–843. [CrossRef]

138. Hallsworth, K.; Thoma, C.; Hollingsworth, K.G.; Cassidy, S.; Anstee, Q.M.; Day, C.P.; Trenell, M.I. Modified high-intensity interval training reduces liver fat and improves cardiac function in non-alcoholic fatty liver disease: A randomized controlled trial. *Clin. Sci.* **2015**, *129*, 1097–1105. [CrossRef]

139. Abdelbasset, W.K.; Tantawy, S.A.; Kamel, D.M.; Alqahtani, B.A.; Soliman, G.S. A randomized controlled trial on the effectiveness of 8-week high-intensity interval exercise on intrahepatic triglycerides, visceral lipids, and health-related quality of life in diabetic obese patients with nonalcoholic fatty liver disease. *Medicine (USA)* **2019**, *98*. [CrossRef]

140. Oh, S.; So, R.; Shida, T.; Matsuo, T.; Kim, B.; Akiyama, K.; Isobe, T.; Okamoto, Y.; Tanaka, K.; Shoda, J. High-intensity aerobic exercise improves both hepatic fat content and stiffness in sedentary obese men with nonalcoholic fatty liver disease. *Sci. Rep.* **2017**, *7*, 1–12. [CrossRef]

141. Hallsworth, K.; Fattakhova, G.; Hollingsworth, K.G.; Thoma, C.; Moore, S.; Taylor, R.; Day, C.P.; Trenell, M.I. Resistance exercise reduces liver fat and its mediators in non-alcoholic fatty liver disease independent of weight loss. *Gut* **2011**, *60*, 1278–1283. [CrossRef]

142. Zelber-Sagi, S.; Buch, A.; Yeshua, H.; Vaisman, N.; Webb, M.; Harari, G.; Kis, O.; Fliss-Isakov, N.; Izkhakov, E.; Halpern, Z.; et al. Effect of resistance training on non-alcoholic fatty-liver disease a randomized-clinical trial. *World J. Gastroenterol.* **2014**, *20*, 4382–4392. [CrossRef]

143. El-Agroudy, N.N.; Kurzbach, A.; Rodionov, R.N.; O'Sullivan, J.; Roden, M.; Birkenfeld, A.L.; Pesta, D.H. Are Lifestyle Therapies Effective for NAFLD Treatment? *Trends Endocrinol. Metab.* **2019**, *30*, 701–709. [CrossRef]

144. Hashida, R.; Kawaguchi, T.; Bekki, M.; Omoto, M.; Matsuse, H.; Nago, T.; Takano, Y.; Ueno, T.; Koga, H.; George, J.; et al. Aerobic vs. resistance exercise in non-alcoholic fatty liver disease: A systematic review. *J. Hepatol.* **2017**, *66*, 142–152. [CrossRef]

145. Katsagoni, C.N.; Georgoulis, M.; Papatheodoridis, G.V.; Panagiotakos, D.B.; Kontogianni, M.D. Effects of lifestyle interventions on clinical characteristics of patients with non-alcoholic fatty liver disease: A meta-analysis. *Metabolism* **2017**, *68*, 119–132. [CrossRef] [PubMed]

146. Golabi, P.; Locklear, C.T.; Austin, P.; Afdhal, S.; Byrns, M.; Gerber, L.; Younossi, Z.M. Effectiveness of exercise in hepatic fat mobilization in nonalcoholic fatty liver disease: Systematic review. *World J. Gastroenterol.* **2016**, *22*, 6318–6327. [CrossRef] [PubMed]

147. Wang, S.-T.; Zheng, J.; Peng, H.-W.; Cai, X.-L.; Pan, X.-T.; Li, H.-Q.; Hong, Q.-Z.; Peng, X.-E. Physical activity intervention for non-diabetic patients with non-alcoholic fatty liver disease: A meta-analysis of randomized controlled trials. *BMC Gastroenterol.* **2020**, *20*, 66. [CrossRef]

148. Yoshioka, N.; Ishigami, M.; Watanabe, Y.; Sumi, H.; Doisaki, M.; Yamaguchi, T.; Ito, T.; Ishizu, Y.; Kuzuya, T.; Honda, T.; et al. Effect of weight change and lifestyle modifications on the development or remission of nonalcoholic fatty liver disease: Sex-specific analysis. *Sci. Rep.* **2020**, *10*. [CrossRef] [PubMed]

149. Romero-Gómez, M.; Zelber-Sagi, S.; Trenell, M. Treatment of NAFLD with diet, physical activity and exercise. *J. Hepatol.* **2017**, *67*, 829–846. [CrossRef] [PubMed]

150. Parry, S.A.; Hodson, L. Managing NAFLD in Type 2 Diabetes: The Effect of Lifestyle Interventions, a Narrative Review. *Adv. Ther.* **2020**, *37*, 1381–1406. [CrossRef] [PubMed]

151. Marchesini, G.; Day, C.P.; Dufour, J.F.; Canbay, A.; Nobili, V.; Ratziu, V.; Tilg, H.; Roden, M.; Gastaldelli, A.; Yki-Jarvinen, H.; et al. EASL-EASD-EASO Clinical Practice Guidelines for the management of non-alcoholic fatty liver disease. *J. Hepatol.* **2016**, *64*, 1388–1402. [CrossRef] [PubMed]

152. Ryan, M.C.; Itsiopoulos, C.; Thodis, T.; Ward, G.; Trost, N.; Hofferberth, S.; O'Dea, K.; Desmond, P.V.; Johnson, N.A.; Wilson, A.M. The Mediterranean diet improves hepatic steatosis and insulin sensitivity in individuals with non-alcoholic fatty liver disease. *J. Hepatol.* **2013**, *59*, 138–143. [CrossRef]

153. Kontogianni, M.D.; Tileli, N.; Margariti, A.; Georgoulis, M.; Deutsch, M.; Tiniakos, D.; Fragopoulou, E.; Zafiropoulou, R.; Manios, Y.; Papatheodoridis, G. Adherence to the Mediterranean diet is associated with the severity of non-alcoholic fatty liver disease. *Clin. Nutr.* **2014**, *33*, 678–683. [CrossRef] [PubMed]

154. Saeed, N.; Nadeau, B.; Shannon, C.; Tincopa, M. Evaluation of dietary approaches for the treatment of non-alcoholic fatty liver disease: A systematic review. *Nutrients* **2019**, *11*, 3064. [CrossRef] [PubMed]

155. Moosavian, S.P.; Arab, A.; Paknahad, Z. The effect of a Mediterranean diet on metabolic parameters in patients with non-alcoholic fatty liver disease: A systematic review of randomized controlled trials. *Clin. Nutr. ESPEN* **2020**, *35*, 40–46. [CrossRef] [PubMed]

156. Vilar-Gomez, E.; Martinez-Perez, Y.; Calzadilla-Bertot, L.; Torres-Gonzalez, A.; Gra-Oramas, B.; Gonzalez-Fabian, L.; Friedman, S.L.; Diago, M.; Romero-Gomez, M. Weight Loss Through Lifestyle Modification Significantly Reduces Features of Nonalcoholic Steatohepatitis. *Gastroenterology* **2015**, *149*, 365–367. [CrossRef]

157. Tendler, D.; Lin, S.; Yancy, W.S.; Mavropoulos, J.; Sylvestre, P.; Rockey, D.C.; Westman, E.C. The effect of a low-carbohydrate, ketogenic diet on nonalcoholic fatty liver disease: A pilot study. *Dig. Dis. Sci.* **2007**, *52*, 589–593. [CrossRef]

158. Luukkonen, P.K.; Dufour, S.; Lyu, K.; Zhang, X.-M.; Hakkarainen, A.; Lehtimäki, T.E.; Cline, G.W.; Petersen, K.F.; Shulman, G.I.; Yki-Järvinen, H. Effect of a ketogenic diet on hepatic steatosis and hepatic mitochondrial metabolism in nonalcoholic fatty liver disease. *Proc. Natl. Acad. Sci. USA* **2020**, *117*, 7347–7354. [CrossRef]

159. Vilar-Gomez, E.; Athinarayanan, S.J.; Adams, R.N.; Hallberg, S.J.; Bhanpuri, N.H.; McKenzie, A.L.; Campbell, W.W.; McCarter, J.P.; Phinney, S.D.; Volek, J.S.; et al. Post hoc analyses of surrogate markers of non-alcoholic fatty liver disease (NAFLD) and liver fibrosis in patients with type 2 diabetes in a digitally supported continuous care intervention: An open-label, non-randomised controlled study. *BMJ Open* **2019**, *9*. [CrossRef]

160. Bergentall, M.; Wu, H.; Bergh, P.-O.; Hakkarainen, A.; Nielsen, J.; Williams, K.J.; Uhlén, M.; Lee, S.; Snyder, M.; Romeo, S.; et al. An Integrated Understanding of the Rapid Metabolic Benefits of a Carbohydrate-Restricted Diet on Hepatic Steatosis in Humans. *Cell Metab.* **2018**, *27*, 559–571.e5. [CrossRef]

161. Watanabe, M.; Tozzi, R.; Risi, R.; Tuccinardi, D.; Mariani, S.; Basciani, S.; Spera, G.; Lubrano, C.; Gnessi, L. Beneficial effects of the ketogenic diet on nonalcoholic fatty liver disease: A comprehensive review of the literature. *Obes. Rev.* **2020**. [CrossRef]

162. Markova, M.; Pivovarova, O.; Hornemann, S.; Sucher, S.; Frahnow, T.; Wegner, K.; Machann, J.; Petzke, K.J.; Hierholzer, J.; Lichtinghagen, R.; et al. Isocaloric Diets High in Animal or Plant Protein Reduce Liver Fat and Inflammation in Individuals With Type 2 Diabetes. *Gastroenterology* **2017**, *152*, 571–585.e8. [CrossRef]

163. Xiao, M.W.; Lin, S.X.; Shen, Z.H.; Luo, W.W.; Wang, X.Y. Systematic review with meta-analysis: The effects of probiotics in nonalcoholic fatty liver disease. *Gastroenterol. Res. Pract.* **2019**, *2019*, 1484598. [CrossRef]

164. Xia, Y.; Zhang, S.; Zhang, Q.; Liu, L.; Meng, G.; Wu, H.; Bao, X.; Gu, Y.; Sun, S.; Wang, X.; et al. Insoluble dietary fibre intake is associated with lower prevalence of newly-diagnosed non-alcoholic fatty liver disease in Chinese men: A large population-based cross-sectional study. *Nutr. Metab.* **2020**, *17*. [CrossRef] [PubMed]

165. Zhao, L.; Zhang, F.; Ding, X.; Wu, G.; Lam, Y.Y.; Wang, X.; Fu, H.; Xue, X.; Lu, C.; Ma, J.; et al. Gut bacteria selectively promoted by dietary fibers alleviate type 2 diabetes. *Science* **2018**, *359*, 1151–1156. [CrossRef] [PubMed]

166. Scorletti, E.; Afolabi, P.R.; Miles, E.A.; Smith, D.E.; Almehmadi, A.; Alshathry, A.; Childs, C.E.; Del Fabbro, S.; Bilson, J.; Moyses, H.E.; et al. Synbiotics Alter Fecal Microbiomes, But Not Liver Fat or Fibrosis, in a Randomized Trial of Patients With Nonalcoholic Fatty Liver Disease. *Gastroenterology* **2020**, *158*, 1597–1610.e7. [CrossRef] [PubMed]

167. George, E.S.; Forsyth, A.; Itsiopoulos, C.; Nicoll, A.J.; Ryan, M.; Sood, S.; Roberts, S.K.; Tierney, A.C. Practical dietary recommendations for the prevention andmanagement of nonalcoholic fatty liver disease in adults. *Adv. Nutr.* **2018**, *9*, 30–40. [CrossRef]

168. Paoli, A.; Tinsley, G.; Bianco, A.; Moro, T. The influence of meal frequency and timing on health in humans: The role of fasting. *Nutrients* **2019**, *11*, 719. [CrossRef]

169. Wong, V.W.S.; Wong, G.L.H.; Chan, R.S.M.; Shu, S.S.T.; Cheung, B.H.K.; Li, L.S.; Chim, A.M.L.; Chan, C.K.M.; Leung, J.K.Y.; Chu, W.C.W.; et al. Beneficial effects of lifestyle intervention in non-obese patients with non-alcoholic fatty liver disease. *J. Hepatol.* **2018**, *69*, 1349–1356. [CrossRef]

170. Meroni, M.; Longo, M.; Rustichelli, A.; Dongiovanni, P. Nutrition and genetics in NAFLD: The perfect binomium. *Int. J. Mol. Sci.* **2020**, *21*, 2986. [CrossRef]

171. Taitano, A.A.; Markow, M.; Finan, J.E.; Wheeler, D.E.; Gonzalvo, J.P.; Murr, M.M. Bariatric Surgery Improves Histological Features of Nonalcoholic Fatty Liver Disease and Liver Fibrosis. *J. Gastrointest. Surg.* **2015**, *19*, 429–437. [CrossRef]

172. Lassailly, G.; Caiazzo, R.; Buob, D.; Pigeyre, M.; Verkindt, H.; Labreuche, J.; Raverdy, V.; Leteurtre, E.; Dharancy, S.; Louvet, A.; et al. Bariatric surgery reduces features of nonalcoholic steatohepatitis in morbidly obese patients. *Gastroenterology* **2015**, *149*, 379–388. [CrossRef]

173. Fakhry, T.K.; Mhaskar, R.; Schwitalla, T.; Muradova, E.; Gonzalvo, J.P.; Murr, M.M. Bariatric surgery improves nonalcoholic fatty liver disease: A contemporary systematic review and meta-analysis. *Surg. Obes. Relat. Dis.* **2019**, *15*, 502–511. [CrossRef] [PubMed]

174. Wirth, K.M.; Sheka, A.C.; Kizy, S.; Irey, R.; Benner, A.; Sieger, G.; Simon, G.; Ma, S.; Lake, J.; Aliferis, C.; et al. Bariatric Surgery is Associated with Decreased Progression of Nonalcoholic Fatty Liver Disease to Cirrhosis. *Ann. Surg.* **2020**, *1*. [CrossRef] [PubMed]

175. Laursen, T.L.; Hagemann, C.A.; Wei, C.; Kazankov, K.; Thomsen, K.L.; Knop, F.K.; Grønbæk, H. Bariatric surgery in patients with non-alcoholic fatty liver disease—From pathophysiology to clinical effects. *World J. Hepatol.* **2019**, *11*, 138–249. [CrossRef] [PubMed]

176. Mathurin, P.; Hollebecque, A.; Arnalsteen, L.; Buob, D.; Leteurtre, E.; Caiazzo, R.; Pigeyre, M.; Verkindt, H.; Dharancy, S.; Louvet, A.; et al. Prospective Study of the Long-Term Effects of Bariatric Surgery on Liver Injury in Patients Without Advanced Disease. *Gastroenterology* **2009**, *137*, 532–540. [CrossRef]

177. Mantovani, A.; Byrne, C.D.; Scorletti, E.; Mantzoros, C.S.; Targher, G. Efficacy and safety of anti-hyperglycaemic drugs in patients with non-alcoholic fatty liver disease with or without diabetes: An updated systematic review of randomized controlled trials. *Diabetes Metab.* **2020**. [CrossRef]

178. Chalasani, N.; Younossi, Z.; Lavine, J.E.; Charlton, M.; Cusi, K.; Rinella, M.; Harrison, S.A.; Brunt, E.M.; Sanyal, A.J. The diagnosis and management of nonalcoholic fatty liver disease: Practice guidance from the American Association for the Study of Liver Diseases. *Hepatology* **2018**, *67*, 328–357. [CrossRef] [PubMed]

179. Chalasani, N.P.; Sanyal, A.J.; Kowdley, K.V.; Robuck, P.R.; Hoofnagle, J.; Kleiner, D.E.; Ünalp, A.; Tonascia, J. Pioglitazone versus vitamin E versus placebo for the treatment of non-diabetic patients with non-alcoholic steatohepatitis: PIVENS trial design. *Contemp. Clin. Trials* **2009**, *30*, 88–96. [CrossRef]

180. Belfort, R.; Harrison, S.A.; Brown, K.; Darland, C.; Finch, J.; Hardies, J.; Balas, B.; Gastaldelli, A.; Tio, F.; Pulcini, J.; et al. A Placebo-Controlled Trial of Pioglitazone in Subjects with Nonalcoholic Steatohepatitis. *N. Engl. J. Med.* **2006**, *355*, 2297–2307. [CrossRef]

181. Cusi, K.; Orsak, B.; Bril, F.; Lomonaco, R.; Hecht, J.; Ortiz-Lopez, C.; Tio, F.; Hardies, J.; Darland, C.; Musi, N.; et al. Long-Term Pioglitazone Treatment for Patients with Nonalcoholic Steatohepatitis and Prediabetes or Type 2 Diabetes Mellitus. *Ann. Intern. Med.* **2016**, *165*, 305. [CrossRef]

182. Brunt, E.M.; Kleiner, D.E.; Wilson, L.A.; Sanyal, A.J.; Neuschwander-Tetri, B.A. Improvements in Histologic Features and Diagnosis Associated with Improvement in Fibrosis in Nonalcoholic Steatohepatitis: Results From the Nonalcoholic Steatohepatitis Clinical Research Network Treatment Trials. *Hepatology* **2019**, *70*, 522–531. [CrossRef]

183. Aithal, G.P.; Thomas, J.A.; Kaye, P.V.; Lawson, A.; Ryder, S.D.; Spendlove, I.; Austin, A.S.; Freeman, J.G.; Morgan, L.; Webber, J. Randomized, Placebo-Controlled Trial of Pioglitazone in Nondiabetic Subjects With Nonalcoholic Steatohepatitis. *Gastroenterology* **2008**, *135*, 1176–1184. [CrossRef] [PubMed]

184. Boettcher, E.; Csako, G.; Pucino, F.; Wesley, R.; Loomba, R. Meta-analysis: Pioglitazone improves liver histology and fibrosis in patients with non-alcoholic steatohepatitis. *Aliment. Pharmacol. Ther.* **2012**, *35*, 66–75. [CrossRef] [PubMed]

185. Bril, F.; Kalavalapalli, S.; Clark, V.C.; Lomonaco, R.; Soldevila-Pico, C.; Liu, I.C.; Orsak, B.; Tio, F.; Cusi, K. Response to Pioglitazone in Patients With Nonalcoholic Steatohepatitis With vs Without Type 2 Diabetes. *Clin. Gastroenterol. Hepatol.* **2018**, *16*, 558–566.e2. [CrossRef]

186. Kawaguchi-Suzuki, M.; Cusi, K.; Bril, F.; Gong, Y.; Langaee, T.; Frye, R.F. A genetic score associates with pioglitazone response in patients with non-alcoholic steatohepatitis. *Front. Pharmacol.* **2018**, *9*. [CrossRef]

187. Shah, P.; Mudaliar, S. Pioglitazone: Side effect and safety profile. *Expert Opin. Drug Saf.* **2010**, *9*, 347–354. [CrossRef]

188. Portillo-Sanchez, P.; Bril, F.; Lomonaco, R.; Barb, D.; Orsak, B.; Bruder, J.M.; Cusi, K. Effect of pioglitazone on bone mineral density in patients with nonalcoholic steatohepatitis: A 36-month clinical trial. *J. Diabetes* **2019**, *11*, 223–231. [CrossRef] [PubMed]

189. Tang, H.; Shi, W.; Fu, S.; Wang, T.; Zhai, S.; Song, Y.; Han, J. Pioglitazone and bladder cancer risk: A systematic review and meta-analysis. *Cancer Med.* **2018**, *7*, 1070–1080. [CrossRef] [PubMed]

190. Ripamonti, E.; Azoulay, L.; Abrahamowicz, M.; Platt, R.W.; Suissa, S. A systematic review of observational studies of the association between pioglitazone use and bladder cancer. *Diabet. Med.* **2019**, *36*, 22–35. [CrossRef]

191. Kersten, S.; Desvergne, B.; Wahli, W. Roles of PPARS in health and disease. *Nature* **2000**, *405*, 421–424. [CrossRef]

192. Czaja, M.J. Pioglitazone: More than just an insulin sensitizer. *Hepatology* **2009**, *49*, 1427–1430. [CrossRef]

193. Soccio, R.E.; Chen, E.R.; Lazar, M.A. Thiazolidinediones and the promise of insulin sensitization in type 2 diabetes. *Cell Metab.* **2014**, *20*, 573–591. [CrossRef]

194. Kalavalapalli, S.; Bril, F.; Koelmel, J.P.; Abdo, K.; Guingab, J.; Andrews, P.; Li, W.-Y.; Jose, D.; Yost, R.A.; Frye, R.F.; et al. Pioglitazone improves hepatic mitochondrial function in a mouse model of nonalcoholic steatohepatitis. *Am. J. Physiol. Metab.* **2018**, *315*. [CrossRef] [PubMed]

195. Hsiao, P.J.; Chiou, H.Y.C.; Jiang, H.J.; Lee, M.Y.; Hsieh, T.J.; Kuo, K.K. Pioglitazone Enhances Cytosolic Lipolysis, β-oxidation and Autophagy to Ameliorate Hepatic Steatosis. *Sci. Rep.* **2017**, *7*. [CrossRef]

196. de Mendonça, M.; dos Santos, B.d.A.C.; de Sousa, É.; Rodrigues, A.C. Adiponectin is required for pioglitazone-induced improvements in hepatic steatosis in mice fed a high-fat diet. *Mol. Cell. Endocrinol.* **2019**, *493*. [CrossRef] [PubMed]

197. Corey, K.E.; Vuppalanchi, R.; Wilson, L.A.; Cummings, O.W.; Chalasani, N. NASH resolution is associated with improvements in HDL and triglyceride levels but not improvement in LDL or non-HDL-C levels. *Aliment. Pharmacol. Ther.* **2015**, *41*, 301–309. [CrossRef]

198. Corey, K.E.; Wilson, L.A.; Altinbas, A.; Yates, K.P.; Kleiner, D.E.; Chung, R.T.; Krauss, R.M.; Chalasani, N.; Bringman, D.; Dasarathy, S.; et al. Relationship between resolution of non-alcoholic steatohepatitis and changes in lipoprotein sub-fractions: A post-hoc analysis of the PIVENS trial. *Aliment. Pharmacol. Ther.* **2019**, *49*, 1205–1213. [CrossRef] [PubMed]

199. Lavine, J.E.; Schwimmer, J.B.; Van Natta, M.L.; Molleston, J.P.; Murray, K.F.; Rosenthal, P.; Abrams, S.H.; Scheimann, A.O.; Sanyal, A.J.; Chalasani, N.; et al. Effect of Vitamin e or metformin for treatment of nonalcoholic fatty liver disease in children and adolescents the tonic randomized controlled trial. *JAMA J. Am. Med. Assoc.* **2011**, *305*, 1659–1668. [CrossRef] [PubMed]

200. Banini, B.A.; Cazanave, S.C.; Yates, K.P.; Asgharpour, A.; Vincent, R.; Mirshahi, F.; Le, P.; Contos, M.J.; Tonascia, J.; Chalasani, N.P.; et al. Haptoglobin 2 Allele is Associated with Histologic Response to Vitamin E in Subjects with Nonalcoholic Steatohepatitis. *J. Clin. Gastroenterol.* **2019**, *53*, 750–758. [CrossRef] [PubMed]

201. Bril, F.; Biernacki, D.M.; Kalavalapalli, S.; Lomonaco, R.; Subbarayan, S.K.; Lai, J.; Tio, F.; Suman, A.; Orsak, B.K.; Hecht, J.; et al. Role of Vitamin E for nonalcoholic steatohepatitis in patients with type 2 diabetes: A randomized controlled trial. *Diabetes Care* **2019**, *42*, 1481–1488. [CrossRef]

202. Tsou, P.; Wu, C.-J. Serum Vitamin E Levels of Adults with Nonalcoholic Fatty Liver Disease: An Inverse Relationship with All-Cause Mortality in Non-Diabetic but Not in Pre-Diabetic or Diabetic Subjects. *J. Clin. Med.* **2019**, *8*, 1057. [CrossRef]

203. Klein, E.A.; Thompson, I.M.; Tangen, C.M.; Crowley, J.J.; Lucia, S.; Goodman, P.J.; Minasian, L.M.; Ford, L.G.; Parnes, H.L.; Gaziano, J.M.; et al. Vitamin E and the risk of prostate cancer: The selenium and vitamin E cancer prevention trial (SELECT). *JAMA J. Am. Med. Assoc.* **2011**, *306*, 1549–1556. [CrossRef] [PubMed]

204. Schürks, M.; Glynn, R.J.; Rist, P.M.; Tzourio, C.; Kurth, T. Effects of vitamin E on stroke subtypes: Meta-analysis of randomised controlled trials. *BMJ* **2010**, *341*, c5702. [CrossRef]

205. Miller, E.R.; Pastor-Barriuso, R.; Dalal, D.; Riemersma, R.A.; Appel, L.J.; Guallar, E. Meta-analysis: High-dosage vitamin E supplementation may increase all-cause mortality. *Ann. Intern. Med.* **2005**, *142*, 37–46. [CrossRef] [PubMed]

206. Abner, E.L.; Schmitt, F.A.; Mendiondo, M.S.; Marcum, J.L.; Kryscio, R.J. Vitamin E and All-Cause Mortality: A Meta-Analysis. *Curr. Aging Sci.* **2012**, *4*, 158–170. [CrossRef] [PubMed]

207. Perumpail, B.; Li, A.; John, N.; Sallam, S.; Shah, N.; Kwong, W.; Cholankeril, G.; Kim, D.; Ahmed, A. The Role of Vitamin E in the Treatment of NAFLD. *Diseases* **2018**, *6*, 86. [CrossRef] [PubMed]

208. He, W.; Xu, Y.; Ren, X.; Xiang, D.; Lei, K.; Zhang, C.; Liu, D. Vitamin E Ameliorates Lipid Metabolism in Mice with Nonalcoholic Fatty Liver Disease via Nrf2/CES1 Signaling Pathway. *Dig. Dis. Sci.* **2019**, *64*, 3182–3191. [CrossRef]

209. Landrier, J.F.; Gouranton, E.; El Yazidi, C.; Malezet, C.; Balaguer, P.; Borel, P.; Amiot, M.J. Adiponectin expression is induced by vitamin E via a peroxisome proliferator-activated receptor γ-dependent mechanism. *Endocrinology* **2009**, *150*, 5318–5325. [CrossRef]

210. Armstrong, M.J.; Gaunt, P.; Aithal, G.P.; Barton, D.; Hull, D.; Parker, R.; Hazlehurst, J.M.; Guo, K.; Abouda, G.; Aldersley, M.A.; et al. Liraglutide safety and efficacy in patients with non-alcoholic steatohepatitis (LEAN): A multicentre, double-blind, randomised, placebo-controlled phase 2 study. *Lancet* **2016**, *387*, 679–690. [CrossRef]

211. Polyzos, S.A.; Kang, E.S.; Boutari, C.; Rhee, E.J.; Mantzoros, C.S. Current and emerging pharmacological options for the treatment of nonalcoholic steatohepatitis. *Metabolism.* **2020**. [CrossRef]

212. Muthiah, M.D.; Sanyal, A.J. Current management of non-alcoholic steatohepatitis. *Liver Int.* **2020**, *40*, 89–95. [CrossRef]

213. Neuschwander-Tetri, B.A. Therapeutic Landscape for NAFLD in 2020. *Gastroenterology* **2020**. [CrossRef] [PubMed]

214. Tanaka, N.; Aoyama, T.; Kimura, S.; Gonzalez, F.J. Targeting nuclear receptors for the treatment of fatty liver disease. *Pharmacol. Ther.* **2017**, *179*, 142–157. [CrossRef] [PubMed]

215. Cave, M.C.; Clair, H.B.; Hardesty, J.E.; Falkner, K.C.; Feng, W.; Clark, B.J.; Sidey, J.; Shi, H.; Aqel, B.A.; McClain, C.J.; et al. Nuclear receptors and nonalcoholic fatty liver disease11This article is part of a Special Issue entitled: Xenobiotic nuclear receptors: New Tricks for An Old Dog, edited by Dr. Wen Xie. *Biochim. Biophys. Acta Gene Regul. Mech.* **2016**, *1859*, 1083–1099. [CrossRef]

216. Yang, X.; Gonzalez, F.J.; Huang, M.; Bi, H. Nuclear receptors and non-alcoholic fatty liver disease: An update. *Liver Res.* **2020**. [CrossRef]

217. Neuschwander-Tetri, B.A.; Loomba, R.; Sanyal, A.J.; Lavine, J.E.; Van Natta, M.L.; Abdelmalek, M.F.; Chalasani, N.; Dasarathy, S.; Diehl, A.M.; Hameed, B.; et al. Farnesoid X nuclear receptor ligand obeticholic acid for non-cirrhotic, non-alcoholic steatohepatitis (FLINT): A multicentre, randomised, placebo-controlled trial. *Lancet* **2015**, *385*, 956–965. [CrossRef]

218. Younossi, Z.M.; Ratziu, V.; Loomba, R.; Rinella, M.; Anstee, Q.M.; Goodman, Z.; Bedossa, P.; Geier, A.; Beckebaum, S.; Newsome, P.N.; et al. Obeticholic acid for the treatment of non-alcoholic steatohepatitis: Interim analysis from a multicentre, randomised, placebo-controlled phase 3 trial. *Lancet* **2019**, *394*, 2184–2196. [CrossRef]

219. Harrison, S.A.; Bashir, M.R.; Guy, C.D.; Zhou, R.; Moylan, C.A.; Frias, J.P.; Alkhouri, N.; Bansal, M.B.; Baum, S.; Neuschwander-Tetri, B.A.; et al. Resmetirom (MGL-3196) for the treatment of non-alcoholic steatohepatitis: A multicentre, randomised, double-blind, placebo-controlled, phase 2 trial. *Lancet* **2019**, *394*, 2012–2024. [CrossRef]

220. Bougarne, N.; Weyers, B.; Desmet, S.J.; Deckers, J.; Ray, D.W.; Staels, B.; De Bosscher, K. Molecular actions of PPARα in lipid metabolism and inflammation. *Endocr. Rev.* **2018**, *39*, 760–802. [CrossRef] [PubMed]

221. Braissant, O.; Foufelle, F.; Scotto, C.; Dauça, M.; Wahli, W. Differential expression of peroxisome proliferator-activated receptors (PPARs): Tissue distribution of PPAR-α, -β, and -γ in the adult rat. *Endocrinology* **1996**, *137*, 354–366. [CrossRef] [PubMed]

222. Patsouris, D.; Reddy, J.K.; Müller, M.; Kersten, S. Peroxisome proliferator-activated receptor α mediates the effects of high-fat diet on hepatic gene expression. *Endocrinology* **2006**, *147*, 1508–1516. [CrossRef] [PubMed]

223. Chakravarthy, M.V.; Pan, Z.; Zhu, Y.; Tordjman, K.; Schneider, J.G.; Coleman, T.; Turk, J.; Semenkovich, C.F. "New" hepatic fat activates PPARα to maintain glucose, lipid, and cholesterol homeostasis. *Cell Metab.* **2005**, *1*, 309–322. [CrossRef]

224. Grygiel-Górniak, B. Peroxisome proliferator-activated receptors and their ligands: Nutritional and clinical implications—A review. *Nutr. J.* **2014**, *13*. [CrossRef]

225. Dubois, V.; Eeckhoute, J.; Lefebvre, P.; Staels, B. Distinct but complementary contributions of PPAR isotypes to energy homeostasis. *J. Clin. Invest.* **2017**, *127*, 1202–1214. [CrossRef] [PubMed]

226. Feige, J.N.; Gelman, L.; Michalik, L.; Desvergne, B.; Wahli, W. From molecular action to physiological outputs: Peroxisome proliferator-activated receptors are nuclear receptors at the crossroads of key cellular functions. *Prog. Lipid Res.* **2006**, *45*, 120–159. [CrossRef]

227. Takeuchi, S.; Matsuda, T.; Kobayashi, S.; Takahashi, T.; Kojima, H. In vitro screening of 200 pesticides for agonistic activity via mouse peroxisome proliferator-activated receptor (PPAR)α and PPARγ and quantitative analysis of in vivo induction pathway. *Toxicol. Appl. Pharmacol.* **2006**, *217*, 235–244. [CrossRef]

228. Xi, Y.; Zhang, Y.; Zhu, S.; Luo, Y.; Xu, P.; Huang, Z. PPAR-Mediated Toxicology and Applied Pharmacology. *Cells* **2020**, *9*, 352. [CrossRef] [PubMed]

229. Casals-Casas, C.; Desvergne, B. Endocrine Disruptors: From Endocrine to Metabolic Disruption. *Annu. Rev. Physiol.* **2011**, *73*, 135–162. [CrossRef] [PubMed]

230. Chen, J.; Montagner, A.; Tan, N.S.; Wahli, W. Insights into the role of PPARβ/δ in NAFLD. *Int. J. Mol. Sci.* **2018**, *19*, 1893. [CrossRef]

231. Ricote, M.; Glass, C.K. PPARs and molecular mechanisms of transrepression. *Biochim. Biophys. Acta Mol. Cell Biol. Lipids* **2007**, *1771*, 926–935. [CrossRef]

232. Brunmeir, R.; Xu, F. Functional regulation of PPARs through post-translational modifications. *Int. J. Mol. Sci.* **2018**, *19*, 1738. [CrossRef]

233. Bookout, A.L.; Jeong, Y.; Downes, M.; Yu, R.T.; Evans, R.M.; Mangelsdorf, D.J.; Yang, X.; Downes, M.; Yu, R.T.; Bookout, A.L.; et al. Nuclear Receptor Expression Links the Circadian Clock to Metabolism. *Cell* **2006**, *126*, 789–799. [CrossRef]

234. Lemberger, T.; Saladin, R.; Vázquez, M.; Assimacopoulos, F.; Staels, B.; Desvergne, B.; Wahli, W.; Auwerx, J. Expression of the peroxisome proliferator-activated receptor α gene is stimulated by stress and follows a diurnal rhythm. *J. Biol. Chem.* **1996**, *271*, 1764–1769. [CrossRef] [PubMed]

235. Montagner, A.; Polizzi, A.; Fouché, E.; Ducheix, S.; Lippi, Y.; Lasserre, F.; Barquissau, V.; Régnier, M.; Lukowicz, C.; Benhamed, F.; et al. Liver PPARα is crucial for whole-body fatty acid homeostasis and is protective against NAFLD. *Gut* **2016**, *65*, 1202–1214. [CrossRef] [PubMed]

236. Liu, S.; Brown, J.D.; Stanya, K.J.; Homan, E.; Leidl, M.; Inouye, K.; Bhargava, P.; Gangl, M.R.; Dai, L.; Hatano, B.; et al. A diurnal serum lipid integrates hepatic lipogenesis and peripheral fatty acid use. *Nature* **2013**, *502*, 550–554. [CrossRef] [PubMed]

237. Régnier, M.; Polizzi, A.; Lippi, Y.; Fouché, E.; Michel, G.; Lukowicz, C.; Smati, S.; Marrot, A.; Lasserre, F.; Naylies, C.; et al. Insights into the role of hepatocyte PPARα activity in response to fasting. *Mol. Cell. Endocrinol.* **2018**, *471*, 75–88. [CrossRef]

238. Kersten, S.; Seydoux, J.; Peters, J.M.; Gonzalez, F.J.; Desvergne, B.; Wahli, W.; Seydoux, J.; Peters, J.M.; Gonzalez, F.J.; Desvergne, B.; et al. Peroxisome proliferator-activated receptor alpha mediates the adaptive response to fasting. *J. Clin. Invest.* **1999**, *103*, 1489–1498. [CrossRef] [PubMed]

239. Lee, S.S.; Pineau, T.; Drago, J.; Lee, E.J.; Owens, J.W.; Kroetz, D.L.; Fernandez-Salguero, P.M.; Westphal, H.; Gonzalez, F.J. Targeted disruption of the alpha isoform of the peroxisome proliferator-activated receptor gene in mice results in abolishment of the pleiotropic effects of peroxisome proliferators. *Mol. Cell. Biol.* **1995**, *15*, 3012–3022. [CrossRef] [PubMed]

240. Pawlak, M.; Baugé, E.; Lalloyer, F.; Lefebvre, P.; Staels, B. Ketone body therapy protects from lipotoxicity and acute liver failure upon Ppar α deficiency. *Mol. Endocrinol.* **2015**, *29*, 1134–1143. [CrossRef]

241. Puchalska, P.; Crawford, P.A. Multi-dimensional Roles of Ketone Bodies in Fuel Metabolism, Signaling, and Therapeutics. *Cell Metab.* **2017**, *25*, 262–284. [CrossRef]

242. Inagaki, T.; Dutchak, P.; Zhao, G.; Ding, X.; Gautron, L.; Parameswara, V.; Li, Y.; Goetz, R.; Mohammadi, M.; Esser, V.; et al. Endocrine Regulation of the Fasting Response by PPARα Mediated Induction of Fibroblast Growth Factor 21. *Cell Metab.* **2007**, *5*, 415–425. [CrossRef]

243. Badman, M.K.; Pissios, P.; Kennedy, A.R.; Koukos, G.; Flier, J.S.; Maratos-Flier, E. Hepatic Fibroblast Growth Factor 21 Is Regulated by PPARα and Is a Key Mediator of Hepatic Lipid Metabolism in Ketotic States. *Cell Metab.* **2007**, *5*, 426–437. [CrossRef] [PubMed]

244. Gälman, C.; Lundåsen, T.; Kharitonenkov, A.; Bina, H.A.; Eriksson, M.; Hafström, I.; Dahlin, M.; Åmark, P.; Angelin, B.; Rudling, M. The Circulating Metabolic Regulator FGF21 Is Induced by Prolonged Fasting and PPARα Activation in Man. *Cell Metab.* **2008**, *8*, 169–174. [CrossRef] [PubMed]

245. Kliewer, S.A.; Mangelsdorf, D.J. A Dozen Years of Discovery: Insights into the Physiology and Pharmacology of FGF21. *Cell Metab.* **2019**, *29*, 246–253. [CrossRef] [PubMed]

246. Kersten, S.; Mandard, S.; Tan, N.S.; Escher, P.; Metzger, D.; Chambon, P.; Gonzalez, F.J.; Desvergne, B.; Wahli, W. Characterization of the fasting-induced adipose factor FIAF, a novel peroxisome proliferator-activated receptor target gene. *J. Biol. Chem.* **2000**, *275*, 28488–28493. [CrossRef]

247. Smati, S.; Régnier, M.; Fougeray, T.; Polizzi, A.; Fougerat, A.; Lasserre, F.; Lukowicz, C.; Tramunt, B.; Guillaume, M.; Burnol, A.F.; et al. Regulation of hepatokine gene expression in response to fasting and feeding: Influence of PPAR-α and insulin-dependent signalling in hepatocytes. *Diabetes Metab.* **2020**, *46*, 129–136. [CrossRef]

248. Lee, J.M.; Wagner, M.; Xiao, R.; Kim, K.H.; Feng, D.; Lazar, M.A.; Moore, D.D. Nutrient-sensing nuclear receptors coordinate autophagy. *Nature* **2014**, *516*, 112–115. [CrossRef] [PubMed]

249. Sinha, R.A.; Rajak, S.; Singh, B.K.; Yen, P.M. Hepatic lipid catabolism via PPARα-lysosomal crosstalk. *Int. J. Mol. Sci.* **2020**, *21*, 2391. [CrossRef]

250. Siong Tan, H.W.; Anjum, B.; Shen, H.M.; Ghosh, S.; Yen, P.M.; Sinha, R.A. Lysosomal inhibition attenuates peroxisomal gene transcription via suppression of PPARA and PPARGC1A levels. *Autophagy* **2019**, *15*, 1455–1459. [CrossRef]

251. Iershov, A.; Nemazanyy, I.; Alkhoury, C.; Girard, M.; Barth, E.; Cagnard, N.; Montagner, A.; Chretien, D.; Rugarli, E.I.; Guillou, H.; et al. The class 3 PI3K coordinates autophagy and mitochondrial lipid catabolism by controlling nuclear receptor PPARα. *Nat. Commun.* **2019**, *10*. [CrossRef]

252. Kersten, S. Peroxisome proliferator activated receptors and lipoprotein metabolism. *PPAR Res.* **2008**, *2008*. [CrossRef]

253. Fernández-Alvarez, A.; Soledad Alvarez, M.; Gonzalez, R.; Cucarella, C.; Muntané, J.; Casado, M. Human SREBP1c expression in liver is directly regulated by Peroxisome Proliferator-activated Receptor α (PPARα). *J. Biol. Chem.* **2011**, *286*, 21466–21477. [CrossRef] [PubMed]

254. Knight, B.L.; Hebbach, A.; Hauton, D.; Brown, A.M.; Wiggins, D.; Patel, D.D.; Gibbons, G.F. A role for PPARα in the control of SREBP activity and lipid synthesis in the liver. *Biochem. J.* **2005**, *389*, 413–421. [CrossRef] [PubMed]

255. Patsouris, D.; Mandard, S.; Voshol, P.J.; Escher, P.; Tan, N.S.; Havekes, L.M.; Koenig, W.; März, W.; Tafuri, S.; Wahli, W.; et al. PPARalpha governs glycerol metabolism. *J. Clin. Invest.* **2004**, *114*, 94–103. [CrossRef] [PubMed]

256. Kersten, S.; Mandard, S.; Escher, P.; Gonzalez, F.J.; Tafuri, S.; Desvergne, B.; Wahli, W. The peroxisome proliferator-activated receptor α regulates amino acid metabolism. *FASEB J.* **2001**, *15*, 1971–1978. [CrossRef] [PubMed]

257. Purushotham, A.; Schug, T.T.; Xu, Q.; Surapureddi, S.; Guo, X.; Li, X. Hepatocyte-Specific Deletion of SIRT1 Alters Fatty Acid Metabolism and Results in Hepatic Steatosis and Inflammation. *Cell Metab.* **2009**, *9*, 327–338. [CrossRef]

258. Muoio, D.M.; Way, J.M.; Tanner, C.J.; Winegar, D.A.; Kliewer, S.A.; Houmard, J.A.; Kraus, W.E.; Lynis Dohm, G. Peroxisome proliferator-activated receptor-α regulates fatty acid utilization in primary human skeletal muscle cells. *Diabetes* **2002**, *51*, 901–909. [CrossRef] [PubMed]

259. Ribet, C.; Montastier, E.; Valle, C.; Bezaire, V.; Mazzucotelli, A.; Mairal, A.; Viguerie, N.; Langin, D. Peroxisome proliferator-activated receptor-α control of lipid and glucose metabolism in human white adipocytes. *Endocrinology* **2010**, *151*, 123–133. [CrossRef] [PubMed]

260. Duncan, J.G.; Bharadwaj, K.G.; Fong, J.L.; Mitra, R.; Sambandam, N.; Courtois, M.R.; Lavine, K.J.; Goldberg, I.J.; Kelly, D.P. Rescue of cardiomyopathy in peroxisome proliferator-activated receptor-α transgenic mice by deletion of lipoprotein lipase identifies sources of cardiac lipids and peroxisome proliferator-activated receptor-α activators. *Circulation* **2010**, *121*, 426–435. [CrossRef] [PubMed]

261. Brocker, C.N.; Patel, D.P.; Velenosi, T.J.; Kim, D.; Yan, T.; Yue, J.; Li, G.; Krausz, K.W.; Gonzalez, F.J. Extrahepatic PPAR modulates fatty acid oxidation and attenuates fasting-induced hepatosteatosis in mice. *J. Lipid Res.* **2018**, *59*, 2140–2152. [CrossRef]

262. Montgomery, M.K.; Bayliss, J.; Keenan, S.; Rhost, S.; Ting, S.B.; Watt, M.J. The role of Ap2a2 in PPARα-mediated regulation of lipolysis in adipose tissue. *FASEB J.* **2019**, *33*, 13267–13279. [CrossRef]

263. Takahashi, H.; Sanada, K.; Nagai, H.; Li, Y.; Aoki, Y.; Ara, T.; Seno, S.; Matsuda, H.; Yu, R.; Kawada, T.; et al. Over-expression of PPARα in obese mice adipose tissue improves insulin sensitivity. *Biochem. Biophys. Res. Commun.* **2017**, *493*, 108–114. [CrossRef] [PubMed]

264. Hondares, E.; Rosell, M.; Diaz-Delfin, J.; Olmos, Y.; Monsalve, M.; Iglesias, R.; Villarroya, F.; Giralt, M. Peroxisome proliferator-activated receptor alpha (PPARalpha) induces PPARgamma coactivator 1alpha (PGC-1alpha) gene expression and contributes to thermogenic activation of brown fat: Involvement of PRDM16 1296. *J. Biol. Chem.* **2011**, *286*, 43112–43122. [CrossRef] [PubMed]

265. Barquissau, V.; Beuzelin, D.; Pisani, D.F.; Beranger, G.E.; Mairal, A.; Montagner, A.; Roussel, B.; Tavernier, G.; Marques, M.A.; Moro, C.; et al. White-to-brite conversion in human adipocytes promotes metabolic reprogramming towards fatty acid anabolic and catabolic pathways. *Mol. Metab.* **2016**, *5*, 352–365. [CrossRef] [PubMed]

266. Defour, M.; Dijk, W.; Ruppert, P.; Nascimento, E.B.M.; Schrauwen, P.; Kersten, S. The Peroxisome Proliferator-Activated Receptor α is dispensable for cold-induced adipose tissue browning in mice. *Mol. Metab.* **2018**, *10*, 39–54. [CrossRef] [PubMed]

267. Lasar, D.; Rosenwald, M.; Kiehlmann, E.; Balaz, M.; Tall, B.; Opitz, L.; Lidell, M.E.; Zamboni, N.; Krznar, P.; Sun, W.; et al. Peroxisome Proliferator Activated Receptor Gamma Controls Mature Brown Adipocyte Inducibility through Glycerol Kinase. *Cell Rep.* **2018**, *22*, 760–773. [CrossRef]

268. Shen, Y.; Su, Y.; Silva, F.J.; Weller, A.H.; Sostre-Colón, J.; Titchenell, P.M.; Steger, D.J.; Seale, P.; Soccio, R.E. Shared PPARα/γ Target Genes Regulate Brown Adipocyte Thermogenic Function. *Cell Rep.* **2020**, *30*, 3079–3091.e5. [CrossRef]

269. Sommars, M.A.; Ramachandran, K.; Senagolage, M.D.; Futtner, C.R.; Germain, D.M.; Allred, A.L.; Omura, Y.; Bederman, I.R.; Barish, G.D. Dynamic repression by BCL6 controls the genome-wide liver response to fasting and steatosis. *Elife* **2019**, *8*. [CrossRef]

270. Li, G.; Brocker, C.N.; Yan, T.; Xie, C.; Krausz, K.W.; Xiang, R.; Gonzalez, F.J. Metabolic adaptation to intermittent fasting is independent of peroxisome proliferator-activated receptor alpha. *Mol. Metab.* **2018**, *7*, 80–89. [CrossRef]

271. Leuenberger, N.; Pradervand, S.; Wahli, W. Sumoylated PPARalpha mediates sex specific gene repression and protects the liver from estrogen induced toxicity in mice. *J. Clin. Invest.* **2009**, *119*, 3138–3148. [CrossRef]

272. Manickam, R.; Wahli, W. Roles of Peroxisome Proliferator-Activated Receptor β/δ in skeletal muscle physiology. *Biochimie* **2017**, *136*, 42–48. [CrossRef]

273. Koh, J.H.; Hancock, C.R.; Terada, S.; Higashida, K.; Holloszy, J.O.; Han, D.H. PPARβ Is Essential for Maintaining Normal Levels of PGC-1α and Mitochondria and for the Increase in Muscle Mitochondria Induced by Exercise. *Cell Metab.* **2017**, *25*, 1176–1185.e5. [CrossRef] [PubMed]

274. Park, J.S.; Holloszy, J.O.; Kim, K.; Koh, J.H. Exercise training-induced PPARβ increases PGC-1α protein stability and improves insulin-induced glucose uptake in rodent muscles. *Nutrients* **2020**, *12*, 652. [CrossRef] [PubMed]

275. Schuler, M.; Ali, F.; Chambon, C.; Duteil, D.; Bornert, J.M.; Tardivel, A.; Desvergne, B.; Wahli, W.; Chambon, P.; Metzger, D. PGC1α expression is controlled in skeletal muscles by PPARβ, whose ablation results in fiber-type switching, obesity, and type 2 diabetes. *Cell Metab.* **2006**, *4*, 407–414. [CrossRef] [PubMed]

276. Wang, Y.X.; Lee, C.H.; Tiep, S.; Yu, R.T.; Ham, J.; Kang, H.; Evans, R.M. Peroxisome-proliferator-activated receptor δ activates fat metabolism to prevent obesity. *Cell* **2003**, *113*, 159–170. [CrossRef]

277. Escher, P.; Braissant, O.; Basu-Modak, S.; Michalik, L.; Wahli, W.; Desvergne, B. Rat PPARs: Quantitative analysis in adult rat tissues and regulation in fasting and refeeding. *Endocrinology* **2001**, *142*, 4195–4202. [CrossRef] [PubMed]

278. Lee, C.H.; Olson, P.; Hevener, A.; Mehl, I.; Chong, L.W.; Olefsky, J.M.; Gonzalez, F.J.; Ham, J.; Kang, H.; Peters, J.M.; et al. PPARδ regulates glucose metabolism and insulin sensitivity *Proc. Natl. Acad. Sci. USA* **2006**, *103*, 3444–3449. [CrossRef] [PubMed]

279. Liu, S.; Hatano, B.; Zhao, M.; Yen, C.C.; Kang, K.; Reilly, S.M.; Gangl, M.R.; Gorgun, C.; Balschi, J.A.; Ntambi, J.M.; et al. Role of peroxisome proliferator-activated receptor δ/β in hepatic metabolic regulation. *J. Biol. Chem.* **2011**, *286*, 1237–1247. [CrossRef]

280. Qin, X.; Xie, X.; Fan, Y.; Tian, J.; Guan, Y.; Wang, X.; Zhu, Y.; Wang, N. Peroxisome proliferator-activated receptor-δ induces insulin-induced gene-1 and suppresses hepatic lipogenesis in obese diabetic mice. *Hepatology* **2008**, *48*, 432–441. [CrossRef]

281. Sanderson, L.M.; Boekschoten, M.V.; Desvergne, B.; Müller, M.; Kersten, S. Transcriptional profiling reveals divergent roles of PPARα and PPARβ/δ in regulation of gene expression in mouse liver. *Physiol. Genomics* **2010**, *41*, 42–52. [CrossRef] [PubMed]

282. Risérus, U.; Sprecher, D.; Johnson, T.; Olson, E.; Hirschberg, S.; Liu, A.; Fang, Z.; Hegde, P.; Richards, D.; Sarov-Blat, L.; et al. Activation of peroxisome proliferator-activated receptor (PPAR)delta promotes reversal of multiple metabolic abnormalities, reduces oxidative stress, and increases fatty acid oxidation in moderately obese men. *Diabetes* **2008**, *57*, 332–339. [CrossRef]

283. Bedu, E.; Wahli, W.; Desvergne, B. Peroxisome proliferator-activated receptor β/δ as a therapeutic target for metabolic diseases. *Expert Opin. Ther. Targets* **2005**, *9*, 861–873. [CrossRef]

284. Zarei, M.; Barroso, E.; Palomer, X.; Dai, J.; Rada, P.; Quesada-López, T.; Escolà-Gil, J.C.; Cedó, L.; Zali, M.R.; Molaei, M.; et al. Hepatic regulation of VLDL receptor by PPARβ/δ and FGF21 modulates non-alcoholic fatty liver disease. *Mol. Metab.* **2018**, *8*, 117–131. [CrossRef] [PubMed]

285. Doktorova, M.; Zwarts, I.; Van Zutphen, T.; Van DIjk, T.H.; Bloks, V.W.; Harkema, L.; De Bruin, A.; Downes, M.; Evans, R.M.; Verkade, H.J.; et al. Intestinal PPARδ protects against diet-induced obesity, insulin resistance and dyslipidemia. *Sci. Rep.* **2017**, *7*. [CrossRef] [PubMed]

286. Wagner, N.; Wagner, K.-D. PPAR Beta/Delta and the Hallmarks of Cancer. *Cells* **2020**, *9*, 1133. [CrossRef]

287. Imai, T.; Takakuwa, R.; Marchand, S.; Dentz, E.; Bornert, J.M.; Messaddeq, N.; Wendling, O.; Mark, M.; Desvergne, B.; Wahli, W.; et al. Peroxisome proliferator-activated receptor γ is required in mature white and brown adipocytes for their survival in the mouse. *Proc. Natl. Acad. Sci. USA* **2004**, *101*, 4543–4547. [CrossRef]

288. Wang, F.; Mullican, S.E.; DiSpirito, J.R.; Peed, L.C.; Lazar, M.A. Lipoatrophy and severe metabolic disturbance in mice with fat-specific deletion of PPARγ. *Proc. Natl. Acad. Sci. USA* **2013**, *110*, 18656–18661. [CrossRef] [PubMed]

289. Jonker, J.W.; Suh, J.M.; Atkins, A.R.; Ahmadian, M.; Li, P.; Whyte, J.; He, M.; Juguilon, H.; Yin, Y.Q.; Phillips, C.T.; et al. A PPARγ-FGF1 axis is required for adaptive adipose remodelling and metabolic homeostasis. *Nature* **2012**, *485*, 391–394. [CrossRef] [PubMed]

290. Blanchard, P.G.; Moreira, R.J.; Castro, É.; Caron, A.; Côté, M.; Andrade, M.L.; Oliveira, T.E.; Ortiz-Silva, M.; Peixoto, A.S.; Dias, F.A.; et al. PPARγ is a major regulator of branched-chain amino acid blood levels and catabolism in white and brown adipose tissues. *Metabolism* **2018**, *89*, 27–38. [CrossRef] [PubMed]

291. Dutchak, P.A.; Katafuchi, T.; Bookout, A.L.; Choi, J.H.; Yu, R.T.; Mangelsdorf, D.J.; Kliewer, S.A. Fibroblast growth factor-21 regulates PPARγ activity and the antidiabetic actions of thiazolidinediones. *Cell* **2012**, *148*, 556–567. [CrossRef] [PubMed]

292. Matsusue, K.; Haluzik, M.; Lambert, G.; Yim, S.-H.; Gavrilova, O.; Ward, J.M.; Brewer, B.; Reitman, M.L.; Gonzalez, F.J. Liver-specific disruption of PPARγ in leptin-deficient mice improves fatty liver but aggravates diabetic phenotypes. *J. Clin. Invest.* **2003**, *111*, 737–747. [CrossRef]

293. Matsusue, K.; Kusakabe, T.; Noguchi, T.; Takiguchi, S.; Suzuki, T.; Yamano, S.; Gonzalez, F.J. Hepatic Steatosis in Leptin-Deficient Mice Is Promoted by the PPARγ Target Gene Fsp27. *Cell Metab.* **2008**, *7*, 302–311. [CrossRef] [PubMed]

294. Wang, W.; Xu, M.J.; Cai, Y.; Zhou, Z.; Cao, H.; Mukhopadhyay, P.; Pacher, P.; Zheng, S.; Gonzalez, F.J.; Gao, B. Inflammation is independent of steatosis in a murine model of steatohepatitis. *Hepatology* **2017**, *66*, 108–123. [CrossRef] [PubMed]

295. Hasenfuss, S.C.; Bakiri, L.; Thomsen, M.K.; Williams, E.G.; Auwerx, J.; Wagner, E.F. Regulation of steatohepatitis and PPARγ signaling by distinct AP-1 dimers. *Cell Metab.* **2014**, *19*, 84–95. [CrossRef] [PubMed]

296. Lee, Y.K.; Park, J.E.; Lee, M.; Hardwick, J.P. Hepatic lipid homeostasis by peroxisome proliferator-activated receptor gamma 2. *Liver Res.* **2018**, *2*, 209–215. [CrossRef] [PubMed]

297. Panasyuk, G.; Espeillac, C.; Chauvin, C.; Pradelli, L.A.; Horie, Y.; Suzuki, A.; Annicotte, J.S.; Fajas, L.; Foretz, M.; Verdeguer, F.; et al. PPARγ contributes to PKM2 and HK2 expression in fatty liver. *Nat. Commun.* **2012**, *3*. [CrossRef]

298. Wang, Y.; Nakajima, T.; Gonzalez, F.J.; Tanaka, N. PPARs as metabolic regulators in the liver: Lessons from liver-specific PPAR-null mice. *Int. J. Mol. Sci.* **2020**, *21*, 2061. [CrossRef]

299. Wahli, W.; Michalik, L. PPARs at the crossroads of lipid signaling and inflammation. *Trends Endocrinol. Metab.* **2012**, *23*, 351–363. [CrossRef]

300. Stienstra, R.; Mandard, S.; Tan, N.S.; Wahli, W.; Trautwein, C.; Richardson, T.A.; Lichtenauer-Kaligis, E.; Kersten, S.; Müller, M. The Interleukin-1 receptor antagonist is a direct target gene of PPARα in liver. *J. Hepatol.* **2007**, *46*, 869–877. [CrossRef]

301. Pawlak, M.; Baugé, E.; Bourguet, W.; De Bosscher, K.; Lalloyer, F.; Tailleux, A.; Lebherz, C.; Lefebvre, P.; Staels, B. The transrepressive activity of peroxisome proliferator-activated receptor alpha is necessary and sufficient to prevent liver fibrosis in mice. *Hepatology* **2014**, *60*, 1593–1606. [CrossRef]

302. Devchand, P.R.; Keller, H.; Peters, J.M.; Vazquez, M.; Gonzalez, F.J.; Wahli, W. The PPARα-leukotriene B4 pathway to inflammation control. *Nature* **1996**, *384*, 39–43. [CrossRef]

303. Jordan, S.; Tung, N.; Casanova-Acebes, M.; Chang, C.; Cantoni, C.; Zhang, D.; Wirtz, T.H.; Naik, S.; Rose, S.A.; Brocker, C.N.; et al. Dietary Intake Regulates the Circulating Inflammatory Monocyte Pool. *Cell* **2019**, *178*, 1102–1114.e17. [CrossRef] [PubMed]

304. Brocker, C.N.; Yue, J.; Kim, D.; Qu, A.; Bonzo, J.A.; Gonzalez, F.J. Hepatocyte-specific PPARA expression exclusively promotes agonist-induced cell proliferation without influence from nonparenchymal cells. *Am. J. Physiol. Gastrointest. Liver Physiol.* **2017**, *312*, G283–G299. [CrossRef] [PubMed]

305. Ip, E.; Farrell, G.; Hall, P.; Robertson, G.; Leclercq, I. Administration of the Potent PPARα Agonist, Wy-14,643, Reverses Nutritional Fibrosis and Steatohepatitis in Mice. *Hepatology* **2004**, *39*, 1286–1296. [CrossRef] [PubMed]

306. Chen, L.; Li, L.; Chen, J.; Li, L.; Zheng, Z.; Ren, J.; Qiu, Y. Oleoylethanolamide, an endogenous PPAR-α ligand, attenuates liver fibrosis targeting hepatic stellate cells. *Oncotarget* **2015**, *6*, 42530–42540. [CrossRef]

307. Odegaard, J.I.; Ricardo-Gonzalez, R.R.; Red Eagle, A.; Vats, D.; Morel, C.R.; Goforth, M.H.; Subramanian, V.; Mukundan, L.; Ferrante, A.W.; Chawla, A. Alternative M2 Activation of Kupffer Cells by PPARδ Ameliorates Obesity-Induced Insulin Resistance. *Cell Metab.* **2008**, *7*, 496–507. [CrossRef]

308. Chen, J.; Zhuang, Y.; Sng, M.K.; Tan, N.S.; Wahli, W. The potential of the FSP1cre-Pparb/d−/− mouse model for studying Juvenile NAFLD. *Int. J. Mol. Sci.* **2019**, *20*, 5115. [CrossRef]

309. Kostadinova, R.; Montagner, A.; Gouranton, E.; Fleury, S.; Guillou, H.; Dombrowicz, D.; Desreumaux, P.; Wahli, W. GW501516-activated PPARβ/δ promotes liver fibrosis via p38-JNK MAPK-induced hepatic stellate cell proliferation. *Cell Biosci.* **2012**, *2*. [CrossRef]

310. Hellemans, K.; Michalik, L.; Dittie, A.; Knorr, A.; Rombouts, K.; De Jong, J.; Heirman, C.; Quartier, E.; Schuit, F.; Wahli, W.; et al. Peroxisome proliferator-activated receptor-β signaling contributes to enhanced proliferation of hepatic stellate cells. *Gastroenterology* **2003**, *124*, 184–201. [CrossRef]

311. Iwaisako, K.; Haimerl, M.; Paik, Y.H.; Taura, K.; Kodama, Y.; Sirlin, C.; Yu, E.; Yu, R.T.; Downes, M.; Evans, R.M.; et al. Protection from liver fibrosis by a peroxisome proliferator-activated receptor δ agonist. *Proc. Natl. Acad. Sci. USA* **2012**, *109*. [CrossRef]

312. Ricote, M.; Li, A.C.; Willson, T.M.; Kelly, C.J.; Glass, C.K. The peroxisome proliferator-activated receptor-γ is a negative regulator of macrophage activation. *Nature* **1998**, *391*, 79–82. [CrossRef]

313. Bouhlel, M.A.; Derudas, B.; Rigamonti, E.; Dièvart, R.; Brozek, J.; Haulon, S.; Zawadzki, C.; Jude, B.; Torpier, G.; Marx, N.; et al. PPARγ Activation Primes Human Monocytes into Alternative M2 Macrophages with Anti-inflammatory Properties. *Cell Metab.* **2007**, *6*, 137–143. [CrossRef]

314. Odegaard, J.I.; Ricardo-Gonzalez, R.R.; Goforth, M.H.; Morel, C.R.; Subramanian, V.; Mukundan, L.; Eagle, A.R.; Vats, D.; Brombacher, F.; Ferrante, A.W.; et al. Macrophage-specific PPARγ controls alternative activation and improves insulin resistance. *Nature* **2007**, *447*, 1116–1120. [CrossRef] [PubMed]

315. Luo, W.; Xu, Q.; Wang, Q.; Wu, H.; Hua, J. Effect of modulation of PPAR-γ activity on Kupffer cells M1/M2 polarization in the development of non-alcoholic fatty liver disease. *Sci. Rep.* **2017**, *7*. [CrossRef]

316. Cipolletta, D.; Feuerer, M.; Li, A.; Kamei, N.; Lee, J.; Shoelson, S.E.; Benoist, C.; Mathis, D. PPAR-γ is a major driver of the accumulation and phenotype of adipose tissue T reg cells. *Nature* **2012**, *486*, 549–553. [CrossRef]

317. Miyahara, T.; Schrum, L.; Rippe, R.; Xiong, S.; Yee, J.; Motomura, K.; Anania, F.A.; Willson, T.M.; Tsukamoto, H. Peroxisome proliferator-activated receptors and hepatic stellate cell activation. *J. Biol. Chem.* **2000**, *275*, 35715–35722. [CrossRef] [PubMed]

318. Hazra, S.; Xiong, S.; Wang, J.; Rippe, R.A.; Chatterjee, V.K.K.; Tsukamoto, H. Peroxisome Proliferator-activated Receptor γ Induces a Phenotypic Switch from Activated to Quiescent Hepatic Stellate Cells. *J. Biol. Chem.* **2004**, *279*, 11392–11401. [CrossRef] [PubMed]

319. Galli, A.; Crabb, D.W.; Ceni, E.; Salzano, R.; Mello, T.; Svegliati-Baroni, G.; Ridolfi, F.; Trozzi, L.; Surrenti, C.; Casini, A. Antidiabetic thiazolidinediones inhibit collagen synthesis and hepatic stellate cell activation in vivo and in vitro. *Gastroenterology* **2002**, *122*, 1924–1940. [CrossRef] [PubMed]

320. Morán-Salvador, E.; Titos, E.; Rius, B.; González-Périz, A.; García-Alonso, V.; López-Vicario, C.; Miquel, R.; Barak, Y.; Arroyo, V.; Clària, J. Cell-specific PPARγ deficiency establishes anti-inflammatory and anti-fibrogenic properties for this nuclear receptor in non-parenchymal liver cells. *J. Hepatol.* **2013**, *59*, 1045–1053. [CrossRef]

321. Mandard, S.; Müller, M.; Kersten, S. Peroxisome proliferator-activated receptor α target genes. *Cell. Mol. Life Sci.* **2004**, *61*, 393–416. [CrossRef]

322. Kersten, S.; Rakhshandehroo, M.; Knoch, B.; Müller, M. Peroxisome proliferator-activated receptor alpha target genes. *PPAR Res.* **2010**, *2010*. [CrossRef]

323. Francque, S.; Verrijken, A.; Caron, S.; Prawitt, J.; Paumelle, R.; Derudas, B.; Lefebvre, P.; Taskinen, M.R.; Van Hul, W.; Mertens, I.; et al. PPARα gene expression correlates with severity and histological treatment response in patients with non-alcoholic steatohepatitis. *J. Hepatol.* **2015**, *63*, 164–173. [CrossRef]

324. Costet, P.; Legendre, C.; Moré, J.; Edgar, A.; Galtier, P.; Pineau, T. Peroxisome proliferator-activated receptor alpha-isoform deficiency leads to progressive dyslipidemia with sexually dimorphic obesity and steatosis. *J. Biol. Chem.* **1998**, *273*, 29577–29585. [CrossRef] [PubMed]

325. Stienstra, R.; Mandard, S.; Patsouris, D.; Maass, C.; Kersten, S.; Müller, M. Peroxisome proliferator-activated receptor α protects against obesity-induced hepatic inflammation. *Endocrinology* **2007**, *148*, 2753–2763. [CrossRef]

326. Abdelmegeed, M.A.; Yoo, S.-H.; Henderson, L.E.; Gonzalez, F.J.; Woodcroft, K.J.; Song, B.-J. PPARα Expression Protects Male Mice from High Fat–Induced Nonalcoholic Fatty Liver. *J. Nutr.* **2011**, *141*, 603–610. [CrossRef] [PubMed]

327. Ip, E.; Farrell, G.C.; Robertson, G.; Hall, P.; Kirsch, R.; Leclercq, I. Central role of PPARα-dependent hepatic lipid turnover in dietary steatohepatitis in mice. *Hepatology* **2003**, *38*, 123–132. [CrossRef] [PubMed]

328. Hu, X.; Tanaka, N.; Guo, R.; Lu, Y.; Nakajima, T.; Gonzalez, F.J.; Aoyama, T. PPARα protects against trans-fatty-acid-containing diet-induced steatohepatitis. *J. Nutr. Biochem.* **2017**, *39*, 77–85. [CrossRef] [PubMed]

329. Régnier, M.; Polizzi, A.; Smati, S.; Lukowicz, C.; Fougerat, A.; Lippi, Y.; Fouché, E.; Lasserre, F.; Naylies, C.; Bétoulières, C.; et al. Hepatocyte-specific deletion of Pparα promotes NAFLD in the context of obesity. *Sci. Rep.* **2020**, *10*. [CrossRef]

330. Liang, N.; Damdimopoulos, A.; Goñi, S.; Huang, Z.; Vedin, L.L.; Jakobsson, T.; Giudici, M.; Ahmed, O.; Pedrelli, M.; Barilla, S.; et al. Hepatocyte-specific loss of GPS2 in mice reduces non-alcoholic steatohepatitis via activation of PPARα. *Nat. Commun.* **2019**, *10*. [CrossRef]

331. Shan, W.; Nicol, C.J.; Ito, S.; Bility, M.T.; Kennett, M.J.; Ward, J.M.; Gonzalez, F.J.; Peters, J.M. Peroxisome proliferator-activated receptor-β/δ protects against chemically induced liver toxicity in mice. *Hepatology* **2008**, *47*, 225–235. [CrossRef] [PubMed]

332. Barak, Y.; Nelson, M.C.; Ong, E.S.; Jones, Y.Z.; Ruiz-Lozano, P.; Chien, K.R.; Koder, A.; Evans, R.M. PPARγ is required for placental, cardiac, and adipose tissue development. *Mol. Cell* **1999**, *4*, 585–595. [CrossRef]

333. Nadra, K.; Anghel, S.I.; Joye, E.; Tan, N.S.; Basu-Modak, S.; Trono, D.; Wahli, W.; Desvergne, B. Differentiation of Trophoblast Giant Cells and Their Metabolic Functions Are Dependent on Peroxisome Proliferator-Activated Receptor β/δ. *Mol. Cell. Biol.* **2006**, *26*, 3266–3281. [CrossRef] [PubMed]

334. Gilardi, F.; Winkler, C.; Quignodon, L.; Diserens, J.G.; Toffoli, B.; Schiffrin, M.; Sardella, C.; Preitner, F.; Desvergne, B. Systemic PPARγ deletion in mice provokes lipoatrophy, organomegaly, severe type 2 diabetes and metabolic inflexibility. *Metabolism* **2019**, *95*, 8–20. [CrossRef] [PubMed]

335. Morán-Salvador, E.; López-Parra, M.; García-Alonso, V.; Titos, E.; Martínez-Clemente, M.; González-Périz, A.; López-Vicario, C.; Barak, Y.; Arroyo, V.; Clària, J. Role for PPARγ in obesity-induced hepatic steatosis as determined by hepatocyte- and macrophage-specific conditional knockouts. *FASEB J.* **2011**, *25*, 2538–2550. [CrossRef] [PubMed]

336. Greenstein, A.W.; Majumdar, N.; Yang, P.; Subbaiah, P.V.; Kineman, R.D.; Cordoba-Chacon, J. Hepatocyte-specific, PPARγ-regulated mechanisms to promote steatosis in adult mice. *J. Endocrinol.* **2017**, *232*, 107–121. [CrossRef] [PubMed]

337. Patitucci, C.; Couchy, G.; Bagattin, A.; Cañeque, T.; De Reyniès, A.; Scoazec, J.Y.; Rodriguez, R.; Pontoglio, M.; Zucman-Rossi, J.; Pende, M.; et al. Hepatocyte nuclear factor 1α suppresses steatosisassociated liver cancer by inhibiting PPARγ transcription. *J. Clin. Invest.* **2017**, *127*, 1873–1888. [CrossRef] [PubMed]

338. Duszka, K.; Ellero-Simatos, S.; Ow, G.S.; Defernez, M.; Paramalingam, E.; Tett, A.; Ying, S.; König, J.; Narbad, A.; Kuznetsov, V.A.; et al. Complementary intestinal mucosa and microbiota responses to caloric restriction. *Sci. Rep.* **2018**, *8*. [CrossRef] [PubMed]

339. Montagner, A.; Korecka, A.; Polizzi, A.; Lippi, Y.; Blum, Y.; Canlet, C.; Tremblay-Franco, M.; Gautier-Stein, A.; Burcelin, R.; Yen, Y.-C.C.; et al. Hepatic circadian clock oscillators and nuclear receptors integrate microbiome-derived signals. *Sci. Rep.* **2016**, *6*, 20127. [CrossRef] [PubMed]

340. Ul Hasan, A.; Rahman, A.; Kobori, H. Interactions between host PPARs and gut microbiota in health and disease. *Int. J. Mol. Sci.* **2019**, *20*, 387. [CrossRef]

341. Choudhary, N.S.; Kumar, N.; Duseja, A. Peroxisome Proliferator-Activated Receptors and Their Agonists in Nonalcoholic Fatty Liver Disease. *J. Clin. Exp. Hepatol.* **2019**, *9*, 731–739. [CrossRef]

342. Boeckmans, J.; Natale, A.; Rombaut, M.; Buyl, K.; Rogiers, V.; De Kock, J.; Vanhaecke, T.; Rodrigues, R.M. Anti-NASH Drug Development Hitches a Lift on PPAR Agonism. *Cells* **2019**, *9*, 37. [CrossRef]

343. Okopień, B.; Bułdak, Ł.; Bołdys, A. Benefits and risks of the treatment with fibrates—a comprehensive summary. *Expert Rev. Clin. Pharmacol.* **2018**, *11*, 1099–1112. [CrossRef]

344. Larter, C.Z.; Yeh, M.M.; van Rooyen, D.M.; Brooling, J.; Ghatora, K.; Farrell, G.C. Peroxisome proliferator-activated receptor-α agonist, Wy 14643, improves metabolic indices, steatosis and ballooning in diabetic mice with non-alcoholic steatohepatitis. *J. Gastroenterol. Hepatol.* **2012**, *27*, 341–350. [CrossRef]

345. Toyama, T.; Nakamura, H.; Harano, Y.; Yamauchi, N.; Morita, A.; Kirishima, T.; Minami, M.; Itoh, Y.; Okanoue, T. PPARα ligands activate antioxidant enzymes and suppress hepatic fibrosis in rats. *Biochem. Biophys. Res. Commun.* **2004**, *324*, 697–704. [CrossRef] [PubMed]

346. Rodríguez-Vilarrupla, A.; Laviña, B.; García-Calderó, H.; Russo, L.; Rosado, E.; Roglans, N.; Bosch, J.; García-Pagán, J.C. PPARα activation improves endothelial dysfunction and reduces fibrosis and portal pressure in cirrhotic rats. *J. Hepatol.* **2012**, *56*, 1033–1039. [CrossRef] [PubMed]

347. Nikam, A.; Patankar, J.V.; Somlapura, M.; Lahiri, P.; Sachdev, V.; Kratky, D.; Denk, H.; Zatloukal, K.; Abuja, P.M. The PPARα Agonist Fenofibrate Prevents Formation of Protein Aggregates (Mallory-Denk bodies) in a Murine Model of Steatohepatitis-like Hepatotoxicity. *Sci. Rep.* **2018**, *8*. [CrossRef] [PubMed]

348. Fabbrini, E.; Mohammed, B.S.; Korenblat, K.M.; Magkos, F.; McCrea, J.; Patterson, B.W.; Klein, S. Effect of Fenofibrate and Niacin on Intrahepatic Triglyceride Content, Very Low-Density Lipoprotein Kinetics, and Insulin Action in Obese Subjects with Nonalcoholic Fatty Liver Disease. *J. Clin. Endocrinol. Metab.* **2010**, *95*, 2727. [CrossRef]

349. Fernández-Miranda, C.; Pérez-Carreras, M.; Colina, F.; López-Alonso, G.; Vargas, C.; Solís-Herruzo, J.A. A pilot trial of fenofibrate for the treatment of non-alcoholic fatty liver disease. *Dig. Liver Dis.* **2008**, *40*, 200–205. [CrossRef]

350. El-Haggar, S.M.; Mostafa, T.M. Comparative clinical study between the effect of fenofibrate alone and its combination with pentoxifylline on biochemical parameters and liver stiffness in patients with non-alcoholic fatty liver disease. *Hepatol. Int.* **2015**, *9*, 471–479. [CrossRef] [PubMed]

351. Oscarsson, J.; Önnerhag, K.; Risérus, U.; Sundén, M.; Johansson, L.; Jansson, P.A.; Moris, L.; Nilsson, P.M.; Eriksson, J.W.; Lind, L. Effects of free omega-3 carboxylic acids and fenofibrate on liver fat content in patients with hypertriglyceridemia and non-alcoholic fatty liver disease: A double-blind, randomized, placebo-controlled study. *J. Clin. Lipidol.* **2018**, *12*, 1390–1403.e4. [CrossRef]

352. Roy, A.; Pahan, K. Gemfibrozil, stretching arms beyond lipid lowering. *Immunopharmacol. Immunotoxicol.* **2009**, *31*, 339–351. [CrossRef]

353. Zhang, X.; Wang, S.; Hu, L.; Wang, J.; Liu, Y.; Shi, P. Gemfibrozil reduces lipid accumulation in smmc-7721 cells via the involvement of pparα and srebp1. *Exp. Ther. Med.* **2019**, *17*, 1282–1289. [CrossRef]

354. Basaranoglu, M.; Acbay, O.; Sonsuz, A. A controlled trial of gemfibrozil in the treatment of patients with nonalcoholic steatohepatitis. *J. Hepatol.* **1999**, *31*, 384. [CrossRef]

355. Yaghoubi, M.; Jafari, S.; Sajedi, B.; Gohari, S.; Akbarieh, S.; Heydari, A.H.; Jameshoorani, M. Comparison of fenofibrate and pioglitazone effects on patients with nonalcoholic fatty liver disease. *Eur. J. Gastroenterol. Hepatol.* **2017**, *29*, 1385–1388. [CrossRef]

356. Garbacz, W.G.; Huang, J.T.J.; Higgins, L.G.; Wahli, W.; Palmer, C.N.A. PPAR Is Required for PPAR δ Action in Regulation of Body Weight and Hepatic Steatosis in Mice. *PPAR Res.* **2015**, *2015*. [CrossRef] [PubMed]

357. Tanaka, T.; Yamamoto, J.; Iwasaki, S.; Asaba, H.; Hamura, H.; Ikeda, Y.; Watanabe, M.; Magoori, K.; Ioka, R.X.; Tachibana, K.; et al. Activation of peroxisome proliferator-activated receptor δ induces fatty acid β-oxidation in skeletal muscle and attenuates metabolic syndrome. *Proc. Natl. Acad. Sci. USA* **2003**, *100*, 15924–15929. [CrossRef]

358. Zarei, M.; Barroso, E.; Palomer, X.; Escolà-Gil, J.C.; Cedó, L.; Wahli, W.; Vázquez-Carrera, M. Pharmacological PPARβ/δ activation upregulates VLDLR in hepatocytes. *Clin. e Investig. en Arterioscler.* **2019**, *31*, 111–118. [CrossRef] [PubMed]

359. Bojic, L.A.; Telford, D.E.; Fullerton, M.D.; Ford, R.J.; Sutherland, B.G.; Edwards, J.Y.; Sawyez, C.G.; Gros, R.; Kemp, B.E.; Steinberg, G.R.; et al. PPAR—Activation attenuates hepatic steatosis in Ldlr Mice by enhanced fat oxidation, reduced lipogenesis, and improved insulin sensitivity. *J. Lipid Res.* **2014**, *55*, 1254–1266. [CrossRef]

360. Chung, C.H.; Lee, M.Y.; Choi, R.; Kim, H.M.; Cho, E.J.; Kim, B.H.; Choi, Y.S.; Naowaboot, J.; Lee, E.Y.; Yang, Y.C.; et al. Peroxisome proliferator-activated receptor δ agonist attenuates hepatic steatosis by anti-inflammatory mechanism. *Exp. Mol. Med.* **2012**, *44*, 578–585. [CrossRef]

361. Silva-Veiga, F.M.; Rachid, T.L.; de Oliveira, L.; Graus-Nunes, F.; Mandarim-de-Lacerda, C.A.; Souza-Mello, V. GW0742 (PPAR-beta agonist) attenuates hepatic endoplasmic reticulum stress by improving hepatic energy metabolism in high-fat diet fed mice. *Mol. Cell. Endocrinol.* **2018**, *474*, 227–237. [CrossRef]

362. Shan, W.; Palkar, P.S.; Murray, I.A.; Mcdevitt, E.I.; Kennett, M.J.; Kang, B.H.; Isom, H.C.; Perdew, G.H.; Gonzalez, F.J.; Peters, J.M. Ligand activation of peroxisome proliferator—Activated receptor β/δ (PPARβ/δ) attenuates carbon tetrachloride hepatotoxicity by downregulating proinflammatory gene expression. *Toxicol. Sci.* **2008**, *105*, 418–428. [CrossRef]

363. Haczeyni, F.; Wang, H.; Barn, V.; Mridha, A.R.; Yeh, M.M.; Haigh, W.G.; Ioannou, G.N.; Choi, Y.-J.; McWherter, C.A.; Teoh, N.C.-H.; et al. The selective peroxisome proliferator-activated receptor-delta agonist seladelpar reverses nonalcoholic steatohepatitis pathology by abrogating lipotoxicity in diabetic obese mice. *Hepatol. Commun.* **2017**, *1*, 663–674. [CrossRef] [PubMed]

364. Daniels, S.J.; Leeming, D.J.; Detlefsen, S.; Bruun, M.F.; Hjuler, S.T.; Henriksen, K.; Hein, P.; Karsdal, M.A.; Brockbank, S.; Cruwys, S. Biochemical and histological characterisation of an experimental rodent model of non-alcoholic steatohepatitis – Effects of a peroxisome proliferator-activated receptor gamma (PPAR-γ) agonist and a glucagon-like peptide-1 analogue. *Biomed. Pharmacother.* **2019**, *111*, 926–933. [CrossRef]

365. Ahn, H.Y.; Kim, H.H.; Hwang, J.Y.; Park, C.; Cho, B.Y.; Park, Y.J. Effects of pioglitazone on nonalcoholic fatty liver disease in the absence of constitutive androstane receptor expression. *PPAR Res.* **2018**, *2018*. [CrossRef] [PubMed]

366. Jia, C.; Huan, Y.; Liu, S.; Hou, S.; Sun, S.; Li, C.; Liu, Q.; Jiang, Q.; Wang, Y.; Shen, Z. Effect of chronic pioglitazone treatment on hepatic gene expression profile in obese C57BL/6J mice. *Int. J. Mol. Sci.* **2015**, *16*, 12213–12229. [CrossRef]

367. Musso, G.; Cassader, M.; Paschetta, E.; Gambino, R. Thiazolidinediones and advanced liver fibrosis in nonalcoholic steatohepatitis: A meta-analysis. *JAMA Intern. Med.* **2017**, *177*, 633–640. [CrossRef]

368. Wu, H.M.; Ni, X.X.; Xu, Q.Y.; Wang, Q.; Li, X.Y.; Hua, J. Regulation of lipid-induced macrophage polarization through modulating peroxisome proliferator-activated receptor-gamma activity affects hepatic lipid metabolism via a Toll-like receptor 4/NF-κB signaling pathway. *J. Gastroenterol. Hepatol.* **2020**. [CrossRef]

369. Nan, Y.M.; Fu, N.; Wu, W.J.; Liang, B.L.; Wang, R.Q.; Zhao, S.X.; Zhao, J.M.; Yu, J. Rosiglitazone prevents nutritional fibrosis and steatohepatitis in mice. *Scand. J. Gastroenterol.* **2009**, *44*, 358–365. [CrossRef]

370. Wei, Z.; Zhao, D.; Zhang, Y.; Chen, Y.; Zhang, S.; Li, Q.; Zeng, P.; Li, X.; Zhang, W.; Duan, Y.; et al. Rosiglitazone ameliorates bile duct ligation-induced liver fibrosis by down-regulating NF-κB-TNF-α signaling pathway in a PPARγ-dependent manner. *Biochem. Biophys. Res. Commun.* **2019**, *519*, 854–860. [CrossRef] [PubMed]

371. Wang, Q.A.; Zhang, F.; Jiang, L.; Ye, R.; An, Y.; Shao, M.; Tao, C.; Gupta, R.K.; Scherer, P.E. Peroxisome Proliferator-Activated Receptor γ and Its Role in Adipocyte Homeostasis and Thiazolidinedione-Mediated Insulin Sensitization. *Mol. Cell. Biol.* **2018**, *38*. [CrossRef] [PubMed]

372. Neuschwander-Tetri, B.A.; Brunt, E.M.; Wehmeier, K.R.; Oliver, D.; Bacon, B.R. Improved nonalcoholic steatohepatitis after 48 weeks of treatment with the PPAR-γ ligand rosiglitazone. *Hepatology* **2003**, *38*, 1008–1017. [CrossRef] [PubMed]

373. Ratziu, V.; Giral, P.; Jacqueminet, S.; Charlotte, F.; Hartemann-Heurtier, A.; Serfaty, L.; Podevin, P.; Lacorte, J.M.; Bernhardt, C.; Bruckert, E.; et al. Rosiglitazone for Nonalcoholic Steatohepatitis: One-Year Results of the Randomized Placebo-Controlled Fatty Liver Improvement With Rosiglitazone Therapy (FLIRT) Trial. *Gastroenterology* **2008**, *135*, 100–110. [CrossRef] [PubMed]

374. Lemoine, M.; Serfaty, L.; Cervera, P.; Capeau, J.; Ratziu, V. Hepatic molecular effects of rosiglitazone in human non-alcoholic steatohepatitis suggest long-term pro-inflammatory damage. *Hepatol. Res.* **2014**, *44*, 1241–1247. [CrossRef] [PubMed]

375. Fruchart, J.C.; Santos, R.D.; Aguilar-Salinas, C.; Aikawa, M.; Al Rasadi, K.; Amarenco, P.; Barter, P.J.; Ceska, R.; Corsini, A.; Després, J.P.; et al. The selective peroxisome proliferator-activated receptor alpha modulator (SPPARMα) paradigm: Conceptual framework and therapeutic potential. *Cardiovasc. Diabetol.* **2019**, *18*, 71. [CrossRef] [PubMed]

376. Kawasaki, M.; Kambe, A.; Yamamoto, Y.; Arulmozhiraja, S.; Ito, S.; Nakagawa, Y.; Tokiwa, H.; Nakano, S.; Shimano, H. Elucidation of molecular mechanism of a selective PPARα modulator, pemafibrate, through combinational approaches of x-ray crystallography, thermodynamic analysis, and first-principle calculations. *Int. J. Mol. Sci.* **2020**, *21*, 361. [CrossRef]

377. Raza-Iqbal, S.; Tanaka, T.; Anai, M.; Inagaki, T.; Matsumura, Y.; Ikeda, K.; Taguchi, A.; Gonzalez, F.J.; Sakai, J.; Kodama, T. Transcriptome analysis of K-877 (A novel selective PPARα modulator (SPPARMα))-regulated genes in primary human hepatocytes and the mouse liver. *J. Atheroscler. Thromb.* **2015**, *22*, 754–772. [CrossRef]

378. Sasaki, Y.; Raza-Iqbal, S.; Tanaka, T.; Murakami, K.; Anai, M.; Osawa, T.; Matsumura, Y.; Sakai, J.; Kodama, T. Gene expression profiles induced by a novel selective peroxisome proliferator-activated receptor α modulator (SPPARMA) pemafibrate. *Int. J. Mol. Sci.* **2019**, *20*, 5682. [CrossRef]

379. Takei, K.; Nakagawa, Y.; Wang, Y.; Han, S.-I.; Satoh, A.; Sekiya, M.; Matsuzaka, T.; Shimano, H. Effects of K-877, a novel selective PPARα modulator, on small intestine contribute to the amelioration of hyperlipidemia in low-density lipoprotein receptor knockout mice. *J. Pharmacol. Sci.* **2017**, *133*, 214–222. [CrossRef]

380. Hennuyer, N.; Duplan, I.; Paquet, C.; Vanhoutte, J.; Woitrain, E.; Touche, V.; Colin, S.; Vallez, E.; Lestavel, S.; Lefebvre, P.; et al. The novel selective PPARα modulator (SPPARMα) pemafibrate improves dyslipidemia, enhances reverse cholesterol transport and decreases inflammation and atherosclerosis. *Atherosclerosis* **2016**, *249*, 200–208. [CrossRef]

381. Sairyo, M.; Kobayashi, T.; Masuda, D.; Kanno, K.; Zhu, Y.; Okada, T.; Koseki, M.; Ohama, T.; Nishida, M.; Sakata, Y.; et al. A novel selective PPARα modulator (SPPARMα), K-877 (pemafibrate), attenuates postprandial hypertriglyceridemia in mice. *J. Atheroscler. Thromb.* **2018**, *25*, 142–152. [CrossRef]

382. Araki, M.; Nakagawa, Y.; Oishi, A.; Han, S.I.; Wang, Y.; Kumagai, K.; Ohno, H.; Mizunoe, Y.; Iwasaki, H.; Sekiya, M.; et al. The peroxisome proliferator-activated receptor α (PPARα) agonist pemafibrate protects against diet-induced obesity in mice. *Int. J. Mol. Sci.* **2018**, *19*, 2148. [CrossRef]

383. Honda, Y.; Kessoku, T.; Ogawa, Y.; Tomeno, W.; Imajo, K.; Fujita, K.; Yoneda, M.; Takizawa, T.; Saito, S.; Nagashima, Y.; et al. Pemafibrate, a novel selective peroxisome proliferator-activated receptor alpha modulator, improves the pathogenesis in a rodent model of nonalcoholic steatohepatitis. *Sci. Rep.* **2017**, *7*, 42477. [CrossRef] [PubMed]

384. Sasaki, Y.; Asahiyama, M.; Tanaka, T.; Yamamoto, S.; Murakami, K.; Kamiya, W.; Matsumura, Y.; Osawa, T.; Anai, M.; Fruchart, J.C.; et al. Pemafibrate, a selective PPARα modulator, prevents non-alcoholic steatohepatitis development without reducing the hepatic triglyceride content. *Sci. Rep.* **2020**, *10*. [CrossRef]

385. Yamashita, S.; Arai, H.; Yokote, K.; Araki, E.; Matsushita, M.; Nojima, T.; Suganami, H.; Ishibashi, S. Efficacy and safety of pemafibrate, a novel selective peroxisome proliferator-activated receptor α modulator (SPPARMα): Pooled analysis of phase 2 and 3 studies in dyslipidemic patients with or without statin combination. *Int. J. Mol. Sci.* **2019**, *20*, 5537. [CrossRef] [PubMed]

386. Ishibashi, S.; Arai, H.; Yokote, K.; Araki, E.; Suganami, H.; Yamashita, S. Efficacy and safety of pemafibrate (K-877), a selective peroxisome proliferator-activated receptor α modulator, in patients with dyslipidemia: Results from a 24-week, randomized, double blind, active-controlled, phase 3 trial. *J. Clin. Lipidol.* **2018**, *12*, 173–184. [CrossRef] [PubMed]

387. Arai, H.; Yamashita, S.; Yokote, K.; Araki, E.; Suganami, H.; Ishibashi, S. Efficacy and safety of pemafibrate versus fenofibrate in patients with high triglyceride and low HDL cholesterol levels: A multicenter, placebo-controlled, double-blind, randomized trial. *J. Atheroscler. Thromb.* **2018**, *25*, 521–538. [CrossRef]

388. Ida, S.; Kaneko, R.; Murata, K. Efficacy and safety of pemafibrate administration in patients with dyslipidemia: A systematic review and meta-analysis. *Cardiovasc. Diabetol.* **2019**, *18*. [CrossRef]

389. Araki, E.; Yamashita, S.; Arai, H.; Yokote, K.; Satoh, J.; Inoguchi, T.; Nakamura, J.; Maegawa, H.; Yoshioka, N.; Tanizawa, Y.; et al. Effects of pemafibrate, a novel selective PPARa modulator, on lipid and glucose metabolism in patients with type 2 diabetes and hypertriglyceridemia: A randomized, double-blind, placebo-controlled, phase 3 trial. *Diabetes Care* **2018**, *41*, 538–546. [CrossRef]

390. Pradhan, A.D.; Paynter, N.P.; Everett, B.M.; Glynn, R.J.; Amarenco, P.; Elam, M.; Ginsberg, H.; Hiatt, W.R.; Ishibashi, S.; Koenig, W.; et al. Rationale and design of the Pemafibrate to Reduce Cardiovascular Outcomes by Reducing Triglycerides in Patients with Diabetes (PROMINENT) study. *Am. Heart J.* **2018**, *206*, 80–93. [CrossRef] [PubMed]

391. Westerouen Van Meeteren, M.J.; Drenth, J.P.H.; Tjwa, E.T.T.L. Elafibranor: A potential drug for the treatment of nonalcoholic steatohepatitis (NASH). *Expert Opin. Investig. Drugs* **2020**, *29*, 117–123. [CrossRef] [PubMed]

392. Staels, B.; Rubenstrunk, A.; Noel, B.; Rigou, G.; Delataille, P.; Millatt, L.J.; Baron, M.; Lucas, A.; Tailleux, A.; Hum, D.W.; et al. Hepatoprotective effects of the dual peroxisome proliferator-activated receptor alpha/delta agonist, GFT505, in rodent models of nonalcoholic fatty liver disease/nonalcoholic steatohepatitis. *Hepatology* **2013**, *58*. [CrossRef] [PubMed]

393. Tølbøl, K.S.; Kristiansen, M.N.; Hansen, H.H.; Veidal, S.S.; Rigbolt, K.T.; Gillum, M.P.; Jelsing, J.; Vrang, N.; Feigh, M. Metabolic and hepatic effects of liraglutide, obeticholic acid and elafibranor in diet-induced obese mouse models of biopsy-confirmed nonalcoholic steatohepatitis. *World J. Gastroenterol.* **2018**, *24*, 179–194. [CrossRef]

394. Roth, J.D.; Veidal, S.S.; Fensholdt, L.K.D.; Rigbolt, K.T.G.; Papazyan, R.; Nielsen, J.C.; Feigh, M.; Vrang, N.; Young, M.; Jelsing, J.; et al. Combined obeticholic acid and elafibranor treatment promotes additive liver histological improvements in a diet-induced ob/ob mouse model of biopsy-confirmed NASH. *Sci. Rep.* **2019**, *9*. [CrossRef] [PubMed]

395. Briand, F.; Heymes, C.; Bonada, L.; Angles, T.; Charpentier, J.; Branchereau, M.; Brousseau, E.; Quinsat, M.; Fazilleau, N.; Burcelin, R.; et al. A 3-week nonalcoholic steatohepatitis mouse model shows elafibranor benefits on hepatic inflammation and cell death. *Clin. Transl. Sci.* **2020**, *13*. [CrossRef] [PubMed]

396. Li, T.H.; Yang, Y.Y.; Huang, C.C.; Liu, C.W.; Tsai, H.C.; Lin, M.W.; Tsai, C.Y.; Huang, S.F.; Wang, Y.W.; Lee, T.Y.; et al. Elafibranor interrupts adipose dysfunction-mediated gut and liver injury in mice with alcoholic steatohepatitis. *Clin. Sci.* **2019**, *133*, 531–544. [CrossRef]

397. Cariou, B.; Hanf, R.; Lambert-Porcheron, S.; Zair, Y.; Sauvinet, V.; Noel, B.; Flet, L.; Vidal, H.; Staels, B.; Laville, M. Dual peroxisome proliferator- activated receptor α/δ agonist gft505 improves hepatic and peripheral insulin sensitivity in abdominally obese subjects. *Diabetes Care* **2013**, *36*, 2923–2930. [CrossRef] [PubMed]

398. Ratziu, V.; Harrison, S.A.; Francque, S.; Bedossa, P.; Lehert, P.; Serfaty, L.; Romero-Gomez, M.; Boursier, J.; Abdelmalek, M.; Caldwell, S.; et al. Elafibranor, an Agonist of the Peroxisome Proliferator-Activated Receptor-α and -δ, Induces Resolution of Nonalcoholic Steatohepatitis Without Fibrosis Worsening. *Gastroenterology* **2016**, *150*, 1147–1159.e5. [CrossRef] [PubMed]

399. Agrawal, R. The First Approved Agent in the Glitazar's Class: Saroglitazar. *Curr. Drug Targets* **2014**, *15*, 151–155. [CrossRef] [PubMed]

400. Jain, M.R.; Giri, S.R.; Trivedi, C.; Bhoi, B.; Rath, A.; Vanage, G.; Vyas, P.; Ranvir, R.; Patel, P.R. Saroglitazar, a novel PPARα/γ agonist with predominant PPARα activity, shows lipid-lowering and insulin-sensitizing effects in preclinical models. *Pharmacol. Res. Perspect.* **2015**, *3*. [CrossRef] [PubMed]

401. Jain, M.R.; Giri, S.R.; Bhoi, B.; Trivedi, C.; Rath, A.; Rathod, R.; Ranvir, R.; Kadam, S.; Patel, H.; Swain, P.; et al. Dual PPARα/γ agonist saroglitazar improves liver histopathology and biochemistry in experimental NASH models. *Liver Int.* **2018**, *38*, 1084–1094. [CrossRef]

402. Kumar, D.; Goand, U.K.; Gupta, S.; Shankar, K.; Varshney, S.; Rajan, S.; Srivastava, A.; Gupta, A.; Vishwakarma, A.L.; Srivastava, A.K.; et al. Saroglitazar reduces obesity and associated inflammatory consequences in murine adipose tissue. *Eur. J. Pharmacol.* **2018**, *822*, 32–42. [CrossRef]

403. Hassan, N.F.; Nada, S.A.; Hassan, A.; El-Ansary, M.R.; Al-Shorbagy, M.Y.; Abdelsalam, R.M. Saroglitazar Deactivates the Hepatic LPS/TLR4 Signaling Pathway and Ameliorates Adipocyte Dysfunction in Rats with High-Fat Emulsion/LPS Model-Induced Non-alcoholic Steatohepatitis. *Inflammation* **2019**, *42*, 1056–1070. [CrossRef]

404. Makled, M.N.; Sharawy, M.H.; El-Awady, M.S. The dual PPAR-α/γ agonist saroglitazar ameliorates thioacetamide-induced liver fibrosis in rats through regulating leptin. *Naunyn. Schmiedebergs. Arch. Pharmacol.* **2019**, *392*, 1569–1576. [CrossRef] [PubMed]

405. Jani, R.H.; Pai, V.; Jha, P.; Jariwala, G.; Mukhopadhyay, S.; Bhansali, A.; Joshi, S. A multicenter, prospective, randomized, double-blind study to evaluate the safety and efficacy of saroglitazar 2 and 4 mg compared with placebo in type 2 diabetes mellitus patients having hypertriglyceridemia not controlled with atorvastatin therapy (PRESS VI). *Diabetes Technol. Ther.* **2014**, *16*, 63–71. [CrossRef]

406. Pai, V.; Paneerselvam, A.; Mukhopadhyay, S.; Bhansali, A.; Kamath, D.; Shankar, V.; Gambhire, D.; Jani, R.H.; Joshi, S.; Patel, P. A multicenter, prospective, randomized, double-blind study to evaluate the safety and efficacy of saroglitazar 2 and 4 mg compared to pioglitazone 45 mg in diabetic dyslipidemia (PRESS V). *J. Diabetes Sci. Technol.* **2014**, *8*, 132–141. [CrossRef] [PubMed]

407. Jain, N.; Bhansali, S.; Kurpad, A.V.; Hawkins, M.; Sharma, A.; Kaur, S.; Rastogi, A.; Bhansali, A. Effect of a Dual PPAR α/γ agonist on Insulin Sensitivity in Patients of Type 2 Diabetes with Hypertriglyceridemia-Randomized double-blind placebo-controlled trial. *Sci. Rep.* **2019**, *9*. [CrossRef]

408. Kaul, U.; Parmar, D.; Manjunath, K.; Shah, M.; Parmar, K.; Patil, K.P.; Jaiswal, A. New dual peroxisome proliferator activated receptor agonist—Saroglitazar in diabetic dyslipidemia and non-alcoholic fatty liver disease: Integrated analysis of the real world evidence. *Cardiovasc. Diabetol.* **2019**, *18*, 80. [CrossRef]

409. Boubia, B.; Poupardin, O.; Barth, M.; Binet, J.; Peralba, P.; Mounier, L.; Jacquier, E.; Gauthier, E.; Lepais, V.; Chatar, M.; et al. Design, Synthesis, and Evaluation of a Novel Series of Indole Sulfonamide Peroxisome Proliferator Activated Receptor (PPAR) $\alpha/\gamma/\delta$ Triple Activators: Discovery of Lanifibranor, a New Antifibrotic Clinical Candidate. *J. Med. Chem.* **2018**, *61*, 2246–2265. [CrossRef]

410. Wettstein, G.; Luccarini, J.-M.; Poekes, L.; Faye, P.; Kupkowski, F.; Adarbes, V.; Defrêne, E.; Estivalet, C.; Gawronski, X.; Jantzen, I.; et al. The new-generation pan-peroxisome proliferator-activated receptor agonist IVA337 protects the liver from metabolic disorders and fibrosis. *Hepatol. Commun.* **2017**, *1*, 524–537. [CrossRef]

411. Lefere, S.; Puengel, T.; Hundertmark, J.; Penners, C.; Frank, A.K.; Guillot, A.; de Muynck, K.; Heymann, F.; Adarbes, V.; Defrêne, E.; et al. Differential effects of selective- and pan-PPAR agonists on experimental steatohepatitis and hepatic macrophages. *J. Hepatol.* **2020**. [CrossRef] [PubMed]

412. Ruzehaji, N.; Frantz, C.; Ponsoye, M.; Avouac, J.; Pezet, S.; Guilbert, T.; Luccarini, J.M.; Broqua, P.; Junien, J.L.; Allanore, Y. Pan PPAR agonist IVA337 is effective in prevention and treatment of experimental skin fibrosis. *Ann. Rheum. Dis.* **2016**, *75*, 2175–2183. [CrossRef]

413. Avouac, J.; Konstantinova, I.; Guignabert, C.; Pezet, S.; Sadoine, J.; Guilbert, T.; Cauvet, A.; Tu, L.; Luccarini, J.M.; Junien, J.L.; et al. Pan-PPAR agonist IVA337 is effective in experimental lung fibrosis and pulmonary hypertension. *Ann. Rheum. Dis.* **2017**, *76*, 1931–1940. [CrossRef]

414. Farrell, G.; Schattenberg, J.M.; Leclercq, I.; Yeh, M.M.; Goldin, R.; Teoh, N.; Schuppan, D. Mouse Models of Nonalcoholic Steatohepatitis: Toward Optimization of Their Relevance to Human Nonalcoholic Steatohepatitis. *Hepatology* **2019**, *69*, 2241–2257. [CrossRef] [PubMed]

415. Kang, Z.; Fan, R. PPARα and NCOR/SMRT corepressor network in liver metabolic regulation. *FASEB J.* **2020**. [CrossRef] [PubMed]

416. Kalliora, C.; Kyriazis, I.D.; Oka, S.I.; Lieu, M.J.; Yue, Y.; Area-Gomez, E.; Pol, C.J.; Tian, Y.; Mizushima, W.; Chin, A.; et al. Dual PPARα/γ activation inhibits SIRT1-PGC1α axis and causes cardiac dysfunction. *JCI Insight* **2019**, *4*. [CrossRef]

417. Li, S.; Yang, B.; Du, Y.; Lin, Y.; Liu, J.; Huang, S.; Zhang, A.; Jia, Z.; Zhang, Y. Targeting PPARα for the treatment and understanding of cardiovascular diseases. *Cell. Physiol. Biochem.* **2019**, *51*, 2760–2775. [CrossRef]

418. Krishnappa, M.; Patil, K.; Parmar, K.; Trivedi, P.; Mody, N.; Shah, C.; Faldu, K.; Maroo, S.; Parmar, D. Effect of saroglitazar 2 mg and 4 mg on glycemic control, lipid profile and cardiovascular disease risk in patients with type 2 diabetes mellitus: A 56-week, randomized, double blind, phase 3 study (PRESS XII study). *Cardiovasc. Diabetol.* **2020**, *19*, 93. [CrossRef] [PubMed]

419. Afroze, S.H.; Munshi, M.K.; Martínez, A.K.; Uddin, M.; Gergely, M.; Szynkarski, C.; Guerrier, M.; Nizamutdinov, D.; Dostal, D.; Glaser, S. Activation of the renin-angiotensin system stimulates biliary hyperplasia during cholestasis induced by extrahepatic bile duct ligation. *Am. J. Physiol. Gastrointest. Liver Physiol.* **2015**, *308*, G691–G701. [CrossRef] [PubMed]

420. Saran, A.R.; Dave, S.; Zarrinpar, A. Circadian Rhythms in the Pathogenesis and Treatment of Fatty Liver Disease. *Gastroenterology* **2020**, *158*. [CrossRef]

Review

PGC-1α in Disease: Recent Renal Insights into a Versatile Metabolic Regulator

Joseph M. Chambers [1],* and Rebecca A. Wingert [2],*

[1] College of Pharmacy, Natural and Health Sciences, Manchester University, Fort Wayne, IN 46845, USA
[2] Department of Biological Sciences, Center for Stem Cells and Regenerative Medicine, Center for Zebrafish Research, Boler-Parseghian Center for Rare and Neglected Diseases, University of Notre Dame, Notre Dame, IN 46556, USA
* Correspondence: JMChambers@manchester.edu (J.M.C.); rwingert@nd.edu (R.A.W.)

Received: 4 July 2020; Accepted: 30 September 2020; Published: 3 October 2020

Abstract: Peroxisome proliferator-activated receptor gamma co-activator 1 alpha (PGC-1α) is perhaps best known as a master regulator of mitochondrial biogenesis and function. However, by virtue of its interactions as a coactivator for numerous nuclear receptors and transcription factors, PGC-1α also regulates many tissue-specific tasks that include adipogenesis, angiogenesis, gluconeogenesis, heme biosynthesis, thermogenesis, and cellular protection against degeneration. Knowledge about these functions continue to be discovered with ongoing research. Unsurprisingly, alterations in PGC-1α expression lead to a range of deleterious outcomes. In this review, we provide a brief background on the PGC-1 family with an overview of PGC-1α's roles as an adaptive link to meet cellular needs and its pathological consequences in several organ contexts. Among the latter, kidney health is especially reliant on PGC-1α. Thus, we discuss here at length how changes in PGC-1α function impact the states of renal cancer, acute kidney injury (AKI) and chronic kidney disease (CKD), as well as emerging data that illuminate pivotal roles for PGC-1α during renal development. We survey a new intriguing association of PGC-1α function with ciliogenesis and polycystic kidney disease (PKD), where recent animal studies revealed that embryonic renal cyst formation can occur in the context of PGC-1α deficiency. Finally, we explore future prospects for PGC-1α research and therapeutic implications for this multifaceted coactivator.

Keywords: PGC-1α; disease; kidney; cancer; AKI; CKD; nephron; PKD; cilia; cystogenesis

1. Introduction

Peroxisome proliferator-activated receptor gamma co-activator 1 alpha (PGC-1α, or alternatively PPARGC1A) serves as a master regulator for mitochondrial biogenesis and function. Mitochondria are classically defined as the powerhouse of the cell, as they function to produce the essential energy needed for cellular processes. PGC-1α was first discovered in the late 1990s as its expression was induced in brown adipose tissue during the response to cold temperature exposure [1]. PGC-1α was found to alter the transcriptional activities of several key mitochondrial genes, which ultimately resulted in increased mitochondrial DNA [1]. Due to its physical interaction with the nuclear hormone receptor peroxisome proliferator-activated receptor gamma (PPARγ) that led to an increase in PPARγ transcriptional activity, this novel co-activator acquired its name, PGC-1α [1]. Interestingly, PGC-1α was shown in the same study to also bind a number of other nuclear hormone receptors including thyroid hormone receptor (THR), retinoid X receptor α (RXRα), and estrogen receptor (ER) [1]. Soon after, the glucocorticoid receptor (GR) was identified as another nuclear receptor binding partner [2]. This was just the beginning of research to uncover the multitude of critical roles that PGC-1α and its family members play during the molecular workings of cellular life [3–7].

It is now appreciated that the PGC-1 family of transcriptional coactivators hold key positions in controlling many cellular pathways, among these figuring prominently in regulating metabolism as well as conducting a host of tissue-specific functions [3–7]. These coactivator proteins do not bind DNA directly but rather all work through interactions with transcription factors and other proteins to influence gene expression. The family consists of three related members: PGC-1α, PGC-1β, and PGC-related coactivator (PRC) [1,8,9]. The PGC-1 family members are conserved across higher vertebrates, including fish, amphibians, birds, and mammals [3–7]. Thought initially to be absent from the genomes of lower eukaryotes such as yeast, insects and worms [2,3], a PGC-1 homologue named Spargel was later identified in *Drosophila*, where it similarly functions in mitochondrial metabolism and regulates insulin signaling [10].

PGC-1 family members share a related overall structure, with functional regions that include an N-terminal activation domain, a central regulatory domain, and a C-terminal RNA recognition motif [2–6]. The activation domain can bind several proteins that possess histone acetyltransferase (HAT) activity and influence chromatin structure [3–7]. The central domain is known to dock various transcription factor targets, but the latter are not limited to interacting with only this region [3–7]. The RNA binding domain enables interactions with proteins of the mediator complex that interacts with the RNA polymerase II machinery [3–7]. Among the family members, PGC-1β shows the closest homology to PGC-1α, and both of these proteins are abundant in tissues with high oxidative metabolism [3–7]. PGC-1α and PGC-1β are the best studied members of this family, and in line with their expression pattern, they both participate in regulating mitochondrial functions [3–7]. In contrast, PRC is expressed nearly ubiquitously [9] but its roles remain comparatively understudied [6,7].

To date, the PGC-1 coactivators have been catalogued to interact with over 20 different transcription factors in sum, with some shared targets and others unique to just one family member [3–7]. These targets are not limited to nuclear receptor family members (such as the previously mentioned PPARγ, THR, ER, RXRα, and GR) but include binding to unliganded nuclear receptors, forkhead/winged helix proteins, zinc-finger proteins, and others [3–7]. Despite large-scale efforts to identify partners such as through genome-wide coactivation screens [10], it is likely that ongoing research will only continue to reveal additional binding partners. Interestingly, the target binding activities of PGC-1α are influenced by numerous post-translational modifications that include acetylation, methylation, phosphorylation and SUMOylation, although comparatively less is known about the influence of such modifications on the other family members [3–7]. Additionally, as we will discuss later in the present work, both PGC-1α and PGC-1β are alternatively spliced, but the functional consequences of this are not yet fully appreciated [3–7].

Today, PGC-1α is known to be specifically expressed in organs with high energy demands such as the heart, kidney, brain, and skeletal muscles, among others [3–7]. Across all of these organs, expression of PGC-1α is highly regulated by a number of signal transduction pathways and hormones to maintain metabolic balance in a tissue specific manner [3–7]. Furthermore, in such locations as brown fat and the liver, dynamic changes in PGC-1α expression are used to respond to fluctuating physiological and environmental stimuli [11]. These changes can consist of cold exposure, exercise, fasting, etc. In the following paragraphs, we provide an overview of some example signaling mechanisms that regulate PGC-1α expression in response to particular stimuli to illustrate the diversity of ways in which PGC-1α can be employed to regulate biological processes.

Cold-induced stress and the cellular response, commonly referred to as adaptive thermogenesis, is essential to maintain body temperature thus ensuring proper organ function. Interestingly, upon subjecting PGC-1α null mice to a standard cold challenge, the animals were unable to maintain core body temperatures and succumbed to hypothermia when exposure was extended beyond 6 h [12]. Defense of body temperature is a functional readout of brown adipose tissue; therefore, this study was one of the first indications that PGC-1α had a molecular role in this cell type [12]. In both brown fat and skeletal muscle, PGC-1α expression is induced by cold and regulates a suite of nuclear and mitochondrial-encoded genes corresponding to subunits of the electron transport chain [7,12].

For example, in brown fat, PGC-1α increases transcriptional activity at the promoter of mitochondrial uncoupling protein-1 (UCP1), where subsequent UCP1 activity results in the production of heat by dissipating the mitochondrial proton gradient [13–16].

Fasting induces PGC-1α expression in the liver through the activation of p38 mitogen-activated protein kinase (p38), where PGC-1α interacts with transcription factors to enable the body to adapt to nutrient deprivation [3,17,18]. PGC-1α is sufficient to activate the processes of gluconeogenesis, fatty acid B-oxidation, ketogenesis, heme biosynthesis, and bile-acid homeostasis. Each of these processes is triggered by PGC-1α coactivation of specific hepatic transcription factors [17–21]. Without PGC-1α, fasted mice develop hypoglycemia and hepatic steatosis [22]. Further, it was recently demonstrated that PGC-1α plays an important role in the hepatic response to insulin during the fasting-to-fed transition by regulating expression of upstream components of the insulin-signaling pathway [23,24].

PGC-1α after exercise-induced cellular stress has been extensively studied in skeletal muscle. In this context, several pathways previously discussed remain relevant, with the overall goal of equilibrating cellular metabolism. In this situation and tissue type, investigators have found several modes of action that result in a PGC-1α-dependent adaptive response. First, nerve stimulation activates the calcium/calmodulin-dependent protein kinase IV (CaMKIV) and calcineurin A (CaN) pathway and increases PGC-1α expression via the myocyte enhancer factor 2C and 2D (MEF2C/2D) [15,19] or by way of the CaMKIV target cAMP response element binding protein (CREB) [15,25–30]. Another previously mentioned pathway involved in both fasting-induced PGC-1α expression and exercise-induced PGC-1α expression is p38 MAPK signaling. Interestingly, multiple factors can be initiated by p38 MAPK that result in downstream PGC-1α functions. Both MEF2 [30] and ATF2 [31] can increase expression of PGC-1α in muscle post exercise.

Much of this information was gathered in the years immediately following the discovery of PGC-1α and laid the groundwork for subsequent studies. This next wave of studies brought to light the extensive network of PGC-1α functionality and its diverse impacts on physiology during health and disease states. Further, recent studies have begun to highlight important developmental roles for PGC-1α. In Section 2 of this review, we will discuss several cellular and molecular roles of PGC-1α during health and disease progression. In Section 3 of this review, we explore the functions of PGC-1α in renal disease and development.

2. PGC-1α in the Context of Human Disease and Development

There has been extensive research regarding the impact of PGC-1α function in human disease (Figure 1). More recently, there is a growing body of work that has begun to reveal insights regarding the roles of PGC-1α in developing tissues. The foundational information presented above illustrates the functions of PGC-1α in the adaptive response to several stimuli including exercise and fasting. Not coincidentally, the relationship between inactivity and poor diet can lead to type 2 diabetes. Type 2 diabetes mellitus is a metabolic disorder caused by abnormal glucose metabolism due to insulin resistance or disrupted B-cell function culminating in abnormal insulin production.

Here, we will highlight part of the relationship between PGC-1α and tissues in terms of type 2 diabetes mellitus. A comprehensive description of these relationships is nicely detailed in another review [30]. As previously discussed, in the liver PGC-1α acts to 1) maintain proper levels of gluconeogenesis with well-studied pathways including CREB, HNF4a, AMPK, Sirtuins, and p38 MAPK 2) control B-cell oxidation and 3) regulate mitochondrial biogenesis by partnering with estrogen related receptors (ERRs) and other factors [17–22,30]. Understandably PGC-1α is also linked with type 2 diabetes mellitus in the pancreas where B-cells are supposed to manufacture and secret insulin. Early studies found PGC-1α overexpression can decrease B-cell insulin secretion [31]. Further research has found roles for PGC-1α in B-cell apoptosis, regeneration, insulin secretion, and mitochondrial metabolism [32–35]. While much of the attention on type 2 diabetes is understandably focused on the liver and pancreas, another largely affected system are skeletal muscles that rely on glucose energetics.

PGC-1α activates the glucose transporter 4 (GLUT4) and in doing so can stimulate glucose uptake. This relationship between PGC-1α and glucose in skeletal muscles is extended to the process of glucose disposal [30,36].

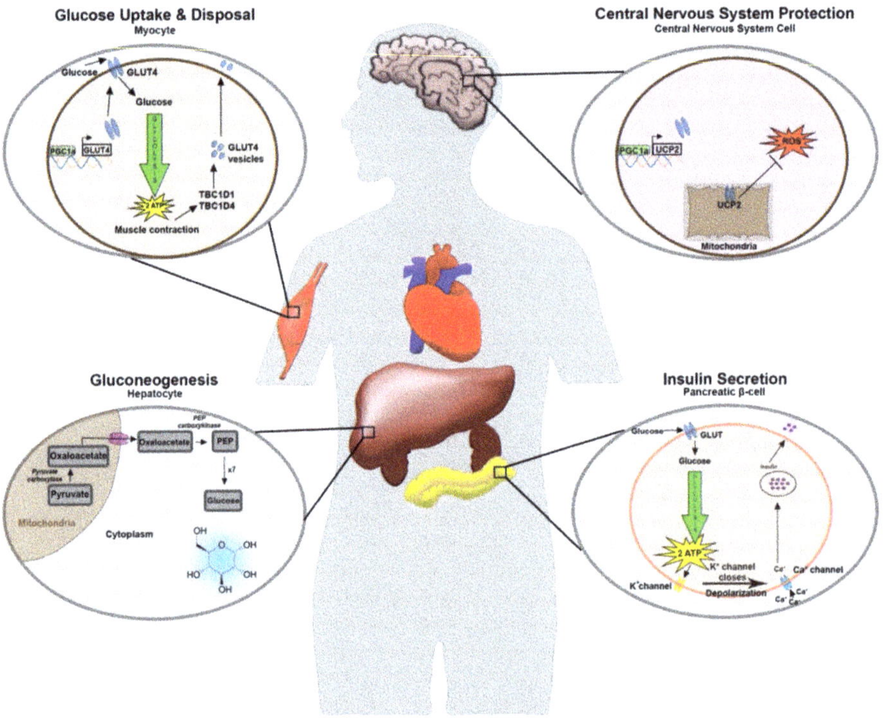

Organ	Disease	Role/Action	Genetic Factors/Partners
Liver	Type 2 Diabetes	Gluconeogenesis; B-cell oxidation; Mitochondrial Biogenesis	CREB; HNF4a; AMPK; Sirtuins; p38 MAPK; ERR
Central Nervous System	Huntington's Disease; ALS; Parkinson's Disease; Alzheimers Disease	Protection from oxidative stress; Oxytocin production; Mitochondrial health	UCP2; SOD2; SIRT1; TFAM; NRF1
Eye	Neovascularization	Retinal angiogenesis	VEGFa
Heart	Cardiomyopathy	Mitochondrial biogenesis; TAC-induced Cardiomyopathy	NRF-1 and -2; ERRs; TFAM
Kidney	Acute Kidney Injury; Chronic Kidney Disease	Protection from Notch-induced fibrosis; reestablishment of renal function following AKI	Notch; HNF1B; Prostaglandin signaling
Multi-organ	Cancer	Tumor development, survival, progression	NF-KB, p65
Pancreas	Type 2 Diabetes	B-cell apoptosis and regeneration; insulin secretion; mitochondrial metabolism	Glucagon; Insulin
Skeletal Muscle	Type 2 Diabetes	Glucose uptake and disposal; inflammatory response; angiogenesis	GLUT4; B-adrenergic; ERR; VEGFa

Figure 1. Systems affected and corresponding disease states in peroxisome proliferator-activated receptor gamma co-activator 1 alpha (PGC-1α) contexts. Schematic depicting the role of PGC-1α in select cellular processes including glucose uptake and disposal via glucose transporter 4 (GLUT4) in skeletal muscle (top, left), central nervous system reactive oxygen species (ROS) protection via PGC-1α uncoupling protein-2 (UCP2) (top, right), gluconeogenesis in the human liver (bottom, left), and insulin secretion from a pancreatic B-cell (bottom, right). Table summarizes key disease states and the associated organ systems in addition to the role PGC-1α plays in those organs and some of the genetic interactions with PGC-1α.

A number of additional studies have demonstrated the importance of PGC-1α in skeletal muscle, perhaps interrelated to type 2 diabetes mellitus. Interestingly, researchers discovered a molecular node connecting the metabolic factors PGC-1α/PGC-1β with immune pathways, specifically the inflammatory response of NF-kB. For example, PGC-1α/PGC-1β repress NF-kB by reducing phosphorylation of p65 in skeletal muscle [37]. Another important association that could impact the relationship between PGC-1α and type 2 diabetes is the discovery of key molecular events connecting exercise to PGC-1α driven angiogenesis in skeletal muscle. This important tie is corroborated by phenotypes observed in PGC-1α deficient mice. As compared to wild-type counterparts, PGC-1α-null mice did not undergo robust angiogenesis post exercise [38].

Further investigation revealed a deeper mechanistic insight, which placed B-adrenergic signaling upstream of PGC-1α in this process and also relies on ERR to activate VEGF signaling to induce angiogenesis [38]. The role PGC-1α plays in angiogenesis is not limited to skeletal muscle as Saint-Geniez et al. (2013) provide evidence PGC-1α controls retinal angiogenesis by inducing VEGFa in multiple retinal cell types, including Muller cells, ganglion cells, photoreceptors, and retinal pigment epithelial cells [39]. Additionally, PGC-1α is highly upregulated in the retina during postnatal development in mice, especially from P5-P17 [39]. Consequently, mice deficient in PGC-1α have signs of reduced retinal angiogenesis in both early developmental stages and adulthood [39]. Further investigation revealed that PGC-1α is involved in pathological neovascularization [39].

A principal area of study has identified roles for PGC-1α in heart development and maturation. A number of compelling findings have been summarized previously [40]. To highlight some of these findings, Lai et al. (2008) show intriguing data indicating a level of redundancy in mice when it comes to the roles of family members PGC-1α and PGC-1b [41]. Mice deficient in one factor develop relatively normally; however, when both factors are knocked out, the compound mutant mice develop a number of heart defects and perish shortly after birth [41]. These redundant layers of regulation are also observed in murine brown adipose tissue [41]. Other studies have suggested timing of the knockout may affect heart function as well [42–45]. As expected from the contents discussed thus far, PGC-1α is responsible for mitochondrial biogenesis in the heart through its relationships with genetic partners, including TFAM, NRF-1, NRF-2, and ERRs [40,43,45,46]. Additionally, loss of PGC-1α results in cardiomyopathy following transverse aortic constriction [44]. Importantly, PGC-1α expression is able to overcome cardiomyopathy, although this depends on the time of induction [43]. Findings indicating the significance of timing and redundancy should elicit a response in the research community to identify the possibility of this relationship in other developmental contexts.

PGC-1α is important for removing reactive oxygen species (ROS). This is especially important for neurological diseases where PGC-1α has been shown to decrease ROS and protect neural cells from oxidative stress by inducing expression of several detoxifying factors such as UCP2 and SOD2 [47]. The link between PGC-1α and neurobiology involves developmental aspects and disease contexts including Alzheimer's disease, Huntington disease, and amyotrophic lateral sclerosis (ALS), among others. In a developmental context, Blechman et al. (2011) found a direct correlation between PGC-1α and oxytocin production in the hypothalamus [48]. The well-studied SIRT1/PGC-1α relationship is very important for mitochondrial recovery following intracerebral hemorrhage in rats [49]. In the case of Huntington disease, PGC-1α and its downstream partners TFAM and NRF-1 are needed to prevent mitochondrial dysfunction that leads to the disease progression [50]. An important finding to both the neuroscience community as well as those interested specifically in PGC-1α came in a 2012 publication that identified several novel PGC-1α isoforms that dictate age of onset for Huntington's disease [51]. These and other tissue and function-specific isoforms are nuances of studying PGC-1α [52]. On a cellular level, PGC-1α impacts ribosomal transcription, which has recently been found to impact Huntington disease progression [53]. Similarly, the development of Parkinson's disease entails central nervous system specific PGC-1α isoforms [54,55]. This same approach is a focus of Alzheimer's disease, where PGC-1α can reduce β-secretase thereby decreasing amyloid-β and the associated neuronal and cognitive loss [56]. A comprehensive understanding of PGC-1α and Alzheimer's disease has been previously reviewed [57].

As cancer studies have shifted to understanding the metabolism of tumors and the tumor microenvironment, there have been several studies focused on how PGC-1α might affect tumorigenesis and metastasis. Currently there is not consensus in the literature regarding PGC-1α levels and the impact on tumorigenesis. Different types of cancers exhibit different PGC-1α expression trends. Further, some cancers such as breast, colon, ovarian, and melanomas have observed both increased and decreased expression of the gene in tumor tissue. The conclusions reached in a very informative review by Mastropasqua et al. (2018) indicate PGC-1α responds to cellular needs in terms of energy demands [58]. This conditional complexity can be mostly explained by the dynamic nature of the PGC-1α gene in combination with the complicated mechanisms leading to cancer progression; therefore, the interface of these two issues presents challenges to conducting research in this area [58]. In addition, PGC-1α has many upstream and perhaps an even greater number of downstream effectors that allow for adaptive responses thus making a simple linear relationship with cancer highly unlikely. An accurate understanding of the role of PGC-1α in cancer will most likely be dependent on the molecular subtype and tissue of origin. The role of PGC-1α cellular metabolism, specifically of cancer cells, in addition to budding therapeutic options, are nicely described in another review by Bost and Kaminski [59]. We will further discuss kidney cancer below.

3. PGC-1α in Kidney Disease and Development

The kidney requires an abundance of energy because of the essential roles it carries out as a vital organ. These functions include blood filtration, ion transport, fluid homeostasis, and waste removal. The nephron is the functional unit of the kidney and is composed of three main components: (1) the glomerulus functions to filter the blood, (2) the segmented tubule that modifies the filtrate, and (3) the collecting duct system that transports waste for excretion [60]. Genetic analyses have illuminated that PGC-1α is necessary for proper nephrogenesis in the zebrafish [61,62]. However, other vertebrate species with PGC-1α loss of function models have not presented with explicit kidney developmental phenotypes in PGC-1α deficient contexts [63]. Interestingly, there is precedence in mice that PGC-1α deficiency can be masked by PGC-1β compensation [41]. Perhaps future studies will shed light on potential for redundant or partially overlapping functions for PGC-1α and PGC-1β in the kidney.

Nevertheless, as one of the most energy demanding organs, it is logical that PGC-1α is highly expressed in the kidney. This expression is conserved across vertebrate species spanning zebrafish, mice, and humans [61,63–67]. Though studies to date have not identified overt loss of function renal phenotypes in mammalian development, there exists a vast amount of literature suggesting a link between PGC-1α and reno-protective properties at later stages. Subjects deficient in PGC-1α experience deleterious outcomes after suffering from acute kidney injury (AKI) as they progressively undergo renal fibrosis and suffer from chronic kidney disease (CKD). In the following subsections, we will detail a number of these studies and key findings, as there is a growing body of evidence revealing critical roles for PGC-1α and regulation of mitochondria biogenesis in kidney development and disease (Figure 2).

3.1. Kidney Cancer

When basal expression of PGC-1α is disrupted in the kidney, the risk of cancer increases. As previously discussed, the exact role of PGC-1α in cancer is not entirely understood, where research has revealed that PGC-1α can be is upregulated or downregulated in a multitude of cancer types [58]. In the context of kidney cancer, the lack of a clear correlation between PGC-1α expression and cancer development and progression is unfortunately only further perplexing. For example, in the context of Birt–Hogg–Dube (BHD) syndrome both PGC-1α and the tumor suppressor FLCN are altered resulting in predisposition to patients developing lung cysts, hair follicle tumors, and renal cancer [68]. FLCN loss induces a metabolic shift toward oxidative phosphorylation and elevated mitochondrial biogenesis in a PGC-1α-dependent manner [68]. Inactivation of PGC-1α in cancer cells rescued hyperplastic phenotypes in FLCN-null kidneys [68]. Together this data indicates that increased FLCN deficiency and subsequent upregulation of PGC-1α enhances energy production and provides malignant tumor cells a growth advantage that fuels renal carcinogenesis [68].

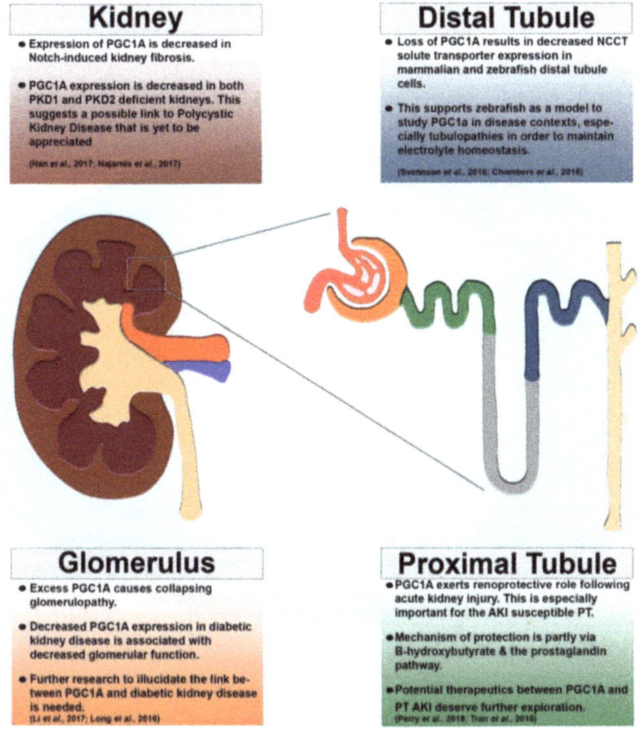

Figure 2. Kidney disease states associated with PGC-1α expression. Renal diagram illustrating general (top, left) and nephron compartment-specific functions of PGC-1α in addition to areas of future research. Kidney (left, maroon), nephron (right), glomerulus (orange), proximal tubule (green), and distal tubule (dark blue). References for cited studies listed within. (PT = proximal tubule, AKI = acute kidney injury).

Contrary to BHD syndrome, the development of clear cell renal carcinoma (ccRCC) involves the downregulation of PGC-1α by a HIF/Dec1-dependent mechanism. The most common genetic driver of ccRCC is the tumor suppressor Von Hippel–Lindau (Vhl), which, when mutated, promotes constitutive expression of HIF-a and promotes metastasis, invasion, angiogenesis, and metabolism [60]. PGC-1α expression in VHL-deficient ccRCC cells restores mitochondrial function, upregulated antioxidant gene expression, suppresses tumor growth, and sensitizes cancer cells to cytotoxic chemotherapy and radiotherapy treatments [69]. ERRα was demonstrated to be an essential co-regulator of PGC-1α in mediating mitochondrial biogenesis in the context of ccRCC [69]. ccRCC is considered a metabolic disease that manifests histological hallmarks such as lipid accumulation and lipid storage, which is thought to support tumor survival. The Cancer Genome Atlas and other data repositories indicate PGC-1α expression is significantly decreased in ccRCC tissues versus healthy kidneys and is associated with poor patient prognosis. A study found supplying ccRCC cells with melatonin restores PGC-1α levels, thereby eliminating abnormal lipid accumulations and inhibiting tumor progression [70]. Furthermore, this melatonin-initiated rescue was driven by a novel 'tumor slimming' mechanism where lipid droplets are broken down via PGC-1α/UCP1-mediated autophagy and lipid browning [70]. Another research group described a novel role for PGC-1α, whereby the PGC-1α/miR-29 axis attenuates the epithelial-to-mesenchymal transition (EMT) program to prevent ccRCC tumor progression [71]. Restoration of PGC-1α in ccRCC cells suppresses collagens associated with invasive and migratory behaviors via activation of miR-29a [71]. PGC-1α-mediated repression of collagen/DDR1 signaling culminates in downregulation of known EMT genes SNAIL1 and SNAIL2, which curbed pro-invasive phenotypes [71].

A genetic loss of function screen revealed MYBBP1A promotes tumorigenesis in absence of glucose by operating a "metabolic switch" involving PGC-1α [72]. In multiple ccRCC lines, downregulation of MYBBP1A induces tumor cell survival and proliferation [72]. In response to decreased MYBBP1A, c-MYB repression is alleviated leading to transcriptional activation of PGC-1α, which elicits a shift to OXPHOS metabolism [72]. This data indicates PGC-1α is a stress sensor of glucose depletion and can provide tumor cells a competitive advantage in restrictive microenvironments similar to previously discussed roles for PGC-1α above [72]. Evaluation of biopsies from 380 ccRCC patients revealed increased MLXIPL and decreased PGC-1α mRNA levels in the tumor microenvironment are significantly correlated with poor overall survival [73]. Further, elevated MlXIPL and PGC-1α were associated with a decline in the number of B cells, CD8+ T cells, macrophages, neutrophils, and dendritic cells [73]. Thus, MlXIPL and PGC-1α signatures are closely related to the degree of immune infiltration, underscoring the potential of monitoring and targeting the tumor microenvironment in ccRCC [73].

3.2. Acute Kidney Injury (AKI)

Acute kidney injury (AKI) is the sudden and rapid decrease in kidney function resulting from a severe insult [74]. This insult can originate from various sources, such as decreased blood flow (ischemia), increased inflammation, introduction of a toxin perhaps in the form of a chemical or antibiotic, or obstructed urine flow. These insults can cause damage to cells throughout the nephron functional units and/or to the interstitium [74]. AKI can be detected clinically by evaluating a number of kidney related outputs including blood urea nitrogen, creatinine, and atypical ion levels.

Over the last decade, a better understanding of the cellular energy defects associated with various types of AKI has understandably focused on the role PGC-1α plays during and after AKI. One of the first studies that examined the specific relationship between AKI and PGC-1α was published over a decade ago. Rasbach and Schnellmann hypothesized that increasing mitochondrial biogenesis could help overcome ischemic injury conditions [75]. They discovered that increasing mitochondrial biogenesis via PGC-1α overexpression helped expedite recovery if applied post-injury [66]. Introducing increased mitochondria via PGC-1α overexpression prior to injury did not prevent injury or promote recovery, suggesting timing and quantity must be finely tuned [75].

Closely following this idea, another research group gathered transcriptional analysis from sepsis-associated AKI, which indicated oxidative phosphorylation genes were downregulated in addition to aberrant mitochondrial function [63]. More specifically, PGC-1α expression was reduced [63]. Global and tissue specific (proximal tubule) PGC-1α knockout mice lines were used in this study and results indicate normal kidney function until AKI where the animals then struggled to recover [63]. PGC-1α action seemingly acts downstream of TNFα induced inflammation as PGC-1α levels decreased after TNFα treatment, and PGC-1α rescued the oxygen consumption of TNFα treated nephron cells [63]. Another study found PGC-1α plays an important role in cellular recovery following an ischemia-reperfusion injury model [76]. Researchers treated with SRT1720 after AKI to agonize the SIRT1 pathway. Upon SIRT1 activation in rescued animals, the group found that PGC-1α had higher rates of deacetylation, an indicator of increased activation [76]. This study further substantiated the thought that PGC-1α could serve as a key therapeutic target for patients suffering AKI.

To this end, potential therapeutics were found in a study that identified trametinib (a MEK1/2 inhibitor) and erlotinib (an EGFR inhibitor) as potentially indirect PGC-1α modifiers that could help in the context of ischemia-reperfusion AKI [77]. In this study, treatment with trametinib blocked ERK1/2 and FOXO3a/1 phosphorylation and resulted in increased PGC-1α expression in kidney cells [77]. Additionally, linking EGFR to this pathway, authors found treatment with erlotinib also blocked

ERK1/2 phosphorylation and exhibited a similar increase in PGC-1α expression as trametinib [77]. This study not only identified potential treatments of AKI, but also pinpointed a pathway including EGFR, ERK1/2, FOXO3a/1, and PGC-1α in the context of mitochondrial response to kidney injury.

Another very interesting relationship between a well-studied pathway and PGC-1α in the kidney has also been documented [66]. Here the researchers accumulated convincing evidence that the disease-causing gene *human nuclear factor 1B* (*HNF1B*) directly controls mitochondria via PGC-1α in kidney cells [66]. Using a similar sepsis-induced AKI model as previously discussed [65], this group found a comparable decrease in PGC-1α and a subsequent decline in other mitochondrial biogenesis genes as well as substantial decreased HNF1B expression and many of its targets [66]. Additional experiments found HNF1B deficiency caused reduced PGC-1α expression and was associated with corresponding mitochondrial morphological defects [66]. A chromatin immunoprecipitation assay found HNF1B binds to the PGC-1α promoter region in mouse kidney cells [66]. The authors believe this is conserved in human kidney cells as an HNF1B-mutant-patient sample had decreased PGC-1α expression [66]. This uniquely connects HNF1B to mitochondrial biogenesis via PGC-1α [66]. Results from this study importantly connect HNF1B to PGC-1α and merit additional research to explore the involvement of PGC-1α in HNF1B associated kidney diseases.

With previous publications finding potential upstream regulators of PGC-1α in response to AKI, a unique study identified the downstream factors of PGC-1α post-AKI [65]. Interestingly, they found post ischemic AKI PGC-1α-deficient mice present with decreased niacinamide (NAM) levels, fat accumulation, and are unable to regain baseline kidney function [65]. Through a number of elegant rescue studies, the authors found PGC-1α regulates renal recovery from AKI by controlling nicotinamide adenine dinucleotide (NAD) biosynthesis via NAM and that NAM was needed for B-hydroxybutyrate to increase prostaglandin E2 production [65]. The authors also found that this NAM relationship to AKI recovery was important during cisplatin induced AKI as well as ischemic induced AKI [65]. Additional research related to this topic discovered the loss of a mitochondrial-related gene-dynamin-related protein 1 (DRP1) benefited proximal tubule cells following ischemic AKI [78]. The benefits included renoprotection via activation of PGC-1α and the associated B-hydroxybutyrate pathway ultimately resulting in decreased inflammation [78]. Further supporting these links, another group found decreased PGC-1α expression in AKI context, specifically toxin-induced AKI [78]. Transcriptomic analysis pointed to PGC-1α acting upstream of many of the genes affected in folic acid induced AKI [79]. They found that decreased PGC-1α resulted in an increase in inflammation (including activation of NF-kB) and cell death [79].

3.3. Chronic Kidney Disease (CKD)

Chronic kidney disease (CDK) is the progressive decrease in renal function due to repeated insults over time [60]. In this section we will discuss both direct and indirect actions of PGC-1α in chronic kidney disease including diabetic nephropathy, glomerular function, and fibrosis.

As previously described, PGC-1α plays a significant role in type 2 diabetes mellitus in multiple tissues including muscle, pancreas, and the kidney. Most of what has been covered up to this point has focused on tissues other than kidney. Here, we will focus on the role of PGC-1α in diabetic nephropathy, also called diabetic kidney disease. Diabetic nephropathy is the leading cause of CKD and too often results in end stage renal disease [80,81]. The glomerulus is the first point of contact for function of the nephron as it filters the blood and passes the filtrate to the nephron tubule for modification. Typically, the function of the glomerulus is affected in diabetic nephropathy as patients exhibit lesions and adverse glomerular filtration rates [80]. Two important studies have indicated a role for PGC-1α and the resulting mitochondrial biogenesis being essential for proper kidney function in the case of diabetic kidney disease [82,83]. Sharma and colleagues (2013) utilized bioinformatics that directed them to discover decreased levels of PGC-1α in diabetic kidneys [82]. The other group also saw decreased mRNA levels of PGC-1α in diabetic kidneys [74]. Furthermore, stimulating PGC-1α expression in diabetic kidneys via AICAR treatment rescued superoxide production in kidneys in addition to a number of clinical outputs [83].

With the evidence supporting an essential role for PGC-1α in the nephron to protect against diabetic nephropathy it is important to understand the possible molecular components regulating the expression of this factor. Upon completing RNA sequencing on glomeruli from a diabetic mouse model, a research team found the long noncoding RNA taurine-upregulated gene 1 (Tug1) was differentially expressed [84]. Additional experiments found Tug1 interacts with PGC-1α to promote PGC-1α expression and resulted in a rescue of downstream PGC-1α targets [84]. Though, another research group found the levels of PGC-1α have to be fine-tuned to allow a beneficial and not a deleterious outcome [85]. There exists a unique balance of PGC-1α expression required for basal glomerulus function [85]. Both mouse and humans suffering from diabetic kidney disease had decreased expression of PGC-1α and the corresponding mitochondrial transcripts, most likely caused by the inhibitory action of TGF-B [85]. Using an inducible nephron-specific PGC-1α overexpression line, the authors found excess PGC-1α causes collapsing glomerulopathy including albuminuria and renal failure/glomerulosclerosis [85]. This suggests a delicate balance of PGC-1α is necessary for proper nephron function.

While these studies strongly indicate a vital role of PGC-1α another group used an inducible nephron specific PGC-1α knockout mouse model to study the physiological role of PGC-1α in the kidney [86]. Results indicated the role of PGC-1α in mitochondrial metabolic systems is observed in the kidney similar to the previously described roles of PGC-1α in other tissues. After inducing the knockout at 12 weeks, researchers noted abnormal sodium excretion, most likely due to the decreased protein expression of sodium transporters NCCT and NKCC2 [86]. These phenotypes ultimately led to mice presenting with renal steatosis, although the authors hypothesize PGC-1α is dispensable for renal physiology due to mild phenotypes and high survival rates [86].

In chronic kidney disease, there exists progressive fibrosis, which leads to the decreased ability of the kidney to perform its tasks as previously functional tissue is converted to nonfunctional connective tissue. Various studies have found PGC-1α plays a direct role in the fibrotic response to repeated insults to the kidney. Renal fibrosis leading to CKD and eventually end stage renal disease can largely be attributed to mis-regulated expression of inflammation and metabolism pathways according to transcriptomics work presented by Kang et al. (2015) [87]. Specifically, they observed fatty acid oxidation components (including PGC-1α specifically) were decreased in the case of fibrosis [87]. This ultimately results in three characteristics of fibrosis: decreased ATP, cell death, and dedifferentiation [87]. Furthermore, they observed molecular connections including evidence that TGFB1 reduced PGC-1α expression via SMAD3 [87]. Interestingly, PGC-1α overexpression was able to rescue a many fibrotic phenotypes leading the field to further pursue a possible targeted treatment for fibrosis progression [87]. This genetic relationship was further supported by a study that also found PGC-1α acts downstream of TGF-B [88]. Once again, there was an adaptive response using PGC-1α in the reaction to stress [88]. Researchers showed that exercise was a possible indicator of more positive outcomes in CKD patients [88]. The mode of action uncovered in this study involves the myokine irisin, which acts to inhibit TGF-B type 1 receptor allowing PGC-1α to decrease the amount of damage to kidneys [88]. PGC-1α expression improves metabolic function and decreases fibrosis in mice kidneys [88].

Understanding renal fibrosis is a key step to formulate innovative therapeutics to combat CKD. Another research team found fatty acid oxidation genes, including PGC-1α were downregulated in fibrotic kidneys due to Notch signaling. In fact, PGC-1α expression was decreased in several types of kidney disease including toxin-induced AKI, physical obstruction induced kidney injury, and a genetic CKD model. The authors found evidence suggesting that the Notch component Hes1 represses PGC-1α in kidney cells. Additionally, PGC-1α expression was able to rescue a number of ailments including kidney physiology, fibrotic gene expression, mitochondrial physiology, and fatty acid oxidation both in vivo and in vitro [67]. A year later, the same group then found a unique role for PGC-1α downstream factor mitochondrial transcription factor A (Tfam) in kidney fibrosis [89]. Using a combination of transcriptomics, loss of function studies, and rescue experiments, the researchers used a number of kidney disease models to support Notch signaling components Jag1 and Notch2 are key players in fibrosis by negatively affecting Tfam, ultimately resulting in aberrant metabolic function [89].

3.4. Future Direction: Polycystic Kidney Disease (PKD) and Cilia

One intriguing area of future studies could be focused on the relationship between PGC-1α and ciliogenesis. There is overlap between kidney diseases and ciliopathies including conditions such as polycystic kidney disease (PKD) and nephronophthisis. Although significant progress has been made in understanding disease progression, the molecular mechanisms and intertwining pathways involved in cilia formation and function is an area of extreme importance as PKD affects over 12 million people worldwide [90]. PKD presents with many fluid-filled sacs (cysts) that develop in the kidneys rendering it more difficult to properly filter the blood and produce urine. PKD is a genetic disorder that affects around 500,000 people in the United States alone [90]. Although two main causative ciliary genes have been identified, there is no cure for PKD and the molecular consequences involving cilia are not well understood. The genetic mechanisms that are known focus on mutations in the genetic drivers polycystin-1 (PKD1) and polycystin-2 (PKD2) that result in cyst formation once a certain threshold of insufficient protein product is reached [91–95]. Both polycystin proteins localize to the ciliary body and function in chemosensing by regulating calcium levels. While this information is important, the current gap in knowledge that needs to be addressed is identifying other genes these polycystin proteins interact with as well as the modes of interaction, either direct or indirect [93–95].

The likely association between PKD and PGC-1α has been supported by researchers who noted that oxidative stress is present early in PKD and mitochondria is involved in pathogenesis of this disease. They discovered a quantitative decrease in PGC-1α mRNA and protein expression that was supported further by histological staining of cyst-lining cells [96]. Additional experiments indicate this PGC-1α dependent mitochondrial dysfunction was caused by decreased calcium via calcineurin/p38-MAPK signaling [96]. The authors were able to rescue superoxide production by supplementing PKD1 mutant cells with MitoQuinone, a mitochondrion antioxidant [96]. The study later proposes that PGC-1α could play a pivotal role in PKD by recapitulating the molecular function of MitoQuinone, where it controls mitochondrial superoxide production and oxidative stress [96]. An interesting link between PGC-1α, its well-studied molecular partners, and PKD has been reported [97]. Here, the authors found evidence that microRNA17 promotes cyst progression in PKD by inhibiting mitochondrial metabolism [97]. In this study, they observe decreased PGC-1α expression in both PKD1 and PKD2 knockout mice [97]. They go on to propose upregulation of c-myc by PKD1/2 results in enhanced activity of miRNA-17, which then functions to inhibit mitochondrial metabolism via the PGC-1α partner PPAR α [97].

While studies of PGC-1α knockouts in mice do not result in cystogenesis, there are a number of reasons to continue investigations. Some of these reasons include: the possibility of PGC-1β compensation [41], the unending connections of PGC-1α pathway components and their relationship to PKD etiology and progression [98] and compelling data from recent studies using zebrafish pronephros where loss of PGC-1α induces cystogenesis in the kidney [99]. Also, PGC-1α has been linked to several PKD/cilia related genes including Hnf1b [66,100,101], Notch signaling [67], prostaglandin signaling [65,99,102,103], and mTOR signaling [26,104]. Specifically, there is mounting evidence mitochondria and the resulting metabolic effects need to strike the balance seen in a number of examples already mentioned in order to properly form cilia [105,106]. The connections between PGC-1α, mitochondria, cilia, and diseases such as PKD and nephronophthisis is an area of research primed to shed light on underlying genetic and cellular mechanisms contributing to disease states.

4. Conclusions

Members of the PGC-1 family are known to be expressed in tissues with oxidative metabolism demands and also dynamically expressed in response to physiological stimuli. PGC-1 coactivators serve diverse functions because they can interact with many binding partners. Among the PGC-1 family members, PGC-1α has incredible versatility to affect cellular processes by virtue of its many transcription factor targets. This versatility also has a role in a multitude of disease contexts throughout the body, including in the liver, nervous system, and kidneys, among others. Here, we have discussed the roles of PGC-1α in a number of contexts including its role as an adaptive response to energy requirements, the current understanding of PGC-1α in disease contexts, with a special focus on the kidneys, and future areas of research relating PGC-1α and mitochondrial deficiencies to PKD. There exists a large challenge when studying PGC-1α knockout models, as recent findings suggest tissue-specific isoforms that may not be detectable by certain analyses (e.g., PCR amplifying or using antibodies for areas of the PGC-1α sequence that is not specific to the tissue of interest) [52]. This variety of transcript variants could result in compensation in knockout murine models, which have historically targeted exons 3-5 [12,46,63], although there are a number of PGC-1α transcript variants that do not include these exons. A fantastic example of this temporal spatial specificity of PGC-1α can be observed in a study that found conditional loss of PGC-1α in adult mice results in decreased dopaminergic neurons in particular regions of the brain [107].

The dynamic expression and roles of PGC-1α and the associated transcript variants give reason to question if PGC-1α may be involved in an even greater number of processes in mammalian physiology. Interestingly, another possibility exists that PGC-1β may compensate for loss of mammalian PGC-1α [41]. It does not appear that this compensation exists in all vertebrates as PGC-1α deficient zebrafish exhibit more severe phenotypes than their mammalian counterparts [61,99]. These differences do not mean there are not also strong similarities as some very specific phenotypes are observed in both mammalian and zebrafish nephrons. Specifically, loss of PGC-1α results in a decrease of expression in the transporter Slc12a3 (NCCT) in zebrafish and mice [61,86]. With the evidence of genetic compensation in Lai et al., (2008) and identification of multiple tissue-specific isoforms [41,51,52,54], future research could seek to identify additional isoforms and ensure complete knockout while also accounting for potential compensatory action of PGC-1β.

Author Contributions: Writing—original draft preparation, J.M.C.; writing—review and editing, R.A.W.; funding acquisition, J.M.C. and R.A.W. All authors have read and agreed to the published version of the manuscript.

Funding: This work was supported in part by the National Institutes of Health (R01DK100237 to R.A.W.) and a 2019 University of Notre Dame Advanced Diagnostics and Therapeutics Graduate Fellowship Award (to J.M.C.). We are grateful to Elizabeth and Michael Gallagher for a generous gift to the University of Notre Dame for the support of stem cell research. The funders had no role in the study design, data collection and analysis, decision to publish, or manuscript preparation.

Acknowledgments: We thank the members of our lab for discussions and insights about this work.

Conflicts of Interest: The authors declare no conflict of interest.

Abbreviations

AKI (acute kidney injury), ALS (amyotrophic lateral sclerosis), AMPK (5′ adenosine monophosphate-activated protein kinase), ATF2 (Activating Transcription Factor 2), BHD (Birt-Hogg-Dube syndrome), CaMKIV (calcium/calmodulin-dependent protein kinase IV), CaN (calcineurin A), ccRCC (clear cell renal carcinoma), CKD (chronic kidney disease), CREB (cAMP response element-binding protein), DRP1 (dynamin-related protein 1), EGFR (epidermal growth factor receptor), EMT (epithelial-to-mesenchymal transition), ER (estrogen receptor), ERK1/2 (Extracellular Signal-Regulated Kinase 1/2), ERR (estrogen related receptor), FLCN (folliculin), FOXO3a/1 (Forkhead Box O3), glucocorticoid receptor (GR), GLUT4 (glucose transporter 4), HAT (histone acetyltransferase), HNF1B (human nuclear factor 1B), HNF4α (hepatocyte nuclear factor-4 alpha), MEF2C/2D (myocyte enhancer factor 2C and 2D), MEK1/2 (Dual specificity mitogen activated protein kinase kinase 1), MLXIPL (MLX Interacting Protein Like), MYBBP1A (MYB Binding Protein 1a), NAD (nicotinamide adenine dinucleotide), NAM (niacinamide), NCCT or Slc12a3 (sodium-chloride symporter), NF-kB (nuclear factor kappa-light-chain-enhancer of activated B cells), NKCC2 (sodium potassium chloride cotransporter), NRF-1 (Nuclear Respiratory Factor 1), NRF-2 (Nuclear Respiratory Factor 2), OXPHOS (Oxidative phosphorylation), p38 (p38 mitogen-activated protein kinase), PGC-1α (Peroxisome proliferator-activated receptor gamma co-activator 1 alpha), PKD (polycystic kidney disease), PKD1 (polycystin-1), PKD2 (polycystin-2), PPARγ (peroxisome proliferator-activated receptor gamma), PPARα (peroxisome proliferator-activated receptor alpha), PRC (PGC-related coactivator), RG (glucocorticoid receptor), ROS (reactive oxygen species), RXRα (retinoid X receptor alpha), SIRT1 (Sirtuin 1),

SMAD3 (Mothers against decapentaplegic homolog 3), SOD2 (Superoxide dismutase 2), TFAM (mitochondrial transcription factor A), TGF-B (Transforming growth factor beta), THR (thyroid hormone receptor), TNFα (Tumor necrosis factor alpha), Tug1 (taurine-upregulated gene 1), UCP1 (uncoupling protein-1), VEGF (Vascular endothelial growth factor), Vhl (Von Hippel-Lindau)

References

1. Puigserver, P.; Wu, Z.; Park, C.W.; Graves, R.; Wright, M.; Spiegelman, B.M. A cold-inducible coactivator of nuclear receptors linked to adaptive thermogenesis. *Cell* **1998**, *92*, 829–839. [CrossRef]
2. Knutti, D.; Kaul, A.; Kralli, A. A tissue-specific coactivator of steroid receptors, identified in a functional genetic screen. *Mol. Cell. Biol.* **2000**, *20*, 2411–2422. [CrossRef]
3. Lin, J.; Handschin, C.; Spiegelman, B.M. Metabolic control through the PGC-1 family of transcription coactivators. *Cell Metab.* **2005**, *1*, 361–370. [CrossRef]
4. Handschin, C.; Spiegelman, B.M. Peroxisome proliferator-activated receptor γ coactivator 1 coactivators, energy homeostasis, and metabolism. *Endocr. Rev.* **2006**, *27*, 728–735. [CrossRef]
5. Fernandez-Marcos, P.J.; Auwerx, J. Regulation of PGC-1α, a nodal regulator of mitochondrial biogenesis. *Am. J. Clin. Nutr.* **2011**, *93*, 884S–890S. [CrossRef]
6. Patten, I.S.; Arany, Z. PGC-1 coactivators in the cardiovascular system. *Trends Endocrinol. Metab.* **2012**, *23*, 90–97. [CrossRef]
7. Liu, C.; Lu, Y. PGC-1 coactivators in the control of energy metabolism. *Acta Biochim. Biophys. Sin.* **2011**, *43*, 248–257. [CrossRef]
8. Lin, J.; Puigserver, P.; Donovan, J.; Tarr, P.; Spiegelman, B.M. Peroxisome proliferator-activated receptor γ coactivator 1β (PGC-1β), a novel PGC-1-related transcription coactivator associated with host cell factor. *J. Biol. Chem.* **2001**, *277*, 1645–1648. [CrossRef]
9. Andersson, U.; Scarpulla, R.C. PGC-1-related coactivator, a novel, serum-inducible coactivator of nuclear respiratory factor 1-dependent transcription in mammalian cells. *Mol. Cell. Biol.* **2001**, *21*, 3738–3749. [CrossRef]
10. Tiefenböck, S.K.; Baltzer, C.; Egli, N.A.; Frei, C. The Drosophila PGC-1 homologue Spargel coordinates mitochondrial activity to insulin signalling. *EMBO J.* **2009**, *29*, 171–183. [CrossRef]
11. Li, S.; Liu, C.; Li, N.; Hao, T.; Han, T.; Hill, D.E.; Vidal, M.; Lu, Y. Genome-wide coactivation analysis of PGC-1α identifies BAF60a as a regulator of hepatic lipid metabolism. *Cell Metab.* **2008**, *8*, 105–117. [CrossRef] [PubMed]
12. Lin, J.; Wu, P.-H.; Tarr, P.T.; Lindenberg, K.S.; St-Pierre, J.; Zhang, C.-Y.; Mootha, V.K.; Jager, S.; Vianna, C.R.; Reznick, R.M.; et al. Defects in adaptive energy metabolism with CNS-linked hyperactivity in PGC-1α null mice. *Cell* **2004**, *119*, 121–135. [CrossRef]
13. Boss, O.; Bachman, E.; Vidal-Puig, A.; Zhang, C.-Y.; Peroni, O.; Lowell, B.B. Role of the β3-adrenergic receptor and/or a putative β4-adrenergic receptor on the expression of uncoupling proteins and peroxisome proliferator-activated receptor-γ coactivator-1. *Biochem. Biophys. Res. Commun.* **1999**, *261*, 870–876. [CrossRef] [PubMed]
14. Gómez-Ambrosi, J.; Frühbeck, G.; Martínez, J.A. Rapid in vivo PGC-1 mRNA upregulation in brown adipose tissue of Wistar rats by a β3-adrenergic agonist and lack of effect of leptin. *Mol. Cell. Endocrinol.* **2001**, *176*, 85–90. [CrossRef]
15. Handschin, C.; Rhee, J.; Lin, J.; Tarr, P.T.; Spiegelman, B.M. An autoregulatory loop controls peroxisome proliferator-activated receptor γ coactivator 1α expression in muscle. *Proc. Natl. Acad. Sci. USA* **2003**, *100*, 7111–7116. [CrossRef]
16. Cao, W.; Daniel, K.W.; Robidoux, J.; Puigserver, P.; Medvedev, A.; Bai, X.; Floering, L.M.; Spiegelman, B.M.; Collins, S. p38 mitogen-activated protein kinase is the central regulator of cyclic AMP-dependent transcription of the brown fat uncoupling protein 1 gene. *Mol. Cell. Biol.* **2004**, *24*, 3057–3067. [CrossRef]
17. Yoon, J.C.; Puigserver, P.; Chen, G.; Donovan, J.; Wu, Z.; Rhee, J.; Adelmant, G.; Stafford, J.M.; Kahn, C.R.; Granner, D.K.; et al. Control of hepatic gluconeogenesis through the transcriptional coactivator PGC-1. *Nature* **2001**, *413*, 131–138. [CrossRef]

18. Estall, J.L.; Kahn, M.; Cooper, M.P.; Fisher, F.M.; Wu, M.K.; Laznik, D.; Qu, L.; Cohen, D.E.; Shulman, G.I.; Spiegelman, B.M. Sensitivity of lipid metabolism and insulin signaling to genetic alterations in hepatic peroxisome proliferator-activated receptor-gamma coactivator-1alpha expression. *Diabetes* **2009**, *58*, 1499–1508. [CrossRef]

19. Herzig, S.; Long, F.; Jhala, U.S.; Hedrick, S.; Quinn, R.; Bauer, A.; Rudolph, D.; Schutz, G.; Yoon, C.; Puigserver, P.; et al. CREB regulates hepatic gluconeogenesis through the coactivator PGC-1. *Nature* **2001**, *413*, 179–183. [CrossRef]

20. Rhee, J.; Inoue, Y.; Yoon, J.C.; Puigserver, P.; Fan, M.; Gonzalez, F.J.; Spiegelman, B.M. Regulation of hepatic fasting response by PPAR coactivator-1 (PGC-1): Requirement for hepatocyte nuclear factor 4 in gluconeogenesis. *Proc. Natl. Acad. Sci. USA* **2003**, *100*, 4012–4017. [CrossRef]

21. Puigserver, P.; Rhee, J.; Donovan, J.; Walkey, C.J.; Yoon, J.C.; Oriente, F.; Kitamura, Y.; Altomonte, J.; Dong, H.H.; Accili, M.; et al. Insulin-regulated hepatic gluconeogenesis through FOXO1–PGC-1α interaction. *Nature* **2003**, *423*, 550–555. [CrossRef] [PubMed]

22. Leone, T.C.; Lehman, J.J.; Finck, B.N.; Schaeffer, P.J.; Wende, A.R.; Boudina, S.; Courtois, M.; Wozniak, D.F.; Sambandam, N.; Bernal-Mizrachi, C.; et al. PGC-1alpha deficiency causes multi-system energy metabolic derangements: Muscle dysfunction, abnormal weight control and hepatic steatosis. *PLoS Biol.* **2005**, *3*, e101. [CrossRef] [PubMed]

23. Cao, W.; Collins, Q.F.; Becker, T.C.; Robidoux, J.; Lupo, E.G.; Xiong, Y.; Daniel, K.W.; Floering, L.; Collins, S. p38 mitogen-activated protein kinase plays a stimulatory role in hepatic gluconeogenesis. *J. Biol. Chem.* **2005**, *280*, 42731–42737. [CrossRef] [PubMed]

24. Besse-Patin, A.; Jeromson, S.; Levesque-Damphousse, P.; Secco, B.; Laplante, M.; Estall, J.L. PGC1A regulates the IRS1:IRS2 ratio during fasting to influence hepatic metabolism downstream of insulin. *Proc. Natl. Acad. Sci. USA* **2019**, *116*, 4285–4290. [CrossRef]

25. Wu, H.; Kanatous, S.B.; Thurmond, F.A.; Gallardo, T.; Isotani, E.; Bassel-Duby, R.; Williams, R.S. Regulation of mitochondrial biogenesis in skeletal muscle by CaMK. *Science* **2002**, *296*, 349–352. [CrossRef]

26. Wu, Z.; Huang, X.; Feng, Y.; Handschin, C.; Gullicksen, P.S.; Bare, O.; Labow, M.; Spiegelman, B.M.; Stevenson, S.C.; Feng, Y.; et al. Transducer of regulated CREB-binding proteins (TORCs) induce PGC-1 transcription and mitochondrial biogenesis in muscle cells. *Proc. Natl. Acad. Sci. USA* **2006**, *103*, 14379–14384. [CrossRef]

27. Akimoto, T.; Sorg, B.S.; Yan, Z. Real-time imaging of peroxisome proliferator-activated receptor-γ coactivator-1α promoter activity in skeletal muscles of living mice. *Am. J. Physiol. Physiol.* **2004**, *287*, C790–C796. [CrossRef]

28. Zhao, M.; New, L.; Kravchenko, V.V.; Kato, Y.; Gram, H.; Di Padova, F.; Olson, E.N.; Ulevitch, R.; Han, J. Regulation of the MEF2 family of transcription factors by p38. *Mol. Cell. Biol.* **1999**, *19*, 21–30. [CrossRef]

29. Akimoto, T.; Pohnert, S.C.; Li, P.; Zhang, M.; Gumbs, C.; Rosenberg, P.B.; Williams, R.S.; Yan, Z. Exercise stimulates PGC-1α transcription in skeletal muscle through activation of the p38 MAPK pathway. *J. Biol. Chem.* **2005**, *280*, 19587–19593. [CrossRef]

30. Wu, H.; Deng, X.; Shi, Y.; Su, Y.; Wei, J.; Duan, H. PGC-1α, glucose metabolism and type 2 diabetes mellitus. *J. Endocrinol.* **2016**, *229*, R99–R115. [CrossRef]

31. Yoon, J.; Xu, G.; Deeney, J.T.; Yang, S.-N.; Rhee, J.; Puigserver, P.; Levens, A.R.; Yang, R.; Zhang, C.-Y.; Lowell, B.B.; et al. Suppression of β cell energy metabolism and insulin release by PGC-1α. *Dev. Cell* **2003**, *5*, 73–83. [CrossRef]

32. Guo, Y.; Xu, M.; Deng, B.; Frontera, J.R.; Kover, K.L.; Aires, D.; Ding, H.; Carlson, S.E.; Turk, J.; Wang, W.; et al. Beta-cell injury in Ncb5or-null mice is exacerbated by consumption of a high-fat diet. *Eur. J. Lipid Sci. Technol.* **2011**, *114*, 233–243. [CrossRef] [PubMed]

33. Kim, J.-W.; Park, S.-Y.; You, Y.-H.; Ham, D.-S.; Park, H.-S.; Lee, S.-H.; Yang, H.K.; Yoon, K.-H. Targeting PGC-1α to overcome the harmful effects of glucocorticoids in porcine neonatal pancreas cell clusters. *Transplantation* **2014**, *97*, 273–279. [CrossRef] [PubMed]

34. De Souza, C.T.; Gasparetti, A.L.; Pereira-Da-Silva, M.; Araújo, E.P.; Carvalheira, J.B.; Saad, M.J.A.; Boschero, A.C.; Carneiro, E.M.; Velloso, L.A. Peroxisome proliferator-activated receptor γ coactivator-1-dependent uncoupling protein-2 expression in pancreatic islets of rats: A novel pathway for neural control of insulin secretion. *Diabetologia* **2003**, *46*, 1522–1531. [CrossRef]

35. Li, N.; Brun, T.; Cnop, M.; Cunha, D.A.; Eizirik, D.L.; Maechler, P. Transient oxidative stress damages mitochondrial machinery inducing persistent β-cell dysfunction. *J. Biol. Chem.* **2009**, *284*, 23602–23612. [CrossRef]

36. Wende, A.R.; Schaeffer, P.J.; Parker, G.J.; Zechner, C.; Han, D.-H.; Chen, M.M.; Hancock, C.R.; Lehman, J.J.; Huss, J.M.; McClain, D.A.; et al. A role for the transcriptional coactivator PGC-1α in muscle refueling. *J. Biol. Chem.* **2007**, *282*, 36642–36651. [CrossRef]

37. Eisele, P.S.; Salatino, S.; Sobek, J.; Hottiger, M.O.; Handschin, C. The peroxisome proliferator-activated receptor γ coactivator 1α/β (PGC-1) coactivators repress the transcriptional activity of NF-κB in skeletal muscle cells*. *J. Biol. Chem.* **2012**, *288*, 2246–2260. [CrossRef]

38. Chinsomboon, J.; Ruas, J.L.; Gupta, R.K.; Thom, R.; Shoag, J.; Rowe, G.C.; Sawada, N.; Raghuram, S.; Arany, Z. The transcriptional coactivator PGC-1α mediates exercise-induced angiogenesis in skeletal muscle. *Proc. Natl. Acad. Sci. USA* **2009**, *106*, 21401–21406. [CrossRef]

39. Saint-Geniez, M.; Jiang, A.; Abend, S.; Liu, L.; Sweigard, H.; Connor, K.M.; Arany, Z. PGC-1α regulates normal and pathological angiogenesis in the retina. *Am. J. Pathol.* **2012**, *182*, 255–265. [CrossRef]

40. Rowe, G.C.; Jiang, A.; Arany, Z. PGC-1 coactivators in cardiac development and disease. *Circ. Res.* **2010**, *107*, 825–838. [CrossRef]

41. Lai, L.; Leone, T.C.; Zechner, C.; Schaeffer, P.J.; Kelly, S.M.; Flanagan, D.P.; Medeiros, D.M.; Kovacs, A.; Kelly, D.P. Transcriptional coactivators PGC-1 and PGC-l control overlapping programs required for perinatal maturation of the heart. *Genes Dev.* **2008**, *22*, 1948–1961. [CrossRef] [PubMed]

42. Martin, O.J.; Lai, L.; Soundarapandian, M.M.; Leone, T.C.; Zorzano, A.; Keller, M.P.; Attie, A.D.; Muoio, D.M.; Kelly, D.P. A role for peroxisome proliferator-activated receptor γ coactivator-1 in the control of mitochondrial dynamics during postnatal cardiac growth. *Circ. Res.* **2014**, *114*, 626–636. [CrossRef] [PubMed]

43. Russell, L.K.; Mansfield, C.M.; Lehman, J.J.; Kovacs, A.; Courtois, M.; Saffitz, J.E.; Medeiros, D.M.; Valencik, M.L.; McDonald, J.A.; Kelly, D.P. Cardiac-specific induction of the transcriptional coactivator peroxisome proliferator-activated receptor γ coactivator-1α promotes mitochondrial biogenesis and reversible cardiomyopathy in a developmental stage-dependent manner. *Circ. Res.* **2004**, *94*, 525–533. [CrossRef]

44. Arany, Z.; Novikov, M.; Chin, S.; Ma, Y.; Rosenzweig, A.; Spiegelman, B.M. Transverse aortic constriction leads to accelerated heart failure in mice lacking PPAR- coactivator 1. *Proc. Natl. Acad. Sci. USA* **2006**, *103*, 10086–10091. [CrossRef] [PubMed]

45. Lehman, J.J.; Barger, P.M.; Kovacs, A.; Saffitz, J.E.; Medeiros, D.M.; Kelly, D.P. Peroxisome proliferator–activated receptor γ coactivator-1 promotes cardiac mitochondrial biogenesis. *J. Clin. Investig.* **2000**, *106*, 847–856. [CrossRef]

46. Arany, Z.; He, H.; Lin, J.; Hoyer, K.; Handschin, C.; Toka, O.; Ahmad, F.; Matsui, T.; Chin, S.; Wu, P.-H.; et al. Transcriptional coactivator PGC-1α controls the energy state and contractile function of cardiac muscle. *Cell Metab.* **2005**, *1*, 259–271. [CrossRef] [PubMed]

47. St-Pierre, J.; Drori, S.; Uldry, M.; Silvaggi, J.M.; Rhee, J.; Jager, S.; Handschin, C.; Zheng, K.; Lin, J.; Yang, W.; et al. Suppression of reactive oxygen species and neurodegeneration by the PGC-1 transcriptional coactivators. *Cell* **2006**, *127*, 397–408. [CrossRef]

48. Blechman, J.; Amir-Zilberstein, L.; Gutnick, A.; Ben-Dor, S.; Levkowitz, G. The metabolic regulator PGC-1 directly controls the expression of the hypothalamic neuropeptide oxytocin. *J. Neurosci.* **2011**, *31*, 14835–14840. [CrossRef]

49. Zhou, Y.; Wang, S.; Li, Y.; Yu, S.; Zhao, Y. SIRT1/PGC-1α signaling promotes mitochondrial functional recovery and reduces apoptosis after intracerebral hemorrhage in rats. *Front. Mol. Neurosci.* **2018**, *10*, 443. [CrossRef]

50. Taherzadeh-Fard, E.; Saft, C.; Akkad, D.A.; Wieczorek, S.; Haghikia, A.; Chan, A.; Epplen, J.T.; Arning, L. PGC-1alpha downstream transcription factors NRF-1 and TFAM are genetic modifiers of Huntington disease. *Mol. Neurodegener.* **2011**, *6*, 32. [CrossRef]

51. Soyal, S.M.; Felder, T.K.; Auer, S.; Hahne, P.; Oberkofler, H.; Witting, A.; Paulmichl, M.; Landwehrmeyer, G.B.; Weydt, P.; Patsch, W.; et al. A greatly extended PPARGC1A genomic locus encodes several new brain-specific isoforms and influences Huntington disease age of onset. *Hum. Mol. Genet.* **2012**, *21*, 3461–3473. [CrossRef] [PubMed]

52. Martinez-Redondo, V.; Pettersson, A.T.; Ruas, J.L. The hitchhiker's guide to PGC-1α isoform structure and biological functions. *Diabetologia* **2015**, *58*, 1969–1977. [CrossRef] [PubMed]

53. Jesse, S.; Bayer, H.; Alupei, M.C.; Zügel, M.; Mulaw, M.A.; Tuorto, F.; Malmsheimer, S.; Singh, K.; Steinacker, J.M.; Schumann, U.; et al. Ribosomal transcription is regulated by PGC-1alpha and disturbed in Huntington's disease. *Sci. Rep.* **2017**, *7*, 1–10. [CrossRef]

54. Soyal, S.M.; Zara, G.; Ferger, B.; Felder, T.K.; Kwik, M.; Nofziger, C.; Dossena, S.; Schwienbacher, C.; Hicks, A.A.; Pramstaller, P.P.; et al. The PPARGC1A locus and CNS-specific PGC-1α isoforms are associated with Parkinson's Disease. *Neurobiol. Dis.* **2019**, *121*, 34–46. [CrossRef]

55. Chandra, G.; Shenoi, R.; Anand, R.; Rajamma, U.; Mohanakumar, K. Reinforcing mitochondrial functions in aging brain: An insight into Parkinson's disease therapeutics. *J. Chem. Neuroanat.* **2019**, *95*, 29–42. [CrossRef]

56. Katsouri, L.; Lim, Y.M.; Blondrath, K.; Eleftheriadou, I.; Lombardero, L.; Birch, A.M.; Mirzaei, N.; Irvine, E.E.; Mazarakis, N.D.; Sastre, M. PPARγ-coactivator-1α gene transfer reduces neuronal loss and amyloid-β generation by reducing β-secretase in an Alzheimer's disease model. *Proc. Natl. Acad. Sci. USA* **2016**, *113*, 12292–12297. [CrossRef]

57. Sweeney, G.; Song, J. The association between PGC-1α and Alzheimer's disease. *Anat. Cell Biol.* **2016**, *49*, 1–6. [CrossRef]

58. Mastropasqua, F.; Girolimetti, G.; Shoshan, M.C. PGC1α: Friend or foe in cancer? *Genes* **2018**, *9*, 48. [CrossRef]

59. Bost, F.; Kaminski, L. The metabolic modulator PGC-1α in cancer. *Am. J. Cancer Res.* **2019**, *9*, 198–211.

60. Li, Y.; Wingert, R.A. Regenerative medicine for the kidney: Stem cell prospects & challenges. *Clin. Transl. Med.* **2013**, *2*, 11. [CrossRef]

61. Chambers, J.M.; Poureetezadi, S.J.; Addiego, A.; Lahne, M.; Wingert, R.A. ppargc1a controls nephron segmentation during zebrafish embryonic kidney ontogeny. *eLife* **2018**, *7*, e40266. [CrossRef] [PubMed]

62. Poureetezadi, S.J.; Cheng, C.N.; Chambers, J.M.; Drummond, B.E.; Wingert, R.A. Prostaglandin signaling regulates nephron segment patterning of renal progenitors during zebrafish kidney development. *eLife* **2016**, *5*, e17551. [CrossRef] [PubMed]

63. Tran, M.; Tam, D.; Bardia, A.; Bhasin, M.K.; Rowe, G.C.; Kher, A.; Zsengeller, Z.K.; Akhavan-Sharif, M.R.; Khankin, E.V.; Saintgeniez, M.; et al. PGC-1α promotes recovery after acute kidney injury during systemic inflammation in mice. *J. Clin. Investig.* **2011**, *121*, 4003–4014. [CrossRef] [PubMed]

64. Fagerberg, L.; Hallström, B.M.; Oksvold, P.; Kampf, C.; Djureinovic, D.; Odeberg, J.; Habuka, M.; Tahmasebpoor, S.; Danielsson, A.; Edlund, K.; et al. Analysis of the human tissue-specific expression by genome-wide integration of transcriptomics and antibody-based proteomics. *Mol. Cell. Proteom.* **2013**, *13*, 397–406. [CrossRef] [PubMed]

65. Tran, M.; Zsengeller, Z.K.; Berg, A.H.; Khankin, E.V.; Bhasin, M.K.; Kim, W.; Clish, C.B.; Stillman, I.E.; Karumanchi, S.A.; Rhee, E.P.; et al. PGC1α drives NAD biosynthesis linking oxidative metabolism to renal protection. *Nature* **2016**, *531*, 528–532. [CrossRef] [PubMed]

66. Casemayou, A.; Fournel, A.; Bagattin, A.; Schanstra, J.P.; Belliere, J.; Decramer, S.; Marsal, D.; Gillet, M.; Chassaing, N.; Huart, A.; et al. Hepatocyte nuclear factor-1β controls mitochondrial respiration in renal tubular cells. *J. Am. Soc. Nephrol.* **2017**, *28*, 3205–3217. [CrossRef] [PubMed]

67. Han, S.H.; Wu, M.-Y.; Nam, B.Y.; Park, J.T.; Yoo, T.-H.; Jhee, J.H.; Park, J.; Chinga, F.; Li, S.-Y.; Susztak, K. PGC-1α protects from notch-induced kidney fibrosis development. *J. Am. Soc. Nephrol.* **2017**, *28*, 3312–3322. [CrossRef]

68. Hasumi, H.; Baba, M.; Hasumi, Y.; Huang, Y.; Oh, H.; Hughes, R.M.; Klein, M.E.; Takikita, S.; Nagashima, K.; Schmidt, L.S.; et al. Regulation of mitochondrial oxidative metabolism by tumor suppressor FLCN. *J. Natl. Cancer Inst.* **2012**, *104*, 1750–1764. [CrossRef]

69. LaGory, E.L.; Wu, C.; Taniguchi, C.M.; Ding, C.-K.C.; Chi, J.-T.; Von Eyben, R.; Scott, D.A.; Richardson, A.D.; Giaccia, A.J. Suppression of PGC-1α is critical for reprogramming oxidative metabolism in renal cell carcinoma. *Cell Rep.* **2015**, *12*, 116–127. [CrossRef]

70. Xiao, W.; Xiong, Z.; Xiong, W.; Yuan, C.; Xiao, H.; Ruan, H.; Song, Z.; Wang, C.; Bao, L.; Cao, Q.; et al. Melatonin/PGC1A/UCP1 promotes tumor slimming and represses tumor progression by initiating autophagy and lipid browning. *J. Pineal Res.* **2019**, *67*, e12607. [CrossRef]

71. Nam, H.; Kundu, A.; Brinkley, G.J.; Chandrashekar, D.S.; Kirkman, R.L.; Chakravarthi, B.V.K.; Orlandella, R.M.; Norian, L.A.; Sonpavde, G.; Ghatalia, P.; et al. PGC1α suppresses kidney cancer progression by inhibiting collagen-induced SNAIL expression. *Matrix Biol.* **2020**, *89*, 43–58. [CrossRef] [PubMed]

72. Felipe-Abrio, B.; Verdugo-Sivianes, E.M.; Carnero, A. c-MYB- and PGC1a-dependent metabolic switch induced by MYBBP1A loss in renal cancer. *Mol. Oncol.* **2019**, *13*, 1519–1533. [CrossRef] [PubMed]

73. Xu, W.-H.; Xu, Y.; Wang, J.; Wan, F.-N.; Wang, H.-K.; Cao, D.-L.; Shi, G.-H.; Qu, Y.-Y.; Zhang, H.-L.; Ye, D.-W. Prognostic value and immune infiltration of novel signatures in clear cell renal cell carcinoma microenvironment. *Aging* **2019**, *11*, 6999–7020. [CrossRef] [PubMed]

74. McCampbell, K.K.; Wingert, R.A. Renal stem cells: Fact or science fiction? *Biochem. J.* **2012**, *444*, 153–168. [CrossRef] [PubMed]

75. Rasbach, K.A.; Schnellmann, R.G. PGC-1α over-expression promotes recovery from mitochondrial dysfunction and cell injury. *Biochem. Biophys. Res. Commun.* **2007**, *355*, 734–739. [CrossRef] [PubMed]

76. Funk, J.A.; Schnellmann, R.G. Accelerated recovery of renal mitochondrial and tubule homeostasis with SIRT1/PGC-1α activation following ischemia–reperfusion injury. *Toxicol. Appl. Pharmacol.* **2013**, *273*, 345–354. [CrossRef] [PubMed]

77. Collier, J.B.; Whitaker, R.M.; Eblen, S.T.; Schnellmann, R.G. Rapid renal regulation of peroxisome proliferator-activated receptor γ coactivator-1α by extracellular signal-regulated kinase 1/2 in physiological and pathological conditions. *J. Biol. Chem.* **2016**, *291*, 26850–26859. [CrossRef] [PubMed]

78. Perry, H.; Huang, L.; Wilson, R.J.; Bajwa, A.; Sesaki, H.; Yan, Z.; Rosin, D.L.; Kashatus, D.F.; Okusa, M.D. Dynamin-related protein 1 deficiency promotes recovery from AKI. *J. Am. Soc. Nephrol.* **2017**, *29*, 194–206. [CrossRef]

79. Fontecha-Barriuso, M.; Martín-Sánchez, D.; Martinez-Moreno, J.M.; Carrasco, S.; Ruiz-Andrés, O.; Monsalve, M.; Sanchez-Ramos, C.; Gómez, M.J.; Ruiz-Ortega, M.; Sánchez-Niño, M.D.; et al. PGC-1α deficiency causes spontaneous kidney inflammation and increases the severity of nephrotoxic AKI. *J. Pathol.* **2019**, *249*, 65–78. [CrossRef]

80. Lim, A.K. Diabetic nephropathy—Complications and treatment. *Int. J. Nephrol. Renov. Dis.* **2014**, *7*, 361–381. [CrossRef]

81. National Institute of Diabetes and Digestive and Kidney Disease NIDDK—National Institute of Health (NIH). Chronic Kidney Disease. 2017. Available online: https://www.niddk.nih.gov/health-information/kidney-disease/chronic-kidney-disease-ckd/what-is-chronic-kidney-disease (accessed on 30 April 2020).

82. Sharma, K.; Karl, B.; Mathew, A.V.; Gangoiti, J.A.; Wassel, C.L.; Saito, R.; Pu, M.; Sharma, S.; You, Y.-H.; Wang, L.; et al. Metabolomics reveals signature of mitochondrial dysfunction in diabetic kidney disease. *J. Am. Soc. Nephrol.* **2013**, *24*, 1901–1912. [CrossRef] [PubMed]

83. Dugan, L.L.; You, Y.-H.; Ali, S.S.; Diamond-Stanic, M.; Miyamoto, S.; Declèves, A.-E.; Andreyev, A.Y.; Quach, T.; Ly, S.; Shekhtman, G.; et al. AMPK dysregulation promotes diabetes-related reduction of superoxide and mitochondrial function. *J. Clin. Investig.* **2013**, *123*, 4888–4899. [CrossRef] [PubMed]

84. Long, J.; Badal, S.S.; Ye, Z.; Wang, Y.; Ayanga, B.A.; Galvan, D.L.; Green, N.H.; Chang, B.H.; Overbeek, P.A.; Danesh, F.R. Long noncoding RNA Tug1 regulates mitochondrial bioenergetics in diabetic nephropathy. *J. Clin. Investig.* **2016**, *126*, 4205–4218. [CrossRef] [PubMed]

85. Li, S.-Y.; Park, J.; Qiu, C.; Han, S.H.; Palmer, M.B.; Arany, Z.; Susztak, K. Increasing the level of peroxisome proliferator-activated receptor γ coactivator-1α in podocytes results in collapsing glomerulopathy. *JCI Insight* **2017**, *2*, e92930. [CrossRef] [PubMed]

86. Svensson, K.; Schnyder, S.; Cardel, B.; Handschin, C. Loss of renal tubular PGC-1α exacerbates diet-induced renal steatosis and age-related urinary sodium excretion in mice. *PLoS ONE* **2016**, *11*, e0158716. [CrossRef]

87. Kang, H.M.; Ahn, S.H.; Choi, P.; Ko, Y.-A.; Han, S.H.; Chinga, F.; Park, A.S.D.; Tao, J.; Sharma, K.; Pullman, J.; et al. Defective fatty acid oxidation in renal tubular epithelial cells has a key role in kidney fibrosis development. *Nat. Med.* **2014**, *21*, 37–46. [CrossRef]

88. Peng, H.; Wang, Q.; Lou, T.; Qin, J.; Jung, S.; Shetty, V.; Li, F.; Wang, Y.; Feng, X.-H.; Mitch, W.E.; et al. Myokine mediated muscle-kidney crosstalk suppresses metabolic reprogramming and fibrosis in damaged kidneys. *Nat. Commun.* **2017**, *8*, 1493. [CrossRef]

89. Huang, S.; Park, J.; Qiu, C.; Chung, K.W.; Li, S.-Y.; Sirin, Y.; Han, S.H.; Taylor, V.; Zimber-Strobl, U.; Susztak, K. Jagged1/Notch2 controls kidney fibrosis via Tfam-mediated metabolic reprogramming. *PLoS Biol.* **2018**, *16*, e2005233. [CrossRef]

90. National Institute of Diabetes and Digestive and Kidney Disease (NIDDKD). Polycystic Kidney Disease. Available online: https://www.niddk.nih.gov/health-information/kidney-disease/polycystic-kidney-disease (accessed on 30 April 2020).

91. Rossetti, S.; Consugar, M.B.; Chapman, A.B.; Torres, V.E.; Guay-Woodford, L.M.; Grantham, J.J.; Bennett, W.M.; Meyers, C.M.; Walker, D.L.; Bae, K.; et al. Comprehensive molecular diagnostics in autosomal dominant polycystic kidney disease. *J. Am. Soc. Nephrol.* **2007**, *18*, 2143–2160. [CrossRef]

92. Chapin, H.C.; Caplan, M.J. The cell biology of polycystic kidney disease. *J. Cell Biol.* **2010**, *191*, 701–710. [CrossRef]

93. Qian, F.; Watnick, T.J.; Onuchic, L.F.; Germino, G.G. The molecular basis of focal cyst formation in human autosomal dominant polycystic kidney disease type I. *Cell* **1996**, *87*, 979–987. [CrossRef]

94. Pei, Y.; Watnick, T.; He, N.; Wang, K.; Liang, Y.; Parfrey, P.; Germino, G.; George-Hyslop, P.S. Somatic PKD2 mutations in individual kidney and liver cysts support a "two-hit" model of cystogenesis in type 2 autosomal dominant polycystic kidney disease. *J. Am. Soc. Nephrol.* **1999**, *10*, 1524–1529. [PubMed]

95. Watnick, T.; He, N.; Wang, K.; Liang, Y.; Parfrey, P.; Hefferton, D.; George-Hyslop, P.S.; Germino, G.G.; Pei, Y. Mutations of PKD1 in ADPKD2 cysts suggest a pathogenic effect of trans-heterozygous mutations. *Nat. Genet.* **2000**, *25*, 143–144. [CrossRef] [PubMed]

96. Ishimoto, Y.; Inagi, R.; Yoshihara, D.; Kugita, M.; Nagao, S.; Shimizu, A.; Takeda, N.; Wake, M.; Honda, K.; Zhou, J.; et al. Mitochondrial abnormality facilitates cyst formation in autosomal dominant polycystic kidney disease. *Mol. Cell. Biol.* **2017**, *37*, e00337-17. [CrossRef]

97. Hajarnis, S.; Lakhia, R.; Yheskel, M.; Williams, D.; Sorourian, M.; Liu, X.; Aboudehen, K.; Zhang, S.; Kersjes, K.; Galasso, R.; et al. microRNA-17 family promotes polycystic kidney disease progression through modulation of mitochondrial metabolism. *Nat. Commun.* **2017**, *8*, 14395. [CrossRef]

98. Li, X. Department of internal medicine epigenetics in ADPKD: Understanding mechanisms and discovering treatment. In *Polycystic Kidney Disease*; Codon Publications: Singapore, 2015; pp. 283–311.

99. Chambers, J.M.; Addiego, A.; Wingert, R.A. Ppargc1a controls ciliated cell development by regulating prostaglandin biosynthesis. CELL-REPORTS-D-19-04964. Available online: http://dx.doi.org/10.2139/ssrn.3509910 (accessed on 27 December 2019).

100. Gresh, L.; Fischer, E.; Reimann, A.; Tanguy, M.; Garbay, S.; Shao, X.; Hiesberger, T.; Fiette, L.; Igarashi, P.; Yaniv, M.; et al. A transcriptional network in polycystic kidney disease. *EMBO J.* **2004**, *23*, 1657–1668. [CrossRef]

101. Hiesberger, T.; Bai, Y.; Shao, X.; McNally, B.T.; Sinclair, A.M.; Tian, X.; Igarashi, P. Mutation of hepatocyte nuclear factor–1β inhibits Pkhd1 gene expression and produces renal cysts in mice. *J. Clin. Invest.* **2004**, *113*, 814–825. [CrossRef]

102. Jin, D.; Ni, T.T.; Sun, J.; Wan, H.; Amack, J.D.; Yu, G.; Fleming, J.; Chiang, C.; Li, W.; Papierniak, A.; et al. Prostaglandin signalling regulates ciliogenesis by modulating intraflagellar transport. *Nat. Cell Biol.* **2014**, *16*, 841–851. [CrossRef]

103. Marra, A.N.; Adeeb, B.D.; Chambers, B.E.; Drummond, B.E.; Ulrich, M.; Addiego, A.; Springer, M.; Poureetezadi, S.J.; Chambers, J.M.; Ronshaugen, M.R.; et al. Prostaglandin signaling regulates renal multiciliated cell specification and maturation. *Proc. Natl. Acad. Sci. USA* **2019**, *116*, 8409–8418. [CrossRef]

104. Boehlke, C.; Kotsis, F.; Patel, V.; Braeg, S.; Voelker, H.; Bredt, S.; Beyer, T.; Janusch, H.; Hamann, C.; Gödel, M.; et al. Primary cilia regulate mTORC1 activity and cell size through Lkb1. *Nat. Cell Biol.* **2010**, *12*, 1115–1122. [CrossRef]

105. Burkhalter, M.D.; Sridhar, A.; Sampaio, P.; Jacinto, R.; Burczyk, M.S.; Donow, C.; Angenendt, M.; Hempel, M.; Walther, P.; Pennekamp, P.; et al. Imbalanced mitochondrial function provokes heterotaxy via aberrant ciliogenesis. *J. Clin. Investig.* **2019**, *129*, 2841–2855. [CrossRef] [PubMed]

106. O'Toole, J.F.; Liu, Y.; Davis, E.E.; Westlake, C.J.; Attanasio, M.; Otto, E.A.; Seelow, D.; Nurnberg, G.; Becker, C.; Nuutinen, M.; et al. Individuals with mutations in XPNPEP3, which encodes a mitochondrial protein, develop a nephronophthisis-like nephropathy. *J. Clin. Investig.* **2010**, *120*, 791–802. [CrossRef] [PubMed]

107. Jiang, H.; Kang, S.U.; Zhang, S.; Karuppagounder, S.; Xu, J.; Lee, Y.-K.; Kang, B.-G.; Lee, Y.; Zhang, J.; Pletnikova, O.; et al. Adult conditional knockout of PGC-1α leads to loss of dopamine neurons. *eNeuro* **2016**, *3*, ENEURO.0183-16.2016. [CrossRef] [PubMed]

Review

Sirt1-PPARS Cross-Talk in Complex Metabolic Diseases and Inherited Disorders of the One Carbon Metabolism

Viola J. Kosgei [1], **David Coelho** [1], **Rosa-Maria Guéant-Rodriguez** [1,2] **and Jean-Louis Guéant** [1,2,*]

[1] UMR Inserm 1256 N-GERE (Nutrition, Génetique et Exposition aux Risques Environmentaux), Université de Lorraine, 54500 Vandoeuvre-lès-Nancy, France; viokos84@yahoo.com (V.J.K.); david.coelho@univ-lorraine.fr (D.C.); rosa-maria.gueant-rodriguez@univ-lorraine.fr (R.-M.G.-R.)

[2] Departments of Digestive Diseases, Nutrition and Endocrinology and Molecular Medicine and National Center of Inborn Errors of Metabolism, University Hospital Center, Université de Lorraine, 54500 Vandoeuvre-lès-Nancy, France

* Correspondence: jean-louis.gueant@univ-lorraine.fr; Tel.: +33-3-72-74-61-35

Received: 20 July 2020; Accepted: 7 August 2020; Published: 11 August 2020

Abstract: Sirtuin1 (Sirt1) has a NAD (+) binding domain and modulates the acetylation status of peroxisome proliferator-activated receptor-γ coactivator-1α (PGC1α) and Fork Head Box O1 transcription factor (Foxo1) according to the nutritional status. Sirt1 is decreased in obese patients and increased in weight loss. Its decreased expression explains part of the pathomechanisms of the metabolic syndrome, diabetes mellitus type 2 (DT2), cardiovascular diseases and nonalcoholic liver disease. Sirt1 plays an important role in the differentiation of adipocytes and in insulin signaling regulated by Foxo1 and phosphatidylinositol 3'-kinase (PI3K) signaling. Its overexpression attenuates inflammation and macrophage infiltration induced by a high fat diet. Its decreased expression plays a prominent role in the heart, liver and brain of rat as manifestations of fetal programming produced by deficit in vitamin B12 and folate during pregnancy and lactation through imbalanced methylation/acetylation of PGC1α and altered expression and methylation of nuclear receptors. The decreased expression of Sirt1 produced by impaired cellular availability of vitamin B12 results from endoplasmic reticulum stress through subcellular mislocalization of ELAVL1/HuR protein that shuttles Sirt1 mRNA between the nucleus and cytoplasm. Preclinical and clinical studies of Sirt1 agonists have produced contrasted results in the treatment of the metabolic syndrome. A preclinical study has produced promising results in the treatment of inherited disorders of vitamin B12 metabolism.

Keywords: Sirtuin1; peroxisome proliferator-activated receptor-γ coactivator-1α; peroxisome proliferator activated receptors; obesity; metabolic syndrome; vitamin B12; folate; fetal programming; inherited metabolic disorders

---

1. Introduction

Sirtuin1 (Sirt1) is one of the seven mammalian proteins belonging to the silent information regulator 2 (Sir2) proteins/Sirtuin family with highly conserved catalytic and nicotinamide adenine dinucleotide (NAD+) binding domain [1]. Sirtuins (Sirt) catalyze histone and non-histone lysine deacetylation in a NAD+ dependent manner [2,3]. Sirt1 has been shown to play an important role in increasing longevity in a number of lower animals like worms and flies [4–6]. Sirt1 is localized in the cytosol and the nucleus, between which it shuttles in response to different pathological and physiological environmental stimuli [2] including cellular stress and metabolic energy dysregulation.

Sirt1 regulates cellular energy metabolism through deacetylation of transcription factors and co-factors of energy metabolism [4,7]. Beside its role in metabolic redox electron transfer, NAD+ is an essential Sirt1-dependent cell sensor of energy metabolism, insulin secretion and adaptation to cell stress [8].

1.1. Role of Sirt1 on the Regulation of Energy Metabolism by Peroxisome Proliferator-Activated Receptor-Γ Coactivator-1α and Peroxisome Proliferator Activated Receptors

Calorie restriction regulates the expression levels of Sirt1 in a tissue-dependent manner [9]. Sirt1 regulates tissue glucose homeostasis, fatty acid beta oxidation and energy metabolism by modulating the acetylation status of peroxisome proliferator-activated receptor-γ coactivator-1α (PGC1α), Fork Head Box O1 transcription factor (Foxo1) and cAMP response element-binding protein (CREB), according to the nutritional status and fasting state. Sirt1 enhances PGC1α activity by deacetylating its lysine residues [10–12]. During fasting, Sirt1 activates PGC1α, which then induces the transcription of genes encoding gluconeogenic enzymes and suppresses the transcription of glycolytic genes in the liver [11]. Deacetylation of Foxo1 by Sirt1 regulates thyroid hormone mediated transcription of gluconeogenic genes like glucose-6-phosphatase and phospho-enoyl pyruvate carboxy kinase in mouse and human hepatic cells [13,14]. Activated Foxo1 binds to insulin response elements (IRE) promoters of these genes to induce their transcription [13,15]. Peroxisome proliferator activated receptors (PPARs) interact with co-regulators, including PGC1α in the regulation of energy homeostasis, insulin sensitivity, inflammation and adipogenesis [16]. Sirt1 regulates lipid metabolism of the liver in response to energy deprivation through deacetylation of PPARα. PPARα target genes encode fatty acid beta oxidation enzymes, carnitine palmitoyltransferase I, medium-chain acyl-CoA dehydrogenase and fatty acyl CoA synthase [17]. Liver specific ablation of Sirt1 induces the decreased transcription of PPARα target genes of fatty acid oxidation [18]. Similarly, the decreased expression of Sirt1 induces dysregulated energy metabolism through imbalanced acetylation and methylation of PGC1α in the myocardium of weaning animals exposed to methyl donor deficiency [19]. Sirt1 inhibits the expression of uncoupling protein gene 2 (UCP2), while Sirt1 silencing produces a mirrored effect in knock out mice [20]. Taken together, these data show that Sirt1 plays a prominent role on the regulation of energy metabolism through the deacetylation of PGC1α and Foxo1.

1.2. The Role of Sirt1 in Metabolic Syndrome and Insulin Resistance

The metabolic syndrome is a global public health problem related to overnutrition. It is defined as a cluster of cardio-metabolic abnormalities that includes obesity, insulin resistance or diabetes mellitus type 2 (DT2), hypertension and dyslipidemia [21,22]. The prevalence and incidence of the metabolic syndrome is increasing dramatically worldwide [21,23]. Systemic overexpression of Sirt1 has been reported to have protective effects against physiological damage in mice exposed to a high fat diet [24]. The decreased expression of Sirt1 explains part of the consequences of fatty acid enriched diets on the metabolic syndrome, DT2, cardiovascular diseases, nonalcoholic liver disease [25–28] and metabolic syndrome-associated cancers [29] (Figure 1). The decreased expression and activity of Sirt1 is involved in the pathogenesis of insulin resistance, DT2 and liver steatosis [30,31]. Decreased Sirt1 expression produces overexpression of miR-180a and impaired insulin signaling in hepatocytes miR-180a targets the 3′ UTR Sirt1 mRNA and is increased in insulin resistant hepatocytes and serum of diabetic patients [32]. Conversely, the overexpression of Sirt1 improves insulin sensitivity of hepatocytes in genetically obese ob/ob mice [33]. Insulin sensitivity is improved through inhibition of endoplasmic reticulum (ER) stress and decreased activity of mammalian/mechanistic target of Rapamycin complex 1 (mTORC1) in these animals [33]. Taken together, these data highlight the decreased activity of Sirt1 as a key component of the molecular mechanisms that produce the outcomes of metabolic syndrome and insulin resistance.

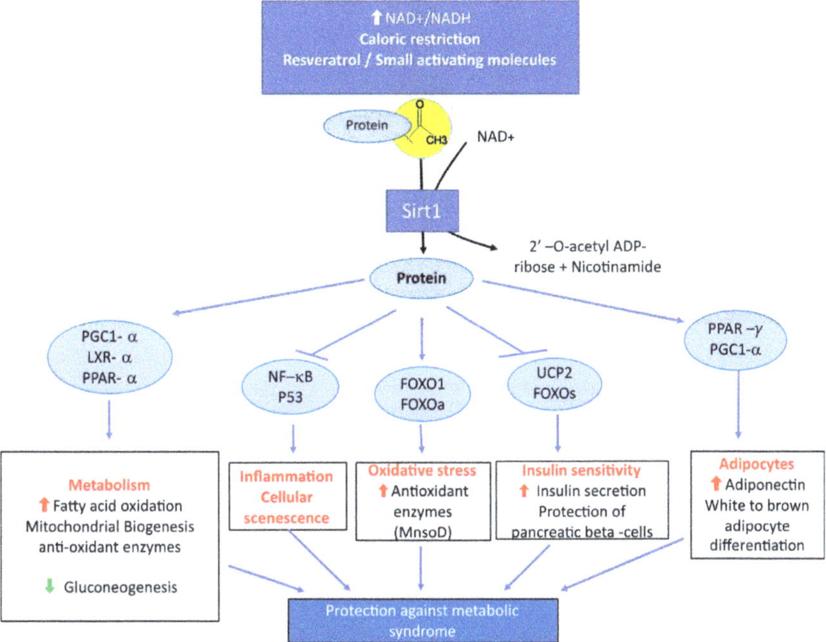

Figure 1. Main metabolic effects of Sirtuin1 related to its protective influence against the manifestations of metabolic syndrome.

1.3. Role of Sirt1 in Pancreatic Beta Cells, Adipose Tissue and Skeletal Muscle

Secretion of insulin by pancreatic beta cells is enhanced by Sirt1 in response to glucose stimulation [34–36]. Conversely, decreased expression of Sirt1 impairs glucose-stimulated insulin secretion and expression of mitochondrial genes that control metabolic coupling. Genetic or pharmacologic activation of Sirt1 protects beta cells against lipotoxicity of circulating lipids [37,38]. Furthermore, the specific deletion of Sirt1 in beta cells decreases the expression of glucose transporters and ER chaperones involved in an unfolded protein response [39]. The treatment of human isolated islets with gamma butyric-acid (GABA) enhances Sirt1 activity and attenuates drug-induced apoptosis, suggesting an anti-apoptotic role of Sirt1 [40].

Clinical studies evidenced decreased Sirt1 expression in adipose tissues of obese patients [41,42] and increased expression in progressive weight loss [43,44]. Adipocyte specific ablation of Sirt1 induces increased adiposity and manifestation of metabolic dysfunction including insulin resistance [25]. Overexpression of Sirt1 in adipose tissue in mice enhances glucose homeostasis and prevents age-induced decline in insulin sensitivity [45]. Chronic exposure to a high fat diet accelerates glucose intolerance and hyperinsulinemia in mice with adipocyte specific knockout of Sirt1 [46]. High fat diet induces cleavage of Sirt1 via Caspase 1 activation in adipose tissues [25]. Sirt1 induces browning remodeling of white adipose tissue by deacetylation of PPARγ [47,48]. PPARγ is highly expressed in adipocytes where it acts as a key regulator of lipid metabolism, insulin sensitivity and adipocyte differentiation [41,42,49]. Adiponectin is an adipocyte hormone associated with obesity and type 1 diabetes [50]. Sirt1 regulates the secretion of adiponectin through PPARγ upregulation of ER oxidoreductase α (Erol-Lα) [51]. Sirt1 also upregulates adiponectin secretion by activating Foxo1 and increasing the interaction between Foxo1 and C/EBP α [52].

Sirt1 plays an important role in the differentiation of skeletal muscle [53]. Insulin signaling is regulated by the Foxo1-Sirt1 pathway in skeletal muscle [54]. Sirt1 enhances insulin sensitivity through

PI3K signaling in response to caloric restriction [55]. Conversely, Sirt1 silencing decreases insulin sensitivity in Sirt1 KO mice. Wild type mice fed with a caloric restriction diet have a dramatic decreased acetylation of p53 and PGC1α, compared to Sirt1 KO mice [55]. In contrast, the overexpression of Sirt1 in skeletal muscle in vivo does not improve insulin resistance and had little impact on mitochondrial metabolism, suggesting that the overexpression of Sirt1 protein is not the single factor involved in insulin sensitization of skeletal muscle [56]. PPARγ- PGC1α couple is crucial in the regulation of energy metabolism in skeletal muscle. Its inhibition induces insulin resistance in C2C12 skeletal muscle cells, while its overexpression attenuates insulin resistance and enhanced glucose uptake [57]. In summary, Sirt1 acts as a sensor of the nutritional status on molecular mechanisms related with metabolic syndrome and insulin resistance, in pancreatic beta cells, adipose tissue and skeletal muscle.

1.4. Anti-Inflammatory Role of Sirt1 and PPAR in Metabolic Syndromes and Related Diseases

Visceral low grade inflammation triggered by adipocytes contributes to the pathogenesis of obesity, insulin resistance and other outcomes of the metabolic syndrome [58–60]. Besides dysregulated energy metabolism, increasing evidences link inflammation to pathogenesis of metabolic syndromes and related disorders, including diabetes and obesity [59–61]. Inflammation of the pancreatic beta cells in the islets is one of the prominent mechanisms of diabetes type 1 (DT1) and DT2. Adipocyte Sirt1 controls systemic insulin sensitivity through its effects on macrophages of adipose tissue [46,62]. Sirt1 knockdown exhibits elevated expression of Tumor Necrosis Factor α (TNF α) in adipocytes, induces macrophage infiltration and inflammation and increases cytokines levels of interleukin 1 beta (IL-1B), Interleukin 10 (IL-10), Interleukin 4 (IL-4) and TNF α in mice fed with high fat diet [63]. Conversely, overexpression of Sirt1 attenuates the adipose tissue inflammation and macrophage infiltration induced by a high fat diet [63]. Sirt1 inhibits inflammation in adipose tissue via mTOR/p70 ribosomal protein kinase 1 (S6k1) and Akt2 interacting pathways [64].

Converging evidences support an antagonistic crosstalk between Sirt1 and nuclear transcription factor Kappa B (NF-κB) [65]. Damaging effect of proinflammatory cytokines on beta cells is attenuated by the overexpression of Sirt1 in isolated rat islets through the deacetylation of P35 subunit of NF-κB [66]. Sirt1 deacetylates lysine 310 in the RelA/P65 subunit of NF-κB [67,68]. As a consequence, deacetylated RelA/P65 impairs methylation of lysine 314 and 315 residues, leading to ubiquitination and degradation [68,69]. Sirt1 knockdown in 3T3-L1 adipocytes activates the NF-κB signaling pathway by increased acetylation of NF-kB components and impaired interaction with promoters of matrix metalloproteinases and monocyte chemoattractant protein 1 (MCP1) [70].

Mice with hepatocyte specific knock out of Sirt1 (Sirt1LKO) exposed to a high fat diet develop hepatic inflammation and ER stress [18]. Liver inflammation of Sirt1 knock out mice results from increased expression of proinflammatory cytokine including TNF-α and IL-1B and macrophage infiltration. Liver ER stress of Sirt1LKO mice results from increased phosphorylation of translation initiation factor (elf2-α) and C-Jun N-terminal (JNK) [18]. PPARα confers protection against cellular stress and inflammation through various mechanisms related to Sirt1 [71]. PPARα agonist fenofibrate upregulates Sirt1 expression, suppresses CD40 and decreases acetylation of NF-κB-P65 in TNF-α treated 3T3-L1 adipocytes [72]. Chronic LPS stimulation of PPARγ deficient macrophages results in increased production of proinflammatory cytokines and decreased expression of anti-inflammatory cytokine IL-10 [73]. Furthermore, PPARγ deficiency causes delayed monocyte kinetic differentiation into macrophages [73]. In addition, PPARγ plays a crucial role in maturation of alternative phenotype in adipose tissues [74]. In summary, Sirt1 exerts protective effects against cellular stress and inflammation through complementary molecular and cellular mechanisms in adipose tissue, pancreatic islets and liver.

1.5. Antioxidant Role of Sirt1 and PPAR against Metabolic Syndromes and Related Diseases

Oxidative stress is associated with pathogenesis of complex metabolic diseases. Sirt1 activates Foxo transcription factors via the feedback loop [75,76]. Activation of Foxo3a and Foxo1a by Sirt1 deacetylation induces the transcription of catalase and manganese superoxide dismutase (MnSOD).

Moderate overexpression of Sirt1 is protective against oxidative stress in the mice heart by upregulating the expression of catalase through Foxo1a-dependent mechanisms [77]. In contrast, high levels of Sirt1 increase oxidative stress and cardiomyopathy [77]. A decreased expression of Sirt1, increased expression of NADPH oxidases (P42Phox) and increased superoxide production is observed in monocytes of DT1 patients [78]. The protective role of Sirt1 against oxidative stress is linked to the deacetylation of check point kinase 2, which increases cell death under oxidative stress [79]. We and others have shown a link between Sirt1 and RNA binding proteins like HUR in response to cellular stress [80,81]. Mitochondrion is one of the main organelle involved in ROS production [82]. PGC1α induces expression of superoxide dismutase 2 (SOD2) and glutathione peroxidase involved in ROS detoxification [12,83]. These data illustrate the important role of Sirt1 in oxidative stress homeostasis.

1.6. Role of Sirt1 in Fetal Programming and Nutritional and Inherited Disorders of Vitamin B12 Metabolism

The fetal programming hypothesis [84], also named "developmental origins of health and disease hypothesis" (DOHaD), proposes that unfavorable intrauterine life, including intrauterine growth restriction (IUGR), predicts the risk of postnatal complex diseases, including insulin resistance, DT2 and other outcomes of pathological obesity. We have shown that the maternal deficiency in methyl donors (MDD, vitamin B12 and folate) during pregnancy and lactation of rodents produces a low birth weight and an epigenomic Sirt1-dependent dysregulation of mitochondrial energy production and fatty acid oxidation in offspring [85,86]. It is noteworthy that the decreased expression of Sirt1 observed in fetal programming of maternal MDD is also a hallmark of overnutrition and pathological obesity (Figure 2). Moreover, population studies have highlighted an association between maternal methyl donor status and manifestations of fetal programming. In India and Nepal, many babies are thin with central obesity. There is a higher prevalence of mothers with low serum vitamin B12, folate deficiency and intrauterine growth restriction, compared to Europe [87]. In these two countries, the most insulin resistant children are born to mothers who have the lowest serum vitamin B12 at the first trimester of pregnancy [88,89]. A variant of adenosyl-methionine decarboxylase is also associated with childhood obesity, in India [90]. Folate seems to influence the metabolic consequences of fetal programming in Europe, despite contrasted results among population studies [91]. In France, a genetic polymorphism (677C > T, relatively common) of methylenetetrahydrofolate reductase (MTHFR) was associated with low birth weight and high insulin resistance in morbidly obese adolescents [92].

The decreased expression of Sirt1 plays a prominent role in the pathomechanisms of fetal programming produced by MDD in rats (vitamin B12 and folate) [19,81,93,94]. Vitamin B12 is metabolized into two active cofactors, methyl-cobalamin (Me-Cbl) and adenosyl-cobalamin (Ado-Cbl), the cofactors for cytoplasmic methionine synthase (MS/MTR) and mitochondrial methyl malonyl CoA mutase, respectively [95]. Me-Cbl and methyl folate are needed for the transmethylation of homocysteine into endogenous methionine, which is catalyzed by methionine synthase. Methionine is the direct precursor of S-adenosyl methionine (SAM), the universal methyl donor needed for transmethylation reactions involved in epigenomic regulatory mechanisms [86].

MDD during pregnancy and lactation induces impaired fatty acid oxidation, reduced activity of complexes I and II, cardiac hypertrophy with enlargement of cardiomyocyte and liver steatosis in weaning rat pups [19,96]. These observations are linked to the hyperacetylation and hypomethylation of PGC1α and dissociation of PGC1α from PPARα through decreased expression of Sirt1 and protein arginine methyltransferase 1 (PMRT1; Figures 2 and 3) [19]. MDD weakens the activator activity of PGC1α for other nuclear receptors, including estrogen receptor-α (ER-α), estrogen-related receptor-α, hepatocyte nuclear factor 4 (HNF-4) and vitamin D receptor (VDR). The liver steatosis of pups born from mothers with MDD during pregnancy and lactation resulted predominantly from hypomethylation of PGC1α, the decreased binding with its partners, including PPARα and HNF4 and the subsequent impaired mitochondrial fatty acid oxidation [96]. The effects of fetal programming on the liver of rats born from MDD mothers are worsened when pups are subsequently subjected to a high fat diet (HF) after d50 [97]. The MDD/HF animals have hallmarks of steato-hepatitis, with increased markers of inflammation and fibrosis, insulin resistance and key genes triggering the pathomechanisms of non-alcoholic steato-hepatitis (NASH; transforming growth factor beta super family, angiotensin and angiotensin receptor type 1). These data show that MDD during pregnancy is a risk factor of NASH in populations subsequently exposed to a HF diet. The deactivation of PGC1α is also involved in the brain manifestations of MDD fetal programming in rats. MDD during gestation and lactation alters the cerebellum plasticity in offspring, with a lower expression of synapsins. The altered neuroplasticity results from decreased expression and methylation of ER-α and subsequent decreased ER-α/PPAR-γ coactivator 1 α (PGC-1α) interaction in the deficiency condition. The impaired ER-α pathway leads to decreased expression of synapsins through a decreased EGR-1/Zif-268 transcription factor and Src-dependent phosphorylation of synapsins [98]. Deficiencies in methyl donors and in vitamin D are independently associated with altered bone development. In young rats, MDD decreases the total body bone mineral density, reduced tibia length and impaired growth plate maturation, and in preosteoblasts, MDD slows cellular proliferation. MDD produces a decreased expression of VDR, estrogen receptor-α, PGC1α, PRMT1 and Sirt1 and decreased nuclear VDR-PGC1α interaction [99]. The weaker VDR-PGC1α interaction is attributed to the reduced expression and imbalanced methylation/acetylation of PGC1α and the nuclear VDR sequestration by heat shock protein 90 (HSP90). These mechanisms together compromise bone development, as reflected by lowered bone alkaline phosphatase and increased proadipogenic PPARγ, adiponectin and estrogen-related receptor-α expression [99].

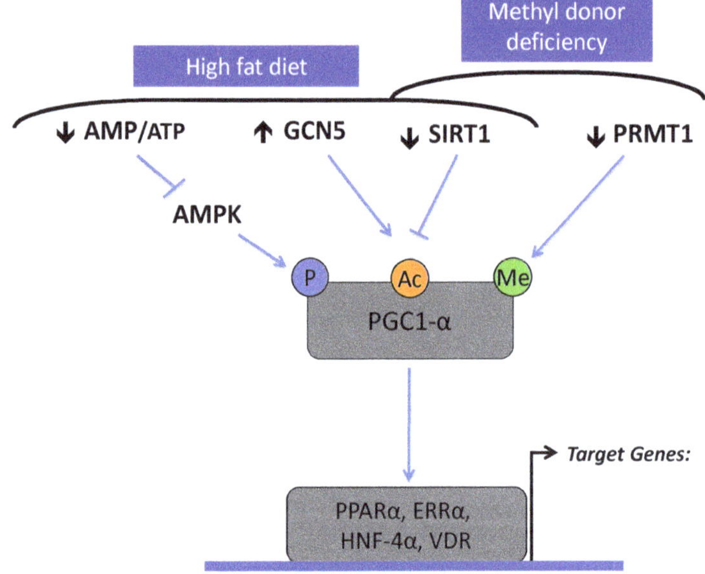

Figure 2. The decreased expression and activity of Sirtuin1 is a common molecular hallmark in a high fat diet and methyl donor deficiency (MDD). Sirtuin1 targets the acetylation of peroxisome proliferator-activated receptor-γ coactivator-1α (PGC1α) and has a complementary role with GCN5 acetylase, AMP kinase and protein arginine methyltransferase (PRMT1) in the regulation of PGC1α activation of nuclear receptors, including PPARα, ERRα, HNF-4α and VDR. PGC1α is phosphorylated and acetylated under the control of AMP kinase (AMPK), GCN5 acetylase and SIRT1 deacetylase. High fat diet and over nutrition decrease activity of AMP kinase, through high intracellular ATP levels, leading to decreased phosphorylation of PGC-1 α. It produces hyperacetylation of PGC-1α through increased expression of GCN5 and decreased expression and activity of SIRT1. MDD produces similar effects through decreased SIRT1 and PRMT1.

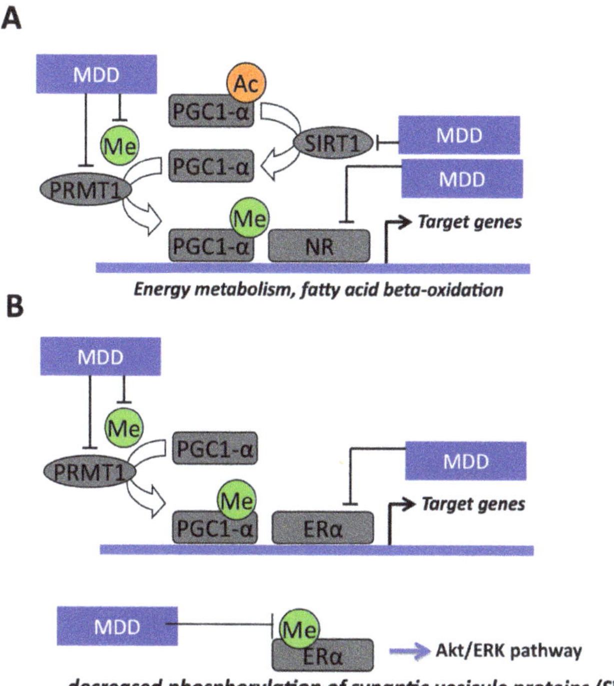

Figure 3. Molecular mechanisms demonstrating the link between methyl donor deficiency (MDD) and fetal programming and energy metabolism in liver and heart and regulation of synapsin expression in neuron. (**A**) The effects of fetal programming on the heart, liver and brain of rats born from methyl donor deficient (MDD) mothers are related to impaired PGC-1α activity through decreased expression of Sirtuin1 (Sirt1) and protein arginine methyltransferase (PRMT1) and decreased synthesis of the universal methyl donor (Me) methyl S- adenosyl methionine (SAM). The decreased activity of PGC-1α results from imbalanced methylation and acetylation. PGC-1α is a regulator of lipid metabolism and fatty acid oxidation through its role as coactivator of PPARα in heart and liver. (**B**) MDD induces decreased phosphorylation of synaptic vesicle proteins through impaired ERα activity linked to decreased SAM levels, PMRT1 expression and PGC-1α activity. PGC-1α is a regulator of synapsin expression and vesicle transport through its role as a coactivator of ERα in the brain.

The impaired cellular availability of vitamin B12 leads to ER stress related to decreased expression of Sirt1. ER stress is evidenced by increased expression and activation of elf2-α and activating transcription factor 6 (ATF6) and may be favored by decreased expression of heat shock proteins (HSP). Decreased Sirt1 impairs the transcription of HSP by increasing the acetylation of heat shock protein factor 1 (HSF1; Figure 4) [93]. Conversely, overexpression of Sirt1 and HSF1 and activation of Sirt1 by SRT1720 as well as addition of vitamin B12 induce a dramatic decrease of ER stress in NIE115 neuronal cells with impaired cellular B12 availability [93]. Similarly, ER stress is increased in fibroblasts of cblC, cblG and cblG* patients with inherited disorders of cellular vitamin B12 metabolism [100]. Some of the molecular mechanisms that underlie the neurological manifestations of patients with inherited disorders of vitamin B12 metabolism are related to transcriptomic changes of genes involved in RNA metabolism and ER stress. The transcriptomic changes result from the subcellular mislocalization of several RNA binding proteins (RBP), including the ELAVL1/HuR protein implicated in neuronal stress and HnRNPA1 and RBM10, in patient fibroblasts and Cd320 knockout mice with impaired cellular uptake of vitamin B12. The decreased interaction of ELAVL1/HuR with the CRM1/exportin protein of the nuclear pore complex and its subsequent mislocalization result from hypomethylation by decreased SAM and protein methyl transferase CARM1 and dephosphorylation by increased protein phosphatase PP2A. The mislocalization of ELAVL1/HuR triggers the decreased expression of Sirt1 deacetylase and other genes involved in brain development, neuroplasticity, myelin formation and brain aging [81,100]. In summary, Sirt1 plays a prominent role in the pathomechanisms produced by MDD through its role on the acetylation of PGC1α, the transcription of HSP and the subcellular localization of RNA binding proteins (RBP).

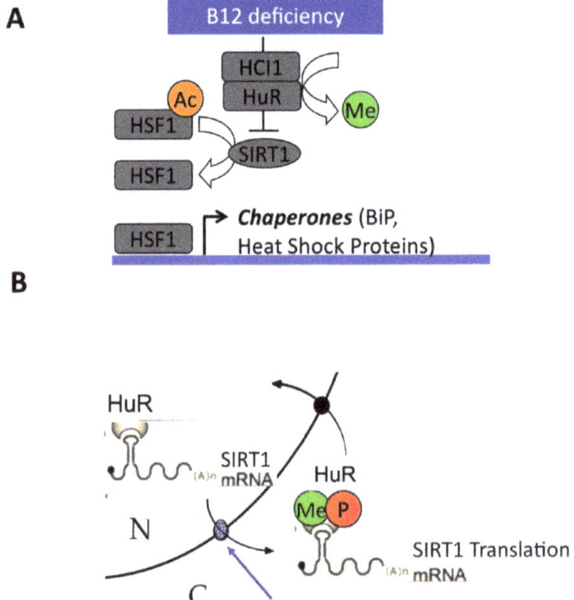

Figure 4. Influence of vitamin B12 cellular availability on endoplasmic reticulum (ER) stress-related decreased expression of Sirtuin1 (Sirt1). (**A**) The decreased cellular availability in B12 activates ER stress pathways and decreases the expression of heat shock proteins through decreased expression of Sirt1 and subsequent hyperacetylation of heat shock factor 1 (HSF1). (**B**) The decreased cellular availability in B12 produces the subcellular mislocalization of the ELAVL1/HuR RNA binding protein implicated in response to ER stress through hypomethylation by decreased synthesis of methyl S-adenosyl methionine (SAM) and dephosphorylation by increased protein phosphatase PP2A. The blue arrow shows the nuclear pore complex. The mislocalization of ELAVL1/HuR triggered the decreased expression of Sirt1 by altered Sirt1 mRNA export from nucleus to cytoplasm.

1.7. Sirt1 Is a Target for the Treatment of Complex and Hereditary Metabolic Diseases

Preclinical and clinical studies in the treatment of complex and inherited metabolic diseases have produced contrasted results in the evaluation of health benefits of activation of Sirt1 by natural and pharmaceutical small activating molecules.

Resveratrol is a natural polyphenolic compound found in red wine and grape and in plants and fruits like berries and peanuts. Resveratrol activates Sirt1 and mimics the caloric restriction status known to be protective against the metabolic syndrome. For decades, the therapeutic use of resveratrol has been considered in regard to its anti-inflammatory, antioxidant, antiaging and anticancer properties [101–103]. It reduces oxidative stress, hepatic steatosis and hypertension in high fat diet induced obesity in rodents and non-human primates [104–108]. In addition, resveratrol protects against high fat diet induced hepatic steatosis and endoplasmic stress [104,108,109] by decreasing the expression of proinflammatory cytokines, including TNF- α, IL-1β and IL-6 and increasing antioxidant enzymes like SOD2 and catalase [110]. Treatment of human monocytes in hyperglycemic condition with resveratrol induces upregulated expression of Sirt1, decreased P42Phox expression and upregulates the expression and activation of Foxo3a, which together lead to decreased production of superoxide [78]. Resveratrol has been proposed as a potential therapeutic agent against cardiovascular diseases [107,111–113]. It reduces hypertension [114,115] and cardiac hypertrophy via LBK1-AMPK1-eNOS signaling pathways in hypertensive animals [114].

Given the health benefits of resveratrol in experimental animal studies, clinical trials have been carried out to evaluate its effects in the prevention and treatment of metabolic syndrome. A preliminary exploratory trial showed that administration of 500 mg resveratrol twice per day for 60 days in type 1 diabetic patients decreased fasting plasma sugar (FPS) and hemoglobin A1C [116]. A prospective open-label randomized controlled trial reported that a 30 days resveratrol supplementation improves hemoglobin A1c, total cholesterol and systolic blood pressure and insulin sensitivity in patients with diabetes type 2 [117,118]. A meta-analysis of nine randomized clinical trials with 283 participants concluded that resveratrol improves insulin sensitivity (HOMA-IR index) in DT2 patients, and had no effects on hemoglobin A1c and the lipid profile [119]. Similar results were observed in another meta-analysis of 29 randomized clinical trials involving 1069 participants [120]. In contrast, a one month randomized clinical double-blinded crossover administration of 150 mg/day of resveratrol had no effect on peripheral and hepatic insulin sensitivity [121]. This contrasted result could be due to the interaction of metformin and resveratrol in the DT2 recruited patients. Resveratrol has been also evaluated in cardiovascular diseases, neurodegenerative diseases and cancers [113,122,123]. For example, it ameliorates the endothelial dysfunction in diabetic and obese mice through sirtuin 1 and peroxisome proliferator-activated receptor δ (PPARδ) [124].

Activators of Sirt1 such as SRT1720, SRT2183 and SRT1460 are 1000-fold more active than resveratrol. SRT1720 and resveratrol improve insulin sensitivity in nutritionally and genetically induced obesity and DT2 in mice [125]. SRT1720 treatment improved insulin sensitivity of liver, skeletal muscle and adipose tissue in insulin resistance Zucker *fa/fa* rats [126]. SRT1720 enhances fatty acid oxidation through direct deacetylation of PGC1α, Foxo1 and indirect activation of the AMP activated protein kinase (AMPK) pathway [126]. Furthermore, SRT1720 extends the lifespan of mice fed a high fat diet, decreases hepatic steatosis, increases insulin sensitivity and reverses inflammation and oxidative stress markers in adult obese mice subjected to a high fat diet [127]. SRT1720 impairs lipopolysaccharide stimulated inflammatory pathways and TNF-α secretion in macrophages and adipose tissues of Zucker fatty rats [70]. SRT1720 also activates AMPK in a Sirt1-independent manner. However, SRT1720 cannot be used in humans because of potential toxicity. Metformin regulates gluconeogenesis in *ob/ob* mice by upregulating hepatic Sirt1 and GCN5 [128]. A computational study confirmed that metformin directly activates Sirt1 [129]. At low NAD+ concentration, leucine and low dose of metformin synergistically activate Sirt1 and AMPK, with enhanced energy metabolism and insulin sensitivity in muscle cells and hepatic cells in vitro [130]. Recently two novel Sirt1 activators with high affinity for Sirt1, SC1C2 and SC1C2.1 attenuated doxorubicin induced DNA damage via deacetylation of P53 and cellular senescence in HepG2 and H9c2 cell lines [131].

Phase 1 clinical trials showed that SRT2104, another small molecule activator of Sirt1 is safe and tolerable in healthy volunteers, including the elderly [132,133]. However, in a prospective double blinded random placebo control crossover study, the daily administration of 2.0 g of SRT2104 for 28 days had a neutral cardiovascular effect in DT2 patients, even if it induced a loss in body weight [134]. Lipid profile of healthy smokers was improved after 28 days SRT2104 administration but again the cardiovascular effects were neutral [134,135].

Some severe forms of inherited disorders of intracellular metabolism of vitamin B12 are resistant to conventional treatments. Decreased Sirt1 activity plays a central role in some of the pathomechanisms of these disorders. We therefore evaluated the effect of Sirt1 agonists in a preclinical study in fibroblasts from patients with cblG and cblC inherited defects of vitamin B12 metabolism and an original transgenic mouse model of methionine synthase deficiency specific to neuronal cells. Patient fibroblasts with cblC and cblG defects of vitamin B12 metabolism presented with endoplasmic reticulum stress, altered subcellular localization of HuR, HnRNPA1 and RBM10, global mRNA mislocalization and increased HnRNPA1-dependent skipping of interferon regulatory factor 3 (IRF3) exons. SRT1720 inhibited ER stress and rescued RBP and mRNA mislocalization and IRF3 splicing. Furthermore, Sirt1 activation by SRT1720 partially restored the methylation and phosphorylation of these RBPs in patients' fibroblasts. Interestingly, SRT1720, vitamin B12 and SAM treatment improved cognitive functions in conditional *MTR*-KO mice with brain specific invalidation of *Mtr* gene encoding MS enzyme [100]. In particular, SRT1720 improved the deficient hippocampo-dependent learning ability of the mice.

In summary, preclinical and clinical studies of Sirt1 agonists have produced contrasted results in the treatment of the metabolic syndrome. A preclinical study has produced promising results in the treatment of inherited disorders of vitamin B12 metabolism.

2. Conclusions

Numerous experimental evidence show a strong link between decreased Sirt1 and the pathological manifestations of metabolic syndrome by synergistic cellular and molecular mechanisms. It is noteworthy that over nutrition and MDD both produce a decrease of Sirt1. There are additive and synergistic effects of the MDD fetal programming and subsequent exposure to a high fat diet in adult life. The therapeutic prospects for using activators of Sirt1 in the treatment of disease outcomes of metabolic syndrome have not been conclusive to date. However, pharmacological activation of Sirt1 opens promising perspectives for the treatment of rare diseases of vitamin B12 metabolism with particular effects on reticulum stress and mislocalization of RBPs.

Author Contributions: Main drafting of the manuscript, V.J.K. and J.-L.G.; partial drafting of specific parts, R.-M.G.-R. and D.C.; critical revision of the manuscript, V.J.K., D.C., R.-M.G.-R. and J.-L.G.; study supervision, J.-L.G. All authors have read and agreed to the published version of the manuscript.

Funding: This research was funded by FHU ARRIMAGE and the French PIA project "Lorraine Université d'Excellence", reference ANR-15-IDEX-04-LUE.

Conflicts of Interest: The authors have declared that no conflict of interest exists.

Abbreviations

SIRT1: Sirtuin 1; **PPARs**, Peroxisome proliferator activated receptors; **PPARα**, Peroxisome proliferator-activated receptor alpha; **PPARγ**, Peroxisome proliferator-activated receptor gamma; **PGC1α**, peroxisome proliferator-activated receptor-γ coactivator-1α; **DT1**, diabetes type 1; **DT2**, diabetes mellitus type 2; **Foxo1**, Forkhead Box O1; **PI3K**, phosphatidylinositol 3′-kinase; **NAD**, *Nicotinamide adenine dinucleotide*; **CREB**, cAMP response element-binding protein; **IRE**, insulin response element; **UCP2**, uncoupling protein gene 2; **ER**, endoplasmic reticulum; **mTORC1**, mammalian/mechanistic target of Rapamycin complex 1; **GABA**, gamma butyric-acid; **Erol-Lα**, ER oxidoreductase α; **C/EBP α**, CCAAT/enhancer binding protein alpha; **TNF α**, Tumor Necrosis Factor alpha, **IL- 1B**, Interleukin 1 beta; **IL- 10**, Interleukin 10; **IL- 4**, Interleukin 4; **S6K1**, p70 ribosomal protein kinase 1; **NF-κB**, nuclear transcription factor Kappa B; **Sirt1LKO**, hepatocyte specific knock out of Sirt1; **MCP1**, monocyte chemoattractant protein 1; **JNK**, C-Jun N-terminal; **LPS**, lipopolysaccharide; **MnSOD**, manganese superoxide dismutase; **DOHaD**, Developmental origins of health and disease hypothesis; **IUGR**, intra-uterine growth restriction, **MDD**, methyl donors deficiency; **MTHFR**, methylenetetrahydrofolate reductase; **Me-Cbl**, methyl-cobalamin; **Ado-Cbl**, adenosyl-cobalamin; **MS/MTR**, methionine synthase, **SAM**, S- adenosyl methionine, **PMRT1**, protein arginine methyltransferase; **HNF-4**, hepatocyte nuclear factor 4; **VDR**, vitamin D receptor **HF**, high fat diet; **NASH**, non-alcoholic steato-hepatitis; **ER-α**,estrogen receptor-α; **ERRα**, estrogen related receptor-α; **HSP**, heat shock protein; **HSF1**, heat shock protein factor 1; **ATF6**, activating transcription factor 6; **RBP**, RNA binding proteins, **AMPK**, AMP activated protein kinase; **FPS**, fasting plasma sugar, **IRF3**, interferon regulatory factor 3.

References

1. Frye, R. Phylogenetic Classification of Prokaryotic and Eukaryotic Sir2-like Proteins. *Biochem. Biophys. Res. Commun.* **2000**, *273*, 793–798. [CrossRef] [PubMed]

2. Tanno, M.; Sakamoto, J.; Miura, T.; Shimamoto, K.; Horio, Y. Nucleocytoplasmic Shuttling of the NAD+-dependent Histone Deacetylase SIRT1. *J. Biol. Chem.* **2006**, *282*, 6823–6832. [CrossRef] [PubMed]

3.   Landry, J.; Sutton, A.; Tafrov, S.T.; Heller, R.C.; Stebbins, J.; Pillus, L.; Sternglanz, R. The silencing protein SIR2 and its homologs are NAD-dependent protein deacetylases. *Proc. Natl. Acad. Sci. USA* **2000**, *97*, 5807–5811. [CrossRef] [PubMed]

4.   Canto, C.; Auwerx, J. Caloric restriction, SIRT1 and longevity. *Trends Endocrinol. Metab.* **2009**, *20*, 325–331. [CrossRef]

5.   Kaeberlein, M.; McVey, M.; Guarente, L. The SIR2/3/4 complex and SIR2 alone promote longevity in Saccharomyces cerevisiae by two different mechanisms. *Genes Dev.* **1999**, *13*, 2570–2580. [CrossRef]

6.   Rogina, B.; Helfand, S.L. Sir2 mediates longevity in the fly through a pathway related to calorie restriction. *Proc. Natl. Acad. Sci. USA* **2004**, *101*, 15998–16003. [CrossRef]

7.   Li, X. SIRT1 and energy metabolism. *Acta Biochim. Biophys. Sin.* **2012**, *45*, 51–60. [CrossRef]

8.   Houtkooper, R.H.; Canto, C.; Wanders, R.J.; Auwerx, J. The secret life of NAD+: An old metabolite controlling new metabolic signaling pathways. *Endocr. Rev.* **2009**, *31*, 194–223. [CrossRef]

9.   Chen, D.; Bruno, J.; Easlon, E.; Lin, S.-J.; Cheng, H.-L.; Alt, F.W.; Guarente, L. Tissue-specific regulation of SIRT1 by calorie restriction. *Genes Dev.* **2008**, *22*, 1753–1757. [CrossRef]

10.   Canto, C.; Auwerx, J. PGC-1α, SIRT1 and AMPK, an energy sensing network that controls energy expenditure. *Curr. Opin. Lipidol.* **2009**, *20*, 98–105. [CrossRef]

11.   Rodgers, J.T.; Lerin, C.; Haas, W.; Gygi, S.P.; Spiegelman, B.M.; Puigserver, P. Nutrient control of glucose homeostasis through a complex of PGC-1α and SIRT1. *Nature* **2005**, *434*, 113–118. [CrossRef]

12.   Nemoto, S.; Fergusson, M.M.; Finkel, T. SIRT1 functionally interacts with the metabolic regulator and transcriptional coactivator PGC-1{alpha}. *J. Biol. Chem.* **2005**, *280*, 16456–16460. [CrossRef]

13.   Nakae, J.; Cao, Y.; Daitoku, H.; Fukamizu, A.; Ogawa, W.; Yano, Y.; Hayashi, Y. The LXXLL motif of murine forkhead transcription factor FoxO1 mediates Sirt1-dependent transcriptional activity. *J. Clin. Investig.* **2006**, *116*, 2473–2483. [CrossRef] [PubMed]

14.   Singh, B.K.; Sinha, R.A.; Zhou, J.; Xie, S.Y.; You, S.-H.; Gauthier, K.; Yen, P.M. FoxO1 Deacetylation Regulates Thyroid Hormone-induced Transcription of Key Hepatic Gluconeogenic Genes*. *J. Biol. Chem.* **2013**, *288*, 30365–30372. [CrossRef] [PubMed]

15.   Durham, S.K.; Suwanichkul, A.; Scheimann, A.O.; Yee, D.; Jackson, J.G.; Barr, F.G.; Powell, D.R. FKHR binds the insulin response element in the insulin-like growth factor binding protein-1 promoter. *Endocrinology* **1999**, *140*, 3140–3146. [CrossRef] [PubMed]

16.   Lamichane, S.; Lamichane, B.D.; Kwon, S.-M. Pivotal Roles of Peroxisome Proliferator-Activated Receptors (PPARs) and Their Signal Cascade for Cellular and Whole-Body Energy Homeostasis. *Int. J. Mol. Sci.* **2018**, *19*, 949. [CrossRef]

17.   Feingold, K.R.; Wang, Y.; Moser, A.; Shigenaga, J.K.; Grunfeld, C. LPS decreases fatty acid oxidation and nuclear hormone receptors in the kidney. *J. Lipid Res.* **2008**, *49*, 2179–2187. [CrossRef]

18.   Purushotham, A.; Schug, T.T.; Xu, Q.; Surapureddi, S.; Guo, X.; Li, X. Hepatocyte-Specific Deletion of SIRT1 Alters Fatty Acid Metabolism and Results in Hepatic Steatosis and Inflammation. *Cell Metab.* **2009**, *9*, 327–338. [CrossRef]

19.   Garcia, M.M.; Guéant-Rodriguez, R.-M.; Pooya, S.; Brachet, P.; Alberto, J.-M.; Jeannesson, E.; Maskali, F.; Gueguen, N.; Marie, P.-Y.; Lacolley, P.; et al. Methyl donor deficiency induces cardiomyopathy through altered methylation/acetylation of PGC-1α by PRMT1 and SIRT1. *J. Pathol.* **2011**, *225*, 324–335. [CrossRef]

20.   Bordone, L.; Motta, M.C.; Picard, F.; Robinson, A.; Jhala, U.S.; Apfeld, J.; McDonagh, T.; Lemieux, M.E.; McBurney, M.; Szilvasi, A.; et al. Sirt1 regulates insulin secretion by repressing UCP2 in pancreatic beta cells. *PLoS Biol.* **2005**, *4*, e31. [CrossRef]

21.   Saklayen, M.G. The Global Epidemic of the Metabolic Syndrome. *Curr. Hypertens. Rep.* **2018**, *20*, 1–8. [CrossRef]

22.   Sigit, F.S.; Tahapary, D.L.; Trompet, S.; Sartono, E.; Van Dijk, K.W.; Rosendaal, F.R.; De Mutsert, R. The prevalence of metabolic syndrome and its association with body fat distribution in middle-aged individuals from Indonesia and the Netherlands: A cross-sectional analysis of two population-based studies. *Diabetol. Metab. Syndr.* **2020**, *12*, 2–11. [CrossRef] [PubMed]

23.   Chooi, Y.C.; Ding, C.; Magkos, F. The epidemiology of obesity. *Metabolism* **2019**, *92*, 6–10. [CrossRef] [PubMed]

24. Guéant, J.; Elakoum, R.; Ziegler, O.; Coelho, D.; Feigerlova, E.; Daval, J.-L.; Gueant-Rodriguez, R.-M. Nutritional models of foetal programming and nutrigenomic and epigenomic dysregulations of fatty acid metabolism in the liver and heart. *Pflüg. Arch. Eur. J. Physiol.* **2013**, *466*, 833–850. [CrossRef] [PubMed]
25. Chalkiadaki, A.; Guarente, L. High-Fat Diet Triggers Inflammation-Induced Cleavage of SIRT1 in Adipose Tissue to Promote Metabolic Dysfunction. *Cell Metab.* **2012**, *16*, 180–188. [CrossRef] [PubMed]
26. Von Frankenberg, A.D.; Marina, A.; Song, X.; Callahan, H.S.; Kratz, M.; Utzschneider, K.M. A high-fat, high-saturated fat diet decreases insulin sensitivity without changing intra-abdominal fat in weight-stable overweight and obese adults. *Eur. J. Nutr.* **2017**, *56*, 431–443. [CrossRef]
27. Hariri, N.; Thibault, L. High-fat diet-induced obesity in animal models. *Nutr. Res. Rev.* **2010**, *23*, 270–299. [CrossRef]
28. Liu, Z.; Gan, L.; Liu, G.; Chen, Y.; Wu, T.; Feng, F.; Chao, S. Sirt1 decreased adipose inflammation by interacting with Akt2 and inhibiting mTOR/S6K1 pathway in mice. *J. Lipid Res.* **2016**, *57*, 1373–1381. [CrossRef]
29. Herranz, D.; Muñoz-Martin, M.; Cañamero, M.; Mulero, F.; Martinez-Pastor, B.; Fernandez-Capetillo, O.; Serrano, M. Sirt1 improves healthy ageing and protects from metabolic syndrome-associated cancer. *Nat. Commun.* **2010**, *1*, 1–8. [CrossRef]
30. Wang, R.H.; Kim, H.S.; Xiao, C.; Xu, X.; Gavrilova, O.; Deng, C.X. Hepatic Sirt1 deficiency in mice impairs mTorc2/Akt signaling and results in hyperglycemia, oxidative damage, and insulin resistance. *J. Clin. Investig.* **2011**, *121*, 4477–4490. [CrossRef]
31. Wang, Y.; Xu, C.; Liang, Y.; Vanhoutte, P.M. SIRT1 in metabolic syndrome: Where to target matters. *Pharmacol. Ther.* **2012**, *136*, 305–318. [CrossRef] [PubMed]
32. Zhou, B.; Li, C.; Qi, W.; Zhang, Y.; Zhang, F.; Wu, J.X.; Hu, Y.N.; Wu, D.M.; Liu, Y.; Yan, T.T.; et al. Downregulation of miR-181a upregulates sirtuin-1 (SIRT1) and improves hepatic insulin sensitivity. *Diabetology* **2012**, *55*, 2032–2043. [CrossRef] [PubMed]
33. Li, Y.; Xu, S.; Giles, A.; Nakamura, K.; Lee, J.W.; Hou, X.; Donmez, G.; Li, J.; Luo, Z.; Walsh, K.; et al. Hepatic overexpression of SIRT1 in mice attenuates endoplasmic reticulum stress and insulin resistance in the liver. *FASEB J.* **2011**, *25*, 1664–1679. [CrossRef] [PubMed]
34. Ramachandran, D.; Roy, U.; Garg, S.; Ghosh, S.; Pathak, S.; Kolthur-Seetharam, U. Sirt1 and mir-9 expression is regulated during glucose-stimulated insulin secretion in pancreatic beta-islets. *FEBS J.* **2011**, *278*, 1167–1174. [CrossRef] [PubMed]
35. Hayashi, C.; Ogawa, O.; Kubo, S.; Mitsuhashi, N.; Onuma, T.; Kawamori, R. Ankle brachial pressure index and carotid intima-media thickness as atherosclerosis markers in Japanese diabetics. *Diabetes Res. Clin. Pract.* **2004**, *66*, 269–275. [CrossRef]
36. Luu, L.; Dai, F.F.; Prentice, K.J.; Huang, X.; Hardy, A.B.; Hansen, J.B.; Liu, Y.; Joseph, J.W.; Wheeler, M.B. The loss of Sirt1 in mouse pancreatic beta cells impairs insulin secretion by disrupting glucose sensing. *Diabetologia* **2013**, *56*, 2010–2020. [CrossRef]
37. Desai, T.; Koulajian, K.; Ivovic, A.; Breen, D.M.; Luu, L.; Tsiani, E.L.; Wheeler, M.B.; Giacca, A. Pharmacologic or genetic activation of SIRT1 attenuates the fat-induced decrease in beta-cell function in vivo. *Nutr. Diabetes* **2019**, *9*, 11. [CrossRef]
38. Wu, L.; Zhou, L.; Lu, Y.; Zhang, J.; Jian, F.; Liu, Y.; Li, F.; Wenyi, L.; Xiao, W.; Li, G. Activation of SIRT1 protects pancreatic beta-cells against palmitate-induced dysfunction. *Biochim. Biophys. Acta* **2012**, *1822*, 1815–18125. [CrossRef]
39. Pinho, A.V.; Bensellam, M.; Wauters, E.; Rees, M.; Giry-Laterriere, M.; Mawson, A.; Ly, L.Q.; Biankin, A.V.; Wu, J.; Laybutt, D.R.; et al. Pancreas-Specific Sirt1-Deficiency in Mice Compromises Beta-Cell Function without Development of Hyperglycemia. *PLoS ONE* **2015**, *10*, e0128012. [CrossRef]
40. Prud'Homme, G.J.; Glinka, Y.; Udovyk, O.; Hasilo, C.; Paraskevas, S.; Wang, Q. GABA protects pancreatic beta cells against apoptosis by increasing SIRT1 expression and activity. *Biochem. Biophys. Res. Commun.* **2014**, *452*, 649–654. [CrossRef]
41. Chawla, A.; Schwarz, E.J.; Dimaculangan, D.D.; Lazar, M.A. Peroxisome proliferator-activated receptor (PPAR) gamma: Adipose-predominant expression and induction early in adipocyte differentiation. *Endocrinology* **1994**, *135*, 798–800. [CrossRef] [PubMed]

42. Medina-Gómez, G.; Gray, S.L.; Yetukuri, L.; Shimomura, K.; Virtue, S.; Campbell, M.; Curtis, R.K.; Jimenez-Liñan, M.; Blount, M.; Yeo, G.S.H.; et al. PPAR gamma 2 Prevents Lipotoxicity by Controlling Adipose Tissue Expandability and Peripheral Lipid Metabolism. *PLoS Genet.* **2007**, *3*, e64. [CrossRef] [PubMed]

43. Mariani, S.; Fiore, D.; Persichetti, A.; Basciani, S.; Lubrano, C.; Poggiogalle, E.; Genco, A.; Donini, L.M.; Gnessi, L. Circulating SIRT1 Increases After Intragastric Balloon Fat Loss in Obese Patients. *Obes. Surg.* **2015**, *26*, 1215–1220. [CrossRef] [PubMed]

44. Rappou, E.; Jukarainen, S.; Rinnankoski-Tuikka, R.; Kaye, S.; Heinonen, S.; Hakkarainen, A.; Lundbom, J.; Lundbom, N.; Saunavaara, V.; Rissanen, A.; et al. Weight Loss Is Associated With Increased NAD(+)/SIRT1 Expression But Reduced PARP Activity in White Adipose Tissue. *J. Clin. Endocrinol. Metab.* **2016**, *101*, 1263–1273. [CrossRef] [PubMed]

45. Xu, C.; Bai, B.; Fan, P.; Cai, Y.; Huang, B.; Law, I.K.; Liu, L.; Xu, A.; Tung, C.; Li, X.; et al. Selective overexpression of human SIRT1 in adipose tissue enhances energy homeostasis and prevents the deterioration of insulin sensitivity with ageing in mice. *Am. J. Transl. Res.* **2013**, *5*, 412–426.

46. Hui, X.; Zhang, M.; Gu, P.; Li, K.; Gao, Y.; Wu, D.; Wang, Y.; Xu, A. Adipocyte SIRT 1 controls systemic insulin sensitivity by modulating macrophages in adipose tissue. *EMBO Rep.* **2017**, *18*, 645–657. [CrossRef]

47. Bargut, T.C.L.; Souza-Mello, V.; Aguila, M.B.; Mandarim-De-Lacerda, C.A. Browning of white adipose tissue: Lessons from experimental models. *Horm. Mol. Biol. Clin. Investig.* **2017**, *31*. [CrossRef]

48. Qiang, L.; Wang, L.; Kon, N.; Zhao, W.; Lee, S.; Zhang, Y.; Rosenbaum, M.; Zhao, Y.; Gu, W.; Farmer, S.R.; et al. Brown remodeling of white adipose tissue by SirT1-dependent deacetylation of Ppargamma. *Cell* **2012**, *150*, 620–632. [CrossRef]

49. Ma, X.; Wang, D.; Zhao, W.; Xu, L. Deciphering the Roles of PPARgamma in Adipocytes via Dynamic Change of Transcription Complex. *Front. Endocrinol.* **2018**, *9*, 473. [CrossRef]

50. Matsuda, M.; Shimomura, I. Roles of adiponectin and oxidative stress in obesity-associated metabolic and cardiovascular diseases. *Rev. Endocr. Metab. Disord.* **2013**, *15*, 1–10. [CrossRef]

51. Qiang, L.; Wang, H.; Farmer, S.R. Adiponectin Secretion Is Regulated by SIRT1 and the Endoplasmic Reticulum Oxidoreductase Ero1-Lα. *Mol. Cell. Biol.* **2007**, *27*, 4698–4707. [CrossRef] [PubMed]

52. Qiao, L.; Shao, J. SIRT1 Regulates Adiponectin Gene Expression through Foxo1-C/Enhancer-binding Protein Transcriptional Complex. *J. Biol. Chem.* **2006**, *281*, 39915–39924. [CrossRef] [PubMed]

53. Fulco, M.; Schiltz, R.; Iezzi, S.; King, M.; Zhao, P.; Kashiwaya, Y.; Hoffman, E.; Veech, R.L.; Sartorelli, V. Sir2 Regulates Skeletal Muscle Differentiation as a Potential Sensor of the Redox State. *Mol. Cell* **2003**, *12*, 51–62. [CrossRef]

54. Sin, T.K.; Yung, B.Y.; Siu, P.M. Modulation of SIRT1-Foxo1 Signaling axis by Resveratrol: Implications in Skeletal Muscle Aging and Insulin Resistance. *Cell. Physiol. Biochem.* **2015**, *35*, 541–552. [CrossRef]

55. Schenk, S.; McCurdy, C.E.; Philp, A.; Chen, M.Z.; Holliday, M.J.; Bandyopadhyay, G.K.; Osborn, O.; Baar, K.; Olefsky, J. Sirt1 enhances skeletal muscle insulin sensitivity in mice during caloric restriction. *J. Clin. Investig.* **2011**, *121*, 4281–4288. [CrossRef] [PubMed]

56. Brandon, A.E.; Tid-Ang, J.; Wright, L.E.; Stuart, E.; Suryana, E.; Bentley, N.; Turner, N.; Cooney, G.J.; Ruderman, N.B.; Kraegen, E.W. Overexpression of SIRT1 in Rat Skeletal Muscle Does Not Alter Glucose Induced Insulin Resistance. *PLoS ONE* **2015**, *10*, e0121959. [CrossRef]

57. Verma, N.K.; Singh, J.; Dey, C.S. PPAR-gamma expression modulates insulin sensitivity in C2C12 skeletal muscle cells. *Br. J. Pharmacol.* **2004**, *143*, 1006–1013. [CrossRef]

58. Donath, M.Y.; Shoelson, S.E. Type 2 diabetes as an inflammatory disease. *Nat. Rev. Immunol.* **2011**, *11*, 98–107. [CrossRef]

59. Fuentes, E.; Fuentes, F.; Vilahur, G.; Badimon, L.; Palomo, I. Mechanisms of Chronic State of Inflammation as Mediators That Link Obese Adipose Tissue and Metabolic Syndrome. *Mediat. Inflamm.* **2013**, *2013*, 1–11. [CrossRef]

60. Tsalamandris, S.; Antonopoulos, A.S.; Oikonomou, E.; Papamikroulis, G.-A.; Vogiatzi, G.; Papaioannou, S.; Deftereos, S.; Tousoulis, D. The Role of Inflammation in Diabetes: Current Concepts and Future Perspectives. *Eur. Cardiol. Rev.* **2019**, *14*, 50–59. [CrossRef]

61. Johnson, A.R.; Milner, J.J.; Makowski, L. The inflammation highway: Metabolism accelerates inflammatory traffic in obesity. *Immunol. Rev.* **2012**, *249*, 218–238. [CrossRef] [PubMed]

62. Perrini, S.; Porro, S.; Nigro, P.; Cignarelli, A.; Caccioppoli, C.; Genchi, V.A.; Martines, G.; De Fazio, M.; Capuano, P.; Natalicchio, A.; et al. Reduced SIRT1 and SIRT2 expression promotes adipogenesis of human visceral adipose stem cells and associates with accumulation of visceral fat in human obesity. *Int. J. Obes.* **2019**, *44*, 307–319. [CrossRef] [PubMed]

63. Gillum, M.P.; Kotas, M.E.; Erion, D.M.; Kursawe, R.; Chatterjee, P.; Nead, K.T.; Muise, E.S.; Hsiao, J.J.; Frederick, D.W.; Yonemitsu, S.; et al. SirT1 Regulates Adipose Tissue Inflammation. *Diabetes* **2011**, *60*, 3235–3245. [CrossRef] [PubMed]

64. Liu, Z.; Patil, I.Y.; Jiang, T.; Sancheti, H.; Walsh, J.P.; Stiles, B.L.; Yin, F.; Cadenas, E. High-Fat Diet Induces Hepatic Insulin Resistance and Impairment of Synaptic Plasticity. *PLoS ONE* **2015**, *10*, e0128274. [CrossRef] [PubMed]

65. Kauppinen, A.; Suuronen, T.; Ojala, J.; Kaarniranta, K.; Salminen, A. Antagonistic crosstalk between NF-kappaB and SIRT1 in the regulation of inflammation and metabolic disorders. *Cell Signal.* **2013**, *25*, 1939–1948. [CrossRef] [PubMed]

66. Lee, J.-H.; Song, M.-Y.; Song, E.-K.; Kim, E.-K.; Moon, W.S.; Han, M.-K.; Park, J.-W.; Kwon, K.-B.; Park, B.-H. Overexpression of SIRT1 Protects Pancreatic β-Cells against Cytokine Toxicity by Suppressing the Nuclear Factor-κB Signaling Pathway. *Diabetes* **2008**, *58*, 344–351. [CrossRef] [PubMed]

67. Chen, L.-F.; Mu, Y.; Greene, W.C. Acetylation of RelA at discrete sites regulates distinct nuclear functions of NF-kappaB. *EMBO J.* **2002**, *21*, 6539–6548. [CrossRef]

68. Yang, X.-D.; Tajkhorshid, E.; Chen, L.-F. Functional Interplay between Acetylation and Methylation of the RelA Subunit of NF-κB. *Mol. Cell. Biol.* **2010**, *30*, 2170–2180. [CrossRef]

69. Yeung, F.; Hoberg, J.E.; Ramsey, C.S.; Keller, M.D.; Jones, D.R.; Frye, R.A.; Mayo, M.W. Modulation of NF-kappaB-dependent transcription and cell survival by the SIRT1 deacetylase. *EMBO J.* **2004**, *23*, 2369–2380. [CrossRef]

70. Yoshizaki, T.; Milne, J.C.; Imamura, T.; Schenk, S.; Sonoda, N.; Babendure, J.L.; Lu, J.-C.; Smith, J.J.; Jirousek, M.R.; Olefsky, J. SIRT1 Exerts Anti-Inflammatory Effects and Improves Insulin Sensitivity in Adipocytes. *Mol. Cell. Biol.* **2008**, *29*, 1363–1374. [CrossRef]

71. Fuentes, E.; Guzman, L.; Moore-Carrasco, R.; Palomo, I. Role of PPARs in inflammatory processes associated with metabolic syndrome (Review). *Mol. Med. Rep.* **2013**, *8*, 1611–1616. [CrossRef] [PubMed]

72. Wang, W.; Lin, Q.; Lin, R.; Zhang, J.; Ren, F.; Zhang, J.; Meixi, J.; Li, Y. PPARalpha agonist fenofibrate attenuates TNF-alpha-induced CD40 expression in 3T3-L1 adipocytes via the SIRT1-dependent signaling pathway. *Exp. Cell Res.* **2013**, *319*, 1523–1533. [CrossRef] [PubMed]

73. Heming, M.; Gran, S.; Jauch, S.L.; Fischer-Riepe, L.; Russo, A.; Klotz, L.; Hermann, S.; Schäfers, M.; Roth, J.; Barczyk-Kahlert, K. Peroxisome Proliferator-Activated Receptor-gamma Modulates the Response of Macrophages to Lipopolysaccharide and Glucocorticoids. *Front. Immunol.* **2018**, *9*, 893. [CrossRef] [PubMed]

74. Odegaard, J.I.; Ricardo-Gonzalez, R.R.; Red Eagle, A.; Vats, D.; Morel, C.R.; Goforth, M.H.; Subramanianet, V.; Mukundan, L.; Ferrante, A.W.; Chawla, A. Alternative M2 activation of Kupffer cells by PPARdelta ameliorates obesity-induced insulin resistance. *Cell Metab.* **2008**, *7*, 496–507. [CrossRef]

75. Brunet, A.; Sweeney, L.B.; Sturgill, J.F.; Chua, K.F.; Greer, P.L.; Lin, Y.; Tran, H.; Ross, S.E.; Mostoslavsky, R.; Cohen, H.Y.; et al. Stress-Dependent Regulation of FOXO Transcription Factors by the SIRT1 Deacetylase. *Science* **2004**, *303*, 2011–2015. [CrossRef]

76. Xiong, S.; Salazar, J.F.A.; Patrushev, N.; Alexander, R.W. FoxO1 Mediates an Autofeedback Loop Regulating SIRT1 Expression. *J. Biol. Chem.* **2010**, *286*, 5289–5299. [CrossRef]

77. Alcendor, R.R.; Gao, S.; Zhai, P.; Zablocki, D.; Holle, E.; Yu, X.; Tian, B.; Wagner, T.; Vatner, S.F.; Sadoshima, J.-I. Sirt1 Regulates Aging and Resistance to Oxidative Stress in the Heart. *Circ. Res.* **2007**, *100*, 1512–1521. [CrossRef]

78. Yun, J.-M.; Chien, A.; Jialal, I.; Devaraj, S. Resveratrol up-regulates SIRT1 and inhibits cellular oxidative stress in the diabetic milieu: Mechanistic insights. *J. Nutr. Biochem.* **2012**, *23*, 699–705. [CrossRef]

79. Kwon, J.; Lee, S.; Kim, Y.-N.; Lee, I.H. Deacetylation of CHK2 by SIRT1 protects cells from oxidative stress-dependent DNA damage response. *Exp. Mol. Med.* **2019**, *51*, 1–9. [CrossRef]

80. Abdelmohsen, K.; Pullmann, R.; Lal, A.; Kim, H.H.; Galban, S.; Yang, X.; Blethrow, J.D.; Walker, M.; Shubert, J.; Gillespie, D.A.; et al. Phosphorylation of HuR by Chk2 Regulates SIRT1 Expression. *Mol. Cell* **2007**, *25*, 543–557. [CrossRef]

81. Battaglia-Hsu, S.-F.; Ghemrawi, R.; Coelho, D.; Dreumont, N.; Mosca, P.; Hergalant, S.; Gauchotte, G.; Sequeira, J.M.; Ndiongue, M.; Houlgatte, R.; et al. Inherited disorders of cobalamin metabolism disrupt nucleocytoplasmic transport of mRNA through impaired methylation/phosphorylation of ELAVL1/HuR. *Nucleic Acids Res.* **2018**, *46*, 7844–7857. [CrossRef] [PubMed]

82. Zhao, R.Z.; Jiang, S.; Zhang, L.; Yu, Z.B. Mitochondrial electron transport chain, ROS generation and uncoupling (Review). *Int. J. Mol. Med.* **2019**, *44*, 3–15. [CrossRef] [PubMed]

83. Pierre, J.S.; Drori, S.; Uldry, M.; Silvaggi, J.M.; Rhee, J.; Jager, S.; Handschin, C.; Zheng, K.; Lin, J.; Yang, W.; et al. Suppression of Reactive Oxygen Species and Neurodegeneration by the PGC-1 Transcriptional Coactivators. *Cell* **2006**, *127*, 397–408. [CrossRef] [PubMed]

84. Godfrey, K.M.; Barker, D.J. Fetal nutrition and adult disease. *Am. J. Clin. Nutr.* **2000**, *71*, 1344S–1352S. [CrossRef]

85. Blaise, S.; Alberto, J.-M.; Audonnet-Blaise, S.; Guéant, J.; Daval, J.-L. Influence of preconditioning-like hypoxia on the liver of developing methyl-deficient rats. *Am. J. Physiol. Metab.* **2007**, *293*, E1492–E1502. [CrossRef]

86. Guéant, J.; Namour, F.; Gueant-Rodriguez, R.-M.; Daval, J.-L. Folate and fetal programming: A play in epigenomics. *Trends Endocrinol. Metab.* **2013**, *24*, 279–289. [CrossRef]

87. Yajnik, C.S.; Deshmukh, U.S. Fetal programming: Maternal nutrition and role of one-carbon metabolism. *Rev. Endocr. Metab. Disord.* **2012**, *13*, 121–127. [CrossRef]

88. Yajnik, C.S.; Deshmukh, U.S. Maternal nutrition, intrauterine programming and consequential risks in the offspring. *Rev. Endocr. Metab. Disord.* **2008**, *9*, 203–211. [CrossRef]

89. Stewart, C.P.; Christian, P.; Schulze, K.J.; Arguello, M.; LeClerq, S.C.; Khatry, S.K.; West, K.P. Low Maternal Vitamin B-12 Status Is Associated with Offspring Insulin Resistance Regardless of Antenatal Micronutrient Supplementation in Rural Nepal. *J. Nutr.* **2011**, *141*, 1912–1917. [CrossRef]

90. Tabassum, R.; Jaiswal, A.; Chauhan, G.; Dwivedi, O.P.; Ghosh, S.; Marwaha, R.K.; Tandon, N.; Bharadwaj, D. Genetic Variant of AMD1 Is Associated with Obesity in Urban Indian Children. *PLoS ONE* **2012**, *7*, e33162. [CrossRef]

91. Lewis, S.J.; Leary, S.; Smith, G.D.; Ness, A. Body composition at age 9 years, maternal folate intake during pregnancy and methyltetrahydrofolate reductase (MTHFR) C677T genotype. *Br. J. Nutr.* **2009**, *102*, 493–496. [CrossRef] [PubMed]

92. Frelut, M.-L.; Nicolas, J.-P.; Guilland, J.-C.; De Courcy, G.P. Methylenetetrahydrofolate reductase 677 C->T polymorphism: A link between birth weight and insulin resistance in obese adolescents. *Pediatr. Obes.* **2011**, *6*, e312–e317. [CrossRef] [PubMed]

93. Ghemrawi, R.; Pooya, S.; Lorentz, S.; Gauchotte, G.; Arnold, C.; Guéant, J.; Battaglia-Hsu, S.-F. Decreased vitamin B12 availability induces ER stress through impaired SIRT1-deacetylation of HSF1. *Cell Death Dis.* **2013**, *4*, e553. [CrossRef] [PubMed]

94. Melhem, H.; Hansmannel, F.; Bressenot, A.; Battaglia-Hsu, S.-F.; Billioud, V.; Alberto, J.M.; Guéant, J.; Peyrin-Biroulet, L. Methyl-deficient diet promotes colitis and SIRT1-mediated endoplasmic reticulum stress. *Gut* **2015**, *65*, 595–606. [CrossRef]

95. Green, R. Vitamin B12 deficiency from the perspective of a practicing hematologist. *Blood* **2017**, *129*, 2603–2611. [CrossRef]

96. Pooya, S.; Blaise, S.; Garcia, M.M.; Giudicelli, J.; Alberto, J.-M.; Gueant-Rodriguez, R.-M.; Jeannesson, E.; Gueguen, N.; Bressenot, A.; Nicolas, B.; et al. Methyl donor deficiency impairs fatty acid oxidation through PGC-1α hypomethylation and decreased ER-α, ERR-α, and HNF-4α in the rat liver. *J. Hepatol.* **2012**, *57*, 344–351. [CrossRef]

97. Bison, A.; Marchal-Bressenot, A.; Li, Z.; Elamouri, I.; Feigerlova, E.; Peng, L.; Houlgatte, R.; Beck, B.; Pourié, G.; Alberto, J.-M.; et al. Foetal programming by methyl donor deficiency produces steato-hepatitis in rats exposed to high fat diet. *Sci. Rep.* **2016**, *6*, 37207. [CrossRef]

98. Pourié, G.; Martin, N.; Bossenmeyer-Pourié, C.; Akchiche, N.; Gueant-Rodriguez, R.-M.; Geoffroy, A.; Jeannesson, E.; Chehadeh, S.E.H.; Mimoun, K.; Brachet, P.; et al. Folate- and vitamin B12–deficient diet during gestation and lactation alters cerebellar synapsin expression via impaired influence of estrogen nuclear receptor α. *FASEB J.* **2015**, *29*, 3713–3725. [CrossRef]

99. Feigerlova, E.; Demarquet, L.; Melhem, H.; Ghemrawi, R.; Battaglia-Hsu, S.F.; Ewu, E.; Alberto, J.M.; Helle, D.; Weryha, G.; Guéant, J.G. Methyl donor deficiency impairs bone development via peroxisome proliferator-activated receptor-gamma coactivator-1alpha-dependent vitamin D receptor pathway. *FASEB J.* **2016**, *30*, 3598–3612. [CrossRef]

100. Ghemrawi, R.; Arnold, C.; Battaglia-Hsu, S.-F.; Pourié, G.; Trinh, I.; Bassila, C.; Rashka, C.; Wiedemann, A.; Flayac, J.; Robert, A.; et al. SIRT1 activation rescues the mislocalization of RNA-binding proteins and cognitive defects induced by inherited cobalamin disorders. *Metabilisem* **2019**, *101*, 153992. [CrossRef]

101. Francioso, A.; Mastromarino, P.; Masci, A.; D'Erme, M.; Mosca, L. Chemistry, Stability and Bioavailability of Resveratrol. *Med. Chem.* **2014**, *10*, 237–245. [CrossRef] [PubMed]

102. Gambini, J.; Inglés, M.; Olaso, G.; Lopez-Grueso, R.; Bonet-Costa, V.; Gimeno-Mallench, L.; Mas-Bargues, C.; Abdelaziz, K.M.; Gomez-Cabrera, M.C.; Vina, J.; et al. Properties of Resveratrol:In VitroandIn VivoStudies about Metabolism, Bioavailability, and Biological Effects in Animal Models and Humans. *Oxidative Med. Cell. Longev.* **2015**, *2015*, 1–13. [CrossRef] [PubMed]

103. Kuršvietienė, L.; Stanevičienė, I.; Mongirdienė, A.; Bernatoniene, J. Multiplicity of effects and health benefits of resveratrol. *Medicina* **2016**, *52*, 148–155. [CrossRef] [PubMed]

104. Cho, S.-J.; Jung, U.J.; Choi, M.-S. Differential effects of low-dose resveratrol on adiposity and hepatic steatosis in diet-induced obese mice. *Br. J. Nutr.* **2012**, *108*, 2166–2175. [CrossRef] [PubMed]

105. Jiménez-Gómez, Y.; Mattison, J.A.; Pearson, K.J.; Martin-Montalvo, A.; Palacios, H.H.; Sossong, A.M.; Ward, T.M.; Younts, C.M.; Lewis, K.; Allard, J.S.; et al. Resveratrol improves adipose insulin signaling and reduces the inflammatory response in adipose tissue of rhesus monkeys on high-fat, high-sugar diet. *Cell Metab.* **2013**, *18*, 533–545. [CrossRef]

106. Kim, S.; Jin, Y.; Choi, Y.; Park, T. Resveratrol exerts anti-obesity effects via mechanisms involving down-regulation of adipogenic and inflammatory processes in mice. *Biochem. Pharmacol.* **2011**, *81*, 1343–1351. [CrossRef]

107. Mattison, J.A.; Wang, M.; Bernier, M.; Zhang, J.; Park, S.-S.; Maudsley, S.; An, S.S.; Santhanam, L.; Martin, B.; Faulkner, S.; et al. Resveratrol prevents high fat/sucrose diet-induced central arterial wall inflammation and stiffening in nonhuman primates. *Cell Metab.* **2014**, *20*, 183–190. [CrossRef]

108. Pan, Q.-R.; Ren, Y.-L.; Liu, W.-X.; Hu, Y.-J.; Zheng, J.-S.; Xu, Y.; Wang, G. Resveratrol prevents hepatic steatosis and endoplasmic reticulum stress and regulates the expression of genes involved in lipid metabolism, insulin resistance, and inflammation in rats. *Nutr. Res.* **2015**, *35*, 576–584. [CrossRef]

109. Andrade, J.M.O.; Paraíso, A.F.; De Oliveira, M.V.M.; Martins, A.M.E.; Neto, J.F.; Guimarães, A.L.S.; De Paula, A.M.; Qureshi, M.; Santos, S.H.S. Resveratrol attenuates hepatic steatosis in high-fat fed mice by decreasing lipogenesis and inflammation. *Nutriens* **2014**, *30*, 915–919. [CrossRef]

110. Bujanda, L.; Hijona, E.; Larzabal, M.; Beraza, M.; Aldazabal, P.; García-Urkia, N.; Sarasqueta, C.; Cosme, A.; Irastorza, B.; Gonzalez, A.; et al. Resveratrol inhibits nonalcoholic fatty liver disease in rats. *BMC Gastroenterol.* **2008**, *8*, 40. [CrossRef]

111. Dolinsky, V.W.; Dyck, J.R. Calorie restriction and resveratrol in cardiovascular health and disease. *Biochim. Biophys. Acta (BBA) Mol. Basis Dis.* **2011**, *1812*, 1477–1489. [CrossRef] [PubMed]

112. Mozafari, M.; Nekooeian, A.A.; Mashghoolozekr, E.; Panjehshahin, M.R. The cardioprotective effects of resveratrol in rats with simultaneous type 2 diabetes and renal hypertension. *Nat. Prod. Commun.* **2015**, *10*, 335–338. [CrossRef] [PubMed]

113. Zordoky, B.N.; Robertson, I.M.; Dyck, J.R. Preclinical and clinical evidence for the role of resveratrol in the treatment of cardiovascular diseases. *Biochim. Biophys. Acta (BBA) Mol. Basis Dis.* **2015**, *1852*, 1155–1177. [CrossRef] [PubMed]

114. Dolinsky, V.W.; Chakrabarti, S.; Pereira, T.J.; Oka, T.; Levasseur, J.; Beker, D.; Zordoky, B.N.; Morton, J.S.; Nagendran, J.; Lopaschuk, G.D.; et al. Resveratrol prevents hypertension and cardiac hypertrophy in hypertensive rats and mice. *Biochim. Biophys. Acta (BBA) Mol. Basis Dis.* **2013**, *1832*, 1723–1733. [CrossRef] [PubMed]

115. Franco, J.G.; Lisboa, P.C.; Lima, N.; Amaral, T.A.; Peixoto-Silva, N.; Resende, A.C.; Oliveira, E.; Passos, M.C.; Moura, E. Resveratrol attenuates oxidative stress and prevents steatosis and hypertension in obese rats programmed by early weaning. *J. Nutr. Biochem.* **2013**, *24*, 960–966. [CrossRef] [PubMed]

116. Movahed, A.; Raj, P.; Nabipour, I.; Mahmoodi, M.; Ostovar, A.; Kalantarhormozi, M.; Netticadan, T. Efficacy and Safety of Resveratrol in Type 1 Diabetes Patients: A Two-Month Preliminary Exploratory Trial. *Nutriens* **2020**, *12*, 161. [CrossRef]

117. Bhatt, J.K.; Thomas, S.; Nanjan, M.J. Resveratrol supplementation improves glycemic control in type 2 diabetes mellitus. *Nutr. Res.* **2012**, *32*, 537–541. [CrossRef]

118. Brasnyó, P.; Molnár, G.A.; Mohás, M.; Markó, L.; Laczy, B.; Cseh, J.; Mikolás, E.; Szijártó, I.A.; Mérei, Á.; Halmai, R.; et al. Resveratrol improves insulin sensitivity, reduces oxidative stress and activates the Akt pathway in type 2 diabetic patients. *Br. J. Nutr.* **2011**, *106*, 383–389. [CrossRef]

119. Zhu, X.; Wu, C.; Qiu, S.; Yuan, X.; Li, L. Effects of resveratrol on glucose control and insulin sensitivity in subjects with type 2 diabetes: Systematic review and meta-analysis. *Nutr. Metab.* **2017**, *14*, 60. [CrossRef]

120. Guo, X.-F.; Li, J.-M.; Tang, J.; Li, D. Effects of resveratrol supplementation on risk factors of non-communicable diseases: A meta-analysis of randomized controlled trials. *Crit. Rev. Food Sci. Nutr.* **2017**, *58*, 3016–3029. [CrossRef]

121. Timmers, S.; De Ligt, M.; Phielix, E.; Van De Weijer, T.; Hansen, J.; Moonen-Kornips, E.; Schaart, G.; Kunz, I.; Hesselink, M.K.; Schrauwen-Hinderling, V.B.; et al. Resveratrol as Add-on Therapy in Subjects With Well-Controlled Type 2 Diabetes: A Randomized Controlled Trial. *Diabetes Care* **2016**, *39*, 2211–2217. [CrossRef] [PubMed]

122. Berman, A.Y.; Motechin, R.A.; Wiesenfeld, M.Y.; Holz, M.K. The therapeutic potential of resveratrol: A review of clinical trials. *NPJ Precis. Oncol.* **2017**, *1*, 35. [CrossRef] [PubMed]

123. Zhang, C.; Yuan, W.; Fang, J.; Wang, W.; He, P.; Lei, J.; Wang, C. Efficacy of Resveratrol Supplementation against Non-Alcoholic Fatty Liver Disease: A Meta-Analysis of Placebo-Controlled Clinical Trials. *PLoS ONE* **2016**, *11*, e0161792. [CrossRef] [PubMed]

124. Cheang, W.S.; Wong, W.T.; Wang, L.; Cheng, C.K.; Lau, C.W.; Ma, R.C.; Xu, A.; Wang, N.; Huang, Y.; Tian, X.Y. Resveratrol ameliorates endothelial dysfunction in diabetic and obese mice through sirtuin 1 and peroxisome proliferator-activated receptor δ. *Pharmacol. Res.* **2019**, *139*, 384–394. [CrossRef] [PubMed]

125. Milne, J.C.; Lambert, P.D.; Schenk, S.; Carney, D.P.; Smith, J.J.; Gagne, D.J.; Jin, L.; Boss, O.; Perni, R.B.; Vu, C.B.; et al. Small molecule activators of SIRT1 as therapeutics for the treatment of type 2 diabetes. *Nature* **2007**, *450*, 712–716. [CrossRef]

126. Feige, J.N.; Lagouge, M.; Canto, C.; Strehle, A.; Houten, S.M.; Milne, J.C.; Lambert, P.D.; Mataki, C.; Elliott, P.J.; Auwerx, J. Specific SIRT1 Activation Mimics Low Energy Levels and Protects against Diet-Induced Metabolic Disorders by Enhancing Fat Oxidation. *Cell Metab.* **2008**, *8*, 347–358. [CrossRef]

127. Minor, R.K.; Baur, J.A.; Gomes, A.P.; Ward, T.M.; Csiszar, A.; Mercken, E.M.; Abdelmohsen, K.; Shin, Y.-K.; Canto, C.; Scheibye-Knudsen, M.; et al. SRT1720 improves survival and healthspan of obese mice. *Sci. Rep.* **2011**, *1*, 70. [CrossRef]

128. Caton, P.W.; Nayuni, N.K.; Kieswich, J.; Khan, N.Q.; Yaqoob, M.M.; Corder, R. Metformin suppresses hepatic gluconeogenesis through induction of SIRT1 and GCN5. *J. Endocrinol.* **2010**, *205*, 97–106. [CrossRef]

129. Cuyàs, E.; Verdura, S.; Llorach-Pares, L.; Fernández-Arroyo, S.; Joven, J.; Martin-Castillo, B.; Bosch-Barrera, J.; Brunet, J.; Nonell-Canals, A.; Sanchez-Martinez, M.; et al. Metformin Is a Direct SIRT1-Activating Compound: Computational Modeling and Experimental Validation. *Front. Endocrinol.* **2018**, *9*. [CrossRef]

130. Banerjee, J.; Bruckbauer, A.; Zemel, M. Activation of the AMPK/Sirt1 pathway by a leucine–metformin combination increases insulin sensitivity in skeletal muscle, and stimulates glucose and lipid metabolism and increases life span in Caenorhabditis elegans. *Metabolisam* **2016**, *65*, 1679–1691. [CrossRef]

131. Scisciola, L.; Sarno, F.; Carafa, V.; Cosconati, S.; Di Maro, S.; Ciuffreda, L.; De Angelis, A.; Stiuso, P.; Feoli, A.; Sbardella, G.; et al. Two novel SIRT1 activators, SCIC2 and SCIC2.1, enhance SIRT1-mediated effects in stress response and senescence. *Epigenetics* **2020**, *15*, 664–683. [CrossRef] [PubMed]

132. Hoffmann, E.; Wald, J.; Lavu, S.; Roberts, J.; Beaumont, C.; Haddad, J.; Elliott, P.; Westphal, C.; Jacobson, E. Pharmacokinetics and tolerability of SRT2104, a first-in-class small molecule activator of SIRT1, after single and repeated oral administration in man. *Br. J. Clin. Pharmacol.* **2012**, *75*, 186–196. [CrossRef] [PubMed]

133. Libri, V.; Brown, A.P.; Gambarota, G.; Haddad, J.; Shields, G.S.; Dawes, H.; Pinato, D.J.; Hoffman, E.; Elliot, P.J.; Vlasuk, G.P.; et al. A Pilot Randomized, Placebo Controlled, Double Blind Phase I Trial of the Novel SIRT1 Activator SRT2104 in Elderly Volunteers. *PLoS ONE* **2012**, *7*, e51395. [CrossRef] [PubMed]

134. Noh, R.M.; Venkatasubramanian, S.; Daga, S.; Langrish, J.; Mills, N.L.; Lang, N.N.; Hoffmann, E.; Waterhouse, B.; Newby, D.E.; Frier, B.M. Cardiometabolic effects of a novel SIRT1 activator, SRT2104, in people with type 2 diabetes mellitus. *Open Heart* **2017**, *4*.

135. Venkatasubramanian, S.; Noh, R.M.; Daga, S.; Langrish, J.P.; Joshi, N.V.; Mills, N.L.; Hoffmann, E.; Jacobson, E.W.; Vlasuk, G.P.; Waterhouse, B.R.; et al. Cardiovascular Effects of a Novel SIRT1 Activator, SRT2104, in Otherwise Healthy Cigarette Smokers. *J. Am. Heart Assoc.* **2013**, *2*, e000042. [CrossRef]

Review

Alzheimer's Disease, a Lipid Story: Involvement of Peroxisome Proliferator-Activated Receptor α

Francisco Sáez-Orellana [1,2,†] , **Jean-Noël Octave** [1,2] and **Nathalie Pierrot** [1,2,*,†]

1   Université Catholique de Louvain, Alzheimer Dementia, Avenue Mounier 53, SSS/IONS/CEMO-Bte B1.53.03, B-1200 Brussels, Belgium; francisco.saez@uclouvain.be (F.S.-O.); jean-noel.octave@uclouvain.be (J.-N.O.)
2   Institute of Neuroscience, Alzheimer Dementia, Avenue Mounier 53, SSS/IONS/CEMO-Bte B1.53.03, B-1200 Brussels, Belgium
*   Correspondence: nathalie.pierrot@uclouvain.be; Tel.: +32-2-764-93-34
†   These authors contributed equally to this work.

Received: 17 April 2020; Accepted: 12 May 2020; Published: 14 May 2020

Abstract: Alzheimer's disease (AD) is the leading cause of dementia in the elderly. Mutations in genes encoding proteins involved in amyloid-β peptide (Aβ) production are responsible for inherited AD cases. The amyloid cascade hypothesis was proposed to explain the pathogeny. Despite the fact that Aβ is considered as the main culprit of the pathology, most clinical trials focusing on Aβ failed and suggested that earlier interventions are needed to influence the course of AD. Therefore, identifying risk factors that predispose to AD is crucial. Among them, the epsilon 4 allele of the *apolipoprotein E* gene that encodes the major brain lipid carrier and metabolic disorders such as obesity and type 2 diabetes were identified as AD risk factors, suggesting that abnormal lipid metabolism could influence the progression of the disease. Among lipids, fatty acids (FAs) play a fundamental role in proper brain function, including memory. Peroxisome proliferator-activated receptor α (PPARα) is a master metabolic regulator that regulates the catabolism of FA. Several studies report an essential role of PPARα in neuronal function governing synaptic plasticity and cognition. In this review, we explore the implication of lipid metabolism in AD, with a special focus on PPARα and its potential role in AD therapy.

Keywords: Alzheimer's; risk factors; PPARs; PPARα; lipids; fatty acids; modulators; cognition; sex; therapy

1. Alzheimer Disease: A Dementing Illness

With a prevalence doubling every 5 years beyond 65, Alzheimer's disease (AD) is a devastating neurodegenerative disorder, which is the most common cause of dementia in the elderly. In 2019, Alzheimer's Disease International estimated that there are over 50 million people living with dementia globally, a figure set to increase to 152 million by 2050 [1].

Memory loss is one of the main clinical features of AD onset, and numerous neuropsychological tests allow for the assessment of the cognitive functions of Alzheimer's patients [2,3]. Along with a decline in cognitive performance, AD is characterized by the coexistence in the brain of two main neuropathological lesions: intraneuronal neurofibrillary tangles composed of hyperphosphorylated microtubule-associated protein tau and extracellular senile plaques containing the amyloid-β (Aβ) peptide generated from the sequential proteolytic processing of its precursor, the amyloid precursor protein (APP). Although the definitive diagnosis of the disease was previously achieved by the postmortem neuropathological brain examination, the detection of specific AD biomarkers in the cerebrospinal fluid, including Aβ and tau, constitutes an early examination and a reliable diagnosis [4]. Moreover, recent non-invasive imaging techniques using Aβ- and tau-PET tracers have led to the preclinical diagnosis of AD, allowing its evolution during the patient's lifetime to be tracked [5–7]. As positron emission tomography (PET) imaging studies have shown that Aβ accumulation occurs long before the onset of clinical AD and given that mutations in the *APP* gene can act as fully penetrant in rare inherited early-onset AD cases (EOAD, about 1% of the cases, (Figure 1)), the amyloid cascade hypothesis was proposed to explain the pathogeny. According to this hypothesis, a gradual accumulation and aggregation of Aβ initiate a neurodegenerative cascade resulting in neurofibrillary tangles formation, cell loss, vascular damage and dementia [8,9]. Although this hypothesis was strengthened by the discovery that mutations in presenilins (Figure 1), the catalytic subunits of the γ-secretase complex, lead to an increase in Aβ production [10], several studies have challenged the amyloid hypothesis over the past ten years [11,12]. Mounting evidence reports that mutations in the *Presenilin 1* (*PSEN1*) gene associated with EOAD have heterogenic manifestations. It was indeed recently shown that the most common mutations found in *PSEN1* decrease the activity of the γ-secretase [13,14] or lead to a loss of its function [15], indicating that in some cases, *PSEN1* mutations either hyper-activate or reduce the activity of the γ-secretase complex. Moreover, γ-secretase activity assessed in brain samples from EOAD and non-demented controls was similar, while it displayed some dysfunctions in a few brain samples from late-onset AD cases (LOAD), which represent the vast majority of AD cases [14]. This suggests that γ-secretase may also play a role in some LOAD cases, in which brain Aβ production levels are similar to those observed in unaffected controls [16].

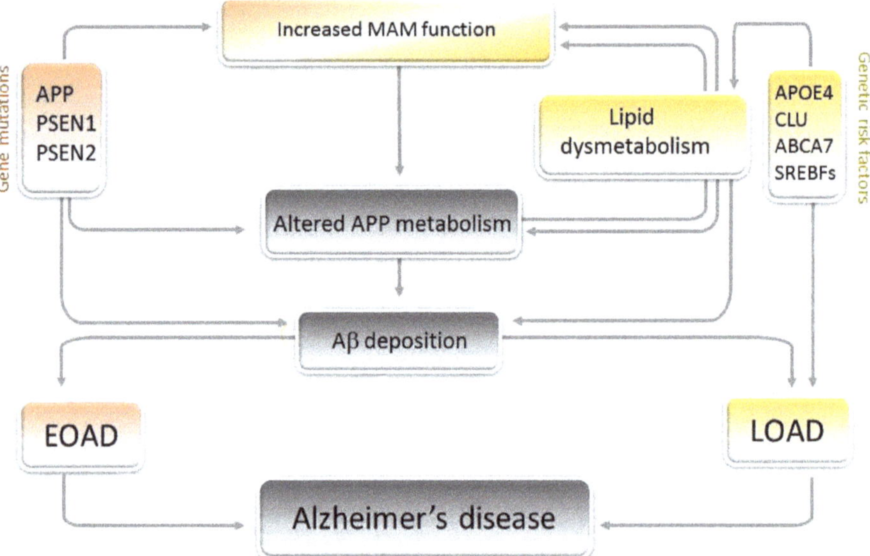

Figure 1. Gene mutations and genetic risk factors linked to lipid dysmetabolism and the progression of Alzheimer's disease (AD). Gene mutations responsible for inherited early-onset AD cases (EOAD, gene mutations) and genetic risk factors for late-onset AD cases (LOAD, genetic risk factors) lead to altered amyloid precursor protein (APP) processing and brain amyloid-β (Aβ) deposition. Disruption of lipid homeostasis induces abnormal lipid composition in rafts and increased mitochondria-associated endoplasmic reticulum membrane (MAM) function in which targeted APP is proteolytically processed into Aβ by presenilins (PSEN). Conversely, cleavage of APP directly affects cellular lipid composition by altering the synthesis of several lipids that are enriched in lipid rafts. Abbreviations: APOE4 (Apolipoprotein E4); CLU (Clusterin); ABCA7 (ATP-binding cassette sub-family A member 7); SREBFs (Sterol regulatory element-binding genes).

Moreover, several studies have shown that humans with Down syndrome, who harbor three copies of the *APP* gene that leads to the overexpression of APP protein, have an age-dependent increased risk for developing AD and develop clinical features and neuropathological changes similar to those observed in AD (for review, see [17]). This aforementioned study suggests that AD could be a combination of different pathologies with diverse etiologies [18] leading to dementia. This is supported by recent findings showing that some pathologies have similar clinical markers and manifestations to those observed in AD, as reported in limbic-predominant, age-related TDP-43 encephalopathy (LATE), in which senile plaques and neurofibrillary tangles that define AD have been brought out in the brain [19].

Since Aβ that builds up in plaques also deposits during normal brain aging, amyloid deposition occurring in the hippocampus and cerebral cortex of AD patients potentially explains deficits in memory and cognitive function observed. Despite the various isoforms of Aβ produced, Aβ toxicity rate is dependent on its state of assembly. Among the three assemblies state of Aβ (monomers, soluble oligomers and insoluble fibrils) (for review, see [20]), soluble Aβ oligomers are organized into different structures ranging from dimers, trimers, tetramers, pentamers, decamers and dodecamers, among others [21–23]. Toxic soluble oligomers have been identified in AD brains [24–26]. However, in some cases, higher aggregates, such as fibrils, showed protective effects in AD models [27], suggesting that there is an inverse correlation between the size of Aβ assembly and its toxicity. Since Aβ dimers form a more stable structure, these dimeric units are described to be the building blocks for toxic aggregates [28]. This supports the idea that disassembling plaques or fibrillar structures could be detrimental if not accompanied by strategies to remove oligomeric aggregates of Aβ. However, it remains difficult to influence the course of AD by removing amyloid deposits. Indeed, some approaches were developed to inhibit secretases activities involved in the release of Aβ or to remove amyloid deposits from the brain using active or passive immunotherapy [29]. While first attempts completely failed in Phase III clinical trials due to the widespread function of secretases and the development of encephalitis in some patients [30], recent attempts have shown that although senile plaques can be effectively removed from the AD brain, cognitive performance is not improved in these patients [31]. Despite the fact that recent results with an immunotherapy clinical trial using aducanumab targeting aggregated forms of Aβ have been encouraging and could prove efficacious [32], this clearly suggests that when structural modifications are found in the brain due to accumulation of abnormal proteins, the proposed treatments arrive too late and are inefficient [33]. Consequently, we must focus on modifications in physiological functions, which could occur long before abnormal protein deposition. Among them, abnormal lipid metabolism could be an important early event in the pathogenesis of AD.

2. Linking Lipids to Alzheimer's Disease

While in EOAD, Aβ accumulation in the brain is caused by gene mutations, the vast majority of AD cases are late-onset AD cases (LOAD, 99% of cases), in which the source of Aβ accumulation in the brain is still unknown. Nevertheless, it is well established that the epsilon 4 allele of the *Apolipoprotein E (APOE)* gene, encoding the main lipid carrier in the brain, is a genetic risk factor for AD. People who are homozygous for this allele have ten times greater risk to develop AD [34]. Therefore, a relationship between AD and lipid metabolism has been established. Furthermore, large-scale genome-wide association studies on AD first confirmed that APOE4 is a major risk factor and provided evidence that at least 20 genetic susceptibility loci in addition to *APOE* genotype are associated with AD [35,36] (Figure 2). Among them are genes encoding Clusterin and ABCA7, two proteins involved in lipid metabolism [37–40] (Figures 1 and 2). Therefore, the identification of these susceptibility loci supports the hypothesis that perturbation of lipid metabolism favors the progression of AD [41]. This hypothesis is sustained by recent reports showing that genetic polymorphisms in *SREBF* genes encoding sterol regulatory element-binding proteins (SREBPs), transcription factors activating lipid metabolism-related genes involved in cholesterol and fatty acids biosynthesis [42,43], were associated with an increased risk of schizophrenia and LOAD [44–46]. Disturbances in the signaling and expression of SREBPs were indeed reported in LOAD cases and, in a rare case of EOAD, harboring a microduplication in the locus of *APP* gene [47,48].

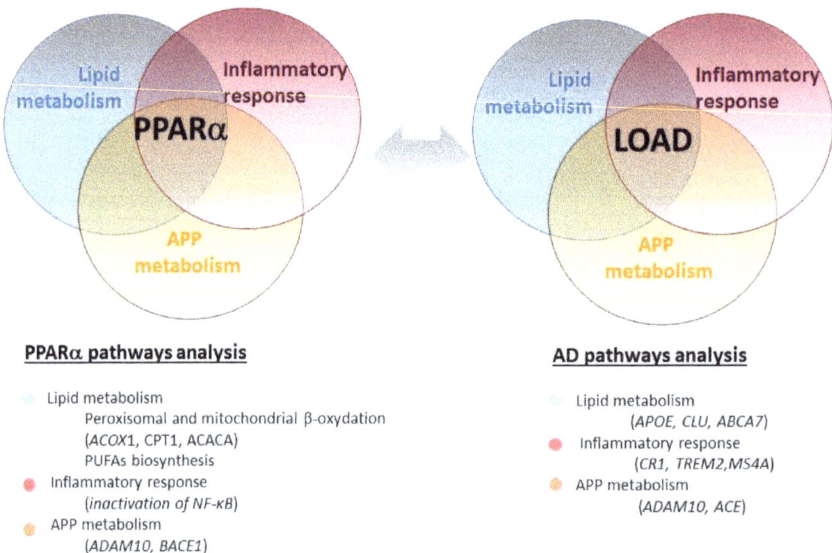

PPARα pathways analysis

Lipid metabolism
    Peroxisomal and mitochondrial β-oxydation
    (*ACOX1*, *CPT1*, *ACACA*)
    PUFAs biosynthesis
Inflammatory response
    (*inactivation of NF-κB*)
APP metabolism
    (*ADAM10, BACE1*)

AD pathways analysis

Lipid metabolism
    (*APOE, CLU, ABCA7*)
Inflammatory response
    (*CR1, TREM2, MS4A*)
APP metabolism
    (*ADAM10, ACE*)

Figure 2. Common pathways regulated by proliferator-activated receptor α (PPARα) and involved in the etiology of AD. Peroxisome proliferator-activated receptor α (PPARα) is a transcription factor that governs pathways involved in the metabolism of lipids, inflammatory response and the metabolism of the amyloid precursor protein (APP), which have been implicated in the etiology of late-onset Alzheimer's disease (LOAD). Genome-wide association studies identified several genetic risk factors for LOAD, which are involved in pathways that are governed by PPARα. Abbreviations: ABCA7 (ATP-binding cassette subfamily A member 7), ACE (Angiotensin-converting enzyme gene), APOE (Apolipoprotein E), APP (Amyloid precursor protein), ACACA (Acetyl-CoA carboxylase), ACOX1 (Acyl-CoA oxidase), ADAM10 (ADAM Metallopeptidase Domain 10), BACE1 (β-Site amyloid precursor protein-cleaving enzyme 1), CLU (Clusterin), CPT1 (Carnitine palmitoyl transferase), CR1 (Complement receptor 1), NF-κB (Nuclear factor κB), MS4A (Membrane-spanning 4A), PUFAs (Polyunsaturated fatty acids), TREM2 (Triggering receptor expressed on myeloid cells 2).

Additional support comes from metabolomic studies that have shown changes in the lipid content of plasma, cerebrospinal fluid and in brain tissue from AD [49–52]. Moreover, perturbations in brain fatty acids profiles observed in brain regions vulnerable to AD pathology [49,53] could influence AD pathogenesis by promoting Aβ accumulation and tau pathology [54–56].

3. Lipids in Alzheimer Disease: Involvement of Fatty Acids in Cognitive Function

The human brain contains the second-highest concentration of lipids (50–60% of its dry weight) after adipose tissue [57]. Due to their structural diversity and their involvement in a wide range of biological processes, lipids play a fundamental role in maintaining brain physiological functions. Phospholipids, sterols, sphingolipids, fatty acids and triacylglycerols are the five main brain lipid classes, which are involved in neuronal differentiation, synaptogenesis, and brain development (for reviews, see [58,59]). At the subcellular level, lipids are basic structural components of cell membranes and are enriched in the myelin sheath surrounding nerve cell axons, which regulates the ability of a neuron to trigger action potentials encoding information [60]. Moreover, lipids and their derivatives modulate membrane fluidity and permeability, which regulate trafficking, localization and function of ion pumps, channels, receptors and transporters at the plasma membrane [61,62].

In particular, any change in lipid homeostasis affects the lipid composition of membrane lipid rafts, which are cholesterol and sphingolipid-enriched microdomains [63] where most of the synaptic-related proteins involved in synaptic transmission and plasticity are embedded [64]. Since lipid rafts from AD brains displayed important changes in their composition [65,66], disturbance in the function of lipid-rafts-associated synaptic proteins could contribute to the development of neuropathological events that favor amyloidogenesis and proteins aggregation [67,68].

One of the main pathological characteristics involved in the pathogenesis of AD implies the APP protein, which is hydrolyzed by β- and γ-secretases, leading to the deposition of Aβ in the brain. APP is a single-pass transmembrane protein with a large extracellular region that contains several domains involved in APP dimerization, protein–protein or metal interactions (e.g., heparin- and copper-binding domains). The APP trans-membrane-helix domain in which the Aβ sequence is inserted and the APP intracellular domain contain cholesterol-binding [69] and YENPTY motifs that regulate the subcellular location, trafficking, and proteolytic processing of APP, respectively [70] (for more details, see [71,72]). These domains and motifs engage APP and its cleavage products in a plethora of physiological functions ranging from synaptic transmission, plasticity, development, neuroprotection, trophic function, cell adhesion, apoptosis, calcium and lipid homeostasis, among others [73–75].

The cholesterol-binding motif found in APP plays an essential role in the interaction of APP with proteins involved in cholesterol metabolism (e.g., SREBP1) [48] and in its location in lipid rafts present in synaptic vesicles and mitochondria-associated endoplasmic reticulum (ER) membranes (MAMs) [76] (Figure 1). MAMs are enriched in cholesterol and sphingomyelin and are points of physical contact between the outer mitochondrial membrane and ER. While they play an essential role in the metabolism of glucose, phospholipids, cholesterol and calcium homeostasis [77,78], MAMs regulate APP processing. Indeed, presenilins and γ-secretase activity, previously localized at the ER, are enriched in MAMs, in which β-cleaved fragment generating Aβ accumulates [79,80]. Interestingly, the activity/function and expression of MAM-associated proteins increase in human and mouse AD brains long before Aβ deposition [81], suggesting a potential role of MAMs in the pathophysiology of AD [81,82]. From these data, the concept of the MAM hypothesis in AD emerged (reviewed in more details in [83,84]).

While rafts are described as noncaveolar lipid microdomains, caveolae are cholesterol-enriched membrane invaginations found in the Golgi network, exocytotic vesicles, ER and plasma membrane in which surface protein markers caveolin are embedded [85,86]. Caveolae are involved in cellular cholesterol transport and are docking sites for signaling proteins and receptors and are therefore considered as hotspots for cell–cell communication [87]. Although caveolae-dependent cell signaling is not yet fully understood, several studies have reported the involvement of caveolin proteins in the pathogenesis of AD [88,89]. Caveolin expression levels are upregulated in the hippocampus and the frontal cortex of AD brain compared to control, suggesting a link between the expression of caveolin and dysregulation of cholesterol homeostasis observed in AD [89–91]. Moreover, increased expression in caveolin promotes oxidative stress and APP processing into Aβ [89,92] that could favor the progression of AD. Although cholesterol and sphingolipid-enriched membrane microdomains could take part in AD physiopathology, fatty acids seem to also contribute to its occurrence.

Fatty acids (FAs) are the major essential monomeric constituents of all lipids [93] and therefore are key components of cellular membranes [94]. They can be unesterified (free) or esterified to plasma membrane phospholipids and are classified based on the length of their carbon chain. FAs are either saturated, monounsaturated, or polyunsaturated (PUFAs) (for review, see [95]). While the brain can produce saturated and monosaturated FAs by de novo lipogenesis, essential PUFAs cannot be synthesized in sufficient quantities [96] and therefore are provided by the diet [94]. Within the brain, palmitic acid and stearic acid are the main saturated FAs, and oleic acid is the main monounsaturated one. Linoleic and α-linolenic essential FAs are transformed into arachidonic and docosahexaenoic acids, the major brain ω-6 and ω-3 PUFAs, respectively [94,95].

PUFAs play a critical role in neurogenesis, synaptic function, inflammation, glucose homeostasis, mood and cognition [94]. As they play a critical role in brain development and functioning [97], high concentrations of dietary saturated long-chain FAs and a decrease in dietary consumption of ω-3 PUFAs have been associated with neurological dysfunction and neuropsychiatric disorders, including neurodegenerative diseases such as AD [98–100]. In addition, diets in the western population are rich in saturated FAs and low in PUFAs [101], which are not only associated with the development of obesity but also to cognitive dysfunction.

Levels of docosahexaenoic acid (DHA), the major brain ω-3 PUFAs, have been reported to be decreased in plasma and post-mortem brains from AD patients [49–51]. Although DHA dietary supplements did not improve memory, cognition or mood [50,102,103], higher dietary intake of DHA is associated with decreased risk of neurological disorders [104] and dementia in elderly individuals [105]. Interestingly, ω-3 PUFAs supplementation in individuals with mild cognitive impairment and in AD patients without the *APOE4* allele has shown benefits [102,106]. While low brain DHA levels were shown to impair behavior in AD mouse models [107,108], the dietary supplementation of ω-3 PUFAs in rodents facilitated hippocampal synaptic plasticity and improved cognitive deficits of aged mice and in several animal models of AD [50,102,109–112]. Moreover, in non-pathological conditions, maternal intake of ω-3 PUFAs increases hippocampal plasticity and cognition in healthy pups rodents [113].

While cellular and molecular mechanisms underlying such effects are poorly understood, more and more studies put forward the involvement of nuclear receptors.

4. RXRs, LXRs and PPARs Nuclear Receptors in AD

4.1. Nuclear Receptors

The nuclear receptors superfamily of ligand-dependent transcription factors regulates energy balance, inflammation, and lipid and glucose metabolism [114]. They control target genes expression through their binding with sequence-specific elements located in gene promoter regions [114]. Structurally, they contain an amino-terminal activation domain needed for the recruitment of coactivators, a carboxyl-terminal ligand- and a DNA-binding domain. Among these receptors, retinoid X receptors (RXRs), liver X receptors (LXRs) and peroxisome proliferator-activated receptors (PPARs) act as master regulators of lipid metabolism by *trans*-activating genes encoding enzymes involved in lipid and fat metabolism. Therefore, they are abundantly expressed in metabolically active tissues, including the brains of rodents and humans [115]. Due to their anti-inflammatory and potential neuroprotective effects, RXRs, LXRs and PPARs activation with specific agonists emerged as promising approaches for treating brain pathologies in several mouse models of Parkinson's, Huntington and Alzheimer's diseases, multiple and amyotrophic lateral sclerosis, stroke and even in a mouse model with physiological brain aging-dependent cognitive decline (reviewed in [116,117]).

4.2. RXRs

Among the three RXR isotypes identified (RXRα, β and γ), RXRα is mainly expressed in the liver, lungs, muscles, kidneys, epidermis and intestine. While RXRβ is expressed ubiquitously, RXRγ is enriched in the brain, heart and muscles. RXRs can be activated by 9-*cis* retinoic acid, linoleic, linolenic and DHA acids, natural RXR ligands [118,119]. As a strong agonist of the RXRs, the retinoid bexarotene synthetic agonist [120], which the U.S. FDA approved for the treatment of cutaneous T-cell lymphoma [121], was described to improve cognitive deficits in AD mouse models [122–127] mainly by inducing the transcription and lipidation of APOE and reducing microglial expression of pro-inflammatory genes among others [128,129]. Although we previously reported that bexarotene improved cognition in a patient with mild AD [130], its efficacy in clinical trials for treating AD pathology has been disappointing [131,132].

4.3. LXRs

As an obligate binding partner of LXRs, RXRs form permissive heterodimers with two LXRs isoforms, LXRα and β [114]. LXRα is abundantly expressed in the liver, intestine, kidney, spleen and adipose tissue, whereas LXRβ is ubiquitously expressed at a lower level but more widely in the brain and mainly in the hippocampus. LXRs are activated by oxysterols, most prominently hydroxylated forms of cholesterol [133,134]. They play therefore a critical role in the control of whole-body cholesterol homeostasis and exert potent anti-inflammatory actions [135]. Once activated, they control the transcription of target genes involved in lipid transport and biosynthesis, such as APOE and SREBP, respectively [136,137]. The expression of the SREBP1 isoform is mediated by LXRs to ensure FAs synthesis needed for the esterification of free cholesterol for protecting cells from a detrimental cholesterol overload. Moreover, unesterified PUFAs exert feedback inhibition on the expression of SREBP1 by antagonizing the oxysterol LXR receptor [138,139].

4.4. PPARs

PPARs were first described for their ability to induce peroxisomal proliferation in the liver in response to xenobiotics [140]. Afterward, they were considered as master metabolic regulators involved in energy homeostasis [141]. They act principally as lipid sensors and regulate whole-body metabolism in response to dietary lipid intake and control their subsequent metabolism and storage by inducing or repressing the expression of genes involved in the metabolism of lipid and glucose [142]. The three PPARs isoforms identified (PPARα, β/δ and γ) have partially overlapping functions and tissue distribution in mammals. PPARα is highly expressed in the liver, heart and kidney but has low levels in the brain [143,144] and more particularly in the hippocampus of rodents and primates [144–149]. PPARα plays an important role in the regulation of FAs catabolism [150] by controlling the expression of genes encoding acyl-CoA oxidase, carnitine palmitoyl transferase and acetyl-CoA carboxylase, enzymes that tightly regulate FAs peroxisomal and mitochondrial β-oxidation, respectively (for reviews, see [151,152]) (Figure 1). Consistent with the first central role of PPARα in FAs catabolism [150], PPARα null mice exhibit greater lipid accumulation [153].

While PPARγ isoform is mainly expressed in white and brown adipose tissue, the large intestine and spleen, in which it regulates adipogenesis, energy balance, lipid biosynthesis and inflammation [154], PPARβ/δ is expressed ubiquitously in all tissues and is the most abundant isoform found in liver, kidney, adipose tissue and skeletal muscle, where it plays mainly a role in FAs oxidation [155].

Although PPARs expression is ubiquitous in the human and mouse brain, PPARα and γ are expressed in both neurons and astrocytes, while PPARβ/δ isoform is exclusively neuronal [115].

Although PPARs were first classified as orphan receptors, many natural and synthetic agonists of PPARs are used in the treatment of glucose and lipid disorders. Several endogenous ligands from dietary lipids and their metabolites were identified, among them the essential FA DHA and eicosanoids [156,157]. Recently, hexadecanamide, 9-octadecanamide and 3-hydroxy-(2,2)-dimethyl butyrate have been identified as endogenous PPARα ligands in mouse brain hippocampus [158]. Moreover, several synthetic ligands are widely used in clinical practices, among them fibrates and thiazolidinediones, PPARα and γ agonists, used in the treatment of hypertriglyceridemia and diabetes mellitus, respectively (for review, see [159]). In addition, PPARs ligands [160] decrease Aβ burden, tau phosphorylation and inflammation and improve behavior in AD mouse models [116,161].

Due to their overlapping expression in all brain cell types from mouse and human and given that they share similarities in their ligand-binding domains [162,163], a tight interconnection between PPARs isoforms was described a couple of years ago [164,165]. The mutual interactions observed between PPARα, β/δ and γ lead to the concept of a "PPAR triad" in the brain (reviewed in [166]). This concept emerged from data reporting that the activation of a PPAR isoform affects the expression of other PPARs due to the low isotype specificity of endogenous PPAR ligands [166,167]. Indeed, the simultaneous activation of different PPARs isoforms was first shown in C6 glioma and lipopolysaccharide (LPS)-stimulated astrocytes, in which the activation of PPARβ/δ increases the expression of PPARγ and to some extent that of PPARα in a positive feedback loop [164,165]. Moreover, such crosstalk between PPARs was also reported in primary cortical neurons and in ischemic rat brain, where PPARγ activation stimulates the interleukin-1 receptor antagonist production through the activation of the PPARβ/δ [168]. Conversely, PPARα agonist reduces the expression of PPARβ/δ in LPS-stimulated primary cultures of astrocytes in a negative feedback loop, leading to the downregulation of the cyclooxygenase (COX)-2 enzyme involved in the synthesis of the endogenous PPAR agonist prostaglandin [164,166]. While the activation of PPARα represses the expression of COX-2, PPARβ/δ activation upregulates the expression of both COX-2 and cytosolic phospholipase A2, producing PUFAs [164,166] (Figure 2). Therefore, the cross-talk between PPAR isoforms highlights a "PPAR network", in which the activation of each PPAR participates in the fine-tuning of genes expression. These interconnections between PPARs should be considered to design appropriate therapeutic strategies for neurodegenerative disorders, including AD.

5. PPARs in Alzheimer's Disease Therapy: The Promising Role of PPARα

5.1. PPARγ and PPARβ/δ in AD

Considering that AD and metabolic diseases such as obesity and type 2 diabetes have overlapping metabolic dysfunctions (e.g., dyslipidemia, glucose metabolism impairment and insulin resistance) and given that PPARs metabolic regulators are expressed in the brain, it is not surprising that changes in PPARs signaling might lead to dementia [169–172].

Since PPARγ and PPARβ/δ regulate both lipid and carbohydrate metabolism and insulin sensitivity, these receptors represent an attractive therapeutic target for AD. The reduction in glucose metabolic rates observed in the AD brain occurs decades before onset of clinical symptoms and supports the idea that metabolic deficits are upstream events, which may influence the course of AD [173]. As a defining feature of AD, brain glucose hypometabolism leads to a decrease in the O-GlcNAcylation (O-GlcNAc) of proteins, including both tau and APP. While an increase in brain O-GlcNAc protects against tau and Aβ peptide toxicity, a decrease in O-GlcNAc promotes neurodegeneration [174].

Moreover, brain insulin resistance promotes AD pathophysiology by disrupting energy homeostasis and insulin signaling pathways [175,176]. Impairment in insulin signaling favors Aβ-mediated oxidative stress, Aβ secretion, brain amyloid deposition and tau pathology (reviewed in [177–179]). Therefore, targeting PPARβ/δ and PPARγ with specific drugs represents an effective strategy to preserve carbohydrate metabolism, insulin-sensitizing pathways and cognitive performance. By far, PPARγ was first considered as a promising target for the treatment of AD. While the thiazolidinedione class of PPARγ agonists has shown improvement in cognitive behavior in murine models of AD [161,180,181], human clinical trials using PPARγ agonists are less encouraging [182,183]. Although the chronic treatment of diabetic patients with the PPARγ agonist pioglitazone reduces dementia risk by 47% [184], Takeda and U.S. partner Zinfandel Pharmaceuticals decided to give up and stop testing a 20-year-old diabetes medicine that fails once more in AD therapy, a lack of success attributed to the low penetrance of glitazones in the brain.

In contrast to PPARγ, PPARβ/δ is highly expressed throughout the brain and therefore represents a new therapeutic target of interest in AD [185]. Indeed, treatments using PPARβ/δ agonists have been reported to decrease brain neuroinflammation, neurodegeneration, amyloid burden and improve cognitive function in several AD mouse and rat models [186–189]. Recently, a Phase IIa clinical study of the dual PPARδ and γ agonist T3D-959 reports plasma metabolome profile changes on lipid, glucose and insulin-related metabolism and improvements of cognitive function (presumably associated with *APOE* genotype) in a small cohort of patients with mild to moderate AD [190].

5.2. PPARα in AD

Although the function of PPARα in the brain remained elusive for a long time, more and more studies indicate that PPARα is involved in physiological and pathological brain functions (e.g., in the sleep-wake cycle [191,192], depression [193–196], epilepsy [197–199], stroke [200–203] and schizophrenia [204]). PPARα modulators (e.g., oleoylethanolamide, a natural PPARα ligand; Wy14643 and fibrates, two synthetic PPARα agonists) regulate dopamine and hippocampal brain-derived neurotrophic factor (BDNF) signaling pathways to rescue depression-related behaviors [193,195,196] and nicotinic acetylcholine receptors and endocannabinoid signaling to alleviate epilepsy and schizophrenia-like effects in mice [197,204].

5.2.1. PPARα Function in Brain Energy Supply

In addition to their anti-inflammatory and potential neuroprotective effects [117,172,205–207], PPARs, in particular PPARα, are master metabolic regulators of energy homeostasis [141]. Several studies report that PPARα plays an essential role in maintaining brain energy supply. Ketone bodies, which are derived from FAs oxidation, are mainly produced in the liver during prolonged fasting or starvation and represent a significant alternative source of fuel to compensate for a lack of glucose in the brain [208–210]. The ketogenic diet has been therefore used in the treatment of several neurological diseases, including Parkinson's and Alzheimer's diseases, traumatic brain injury and epilepsy (reviewed in [211]). More and more evidence indicates that the ketogenic diet shows benefits in both in vitro and in vivo AD models. Treatment with the ketone body d-β-hydroxybutyrate protects hippocampal neurons from Aβ toxicity [212] and ketogenic diet decreases brain amyloid pathology in a mouse model of AD [213]. Moreover, the oral administration of the ketogenic compound AC-1202 reduces oxidative stress and inflammation and improves cognitive function in mild to moderate AD patients [214].

5.2.2. PPARα and Cognitive Function

More recently, an essential role of PPARα in cognition has emerged. By using a passive-avoidance task, Mazzola et al. reported that memory acquisition is enhanced in rats treated with the PPARα agonist Wy14643 [215]. Moreover, treatment of mice with the Wy14643 attenuates cognitive impairments induced by scopolamine, a muscarinic acetylcholine receptor antagonist [216]. Consistent with the potential role of PPARα in cognition and memory, PPARα-deficient mice have spatial learning and long-term memory deficits [149], indicating that PPARα is required for normal cognitive function [217]. Roy et al. have shown that PPARα, and not PPARγ and PPARβ/δ isoforms, regulates the expression of a set of synaptic-related proteins involved in excitatory neurotransmission, including BDNF, GluN2A and GluN2B subunits containing N-methyl-D-aspartate receptors (NMDARs) and GluA1 subunit containing alpha-amino-3-hydroxy-5-methyl-4-isoxazolepropionic acid receptors (AMPARs) [149]. In agreement with this, we have recently reported that the absence of PPARα severely impairs hippocampal long-term potentiation (LTP), which is defined as an activity-dependent enhancement of synaptic strength involved in memory processing [218], and GluA1 expression in male mice [125].

Moreover, Roy et al. identified a PPAR-responsive element in the promoter of genes encoding the cAMP response element-binding (CREB) protein and therefore identified it as a PPARα target [149]. Interestingly, recent data indicate that RXR activation induces neuronal CREB signaling and increases dendritic complexity and branching of neurons promoting their differentiation and development [219, 220]. In addition, activation of RXRs upregulates the expression of synaptic markers and improves cognition in a mouse model of AD [124]. Altogether, these data indicate that effects mediated by RXR activation on the expression of synaptic-related proteins and cognition could be PPARα-mediated.

It is also interesting to note that Chikahisa et al. recently reported that PPARα-null mice exhibit an enhancement of fear learning [221]. This enhancement results from an increase in levels of dopamine and its metabolites in the amygdala [221], suggesting that PPARα is likewise involved in the regulation of emotional memory via the dopamine pathway in the amygdala.

5.2.3. Potential Link between PPARα and AD

The relevance of a potential beneficial effect of PPARα for dementia is supported by some studies showing that polymorphisms in *PPARA* gene encoding PPARα were associated with an increased risk of AD. In 2003, Brune et al. were the first to report an association of the *PPARA* L162V polymorphism with the AD risk [222]. They indicate that this risk is even higher in carriers harboring a polymorphism in *INS* gene encoding insulin [222]. The interaction of *INS* and *PPARA* genes in AD was thereafter investigated by Kölsch and colleagues [223]. In their study, they report an interaction on AD risk between *PPARA* L162V and *INS* in Northern Europeans, in whom Aβ42 and pro-inflammatory cytokines levels were increased in the cerebrospinal fluid (CSF) [223,224]. However, Sjölander et al. later reported a lack of replication of these studies [225]. They did not find significant differences in genotype or allele distributions between AD patients and controls and found no influence of *PPARA* variants on CSF markers [225]. Although these conflicting results question the promising role of PPARα in AD therapy, previous results indicate that expression levels of PPARα and β/δ are significantly reduced, whereas PPARγ expression is selectively increased in AD brains [226], indicating that PPARs function is impaired in AD and therefore may contribute to the progression of the disease.

5.2.4. PPARα Ligands and AD

More and more studies report the beneficial effects of several PPARα synthetic agonists on cognitive behavior in several AD mouse models. Among them, fibrates (e.g., fenofibrate, bezafibrate, ciprofibrate and gemfibrozil) are a class of lipid-lowering drugs used in the treatment of metabolic syndromes, including hypertriglyceridemia, obesity and type 2 diabetes, which prevents the progression of atherosclerotic lesions, cardiovascular events and non-alcoholic fatty liver disease (reviewed in [227,228]). Among fibrates, fenofibrate has been widely used, but its relatively low efficiency as PPARα agonist [229,230] leads to the development of pemafibrate, a more potent and selective agonist for PPARα [231–233]. Recent results from two Japanese Phase III clinical studies indicate that pemafibrate improves lipid profiles in patients with type 2 diabetes and hypertriglyceridemia [234] and was useful for dyslipidemia, with a much higher efficacy than fenofibrate [235].

The salutary effects of fibrates on memory have been reported in several preclinical AD models. It was recently demonstrated that administration of the PPARα activator gemfibrozil decreases amyloid plaque burden, microgliosis and astrogliosis in the hippocampus and cortex of 5XFAD mice [236], a well-characterized transgenic mouse model of AD, in which age-dependent synaptic and cognitive deficits occur [237]. Although a decrease in the expression of PPARα was observed in the brain of 5XFAD mice [125,236], oral administration of gemfibrozil or pemafibrate improves spatial learning, memory and hippocampal LTP, respectively, in these mice [125,236].

More recently, Luo et al. reported that amyloid pathology, memory deficits and anxiety were reversed in the APP-PSEN1ΔE9 mouse model of AD treated with either gemfibrozil or Wy14643 [238]. The effects observed were mediated by a PPARα-dependent enhancement of autophagosome biogenesis [238].

In addition, the activation of PPARα with non-conventional ligands such as statins or aspirin, cholesterol-lowering and nonsteroidal anti-inflammatory drugs, improves [239] hippocampal plasticity and memory in 5XFAD but not in 5XFAD/*Ppara*-null mice by mediating the transcriptional activation of BDNF and CREB, respectively [239,240]. Moreover, oral administration of low-dose aspirin decreased amyloid plaque pathology in 5xFAD mice by stimulating PPARα-mediated lysosomal biogenesis [241].

Overall, these results support the potential for using PPARα ligands as a promising strategy for the treatment of AD.

Among PPARα ligands, gemfibrozil has recently been assessed as a possible treatment for AD. Although gemfibrozil has first been tested in a Phase I clinical trial (NCT00966966 [242]) in healthy volunteers to evaluate its safety and absorption (unpublished data), a second early Phase I trial (NCT02045056 [243]) is testing its efficiency to prevent AD by evaluating its ability to increase microRNA107 (mir-107) levels in participants with either no or mild cognitive impairment. It was previously shown that expression of mir-107, a small noncoding RNA involved in the regulation of gene expression [244], is reduced in AD and may accelerate disease progression through the regulation of β-Site amyloid precursor protein-cleaving enzyme 1 (BACE1) [245,246], an endopeptidase that cleaves APP to generate Aβ [247,248]. Moreover, gemfibrozil-mediated activation of PPARα has been reported to promote the non-amyloidogenic APP processing [249]. In 2015, Corbett et al. identified PPAR-responsive elements in the promoter of the gene encoding the α-secretase "a disintegrin and metalloproteinase domain-containing protein 10" (ADAM10), a new PPARα target [249] (Figure 2). APP cleavage by ADAM10 precludes Aβ generation and results in the release of a soluble APP α (sAPPα) fragment which exerts neurotrophic and neuroprotective properties involved in the maintenance of dendritic integrity in the hippocampus [250]. Treatment of wild-type mouse hippocampal neurons with gemfibrozil increases sAPPα and decreases Aβ production [249]. Moreover, the production of brain endogenous Aβ is increased in PPARα-deficient mice and exacerbated in 5XFAD/*Ppara*-null when compared to wild-type and 5xFAD respective littermates [249].

Although PPARα is indubitably involved in the non-amyloidogenic processing of APP, PPARγ was previously demonstrated to also regulate Aβ production by controlling the expression of *Bace1* gene. PPARγ activation with specific agonists (e.g., thiazolidinediones and non-steroidal anti-inflammatory drugs including ibuprofen) decreases the expression of BACE1 [251], whereas a lack of PPARγ has an opposite effect in cultured mouse embryonic fibroblasts [252], suggesting that PPARγ is a repressor of *Bace1*.

5.2.5. PPARα and Sex

While sexual dimorphisms of PPARγ agonist rosiglitazone were previously reported on insulin sensitization and glucose in mice [253], most in vivo studies have analyzed the potential effects of PPARα modulators on cognition mainly in male and not in female rodents. We have previously reported a sex-regulated gene dosage effect of PPARα on synaptic plasticity [125]. PPARα activation improves synaptic plasticity only in male but not in female 5XFAD mice [125]. These observations were concomitant with a higher expression of PPARα in brains of males as compared to females [125]. Such differences in PPARα expression between male and female rodents were previously reported in liver [254], lymphocytes [255] and ischemic brain [202]. Moreover, a sexual dimorphism was also observed in hippocampus-dependent behaviors. Numerous studies have previously reported that the magnitude and maintenance of LTP were larger in males than in females, not only at CA3-CA1 synapses but also in the dentate gyrus-CA3 and temproammonic-CA1 synapses of the hippocampus [256–261].

The most obvious difference between males and females is sexual hormones. Hormones are known to influence the expression of PPARα in a sex-specific manner since gonadectomy of male rats decreases PPARα expression [254]. Estrogens, such as estradiol, are known to improve synaptic plasticity [262,263], and behavior is affected in ovariectomized female rats [264–266]. In humans, cognitive impairments in older women have long been attributed to the decrease in circulating estradiol levels after menopause. Exogenous restitution of this hormone during the perimenopausal period ameliorates such impairments [267,268]. In addition, estrogen replacement therapy in women in a specific time window is associated with reduced incidence of AD (reviewed in [269]). Although no differences in PPARα expression were reported between men and women in skeletal muscles [270], several studies indicate that human circulating mononuclear and T cells exhibit sex differences driven by the expression of PPARα and PPARγ [271,272]. In their study, Zhang et al. showed that the treatment of T cells with androgens increases PPARα and decreases PPARγ expression [272].

It is known that women are at a higher risk for AD (two-thirds of those with AD are women). This results partly from differences between men and women in life-expectation and biology (e.g., sex-specific differences in gene expression, hormone levels, brain structure and function and in inflammatory response) (reviewed in [273,274]). Such differences are not exclusively related to AD but are also observed in cardiovascular diseases, metabolic syndromes and diabetes, where postmenopausal women or women with endocrine disorders (e.g., in Polycystic Ovary Syndrome or Primary Ovarian

Insufficiency, in which levels of androgens or estrogens are increased or decreased, respectively), are at higher risk to develop these pathologies when compared to non-affected women (reviewed in [275]). This suggests a potential role for estrogens in metabolic function and in particular in brain metabolism. Indeed, impairment in estrogenic regulation of brain glucose metabolism was previously reported during perimenopause [276,277], and brain hypometabolism reported in women in menopausal transition is associated with cognitive dysfunctions [278,279]. This could result from a decrease in the activation of estrogen receptors (members of the superfamily class of nuclear receptors) in brain areas involved in learning and memory processes, including the prefrontal cortex and hippocampus [280]. Moreover, impairments in the estrogenic regulation of mitochondrial bioenergetics [281] could lead to subsequent oxidative stress, promoting Aß accumulation and neuronal dysfunction [282]. Furthermore, sex differences in PPARs expression and function reported in rodent and human brain could result from changes in sex hormones levels and cross-talks between estrogen receptors and PPARs network (reviewed in [283]), suggesting a role for these receptors in sexual dimorphisms observed in metabolism, inflammatory response and brain function. Moreover, sex differences of the plasma and brain lipidome have been mentioned in humans [284–287], supporting a potential role of PPARα in this aspect.

6. Conclusions

AD is a multifactorial neurodegenerative disorder in which cognitive deficits occurred. AD is influenced by genotype and environmental factors. Among risk factors identified, genomic loci encoding proteins involved in lipid metabolism and altered lipidome of AD brain suggest that brain lipid metabolism is impaired in AD. Moreover, obesity and type 2 diabetes metabolic disorders, identified as AD risk factors, support the essential role of lipid homeostasis in the etiology of AD. Interestingly, PPARs are nuclear transcription factors that govern pathways implicated in the etiology of AD, including lipid metabolism and inflammatory response (Figure 2). Among them, PPARα involved in the catabolism of FAs plays a crucial role in cognitive brain function. While disease-modifying treatments for AD are seeking to interfere with the pathogenic steps responsible for clinical symptoms, PPARs modulators are a promising target in AD therapy. Among PPARs, PPARα has a particular interest since it is the only PPAR isoform described to have neuronal functions involved in memory processes.

As the expression of PPARs is modified in the AD brain, the characterization of new synthetic molecules able to activate several PPARs isoforms could be needed for an efficient treatment for AD. Alternative strategies could be therefore to design novel pan-agonists that can simultaneously activate PPARα, PPARβ and PPARγ. Their efficiencies were previously demonstrated in preclinical mouse model of AD and therefore deserve further investigation.

Since a sexual dimorphism of PPARα agonist was observed in mechanisms underlying memory processes in vivo, sex differences should be considered in therapy aiming to use PPARα modulators. In humans, the influence of sex on the incidence, manifestation and treatment of numerous neurological and psychiatric diseases is well documented. Therefore, it is of crucial importance to decipher sex differences in AD, in which complex cognitive and neuropsychiatric symptoms occur, in order to define novel PPARα sex-specific therapeutic strategies.

Author Contributions: F.S.-O. and N.P. designed and wrote the manuscript. J.-N.O. critically evaluated and approved the review. All authors have read and agreed to the published version of the manuscript.

Funding: This review received no external funding.

Acknowledgments: Research in the area addressed in this review was supported by the Fondation Louvain, the Belgian Fonds pour la Recherche Scientifique, Interuniversity Attraction Poles Program-Belgian State-Belgian Science Policy, The Belgian Fonds de la Recherche Scientifique Médicale, the Queen Elisabeth Medical Foundation and the Fondation pour la Recherche sur la Maladie d'Alzheimer, CONICYT Becas Chile Program.

Conflicts of Interest: The authors declare no conflict of interest related to the material presented in this review.

References

1. Alzheimer's Disease International. *World Alzheimer Report 2019, Attitudes to Dementia*; Alzheimer's Disease International: London, UK, 2019.
2. Feldman, H.H.; Jacova, C.; Robillard, A.; Garcia, A.; Chow, T.; Borrie, M.; Schipper, H.M.; Blair, M.; Kertesz, A.; Chertkow, H. Diagnosis and treatment of dementia: 2. Diagnosis. *Can. Med. Assoc. J.* **2008**, *178*, 825–836. [CrossRef]
3. Kuslansky, G.; Katz, M.; Verghese, J.; Hall, C.B.; Lapuerta, P.; LaRuffa, G.; Lipton, R.B. Detecting dementia with the Hopkins Verbal Learning Test and the Mini-Mental State Examination. *Arch. Clin. Neuropsychol.* **2004**, *19*, 89–104. [CrossRef] [PubMed]
4. Parnetti, L.; Lanari, A.; Silvestrelli, G.; Saggese, E.; Reboldi, P. Diagnosing prodromal Alzheimer's disease: Role of CSF biochemical markers. *Mech. Ageing Dev.* **2006**, *127*, 129–132. [CrossRef] [PubMed]
5. Marcus, C.; Mena, E.; Subramaniam, R.M. Brain PET in the diagnosis of Alzheimer's disease. *Clin. Nucl. Med.* **2014**, *39*, e413–e426. [CrossRef] [PubMed]
6. Iaccarino, L.; Sala, A.; Caminiti, S.; Perani, D. The emerging role of PET imaging in dementia [version 1; peer review: 3 approved]. *F1000 Res.* **2017**, *6*. [CrossRef]
7. Villemagne, V.L.; Doré, V.; Burnham, S.C.; Masters, C.L.; Rowe, C.C. Imaging tau and amyloid-β proteinopathies in Alzheimer disease and other conditions. *Nat. Rev. Neurol.* **2018**, *14*, 225–236. [CrossRef] [PubMed]
8. Hardy, J.; Allsop, D. Amyloid deposition as the central event in the aetiology of Alzheimer's disease. *Trends Pharmacol. Sci.* **1991**, *12*, 383–388. [CrossRef]
9. Selkoe, D.J.; Hardy, J. The amyloid hypothesis of Alzheimer's disease at 25 years. *Embo Mol. Med.* **2016**, *8*, 595–608. [CrossRef]
10. Bergmans, B.A.; De Strooper, B. γ-secretases: From cell biology to therapeutic strategies. *Lancet Neurol.* **2010**, *9*, 215–226. [CrossRef]
11. Hillen, H. The Beta Amyloid Dysfunction (BAD) Hypothesis for Alzheimer's Disease. *Front. Neurosci.* **2019**, *13*. [CrossRef]
12. Castellani, R.J.; Plascencia-Villa, G.; Perry, G. The amyloid cascade and Alzheimer's disease therapeutics: Theory versus observation. *Lab. Investig.* **2019**, *99*, 958–970. [CrossRef] [PubMed]
13. Sun, L.; Zhou, R.; Yang, G.; Shi, Y. Analysis of 138 pathogenic mutations in presenilin-1 on the in vitro production of Aβ42 and Aβ40 peptides by γ-secretase. *Proc. Natl. Acad. Sci. USA* **2017**, *114*, E476–E485. [CrossRef] [PubMed]
14. Ben-Gedalya, T.; Moll, L.; Bejerano-Sagie, M.; Frere, S.; Cabral, W.A.; Friedmann-Morvinski, D.; Slutsky, I.; Burstyn-Cohen, T.; Marini, J.C.; Cohen, E. Alzheimer's disease-causing proline substitutions lead to presenilin 1 aggregation and malfunction. *EMBO J.* **2015**, *34*, 2820–2839. [CrossRef] [PubMed]
15. Xia, D.; Watanabe, H.; Wu, B.; Lee, S.H.; Li, Y.; Tsvetkov, E.; Bolshakov, V.Y.; Shen, J.; Kelleher, R.J. Presenilin-1 Knockin Mice Reveal Loss-of-Function Mechanism for Familial Alzheimer's Disease. *Neuron* **2015**, *85*, 967–981. [CrossRef] [PubMed]
16. Szaruga, M.; Veugelen, S.; Benurwar, M.; Lismont, S.; Sepulveda-Falla, D.; Lleo, A.; Ryan, N.S.; Lashley, T.; Fox, N.C.; Murayama, S.; et al. Qualitative changes in human γ-secretase underlie familial Alzheimer's disease. *J. Exp. Med.* **2015**, *212*, 2003–2013. [CrossRef]
17. Lott, I.T.; Head, E. Dementia in Down syndrome: Unique insights for Alzheimer disease research. *Nat. Rev. Neurol.* **2019**, *15*, 135–147. [CrossRef]
18. Gong, C.-X.; Liu, F.; Iqbal, K. Multifactorial Hypothesis and Multi-Targets for Alzheimer's Disease. *J. Alzheimer's Dis.* **2018**, *64*, S107–S117. [CrossRef]
19. Nelson, P.T.; Dickson, D.W.; Trojanowski, J.Q.; Jack, C.R.; Boyle, P.A.; Arfanakis, K.; Rademakers, R.; Alafuzoff, I.; Attems, J.; Brayne, C.; et al. Limbic-predominant age-related TDP-43 encephalopathy (LATE): Consensus working group report. *Brain* **2019**, *142*, 1503–1527. [CrossRef]
20. Goure, W.F.; Krafft, G.A.; Jerecic, J.; Hefti, F. Targeting the proper amyloid-beta neuronal toxins: A path forward for Alzheimer's disease immunotherapeutics. *Alzheimer's Res. Ther.* **2014**, *6*, 42. [CrossRef]
21. Walsh, D.M.; Tseng, B.P.; Rydel, R.E.; Podlisny, M.B.; Selkoe, D.J. The Oligomerization of Amyloid β-Protein Begins Intracellularly in Cells Derived from Human Brain. *Biochemistry* **2000**, *39*, 10831–10839. [CrossRef]

22. Chen, Y.-R.; Glabe, C.G. Distinct Early Folding and Aggregation Properties of Alzheimer Amyloid-β Peptides Aβ40 and Aβ42: STABLE TRIMER OR TETRAMER FORMATION BY Aβ42. *J. Biol. Chem.* **2006**, *281*, 24414–24422. [CrossRef] [PubMed]

23. Ahmed, M.; Davis, J.; Aucoin, D.; Sato, T.; Ahuja, S.; Aimoto, S.; Elliott, J.I.; Van Nostrand, W.E.; Smith, S.O. Structural conversion of neurotoxic amyloid-β1–42 oligomers to fibrils. *Nat. Struct. Mol. Biol.* **2010**, *17*, 561–567. [CrossRef] [PubMed]

24. Shankar, G.M.; Li, S.; Mehta, T.H.; Garcia-Munoz, A.; Shepardson, N.E.; Smith, I.; Brett, F.M.; Farrell, M.A.; Rowan, M.J.; Lemere, C.A.; et al. Amyloid-β protein dimers isolated directly from Alzheimer's brains impair synaptic plasticity and memory. *Nat. Med.* **2008**, *14*, 837–842. [CrossRef] [PubMed]

25. Lesné, S.E.; Sherman, M.A.; Grant, M.; Kuskowski, M.; Schneider, J.A.; Bennett, D.A.; Ashe, K.H. Brain amyloid-β oligomers in ageing and Alzheimer's disease. *Brain* **2013**, *136*, 1383–1398. [CrossRef] [PubMed]

26. Kayed, R.; Head, E.; Thompson, J.L.; McIntire, T.M.; Milton, S.C.; Cotman, C.W.; Glabe, C.G. Common Structure of Soluble Amyloid Oligomers Implies Common Mechanism of Pathogenesis. *Science* **2003**, *300*, 486–489. [CrossRef] [PubMed]

27. Cohen, E.; Paulsson, J.F.; Blinder, P.; Burstyn-Cohen, T.; Du, D.; Estepa, G.; Adame, A.; Pham, H.M.; Holzenberger, M.; Kelly, J.W.; et al. Reduced IGF-1 Signaling Delays Age-Associated Proteotoxicity in Mice. *Cell* **2009**, *139*, 1157–1169. [CrossRef]

28. O'Nuallain, B.; Freir, D.B.; Nicoll, A.J.; Risse, E.; Ferguson, N.; Herron, C.E.; Collinge, J.; Walsh, D.M. Amyloid β-Protein Dimers Rapidly Form Stable Synaptotoxic Protofibrils. *J. Neurosci.* **2010**, *30*, 14411–14419. [CrossRef]

29. Canter, R.G.; Penney, J.; Tsai, L.-H. The road to restoring neural circuits for the treatment of Alzheimer's disease. *Nature* **2016**, *539*, 187–196. [CrossRef]

30. Eisai. Eisai And Biogen To Discontinue Phase III Clinical Studies Of BACE Inhibitor Elenbecestat In Early Alzheimer's Disease. Available online: http://eisai.mediaroom.com/2019-09-13-Eisai-And-Biogen-To-Discontinue-Phase-III-Clinical-Studies-Of-BACE-Inhibitor-Elenbecestat-In-Early-Alzheimers-Disease (accessed on 1 April 2020).

31. Liu, P.-P.; Xie, Y.; Meng, X.-Y.; Kang, J.-S. History and progress of hypotheses and clinical trials for Alzheimer's disease. *Signal Transduct Target* **2019**, *4*, 29. [CrossRef]

32. Schneider, L. A resurrection of aducanumab for Alzheimer's disease. *Lancet Neurol.* **2020**, *19*, 111–112. [CrossRef]

33. Huang, L.-K.; Chao, S.-P.; Hu, C.-J. Clinical trials of new drugs for Alzheimer disease. *J. Biomed. Sci.* **2020**, *27*, 18. [CrossRef] [PubMed]

34. Yu, J.-T.; Tan, L.; Hardy, J. Apolipoprotein E in Alzheimer's Disease: An Update. *Annu. Rev. Neurosci.* **2014**, *37*, 79–100. [CrossRef] [PubMed]

35. Kunkle, B.W.; Grenier-Boley, B.; Sims, R.; Bis, J.C.; Damotte, V.; Naj, A.C.; Boland, A.; Vronskaya, M.; van der Lee, S.J.; Amlie-Wolf, A.; et al. Genetic meta-analysis of diagnosed Alzheimer's disease identifies new risk loci and implicates Aβ, tau, immunity and lipid processing. *Nat. Genet.* **2019**, *51*, 414–430. [CrossRef] [PubMed]

36. Cuyvers, E.; Sleegers, K. Genetic variations underlying Alzheimer's disease: Evidence from genome-wide association studies and beyond. *Lancet Neurol.* **2016**, *15*, 857–868. [CrossRef]

37. De Roeck, A.; Van Broeckhoven, C.; Sleegers, K. The role of ABCA7 in Alzheimer's disease: Evidence from genomics, transcriptomics and methylomics. *Acta Neuropathol.* **2019**, *138*, 201–220. [CrossRef] [PubMed]

38. Harold, D.; Abraham, R.; Hollingworth, P.; Sims, R.; Gerrish, A.; Hamshere, M.L.; Pahwa, J.S.; Moskvina, V.; Dowzell, K.; Williams, A.; et al. Genome-wide association study identifies variants at CLU and PICALM associated with Alzheimer's disease. *Nat. Genet.* **2009**, *41*, 1088–1093. [CrossRef]

39. Hollingworth, P.; Harold, D.; Sims, R.; Gerrish, A.; Lambert, J.-C.; Carrasquillo, M.M.; Abraham, R.; Hamshere, M.L.; Pahwa, J.S.; Moskvina, V.; et al. Common variants at ABCA7, MS4A6A/MS4A4E, EPHA1, CD33 and CD2AP are associated with Alzheimer's disease. *Nat. Genet.* **2011**, *43*, 429–435. [CrossRef]

40. Lambert, J.-C.; Heath, S.; Even, G.; Campion, D.; Sleegers, K.; Hiltunen, M.; Combarros, O.; Zelenika, D.; Bullido, M.J.; Tavernier, B.; et al. Genome-wide association study identifies variants at CLU and CR1 associated with Alzheimer's disease. *Nat. Genet.* **2009**, *41*, 1094–1099. [CrossRef]

41. Di Paolo, G.; Kim, T.-W. Linking lipids to Alzheimer's disease: Cholesterol and beyond. *Nat. Rev. Neurosci.* **2011**, *12*, 284–296. [CrossRef]

42. Shimano, H.; Shimomura, I.; Hammer, R.E.; Herz, J.; Goldstein, J.L.; Brown, M.S.; Horton, J.D. Elevated levels of SREBP-2 and cholesterol synthesis in livers of mice homozygous for a targeted disruption of the SREBP-1 gene. *J. Clin. Investig.* **1997**, *100*, 2115–2124. [CrossRef]

43. Horton, J.D. Sterol regulatory element-binding proteins: Transcriptional activators of lipid synthesis. *Biochem. Soc. Trans.* **2002**, *30*, 1091–1095. [CrossRef]

44. Chen, Y.; Bang, S.; McMullen, M.F.; Kazi, H.; Talbot, K.; Ho, M.-X.; Carlson, G.; Arnold, S.E.; Ong, W.-Y.; Kim, S.F. Neuronal Activity-Induced Sterol Regulatory Element Binding Protein-1 (SREBP1) is Disrupted in Dysbindin-Null Mice—Potential Link to Cognitive Impairment in Schizophrenia. *Mol. Neurobiol.* **2017**, *54*, 1699–1709. [CrossRef] [PubMed]

45. Picard, C.; Julien, C.; Frappier, J.; Miron, J.; Théroux, L.; Dea, D.; Breitner, J.C.S.; Poirier, J. Alterations in cholesterol metabolism–related genes in sporadic Alzheimer's disease. *Neurobiol. Aging* **2018**, *66*, e181–e189. [CrossRef] [PubMed]

46. Steen, V.M.; Skrede, S.; Polushina, T.; López, M.; Andreassen, O.A.; Fernø, J.; Hellard, S.L. Genetic evidence for a role of the SREBP transcription system and lipid biosynthesis in schizophrenia and antipsychotic treatment. *Eur. Neuropsychopharmacol.* **2017**, *27*, 589–598. [CrossRef] [PubMed]

47. Wang, C.; Zhao, F.; Shen, K.; Wang, W.; Siedlak, S.L.; Lee, H.-g.; Phelix, C.F.; Perry, G.; Shen, L.; Tang, B.; et al. The sterol regulatory element-binding protein 2 is dysregulated by tau alterations in Alzheimer disease. *Brain Pathol.* **2019**, *29*, 530–543. [CrossRef] [PubMed]

48. Pierrot, N.; Tyteca, D.; D'auria, L.; Dewachter, I.; Gailly, P.; Hendrickx, A.; Tasiaux, B.; Haylani, L.E.; Muls, N.; N'Kuli, F.; et al. Amyloid precursor protein controls cholesterol turnover needed for neuronal activity. *Embo Mol. Med.* **2013**, *5*, 608–625. [CrossRef]

49. Cunnane, S.C.; Schneider, J.A.; Tangney, C.; Tremblay-Mercier, J.; Fortier, M.; Bennett, D.A.; Morris, M.C. Plasma and Brain Fatty Acid Profiles in Mild Cognitive Impairment and Alzheimer's Disease. *J. Alzheimer's Dis.* **2012**, *29*, 691–697. [CrossRef]

50. Joffre, C.; Nadjar, A.; Lebbadi, M.; Calon, F.; Laye, S. n-3 LCPUFA improves cognition: The young, the old and the sick. *ProstaglandinsLeukot. Essent. Fat. Acids* **2014**, *91*, 1–20. [CrossRef]

51. Lukiw, W.J.; Pappolla, M.; Pelaez, R.P.; Bazan, N.G. Alzheimer's Disease—A Dysfunction in Cholesterol and Lipid Metabolism. *Cell. Mol. Neurobiol.* **2005**, *25*, 475–483. [CrossRef]

52. Proitsi, P.; Kim, M.; Whiley, L.; Pritchard, M.; Leung, R.; Soininen, H.; Kloszewska, I.; Mecocci, P.; Tsolaki, M.; Vellas, B.; et al. Plasma lipidomics analysis finds long chain cholesteryl esters to be associated with Alzheimer's disease. *Transl. Psychiatry* **2015**, *5*, e494. [CrossRef]

53. Snowden, S.G.; Ebshiana, A.A.; Hye, A.; An, Y.; Pletnikova, O.; O'Brien, R.; Troncoso, J.; Legido-Quigley, C.; Thambisetty, M. Association between fatty acid metabolism in the brain and Alzheimer disease neuropathology and cognitive performance: A nontargeted metabolomic study. *PLoS Med.* **2017**, *14*, e1002266. [CrossRef]

54. Amtul, Z.; Uhrig, M.; Wang, L.; Rozmahel, R.F.; Beyreuther, K. Detrimental effects of arachidonic acid and its metabolites in cellular and mouse models of Alzheimer's disease: Structural insight. *Neurobiol. Aging* **2012**, *33*, e821–e831. [CrossRef] [PubMed]

55. Barbero-Camps, E.; Fernández, A.; Martínez, L.; Fernández-Checa, J.C.; Colell, A. APP/PS1 mice overexpressing SREBP-2 exhibit combined Aβ accumulation and tau pathology underlying Alzheimer's disease. *Hum. Mol. Genet.* **2013**, *22*, 3460–3476. [CrossRef] [PubMed]

56. Grimm, M.O.W.; Kuchenbecker, J.; Grösgen, S.; Burg, V.K.; Hundsdörfer, B.; Rothhaar, T.L.; Friess, P.; de Wilde, M.C.; Broersen, L.M.; Penke, B.; et al. Docosahexaenoic Acid Reduces Amyloid β Production via Multiple Pleiotropic Mechanisms. *J. Biol. Chem.* **2011**, *286*, 14028–14039. [CrossRef] [PubMed]

57. Hamilton, J.A.; Hillard, C.J.; Spector, A.A.; Watkins, P.A. Brain Uptake and Utilization of Fatty Acids, Lipids and Lipoproteins: Application to Neurological Disorders. *J. Mol. Neurosci.* **2007**, *33*, 2–11. [CrossRef]

58. Grimm, M.O.W.; Michaelson, D.M.; Hartmann, T. Omega-3 fatty acids, lipids, and apoE lipidation in Alzheimer's disease: A rationale for multi-nutrient dementia prevention. *J. Lipid Res.* **2017**, *58*, 2083–2101. [CrossRef]

59. Hussain, G.; Wang, J.; Rasul, A.; Anwar, H.; Imran, A.; Qasim, M.; Zafar, S.; Kamran, S.K.S.; Razzaq, A.; Aziz, N.; et al. Role of cholesterol and sphingolipids in brain development and neurological diseases. *Lipids Health Dis.* **2019**, *18*, 26. [CrossRef]

60. Fields, R.D. A new mechanism of nervous system plasticity: Activity-dependent myelination. *Nat. Rev. Neurosci.* **2015**, *16*, 756–767. [CrossRef]

61. Duncan, A.L.; Song, W.; Sansom, M.S.P. Lipid-Dependent Regulation of Ion Channels and G Protein–Coupled Receptors: Insights from Structures and Simulations. *Annu. Rev. Pharmacol. Toxicol.* **2020**, *60*, 31–50. [CrossRef]

62. Cordero-Morales, J.F.; Vásquez, V. How lipids contribute to ion channel function, a fat perspective on direct and indirect interactions. *Curr. Opin. Struct. Biol.* **2018**, *51*, 92–98. [CrossRef]

63. Munro, S. Lipid Rafts: Elusive or Illusive? *Cell* **2003**, *115*, 377–388. [CrossRef]

64. Sebastião, A.M.; Colino-Oliveira, M.; Assaife-Lopes, N.; Dias, R.B.; Ribeiro, J.A. Lipid rafts, synaptic transmission and plasticity: Impact in age-related neurodegenerative diseases. *Neuropharmacology* **2013**, *64*, 97–107. [CrossRef] [PubMed]

65. Martín, V.; Fabelo, N.; Santpere, G.; Puig, B.; Marín, R.; Ferrer, I.; Díaz, M. Lipid Alterations in Lipid Rafts from Alzheimer's Disease Human Brain Cortex. *J. Alzheimer's Dis.* **2010**, *19*, 489–502. [CrossRef]

66. Raquel, M.; Noemí, F.; Cecilia, F.-E.; Ana, C.-A.; Deiene, R.-B.; David, Q.-A.; Fátima, M.-H.; Mario, D. Lipid Raft Alterations in Aged-Associated Neuropathologies. *Curr. Alzheimer Res.* **2016**, *13*, 973–984. [CrossRef]

67. Hicks, D.; Nalivaeva, N.; Turner, A. Lipid Rafts and Alzheimer's Disease: Protein-Lipid Interactions and Perturbation of Signaling. *Front. Physiol.* **2012**, *3*. [CrossRef]

68. Dart, C. SYMPOSIUM REVIEW: Lipid microdomains and the regulation of ion channel function. *J. Physiol.* **2010**, *588*, 3169–3178. [CrossRef]

69. Barrett, P.J.; Song, Y.; Van Horn, W.D.; Hustedt, E.J.; Schafer, J.M.; Hadziselimovic, A.; Beel, A.J.; Sanders, C.R. The Amyloid Precursor Protein Has a Flexible Transmembrane Domain and Binds Cholesterol. *Science* **2012**, *336*, 1168–1171. [CrossRef]

70. Coburger, I.; Hoefgen, S.; Than, M.E. The structural biology of the amyloid precursor protein APP—A complex puzzle reveals its multi-domain architecture. *Biol. Chem.* **2014**, *395*, 485–498. [CrossRef]

71. Van der Kant, R.; Goldstein, L.S.B. Cellular Functions of the Amyloid Precursor Protein from Development to Dementia. *Dev. Cell* **2015**, *32*, 502–515. [CrossRef]

72. Müller, U.C.; Deller, T.; Korte, M. Not just amyloid: Physiological functions of the amyloid precursor protein family. *Nat. Rev. Neurosci.* **2017**, *18*, 281–298. [CrossRef]

73. Ludewig, S.; Korte, M. Novel Insights into the Physiological Function of the APP (Gene) Family and Its Proteolytic Fragments in Synaptic Plasticity. *Front. Mol. Neurosci.* **2017**, *9*. [CrossRef]

74. Grimm, M.O.W.; Mett, J.; Grimm, H.S.; Hartmann, T. APP Function and Lipids: A Bidirectional Link. *Front. Mol. Neurosci.* **2017**, *10*. [CrossRef]

75. Hefter, D.; Ludewig, S.; Draguhn, A.; Korte, M. Amyloid, APP, and Electrical Activity of the Brain. *Neuroscientist* **2020**, *26*, 231–251. [CrossRef] [PubMed]

76. DelBove, C.E.; Strothman, C.E.; Lazarenko, R.M.; Huang, H.; Sanders, C.R.; Zhang, Q. Reciprocal modulation between amyloid precursor protein and synaptic membrane cholesterol revealed by live cell imaging. *Neurobiol. Dis.* **2019**, *127*, 449–461. [CrossRef] [PubMed]

77. Area-Gomez, E.; Schon, E.A. Mitochondria-associated ER membranes and Alzheimer disease. *Curr. Opin. Genet. Dev.* **2016**, *38*, 90–96. [CrossRef] [PubMed]

78. Janikiewicz, J.; Szymański, J.; Malinska, D.; Patalas-Krawczyk, P.; Michalska, B.; Duszyński, J.; Giorgi, C.; Bonora, M.; Dobrzyn, A.; Wieckowski, M.R. Mitochondria-associated membranes in aging and senescence: Structure, function, and dynamics. *Cell Death Dis.* **2018**, *9*, 332. [CrossRef] [PubMed]

79. Area-Gomez, E.; de Groof, A.J.C.; Boldogh, I.; Bird, T.D.; Gibson, G.E.; Koehler, C.M.; Yu, W.H.; Duff, K.E.; Yaffe, M.P.; Pon, L.A.; et al. Presenilins Are Enriched in Endoplasmic Reticulum Membranes Associated with Mitochondria. *Am. J. Pathol.* **2009**, *175*, 1810–1816. [CrossRef]

80. Pera, M.; Larrea, D.; Guardia-Laguarta, C.; Montesinos, J.; Velasco, K.R.; Agrawal, R.R.; Xu, Y.; Chan, R.B.; Di Paolo, G.; Mehler, M.F.; et al. Increased localization of APP-C99 in mitochondria-associated ER membranes causes mitochondrial dysfunction in Alzheimer disease. *Embo J.* **2017**, *36*, 3356–3371. [CrossRef]

81. Hedskog, L.; Pinho, C.M.; Filadi, R.; Rönnbäck, A.; Hertwig, L.; Wiehager, B.; Larssen, P.; Gellhaar, S.; Sandebring, A.; Westerlund, M.; et al. Modulation of the endoplasmic reticulum–mitochondria interface in Alzheimer's disease and related models. *Proc. Natl. Acad. Sci. USA* **2013**, *110*, 7916–7921. [CrossRef]

82. Area-Gomez, E.; del Carmen Lara Castillo, M.; Tambini, M.D.; Guardia-Laguarta, C.; de Groof, A.J.C.; Madra, M.; Ikenouchi, J.; Umeda, M.; Bird, T.D.; Sturley, S.L.; et al. Upregulated function of mitochondria-associated ER membranes in Alzheimer disease. *Embo J.* **2012**, *31*, 4106–4123. [CrossRef]
83. Area-Gomez, E.; de Groof, A.; Bonilla, E.; Montesinos, J.; Tanji, K.; Boldogh, I.; Pon, L.; Schon, E.A. A key role for MAM in mediating mitochondrial dysfunction in Alzheimer disease. *Cell Death Dis.* **2018**, *9*, 335. [CrossRef] [PubMed]
84. Area-Gomez, E.; Schon, E.A. On the Pathogenesis of Alzheimer's Disease: The MAM Hypothesis. *Faseb J.* **2017**, *31*, 864–867. [CrossRef] [PubMed]
85. McMahon, H.T.; Boucrot, E. Membrane curvature at a glance. *J. Cell Sci.* **2015**, *128*, 1065–1070. [CrossRef] [PubMed]
86. Simons, K.; Toomre, D. Lipid rafts and signal transduction. *Nat. Rev. Mol. Cell Biol.* **2000**, *1*, 31–39. [CrossRef] [PubMed]
87. Srivastav, R.K.; Ansari, T.M.; Prasad, M.; Vishwakarma, V.K.; Bagga, P.; Ahsan, F. Caveolins; An Assailant or An Ally of Various Cellular Disorders. *Drug Res (Stuttg)* **2019**, *69*, 419–427. [CrossRef]
88. Chauhan, N.B. Membrane dynamics, cholesterol homeostasis, and Alzheimer's disease. *J. Lipid Res.* **2003**, *44*, 2019–2029. [CrossRef]
89. Gaudreault, S.B.; Dea, D.; Poirier, J. Increased caveolin-1 expression in Alzheimer's disease brain. *Neurobiol. Aging* **2004**, *25*, 753–759. [CrossRef]
90. Alsaqati, M.; Thomas, R.S.; Kidd, E.J. Proteins Involved in Endocytosis Are Upregulated by Ageing in the Normal Human Brain: Implications for the Development of Alzheimer's Disease. *J. Gerontol.: Ser. A* **2017**, *73*, 289–298. [CrossRef]
91. Cameron, P.L.; Ruffin, J.W.; Bollag, R.; Rasmussen, H.; Cameron, R.S. Identification of Caveolin and Caveolin-Related Proteins in the Brain. *J. Neurosci.* **1997**, *17*, 9520–9535. [CrossRef]
92. Head, B.P.; Peart, J.N.; Panneerselvam, M.; Yokoyama, T.; Pearn, M.L.; Niesman, I.R.; Bonds, J.A.; Schilling, J.M.; Miyanohara, A.; Headrick, J.; et al. Loss of Caveolin-1 Accelerates Neurodegeneration and Aging. *PLoS ONE* **2010**, *5*, e15697. [CrossRef]
93. Chang, C.-Y.; Ke, D.-S.; Chen, J.-Y. Essential Fatty Acids and Human Brain. *Acta Neurol. Taiwanica* **2009**, *18*, 231–241.
94. Bazinet, R.P.; Layé, S. Polyunsaturated fatty acids and their metabolites in brain function and disease. *Nat. Rev. Neurosci.* **2014**, *15*, 771–785. [CrossRef] [PubMed]
95. Tracey, T.J.; Steyn, F.J.; Wolvetang, E.J.; Ngo, S.T. Neuronal Lipid Metabolism: Multiple Pathways Driving Functional Outcomes in Health and Disease. *Front. Mol. Neurosci.* **2018**, *11*. [CrossRef] [PubMed]
96. Moore, S.A. Polyunsaturated fatty acid synthesis and release by brain-derived cells in vitro. *J. Mol. Neurosci.* **2001**, *16*, 195–200. [CrossRef]
97. Luchtman, D.W.; Song, C. Cognitive enhancement by omega-3 fatty acids from child-hood to old age: Findings from animal and clinical studies. *Neuropharmacology* **2013**, *64*, 550–565. [CrossRef]
98. Hussain, G.; Schmitt, F.; Loeffler, J. P.; Gonzalez De Aguilar, J. Fatting the brain: A brief of recent research. *Front. Cell. Neurosci.* **2013**, *7*. [CrossRef] [PubMed]
99. Conquer, J.A.; Tierney, M.C.; Zecevic, J.; Bettger, W.J.; Fisher, R.H. Fatty acid analysis of blood plasma of patients with Alzheimer's disease, other types of dementia, and cognitive impairment. *Lipids* **2000**, *35*, 1305–1312. [CrossRef]
100. Schaefer, E.J.; Bongard, V.; Beiser, A.S.; Lamon-Fava, S.; Robins, S.J.; Au, R.; Tucker, K.L.; Kyle, D.J.; Wilson, P.W.F.; Wolf, P.A. Plasma Phosphatidylcholine Docosahexaenoic Acid Content and Risk of Dementia and Alzheimer Disease: The Framingham Heart Study. *Arch. Neurol.* **2006**, *63*, 1545–1550. [CrossRef]
101. Odermatt, A. The Western-style diet: A major risk factor for impaired kidney function and chronic kidney disease. *Am. J. Physiol.-Ren. Physiol.* **2011**, *301*, F919–F931. [CrossRef]
102. Quinn, J.F.; Raman, R.; Thomas, R.G.; Yurko-Mauro, K.; Nelson, E.B.; Van Dyck, C.; Galvin, J.E.; Emond, J.; Jack, C.R.; Weiner, M.; et al. Docosahexaenoic Acid Supplementation and Cognitive Decline in Alzheimer Disease: A Randomized Trial. *JAMA* **2010**, *304*, 1903–1911. [CrossRef]
103. Mazereeuw, G.; Lanctôt, K.L.; Chau, S.A.; Swardfager, W.; Herrmann, N. Effects of omega-3 fatty acids on cognitive performance: A meta-analysis. *Neurobiol. Aging* **2012**, *33*, e1417–e1429. [CrossRef]
104. Layé, S. Polyunsaturated fatty acids, neuroinflammation and well being. *ProstaglandinsLeukot. Essent. Fat. Acids (Plefa)* **2010**, *82*, 295–303. [CrossRef] [PubMed]

105. Féart, C.; Peuchant, E.; Letenneur, L.; Samieri, C.; Montagnier, D.; Fourrier-Reglat, A.; Barberger-Gateau, P. Plasma eicosapentaenoic acid is inversely associated with severity of depressive symptomatology in the elderly: Data from the Bordeaux sample of the Three-City Study. *Am. J. Clin. Nutr.* **2008**, *87*, 1156–1162. [CrossRef] [PubMed]

106. Freund-Levi, Y.; Eriksdotter-Jönhagen, M.; Cederholm, T.; Basun, H.; Faxén-Irving, G.; Garlind, A.; Vedin, I.; Vessby, B.; Wahlund, L.-O.; Palmblad, J. ω-3 Fatty Acid Treatment in 174 Patients With Mild to Moderate Alzheimer Disease: OmegAD Study: A Randomized Double-blind Trial. *Arch. Neurol.* **2006**, *63*, 1402–1408. [CrossRef] [PubMed]

107. Calon, F.; Lim, G.P.; Yang, F.; Morihara, T.; Teter, B.; Ubeda, O.; Rostaing, P.; Triller, A.; Salem, N.; Ashe, K.H.; et al. Docosahexaenoic Acid Protects from Dendritic Pathology in an Alzheimer's Disease Mouse Model. *Neuron* **2004**, *43*, 633–645. [CrossRef]

108. Calon, F.; Cole, G. Neuroprotective action of omega-3 polyunsaturated fatty acids against neurodegenerative diseases: Evidence from animal studies. *ProstaglandinsLeukot. Essent. Fat. Acids* **2007**, *77*, 287–293. [CrossRef]

109. Connor, S.; Tenorio, G.; Clandinin, M.T.; Sauvé, Y. DHA supplementation enhances high-frequency, stimulation-induced synaptic transmission in mouse hippocampus. *Appl. Physiol. Nutr. Metab.* **2012**, *37*, 880–887. [CrossRef]

110. Labrousse, V.F.; Nadjar, A.; Joffre, C.; Costes, L.; Aubert, A.; Grégoire, S.; Bretillon, L.; Layé, S. Short-Term Long Chain Omega3 Diet Protects from Neuroinflammatory Processes and Memory Impairment in Aged Mice. *PLoS ONE* **2012**, *7*, e36861. [CrossRef] [PubMed]

111. Hashimoto, M.; Hossain, S.; Shimada, T.; Sugioka, K.; Yamasaki, H.; Fujii, Y.; Ishibashi, Y.; Oka, J.-I.; Shido, O. Docosahexaenoic acid provides protection from impairment of learning ability in Alzheimer's disease model rats. *J. Neurochem.* **2002**, *81*, 1084–1091. [CrossRef]

112. Cole, G.M.; Frautschy, S.A. Docosahexaenoic Acid Protects from Amyloid and Dendritic Pathology in an Alzheimer's Disease Mouse Model. *Nutr. Health* **2006**, *18*, 249–259. [CrossRef]

113. Kavraal, S.; Oncu, S.K.; Bitiktas, S.; Artis, A.S.; Dolu, N.; Gunes, T.; Suer, C. Maternal intake of Omega-3 essential fatty acids improves long term potentiation in the dentate gyrus and Morris water maze performance in rats. *Brain Res.* **2012**, *1482*, 32–39. [CrossRef] [PubMed]

114. Evans, R.M.; Mangelsdorf, D.J. Nuclear Receptors, RXR, and the Big Bang. *Cell* **2014**, *157*, 255–266. [CrossRef] [PubMed]

115. Warden, A.; Truitt, J.; Merriman, M.; Ponomareva, O.; Jameson, K.; Ferguson, L.B.; Mayfield, R.D.; Harris, R.A. Localization of PPAR isotypes in the adult mouse and human brain. *Sci. Rep.* **2016**, *6*, 27618. [CrossRef] [PubMed]

116. Moutinho, M.; Landreth, G.E. Therapeutic potential of nuclear receptor agonists in Alzheimer's disease. *J. Lipid Res.* **2017**, *58*, 1937–1949. [CrossRef] [PubMed]

117. Zolezzi, J.M.; Santos, M.J.; Bastías-Candia, S.; Pinto, C.; Godoy, J.A.; Inestrosa, N.C. PPARs in the central nervous system: Roles in neurodegeneration and neuroinflammation. *Biol. Rev.* **2017**, *92*, 2046–2069. [CrossRef]

118. Wolf, G. Is 9-Cis-Retinoic Acid the Endogenous Ligand for the Retinoic Acid-X Receptor? *Nutr. Rev.* **2006**, *64*, 532–538. [CrossRef]

119. Heyman, R.A.; Mangelsdorf, D.J.; Dyck, J.A.; Stein, R.B.; Eichele, G.; Evans, R.M.; Thaller, C. 9-cis retinoic acid is a high affinity ligand for the retinoid X receptor. *Cell* **1992**, *68*, 397–406. [CrossRef]

120. Chitranshi, N.; Dheer, Y.; Kumar, S.; Graham, S.L.; Gupta, V. Molecular docking, dynamics, and pharmacology studies on bexarotene as an agonist of ligand-activated transcription factors, retinoid X receptors. *J. Cell. Biochem.* **2019**, *120*, 11745–11760. [CrossRef]

121. Zhang, C.; Duvic, M. Treatment of cutaneous T-cell lymphoma with retinoids. *Dermatol. Ther.* **2006**, *19*, 264–271. [CrossRef] [PubMed]

122. Jiang, Q.; Lee, C.Y.D.; Mandrekar, S.; Wilkinson, B.; Cramer, P.; Zelcer, N.; Mann, K.; Lamb, B.; Willson, T.M.; Collins, J.L.; et al. ApoE Promotes the Proteolytic Degradation of Aβ. *Neuron* **2008**, *58*, 681–693. [CrossRef]

123. Cramer, P.E.; Cirrito, J.R.; Wesson, D.W.; Lee, C.Y.D.; Karlo, J.C.; Zinn, A.E.; Casali, B.T.; Restivo, J.L.; Goebel, W.D.; James, M.J.; et al. ApoE-Directed Therapeutics Rapidly Clear β-Amyloid and Reverse Deficits in AD Mouse Models. *Science* **2012**, *335*, 1503–1506. [CrossRef] [PubMed]

124. Mariani, M.M.; Malm, T.; Lamb, R.; Jay, T.R.; Neilson, L.; Casali, B.; Medarametla, L.; Landreth, G.E. Neuronally-directed effects of RXR activation in a mouse model of Alzheimer's disease. *Sci. Rep.* **2017**, *7*, 42270. [CrossRef] [PubMed]

125. Pierrot, N.; Ris, L.; Stancu, I.-C.; Doshina, A.; Ribeiro, F.; Tyteca, D.; Baugé, E.; Lalloyer, F.; Malong, L.; Schakman, O.; et al. Sex-regulated gene dosage effect of PPARα on synaptic plasticity. *Life Sci. Alliance* **2019**, *2*, e201800262. [CrossRef] [PubMed]

126. Savage, J.C.; Jay, T.; Goduni, E.; Quigley, C.; Mariani, M.M.; Malm, T.; Ransohoff, R.M.; Lamb, B.T.; Landreth, G.E. Nuclear Receptors License Phagocytosis by Trem2$^+$ Myeloid Cells in Mouse Models of Alzheimer's Disease. *J. Neurosci.* **2015**, *35*, 6532–6543. [CrossRef] [PubMed]

127. Yamanaka, M.; Ishikawa, T.; Griep, A.; Axt, D.; Kummer, M.P.; Heneka, M.T. PPARγ/RXRα-Induced and CD36-Mediated Microglial Amyloid-β Phagocytosis Results in Cognitive Improvement in Amyloid Precursor Protein/Presenilin 1 Mice. *J. Neurosci.* **2012**, *32*, 17321–17331. [CrossRef]

128. Liang, Y.; Lin, S.; Beyer, T.P.; Zhang, Y.; Wu, X.; Bales, K.R.; DeMattos, R.B.; May, P.C.; Li, S.D.; Jiang, X.-C.; et al. A liver X receptor and retinoid X receptor heterodimer mediates apolipoprotein E expression, secretion and cholesterol homeostasis in astrocytes. *J. Neurochem.* **2004**, *88*, 623–634. [CrossRef]

129. Zhao, J.; Fu, Y.; Liu, C.-C.; Shinohara, M.; Nielsen, H.M.; Dong, Q.; Kanekiyo, T.; Bu, G. Retinoic Acid Isomers Facilitate Apolipoprotein E Production and Lipidation in Astrocytes through the Retinoid X Receptor/Retinoic Acid Receptor Pathway. *J. Biol. Chem.* **2014**, *289*, 11282–11292. [CrossRef]

130. Pierrot, N.; Lhommel, R.; Quenon, L.; Hanseeuw, B.; Dricot, L.; Sindic, C.; Maloteaux, J.-M.; Octavea, J.-N.; Ivanoiu, A. Targretin Improves Cognitive and Biological Markers in a Patient with Alzheimer's Disease. *J. Alzheimer's Dis.* **2016**, *49*, 271–276. [CrossRef]

131. Ghosal, K.; Haag, M.; Verghese, P.B.; West, T.; Veenstra, T.; Braunstein, J.B.; Bateman, R.J.; Holtzman, D.M.; Landreth, G.E. A randomized controlled study to evaluate the effect of bexarotene on amyloid-β and apolipoprotein E metabolism in healthy subjects. *Alzheimers Dement (N Y)* **2016**, *2*, 110–120. [CrossRef]

132. Cummings, J.L.; Zhong, K.; Kinney, J.W.; Heaney, C.; Moll-Tudla, J.; Joshi, A.; Pontecorvo, M.; Devous, M.; Tang, A.; Bena, J. Double-blind, placebo-controlled, proof-of-concept trial of bexarotene in moderate Alzheimer's disease. *Alzheimer's Res. Ther.* **2016**, *8*, 4. [CrossRef]

133. George, J.; Schroepfer, J. Oxysterols: Modulators of Cholesterol Metabolism and Other Processes. *Physiol. Rev.* **2000**, *80*, 361–554. [CrossRef]

134. Ma, L.; Nelson, E.R. Oxysterols and nuclear receptors. *Mol. Cell. Endocrinol.* **2019**, *484*, 42–51. [CrossRef] [PubMed]

135. Bensinger, S.J.; Tontonoz, P. Integration of metabolism and inflammation by lipid-activated nuclear receptors. *Nature* **2008**, *454*, 470–477. [CrossRef] [PubMed]

136. Gan, X.; Kaplan, R.; Menke, J.G.; MacNaul, K.; Chen, Y.; Sparrow, C.P.; Zhou, G.; Wright, S.D.; Cai, T.-Q. Dual Mechanisms of ABCA1 Regulation by Geranylgeranyl Pyrophosphate. *J. Biol. Chem.* **2001**, *276*, 48702–48708. [CrossRef]

137. Langmann, T.; Liebisch, G.; Moehle, C.; Schifferer, R.; Dayoub, R.; Heiduczek, S.; Grandl, M.; Dada, A.; Schmitz, G. Gene expression profiling identifies retinoids as potent inducers of macrophage lipid efflux. *Biochim. Biophys. Acta (Bba)-Mol. Basis Dis.* **2005**, *1740*, 155–161. [CrossRef]

138. Ou, J.; Tu, H.; Shan, B.; Luk, A.; DeBose-Boyd, R.A.; Bashmakov, Y.; Goldstein, J.L.; Brown, M.S. Unsaturated fatty acids inhibit transcription of the sterol regulatory element-binding protein-1c (SREBP-1c) gene by antagonizing ligand-dependent activation of the LXR. *Proc. Natl. Acad. Sci. USA* **2001**, *98*, 6027–6032. [CrossRef]

139. Yoshikawa, T.; Shimano, H.; Yahagi, N.; Ide, T.; Amemiya-Kudo, M.; Matsuzaka, T.; Nakakuki, M.; Tomita, S.; Okazaki, H.; Tamura, Y.; et al. Polyunsaturated Fatty Acids Suppress Sterol Regulatory Element-binding Protein 1c Promoter Activity by Inhibition of Liver X Receptor (LXR) Binding to LXR Response Elements. *J. Biol. Chem.* **2002**, *277*, 1705–1711. [CrossRef]

140. Reddy, J.K.; Rao, M.S. *Xenobiotic-Induced Peroxisome Proliferation: Role of Tissue Specificity and Species Differences in Response in the Evaluation of the Implications for Human Health*; Springer: Berlin/Heidelberg, Germany, 1987; pp. 43–53.

141. Wang, Y.-X. PPARs: Diverse regulators in energy metabolism and metabolic diseases. *Cell Res.* **2010**, *20*, 124–137. [CrossRef]

142. Michalik, L.; Auwerx, J.; Berger, J.P.; Chatterjee, V.K.; Glass, C.K.; Gonzalez, F.J.; Grimaldi, P.A.; Kadowaki, T.; Lazar, M.A.; O'Rahilly, S.; et al. International Union of Pharmacology. LXI. Peroxisome Proliferator-Activated Receptors. *Pharmacol. Rev.* **2006**, *58*, 726–741. [CrossRef]

143. Auboeuf, D.; Rieusset, J.; Fajas, L.; Vallier, P.; Frering, V.; Riou, J.P.; Staels, B.; Auwerx, J.; Laville, M.; Vidal, H. Tissue Distribution and Quantification of the Expression of mRNAs of Peroxisome Proliferator–Activated Receptors and Liver X Receptor-α in Humans: No Alteration in Adipose Tissue of Obese and NIDDM Patients. *Diabetes* **1997**, *46*, 1319–1327. [CrossRef]

144. Braissant, O.; Foufelle, F.; Scotto, C.; Dauça, M.; Wahli, W. Differential expression of peroxisome proliferator-activated receptors (PPARs): Tissue distribution of PPAR-alpha, -beta, and -gamma in the adult rat. *Endocrinology* **1996**, *137*, 354–366. [CrossRef] [PubMed]

145. Rivera, P.; Arrabal, S.; Vargas, A.; Blanco Calvo, E.; Serrano, A.; Pavon, F.J.; Rodriguez de Fonseca, F.; Suarez, J. Localization of peroxisome proliferator-activated receptor alpha (PPARα) and N-acyl phosphatidylethanolamine phospholipase D (NAPE-PLD) in cells expressing the Ca2+-binding proteins calbindin, calretinin, and parvalbumin in the adult rat hippocampus. *Front. Neuroanat.* **2014**, *8*. [CrossRef]

146. Moreno, S.; Farioli-Vecchioli, S.; Cerù, M.P. Immunolocalization of peroxisome proliferator-activated receptors and retinoid x receptors in the adult rat CNS. *Neuroscience* **2004**, *123*, 131–145. [CrossRef] [PubMed]

147. Gofflot, F.; Chartoire, N.; Vasseur, L.; Heikkinen, S.; Dembele, D.; Le Merrer, J.; Auwerx, J. Systematic Gene Expression Mapping Clusters Nuclear Receptors According to Their Function in the Brain. *Cell* **2007**, *131*, 405–418. [CrossRef] [PubMed]

148. Cullingford, T.E.; Bhakoo, K.; Peuchen, S.; Dolphin, C.T.; Patel, R.; Clark, J.B. Distribution of mRNAs Encoding the Peroxisome Proliferator-Activated Receptor α, β, and γ and the Retinoid X Receptor α, β, and γ in Rat Central Nervous System. *J. Neurochem.* **1998**, *70*, 1366–1375. [CrossRef]

149. Roy, A.; Jana, M.; Corbett, G.T.; Ramaswamy, S.; Kordower, J.H.; Gonzalez, F.J.; Pahan, K. Regulation of Cyclic AMP Response Element Binding and Hippocampal Plasticity-Related Genes by Peroxisome Proliferator-Activated Receptor α. *Cell Rep.* **2013**, *4*, 724–737. [CrossRef]

150. Staels, B.; Dallongeville, J.; Auwerx, J.; Schoonjans, K.; Leitersdorf, E.; Fruchart, J.-C. Mechanism of Action of Fibrates on Lipid and Lipoprotein Metabolism. *Circulation* **1998**, *98*, 2088–2093. [CrossRef]

151. and, J.K.R.; Hashimoto, T. PEROXISOMAL β-OXIDATION AND PEROXISOME PROLIFERATOR–ACTIVATED RECEPTOR α: An Adaptive Metabolic System. *Annu. Rev. Nutr.* **2001**, *21*, 193–230. [CrossRef]

152. Schreurs, M.; Kuipers, F.; Van Der Leij, F.R. Regulatory enzymes of mitochondrial β-oxidation as targets for treatment of the metabolic syndrome. *Obes. Rev.* **2010**, *11*, 380–388. [CrossRef]

153. Chung, K.W.; Lee, E.K.; Lee, M.K.; Oh, G.T.; Yu, B.P.; Chung, H.Y. Impairment of PPARα and the Fatty Acid Oxidation Pathway Aggravates Renal Fibrosis during Aging. *J. Am. Soc. Nephrol.* **2018**, *29*, 1223–1237. [CrossRef]

154. Semple, R.K.; Chatterjee, V.K.K.; O'Rahilly, S. PPARγ and human metabolic disease. *J. Clin. Investig.* **2006**, *116*, 581–589. [CrossRef] [PubMed]

155. Barish, G.D.; Narkar, V.A.; Evans, R.M. PPARδ: A dagger in the heart of the metabolic syndrome. *J. Clin. Investig.* **2006**, *116*, 590–597. [CrossRef] [PubMed]

156. Kliewer, S.A.; Sundseth, S.S.; Jones, S.A.; Brown, P.J.; Wisely, G.B.; Koble, C.S.; Devchand, P.; Wahli, W.; Willson, T.M.; Lenhard, J.M.; et al. Fatty acids and eicosanoids regulate gene expression through direct interactions with peroxisome proliferator-activated receptors α and γ. *Proc. Natl. Acad. Sci. USA* **1997**, *94*, 4318–4323. [CrossRef] [PubMed]

157. Desvergne, B. PPARs special issue: Anchoring the present to explore the future. *Biochim. Biophys. Acta (Bba) Mol. Cell Biol. Lipids* **2007**, *1771*, 913–914. [CrossRef]

158. Roy, A.; Kundu, M.; Jana, M.; Mishra, R.K.; Yung, Y.; Luan, C.-H.; Gonzalez, F.J.; Pahan, K. Identification and characterization of PPARα ligands in the hippocampus. *Nat. Chem. Biol.* **2016**, *12*, 1075–1083. [CrossRef]

159. Cheng, H.S.; Tan, W.R.; Low, Z.S.; Marvalim, C.; Lee, J.Y.H.; Tan, N.S. Exploration and Development of PPAR Modulators in Health and Disease: An Update of Clinical Evidence. *Int. J. Mol. Sci.* **2019**, *20*, 5055. [CrossRef]

160. Burris, T.P. PPARα ligands make memories. *Nat. Chem. Biol.* **2016**, *12*, 993–994. [CrossRef]

161. Heneka, M.T.; Reyes-Irisarri, E.; Hüll, M.; Kummer, M.P. Impact and Therapeutic Potential of PPARs in Alzheimer's Disease. *Curr. Neuropharmacol.* **2011**, *9*, 643–650. [CrossRef]

162. Korbecki, J.; Bobiński, R.; Dutka, M. Self-regulation of the inflammatory response by peroxisome proliferator-activated receptors. *Inflamm. Res.* **2019**, *68*, 443–458. [CrossRef]
163. Marco, F.; Francesca, F.; Maria Paola, C.; Sandra, M. Neuroprotective Properties of Peroxisome Proliferator-Activated Receptor Alpha (PPARα) and its Lipid Ligands. *Curr. Med. Chem.* **2014**, *21*, 2803–2821. [CrossRef]
164. Aleshin, S.; Grabeklis, S.; Hanck, T.; Sergeeva, M.; Reiser, G. Peroxisome Proliferator-Activated Receptor (PPAR)-γ Positively Controls and PPARα Negatively Controls Cyclooxygenase-2 Expression in Rat Brain Astrocytes through a Convergence on PPARβ/δ via Mutual Control of PPAR Expression Levels. *Mol. Pharmacol.* **2009**, *76*, 414–424. [CrossRef] [PubMed]
165. Leisewitz, A.V.; Urrutia, C.R.; Martinez, G.R.; Loyola, G.; Bronfman, M. A PPARs cross-talk concertedly commits C6 glioma cells to oligodendrocytes and induces enzymes involved in myelin synthesis. *J. Cell. Physiol.* **2008**, *217*, 367–376. [CrossRef] [PubMed]
166. Aleshin, S.; Reiser, G. Role of the peroxisome proliferator-activated receptors (PPAR)-α, β/δ and γ triad in regulation of reactive oxygen species signaling in brain. *Biol. Chem.* **2013**, *394*, 1553. [CrossRef] [PubMed]
167. Aleshin, S.; Strokin, M.; Sergeeva, M.; Reiser, G. Peroxisome proliferator-activated receptor (PPAR)β/δ, a possible nexus of PPARα- and PPARγ-dependent molecular pathways in neurodegenerative diseases: Review and novel hypotheses. *Neurochem. Int.* **2013**, *63*, 322–330. [CrossRef] [PubMed]
168. Glatz, T.; Stöck, I.; Nguyen-Ngoc, M.; Gohlke, P.; Herdegen, T.; Culman, J.; Zhao, Y. Peroxisome-proliferator-activated receptors γ and peroxisome-proliferator-activated receptors β/δ and the regulation of interleukin 1 receptor antagonist expression by pioglitazone in ischaemic brain. *J. Hypertens.* **2010**, *28*, 1488–1497. [CrossRef] [PubMed]
169. Arnold, S.E.; Arvanitakis, Z.; Macauley-Rambach, S.L.; Koenig, A.M.; Wang, H.-Y.; Ahima, R.S.; Craft, S.; Gandy, S.; Buettner, C.; Stoeckel, L.E.; et al. Brain insulin resistance in type 2 diabetes and Alzheimer disease: Concepts and conundrums. *Nat. Rev. Neurol.* **2018**, *14*, 168–181. [CrossRef]
170. Mazon, J.N.; de Mello, A.H.; Ferreira, G.K.; Rezin, G.T. The impact of obesity on neurodegenerative diseases. *Life Sci.* **2017**, *182*, 22–28. [CrossRef]
171. Procaccini, C.; Santopaolo, M.; Faicchia, D.; Colamatteo, A.; Formisano, L.; de Candia, P.; Galgani, M.; De Rosa, V.; Matarese, G. Role of metabolism in neurodegenerative disorders. *Metabolism* **2016**, *65*, 1376–1390. [CrossRef]
172. Wójtowicz, S.; Strosznajder, A.K.; Jeżyna, M.; Strosznajder, J.B. The Novel Role of PPAR Alpha in the Brain: Promising Target in Therapy of Alzheimer's Disease and Other Neurodegenerative Disorders. *Neurochem. Res.* **2020**, *45*, 972–988. [CrossRef]
173. Mosconi, L. Glucose metabolism in normal aging and Alzheimer's disease: Methodological and physiological considerations for PET studies. *Clin. Transl. Imaging* **2013**, *1*, 217–233. [CrossRef]
174. Gong, C.-X.; Liu, F.; Iqbal, K. O-GlcNAcylation: A regulator of tau pathology and neurodegeneration. *Alzheimer's Dement.* **2016**, *12*, 1078–1089. [CrossRef] [PubMed]
175. Kandimalla, R.; Thirumala, V.; Reddy, P.H. Is Alzheimer's disease a Type 3 Diabetes? A critical appraisal. *Biochim. Biophys. Acta (Bba)-Mol. Basis Dis.* **2017**, *1863*, 1078–1089. [CrossRef] [PubMed]
176. de la Monte, S.M.; Wands, J.R. Review of insulin and insulin-like growth factor expression, signaling, and malfunction in the central nervous system: Relevance to Alzheimer's disease. *J. Alzheimer's Dis.* **2005**, *7*, 45–61. [CrossRef] [PubMed]
177. De Felice, F.G. Alzheimer's disease and insulin resistance: Translating basic science into clinical applications. *J. Clin. Investig.* **2013**, *123*, 531–539. [CrossRef] [PubMed]
178. Mullins, R.J.; Diehl, T.C.; Chia, C.W.; Kapogiannis, D. Insulin Resistance as a Link between Amyloid-Beta and Tau Pathologies in Alzheimer's Disease. *Front. Aging Neurosci.* **2017**, *9*. [CrossRef] [PubMed]
179. Gonçalves, R.A.; Wijesekara, N.; Fraser, P.E.; De Felice, F.G. The Link Between Tau and Insulin Signaling: Implications for Alzheimer's Disease and Other Tauopathies. *Front. Cell. Neurosci.* **2019**, *13*. [CrossRef]
180. Skerrett, R.; Pellegrino, M.P.; Casali, B.T.; Taraboanta, L.; Landreth, G.E. Combined Liver X Receptor/Peroxisome Proliferator-activated Receptor γ Agonist Treatment Reduces Amyloid β Levels and Improves Behavior in Amyloid Precursor Protein/Presenilin 1 Mice. *J. Biol. Chem.* **2015**, *290*, 21591–21602. [CrossRef]
181. Escribano, L.; Simón, A.-M.; Gimeno, E.; Cuadrado-Tejedor, M.; López de Maturana, R.; García-Osta, A.; Ricobaraza, A.; Pérez-Mediavilla, A.; Del Río, J.; Frechilla, D. Rosiglitazone Rescues Memory Impairment

in Alzheimer's Transgenic Mice: Mechanisms Involving a Reduced Amyloid and Tau Pathology. *Neuropsychopharmacology* **2010**, *35*, 1593–1604. [CrossRef]

182. Gold, M.; Gold, M.; Alderton, C.; Zvartau-Hind, M.; Egginton, S.; Saunders, A.M.; Irizarry, M.; Craft, S.; Landreth, G.; Linnamägi, Ü.; et al. Rosiglitazone Monotherapy in Mild-to-Moderate Alzheimer's Disease: Results from a Randomized, Double-Blind, Placebo-Controlled Phase III Study. *Dement. Geriatr. Cogn. Disord.* **2010**, *30*, 131–146. [CrossRef]

183. Galimberti, D.; Scarpini, E. Pioglitazone for the treatment of Alzheimer's disease. *Expert Opin. Investig. Drugs* **2017**, *26*, 97–101. [CrossRef]

184. Heneka, M.T.; Fink, A.; Doblhammer, G. Effect of pioglitazone medication on the incidence of dementia. *Ann. Neurol.* **2015**, *78*, 284–294. [CrossRef] [PubMed]

185. Woods, J.W.; Tanen, M.; Figueroa, D.J.; Biswas, C.; Zycband, E.; Moller, D.E.; Austin, C.P.; Berger, J.P. Localization of PPARδ in murine central nervous system: Expression in oligodendrocytes and neurons. *Brain Res.* **2003**, *975*, 10–21. [CrossRef]

186. Sergey, K.; Jill, C.R.; Douglas, L.F. A PPARdelta Agonist Reduces Amyloid Burden and Brain Inflammation in a Transgenic Mouse Model of Alzheimers Disease. *Curr. Alzheimer Res.* **2009**, *6*, 431–437. [CrossRef]

187. Tong, M.; Deochand, C.; Didsbury, J.; de la Monte, S.M. T3D-959: A Multi-Faceted Disease Remedial Drug Candidate for the Treatment of Alzheimer's Disease. *J. Alzheimer's Dis.* **2016**, *51*, 123–138. [CrossRef]

188. Reich, D.; Gallucci, G.; Tong, M.; de la Monte, S.M. Therapeutic Advantages of Dual Targeting of PPAR-δ and PPAR-γ in an Experimental Model of Sporadic Alzheimer's Disease. *J. Parkinson's Dis. Alzheimer's Dis.* **2018**, *5*. [CrossRef]

189. de la Monte, S.M.; Tong, M.; Schiano, I.; Didsbury, J. Improved Brain Insulin/IGF Signaling and Reduced Neuroinflammation with T3D-959 in an Experimental Model of Sporadic Alzheimer's Disease. *J. Alzheimer's Dis.* **2017**, *55*, 849–864. [CrossRef]

190. Chamberlain, S.; Gabriel, H.; Strittmatter, W.; Didsbury, J. An Exploratory Phase IIa Study of the PPAR delta/gamma Agonist T3D-959 Assessing Metabolic and Cognitive Function in Subjects with Mild to Moderate Alzheimer's Disease. *J. Alzheimer's Dis.* **2020**, *73*, 1085–1103. [CrossRef]

191. Murillo-Rodríguez, E.; Guzmán, K.; Arankowsky-Sandoval, G.; Salas-Crisóstomo, M.; Jiménez-Moreno, R.; Arias-Carrión, O. Evidence that activation of nuclear peroxisome proliferator-activated receptor alpha (PPARα) modulates sleep homeostasis in rats. *Brain Res. Bull.* **2016**, *127*, 156–163. [CrossRef]

192. Mijangos-Moreno, S.; Poot-Aké, A.; Guzmán, K.; Arankowsky-Sandoval, G.; Arias-Carrión, O.; Zaldívar-Rae, J.; Sarro-Ramírez, A.; Murillo-Rodríguez, E. Sleep and neurochemical modulation by the nuclear peroxisome proliferator-activated receptor α (PPAR-α) in rat. *Neurosci. Res.* **2016**, *105*, 65–69. [CrossRef]

193. Jiang, B.; Wang, Y.-J.; Wang, H.; Song, L.; Huang, C.; Zhu, Q.; Wu, F.; Zhang, W. Antidepressant-like effects of fenofibrate in mice via the hippocampal brain-derived neurotrophic factor signalling pathway. *Br. J. Pharmacol.* **2017**, *174*, 177–194. [CrossRef]

194. Li, M.; Wang, D.; Bi, W.; Jiang, Z.-e.; Piao, R.; Yu, H. Palmitoylethanolamide Exerts Antidepressant-Like Effects in Rats: Involvement of PPAR Pathway in the Hippocampus. *J. Pharmacol. Exp. Ther.* **2019**, *369*, 163–172. [CrossRef] [PubMed]

195. Scheggi, S.; Melis, M.; De Felice, M.; Aroni, S.; Muntoni, A.L.; Pelliccia, T.; Gambarana, C.; De Montis, M.G.; Pistis, M. PPARα modulation of mesolimbic dopamine transmission rescues depression-related behaviors. *Neuropharmacology* **2016**, *110*, 251–259. [CrossRef] [PubMed]

196. Jiang, B.; Huang, C.; Zhu, Q.; Tong, L.-J.; Zhang, W. WY14643 produces anti-depressant-like effects in mice via the BDNF signaling pathway. *Psychopharmacology* **2015**, *232*, 1629–1642. [CrossRef] [PubMed]

197. Puligheddu, M.; Pillolla, G.; Melis, M.; Lecca, S.; Marrosu, F.; De Montis, M.G.; Scheggi, S.; Carta, G.; Murru, E.; Aroni, S.; et al. PPAR-Alpha Agonists as Novel Antiepileptic Drugs: Preclinical Findings. *PLoS ONE* **2013**, *8*, e64541. [CrossRef] [PubMed]

198. Citraro, R.; Russo, E.; Leo, A.; Russo, R.; Avagliano, C.; Navarra, M.; Calignano, A.; De Sarro, G. Pharmacokinetic-pharmacodynamic influence of N-palmitoylethanolamine, arachidonyl-2′-chloroethylamide and WIN 55,212-2 on the anticonvulsant activity of antiepileptic drugs against audiogenic seizures in DBA/2 mice. *Eur. J. Pharmacol.* **2016**, *791*, 523–534. [CrossRef] [PubMed]

199. Porta, N.; Vallée, L.; Lecointe, C.; Bouchaert, E.; Staels, B.; Bordet, R.; Auvin, S. Fenofibrate, a peroxisome proliferator–activated receptor-α agonist, exerts anticonvulsive properties. *Epilepsia* **2009**, *50*, 943–948. [CrossRef]

200. Zhou, Y.; Yang, L.; Ma, A.; Zhang, X.; Li, W.; Yang, W.; Chen, C.; Jin, X. Orally administered oleoylethanolamide protects mice from focal cerebral ischemic injury by activating peroxisome proliferator-activated receptor α. *Neuropharmacology* **2012**, *63*, 242–249. [CrossRef]

201. Sun, Y.; Alexander, S.P.H.; Garle, M.J.; Gibson, C.L.; Hewitt, K.; Murphy, S.P.; Kendall, D.A.; Bennett, A.J. Cannabinoid activation of PPAR alpha; a novel neuroprotective mechanism. *Br. J. Pharmacol.* **2007**, *152*, 734–743. [CrossRef]

202. Dotson, A.L.; Wang, J.; Chen, Y.; Manning, D.; Nguyen, H.; Saugstad, J.A.; Offner, H. Sex differences and the role of PPAR alpha in experimental stroke. *Metab. Brain Dis.* **2016**, *31*, 539–547. [CrossRef]

203. Dotson, A.L.; Wang, J.; Liang, J.; Nguyen, H.; Manning, D.; Saugstad, J.A.; Offner, H. Loss of PPARα perpetuates sex differences in stroke reflected by peripheral immune mechanisms. *Metab. Brain Dis.* **2016**, *31*, 683–692. [CrossRef]

204. Costa, M.; Squassina, A.; Congiu, D.; Chillotti, C.; Niola, P.; Galderisi, S.; Pistis, M.; Del Zompo, M. Investigation of endocannabinoid system genes suggests association between peroxisome proliferator activator receptor-α gene (PPARA) and schizophrenia. *Eur. Neuropsychopharmacol.* **2013**, *23*, 749–759. [CrossRef] [PubMed]

205. D'Agostino, G.; La Rana, G.; Russo, R.; Sasso, O.; Iacono, A.; Esposito, E.; Raso, G.M.; Cuzzocrea, S.; Lo Verme, J.; Piomelli, D.; et al. Acute Intracerebroventricular Administration of Palmitoylethanolamide, an Endogenous Peroxisome Proliferator-Activated Receptor-α Agonist, Modulates Carrageenan-Induced Paw Edema in Mice. *J. Pharmacol. Exp. Ther.* **2007**, *322*, 1137–1143. [CrossRef] [PubMed]

206. Cuzzocrea, S.; Mazzon, E.; Di Paola, R.; Peli, A.; Bonato, A.; Britti, D.; Genovese, T.; Muià, C.; Crisafulli, C.; Caputi, A.P. The role of the peroxisome proliferator-activated receptor-α (PPAR-α) in the regulation of acute inflammation. *J. Leukoc. Biol.* **2006**, *79*, 999–1010. [CrossRef] [PubMed]

207. D'Orio, B.; Fracassi, A.; Ceru, M.P.; Moreno, S. Targeting PPARalpha in Alzheimer's Disease. *Curr. Alzheimer Res.* **2018**, *15*, 345–354. [CrossRef]

208. Kersten, S.; Stienstra, R. The role and regulation of the peroxisome proliferator activated receptor alpha in human liver. *Biochimie* **2017**, *136*, 75–84. [CrossRef]

209. Bougarne, N.; Weyers, B.; Desmet, S.J.; Deckers, J.; Ray, D.W.; Staels, B.; De Bosscher, K. Molecular Actions of PPARα in Lipid Metabolism and Inflammation. *Endocr. Rev.* **2018**, *39*, 760–802. [CrossRef]

210. Cunnane, S.C.; Courchesne-Loyer, A.; Vandenberghe, C.; St-Pierre, V.; Fortier, M.; Hennebelle, M.; Croteau, E.; Bocti, C.; Fulop, T.; Castellano, C.-A. Can Ketones Help Rescue Brain Fuel Supply in Later Life? Implications for Cognitive Health during Aging and the Treatment of Alzheimer's Disease. *Front. Mol. Neurosci.* **2016**, *9*. [CrossRef]

211. Yang, H.; Shan, W.; Zhu, F.; Wu, J.; Wang, Q. Ketone Bodies in Neurological Diseases: Focus on Neuroprotection and Underlying Mechanisms. *Front. Neurol.* **2019**, *10*. [CrossRef]

212. Kashiwaya, Y.; Takeshima, T.; Mori, N.; Nakashima, K.; Clarke, K.; Veech, R.L. D-β-Hydroxybutyrate protects neurons in models of Alzheimer's and Parkinson's disease. *Proc. Natl. Acad. Sci. USA* **2000**, *97*, 5440–5444. [CrossRef]

213. Yao, J.; Chen, S.; Mao, Z.; Cadenas, E.; Brinton, R.D. 2-Deoxy-D-Glucose Treatment Induces Ketogenesis, Sustains Mitochondrial Function, and Reduces Pathology in Female Mouse Model of Alzheimer's Disease. *PLoS ONE* **2011**, *6*, e21788. [CrossRef]

214. Henderson, S.T.; Vogel, J.L.; Barr, L.J.; Garvin, F.; Jones, J.J.; Costantini, L.C. Study of the ketogenic agent AC-1202 in mild to moderate Alzheimer's disease: A randomized, double-blind, placebo-controlled, multicenter trial. *Nutr. Metab.* **2009**, *6*, 31. [CrossRef]

215. Mazzola, C.; Medalie, J.; Scherma, M.; Panlilio, L.V.; Solinas, M.; Tanda, G.; Drago, F.; Cadet, J.L.; Goldberg, S.R.; Yasar, S. Fatty acid amide hydrolase (FAAH) inhibition enhances memory acquisition through activation of PPAR-alpha nuclear receptors. *Learn Mem* **2009**, *16*, 332–337. [CrossRef]

216. Xu, H.; You, Z.; Wu, Z.; Zhou, L.; Shen, J.; Gu, Z. WY14643 Attenuates the Scopolamine-Induced Memory Impairments in Mice. *Neurochem. Res.* **2016**, *41*, 2868–2879. [CrossRef] [PubMed]

217. D'Agostino, G.; Cristiano, C.; Lyons, D.J.; Citraro, R.; Russo, E.; Avagliano, C.; Russo, R.; Raso, G.M.; Meli, R.; De Sarro, G.; et al. Peroxisome proliferator-activated receptor alpha plays a crucial role in behavioral repetition and cognitive flexibility in mice. *Mol. Metab.* **2015**, *4*, 528–536. [CrossRef] [PubMed]

218. Bliss, T.V.P.; Collingridge, G.L. A synaptic model of memory: Long-term potentiation in the hippocampus. *Nature* **1993**, *361*, 31–39. [CrossRef] [PubMed]

219. Mounier, A.; Georgiev, D.; Nam, K.N.; Fitz, N.F.; Castranio, E.L.; Wolfe, C.M.; Cronican, A.A.; Schug, J.; Lefterov, I.; Koldamova, R. Bexarotene-Activated Retinoid X Receptors Regulate Neuronal Differentiation and Dendritic Complexity. *J. Neurosci.* **2015**, *35*, 11862–11876. [CrossRef] [PubMed]

220. Nam, K.N.; Mounier, A.; Fitz, N.F.; Wolfe, C.; Schug, J., Lefterov, I.; Koldamova, R. RXR controlled regulatory networks identified in mouse brain counteract deleterious effects of Aβ oligomers. *Sci. Rep.* **2016**, *6*, 24048. [CrossRef]

221. Chikahisa, S.; Chida, D.; Shiuchi, T.; Harada, S.; Shimizu, N.; Otsuka, A.; Tanioka, D.; Séi, H. Enhancement of fear learning in PPARα knockout mice. *Behav. Brain Res.* **2019**, *359*, 664–670. [CrossRef]

222. Brune, S.; Kölsch, H.; Ptok, U.; Majores, M.; Schulz, A.; Schlosser, R.; Rao, M.L.; Maier, W.; Heun, R. Polymorphism in the peroxisome proliferator-activated receptor α gene influences the risk for Alzheimer's disease. *J. Neural Transm.* **2003**, *110*, 1041–1050. [CrossRef]

223. Kölsch, H.; Lehmann, D.J.; Ibrahim-Verbaas, C.A.; Combarros, O.; van Duijn, C.M.; Hammond, N.; Belbin, O.; Cortina-Borja, M.; Lehmann, M.G.; Aulchenko, Y.S.; et al. Interaction of insulin and PPAR-α genes in Alzheimer's disease: The Epistasis Project. *J. Neural Transm.* **2012**, *119*, 473–479. [CrossRef]

224. Heun, R.; Kölsch, H.; Ibrahim-Verbaas, C.A.; Combarros, O.; Aulchenko, Y.S.; Breteler, M.; Schuur, M.; van Duijn, C.M.; Hammond, N.; Belbin, O.; et al. Interactions between PPAR-α and inflammation-related cytokine genes on the development of Alzheimer's disease, observed by the Epistasis Project. *Int. J. Mol. Epidemiol. Genet.* **2012**, *3*, 39–47. [PubMed]

225. Sjölander, A.; Minthon, L.; Bogdanovic, N.; Wallin, A.; Zetterberg, H.; Blennow, K. The PPAR-α gene in Alzheimer's disease: Lack of replication of earlier association. *Neurobiol. Aging* **2009**, *30*, 666–668. [CrossRef] [PubMed]

226. de la Monte, S.M.; Wands, J.R. Molecular indices of oxidative stress and mitochondrial dysfunction occur early and often progress with severity of Alzheimer's disease. *J. Alzheimer's Dis.* **2006**, *9*, 167–181. [CrossRef] [PubMed]

227. Staels, B.; Maes, M.; Zambon, A. Fibrates and future PPARα agonists in the treatment of cardiovascular disease. *Nat. Clin. Pract. Cardiovasc. Med.* **2008**, *5*, 542–553. [CrossRef]

228. Gross, B.; Pawlak, M.; Lefebvre, P.; Staels, B. PPARs in obesity-induced T2DM, dyslipidaemia and NAFLD. *Nat. Rev. Endocrinol.* **2017**, *13*, 36–49. [CrossRef]

229. Brown, W.V.; Dujovne, C.A.; Farquhar, J.W.; Feldman, E.B.; Grundy, S.M.; Knopp, R.H.; Lasser, N.L.; Mellies, M.J.; Palmer, R.H.; Samuel, P. Effects of fenofibrate on plasma lipids. Double-blind, multicenter study in patients with type IIA or IIB hyperlipidemia. *Arterioscler.: Off. J. Am. Heart Assoc. Inc.* **1986**, *6*, 670–678. [CrossRef]

230. Brown, W.V. Focus on Fenofibrate. *Hosp. Pract.* **1988**, *23*, 31–40. [CrossRef]

231. Hennuyer, N.; Duplan, I.; Paquet, C.; Vanhoutte, J.; Woitrain, E.; Touche, V.; Colin, S.; Vallez, E.; Lestavel, S.; Lefebvre, P.; et al. The novel selective PPARα modulator (SPPARMα) pemafibrate improves dyslipidemia, enhances reverse cholesterol transport and decreases inflammation and atherosclerosis. *Atherosclerosis* **2016**, *249*, 200–208. [CrossRef]

232. Yamazaki, Y.; Abe, K.; Toma, T.; Nishikawa, M.; Ozawa, H.; Okuda, A.; Araki, T.; Oda, S.; Inoue, K.; Shibuya, K.; et al. Design and synthesis of highly potent and selective human peroxisome proliferator-activated receptor α agonists. *Bioorg. Med. Chem. Lett.* **2007**, *17*, 4689–4693. [CrossRef] [PubMed]

233. Fruchart, J.-C. Selective peroxisome proliferator-activated receptorα modulators (SPPARMα): The next generation of peroxisome proliferator-activated receptor α-agonists. *Cardiovasc. Diabetol.* **2013**, *12*, 82. [CrossRef]

234. Araki, E.; Yamashita, S.; Arai, H.; Yokote, K.; Satoh, J.; Inoguchi, T.; Nakamura, J.; Maegawa, H.; Yoshioka, N.; Tanizawa, Y.; et al. Effects of Pemafibrate, a Novel Selective PPARα Modulator, on Lipid and Glucose Metabolism in Patients With Type 2 Diabetes and Hypertriglyceridemia: A Randomized, Double-Blind, Placebo-Controlled, Phase 3 Trial. *Diabetes Care* **2018**, *41*, 538–546. [CrossRef] [PubMed]

235. Ishibashi, S.; Arai, H.; Yokote, K.; Araki, E.; Suganami, H.; Yamashita, S. Efficacy and safety of pemafibrate (K-877), a selective peroxisome proliferator-activated receptor α modulator, in patients with dyslipidemia: Results from a 24-week, randomized, double blind, active-controlled, phase 3 trial. *J. Clin. Lipidol.* **2018**, *12*, 173–184. [CrossRef] [PubMed]

236. Chandra, S.; Roy, A.; Jana, M.; Pahan, K. Cinnamic acid activates PPARα to stimulate Lysosomal biogenesis and lower Amyloid plaque pathology in an Alzheimer's disease mouse model. *Neurobiol. Dis.* **2019**, *124*, 379–395. [CrossRef] [PubMed]

237. Oakley, H.; Cole, S.L.; Logan, S.; Maus, E.; Shao, P.; Craft, J.; Guillozet-Bongaarts, A.; Ohno, M.; Disterhoft, J.; Van Eldik, L.; et al. Intraneuronal β-Amyloid Aggregates, Neurodegeneration, and Neuron Loss in Transgenic Mice with Five Familial Alzheimer's Disease Mutations: Potential Factors in Amyloid Plaque Formation. *J. Neurosci.* **2006**, *26*, 10129–10140. [CrossRef] [PubMed]

238. Luo, R.; Su, L.-Y.; Li, G.; Yang, J.; Liu, Q.; Yang, L.-X.; Zhang, D.-F.; Zhou, H.; Xu, M.; Fan, Y.; et al. Activation of PPARA-mediated autophagy reduces Alzheimer disease-like pathology and cognitive decline in a murine model. *Autophagy* **2020**, *16*, 52–69. [CrossRef] [PubMed]

239. Roy, A.; Jana, M.; Kundu, M.; Corbett, G.T.; Rangaswamy, S.B.; Mishra, R.K.; Luan, C.-H.; Gonzalez, F.J.; Pahan, K. HMG-CoA Reductase Inhibitors Bind to PPARα to Upregulate Neurotrophin Expression in the Brain and Improve Memory in Mice. *Cell Metab.* **2015**, *22*, 253–265. [CrossRef]

240. Patel, D.; Roy, A.; Kundu, M.; Jana, M.; Luan, C.-H.; Gonzalez, F.J.; Pahan, K. Aspirin binds to PPARα to stimulate hippocampal plasticity and protect memory. *Proc. Natl. Acad. Sci. USA* **2018**, *115*, E7408–E7417. [CrossRef]

241. Chandra, S.; Jana, M.; Pahan, K. Aspirin Induces Lysosomal Biogenesis and Attenuates Amyloid Plaque Pathology in a Mouse Model of Alzheimer's Disease via PPARα. *J. Neurosci.* **2018**, *38*, 6682–6699. [CrossRef]

242. Medicine, N.L.O. Study Evaluating Potential Interaction Between SAM-531 And Gemfibrozil When Co-Administered. Available online: https://clinicaltrials.gov/ct2/show/NCT00966966 (accessed on 10 April 2020).

243. Medicine, N.L.O. Modulation of Micro-RNA Pathways by Gemfibrozil in Predementia Alzheimer Disease. Available online: https://clinicaltrials.gov/ct2/show/NCT02045056 (accessed on 10 April 2020).

244. Krol, J.; Loedige, I.; Filipowicz, W. The widespread regulation of microRNA biogenesis, function and decay. *Nat. Rev. Genet.* **2010**, *11*, 597–610. [CrossRef]

245. Wang, W.-X.; Rajeev, B.W.; Stromberg, A.J.; Ren, N.; Tang, G.; Huang, Q.; Rigoutsos, I.; Nelson, P.T. The Expression of MicroRNA miR-107 Decreases Early in Alzheimer's Disease and May Accelerate Disease Progression through Regulation of β-Site Amyloid Precursor Protein-Cleaving Enzyme 1. *J. Neurosci.* **2008**, *28*, 1213–1223. [CrossRef]

246. Nelson, P.T.; Wang, W.-X. MiR-107 is Reduced in Alzheimer's Disease Brain Neocortex: Validation Study. *J. Alzheimer's Dis.* **2010**, *21*, 75–79. [CrossRef]

247. Vassar, R.; Bennett, B.D.; Babu-Khan, S.; Kahn, S.; Mendiaz, E.A.; Denis, P.; Teplow, D.B.; Ross, S.; Amarante, P.; Loeloff, R.; et al. β-Secretase Cleavage of Alzheimer's Amyloid Precursor Protein by the Transmembrane Aspartic Protease BACE. *Science* **1999**, *286*, 735–741. [CrossRef] [PubMed]

248. Haniu, M.; Denis, P.; Young, Y.; Mendiaz, E.A.; Fuller, J.; Hui, J.O.; Bennett, B.D.; Kahn, S.; Ross, S.; Burgess, T.; et al. Characterization of Alzheimer's β-Secretase Protein BACE: A PEPSIN FAMILY MEMBER WITH UNUSUAL PROPERTIES. *J. Biol. Chem.* **2000**, *275*, 21099–21106. [CrossRef] [PubMed]

249. Corbett, G.T.; Gonzalez, F.J.; Pahan, K. Activation of peroxisome proliferator-activated receptor α stimulates ADAM10-mediated proteolysis of APP. *Proc. Natl. Acad. Sci. USA* **2015**, *112*, 8445–8450. [CrossRef] [PubMed]

250. Yuan, X.-Z.; Sun, S.; Tan, C.-C.; Yu, J.-T.; Tan, L. The Role of ADAM10 in Alzheimer's Disease. *J. Alzheimer's Dis.* **2017**, *58*, 303–322. [CrossRef]

251. Sastre, M.; Dewachter, I.; Landreth, G.E.; Willson, T.M.; Klockgether, T.; van Leuven, F.; Heneka, M.T. Nonsteroidal Anti-Inflammatory Drugs and Peroxisome Proliferator-Activated Receptor-γ Agonists Modulate Immunostimulated Processing of Amyloid Precursor Protein through Regulation of β-Secretase. *J. Neurosci.* **2003**, *23*, 9796–9804. [CrossRef]

252. Sastre, M.; Dewachter, I.; Rossner, S.; Bogdanovic, N.; Rosen, E.; Borghgraef, P.; Evert, B.O.; Dumitrescu-Ozimek, L.; Thal, D.R.; Landreth, G.; et al. Nonsteroidal anti-inflammatory drugs repress β-secretase gene promoter activity by the activation of PPARγ. *Proc. Natl. Acad. Sci. USA* **2006**, *103*, 443–448. [CrossRef]

253. Duan, S.Z.; Usher, M.G.; Foley, E.L.; Milstone, D.S.; Brosius, F.C.; Mortensen, R.M. Sex dimorphic actions of rosiglitazone in generalised peroxisome proliferator-activated receptor-γ (PPAR-γ)-deficient mice. *Diabetologia* **2010**, *53*, 1493–1505. [CrossRef]

254. Jalouli, M.; Carlsson, L.; Améen, C.; Lindén, D.; Ljungberg, A.; Michalik, L.; Edén, S.; Wahli, W.; Oscarsson, J. Sex Difference in Hepatic Peroxisome Proliferator-Activated Receptor α Expression: Influence of Pituitary and Gonadal Hormones. *Endocrinology* **2003**, *144*, 101–109. [CrossRef]

255. Dunn, S.E.; Ousman, S.S.; Sobel, R.A.; Zuniga, L.; Baranzini, S.E.; Youssef, S.; Crowell, A.; Loh, J.; Oksenberg, J.; Steinman, L. Peroxisome proliferator–activated receptor (PPAR)α expression in T cells mediates gender differences in development of T cell–mediated autoimmunity. *J. Exp. Med.* **2007**, *204*, 321–330. [CrossRef]

256. Maren, S.; De Oca, B.; Fanselow, M.S. Sex differences in hippocampal long-term potentiation (LTP) and Pavlovian fear conditioning in rats: Positive correlation between LTP and contextual learning. *Brain Res.* **1994**, *661*, 25–34. [CrossRef]

257. Qi, X.; Zhang, K.; Xu, T.; Yamaki, V.N.; Wei, Z.; Huang, M.; Rose, G.M.; Cai, X. Sex Differences in Long-Term Potentiation at Temporoammonic-CA1 Synapses: Potential Implications for Memory Consolidation. *PLoS ONE* **2016**, *11*, e0165891. [CrossRef]

258. Monfort, P.; Gomez-Gimenez, B.; Llansola, M.; Felipo, V. Gender Differences in Spatial Learning, Synaptic Activity, and Long-Term Potentiation in the Hippocampus in Rats: Molecular Mechanisms. *ACS Chem. Neurosci.* **2015**, *6*, 1420–1427. [CrossRef] [PubMed]

259. Yang, D.-W.; Pan, B.; Han, T.-Z.; Xie, W. Sexual dimorphism in the induction of LTP: Critical role of tetanizing stimulation. *Life Sci.* **2004**, *75*, 119–127. [CrossRef]

260. Ris, L.; Angelo, M.; Plattner, F.; Capron, B.; Errington, M.L.; Bliss, T.V.P.; Godaux, E.; Giese, K.P. Sexual dimorphisms in the effect of low-level p25 expression on synaptic plasticity and memory. *Eur. J. Neurosci.* **2005**, *21*, 3023–3033. [CrossRef]

261. Harte-Hargrove, L.C.; Varga-Wesson, A.; Duffy, A.M.; Milner, T.A.; Scharfman, H.E. Opioid Receptor-Dependent Sex Differences in Synaptic Plasticity in the Hippocampal Mossy Fiber Pathway of the Adult Rat. *J. Neurosci.* **2015**, *35*, 1723–1738. [CrossRef]

262. Córdoba Montoya, D.A.; Carrer, H.F. Estrogen facilitates induction of long term potentiation in the hippocampus of awake rats1First published on the World Wide Web on 4 November 1997.1. *Brain Res.* **1997**, *778*, 430–438. [CrossRef]

263. Good, M.; Day, M.; Muir, J.L. Cyclical changes in endogenous levels of oestrogen modulate the induction of LTD and LTP in the hippocampal CA1 region. *Eur. J. Neurosci.* **1999**, *11*, 4476–4480. [CrossRef] [PubMed]

264. Daniel, J.M.; Lee, C.D. Estrogen replacement in ovariectomized rats affects strategy selection in the Morris water maze. *Neurobiol. Learn. Mem.* **2004**, *82*, 142–149. [CrossRef]

265. Korol, D.L.; Malin, E.L.; Borden, K.A.; Busby, R.A.; Couper-Leo, J. Shifts in preferred learning strategy across the estrous cycle in female rats. *Horm. Behav.* **2004**, *45*, 330–338. [CrossRef]

266. Gupta, R.R.; Sen, S.; Diepenhorst, L.L.; Rudick, C.N.; Maren, S. Estrogen modulates sexually dimorphic contextual fear conditioning and hippocampal long-term potentiation (LTP) in rats11Published on the World Wide Web on 1 December 2000. *Brain Res.* **2001**, *888*, 356–365. [CrossRef]

267. Joffe, H.; Hall, J.E.; Gruber, S.; Sarmiento, I.A.; Cohen, L.S.; Yurgelun-Todd, D.; Martin, K.A. Estrogen therapy selectively enhances prefrontal cognitive processes: A randomized, double-blind, placebo-controlled study with functional magnetic resonance imaging in perimenopausal and recently postmenopausal women. *Menopause* **2006**, *13*, 411–422. [CrossRef] [PubMed]

268. Scott, E.; Zhang, Q.-g.; Wang, R.; Vadlamudi, R.; Brann, D. Estrogen neuroprotection and the critical period hypothesis. *Front. Neuroendocrinol.* **2012**, *33*, 85–104. [CrossRef] [PubMed]

269. Merlo, S.; Spampinato, S.F.; Sortino, M.A. Estrogen and Alzheimer's disease: Still an attractive topic despite disappointment from early clinical results. *Eur. J. Pharmacol.* **2017**, *817*, 51–58. [CrossRef]

270. Maher, A.C.; Akhtar, M.; Vockley, J.; Tarnopolsky, M.A. Women Have Higher Protein Content of β-Oxidation Enzymes in Skeletal Muscle than Men. *PLoS ONE* **2010**, *5*, e12025. [CrossRef]

271. Wege, N.; Schutkowski, A.; Boenn, M.; Bialek, J.; Schlitt, A.; Noack, F.; Grosse, I.; Stangl, G.I. Men and women differ in their diurnal expression of monocyte peroxisome proliferator-activated receptor-α in the fed but not in the fasted state. *FASEB J.* **2015**, *29*, 2905–2911. [CrossRef]

272. Zhang, M.A.; Rego, D.; Moshkova, M.; Kebir, H.; Chruscinski, A.; Nguyen, H.; Akkermann, R.; Stanczyk, F.Z.; Prat, A.; Steinman, L.; et al. Peroxisome proliferator-activated receptor (PPAR)α and -γ regulate IFNγ and

IL-17A production by human T cells in a sex-specific way. *Proc. Natl. Acad. Sci. USA* **2012**, *109*, 9505–9510. [CrossRef]

273. Fisher, D.W.; Bennett, D.A.; Dong, H. Sexual dimorphism in predisposition to Alzheimer's disease. *Neurobiol. Aging* **2018**, *70*, 308–324. [CrossRef]

274. Riedel, B.C.; Thompson, P.M.; Brinton, R.D. Age, APOE and sex: Triad of risk of Alzheimer's disease. *J. Steroid Biochem. Mol. Biol.* **2016**, *160*, 134–147. [CrossRef]

275. Appelman, Y.; van Rijn, B.B.; ten Haaf, M.E.; Boersma, E.; Peters, S.A.E. Sex differences in cardiovascular risk factors and disease prevention. *Atherosclerosis* **2015**, *241*, 211–218. [CrossRef]

276. Brinton, R.D. Estrogen-induced plasticity from cells to circuits: Predictions for cognitive function. *Trends Pharmacol. Sci.* **2009**, *30*, 212–222. [CrossRef]

277. Liu, F.; Day, M.; Muñiz, L.C.; Bitran, D.; Arias, R.; Revilla-Sanchez, R.; Grauer, S.; Zhang, G.; Kelley, C.; Pulito, V.; et al. Activation of estrogen receptor-β regulates hippocampal synaptic plasticity and improves memory. *Nat. Neurosci.* **2008**, *11*, 334–343. [CrossRef] [PubMed]

278. Scheyer, O.; Rahman, A.; Hristov, H.; Berkowitz, C.; Isaacson, R.S.; Diaz Brinton, R.; Mosconi, L. Female Sex and Alzheimer's Risk: The Menopause Connection. *J. Prev. Alzheimer's Dis.* **2018**, *5*, 225–230. [CrossRef] [PubMed]

279. Rahman, A.; Jackson, H.; Hristov, H.; Isaacson, R.S.; Saif, N.; Shetty, T.; Etingin, O.; Henchcliffe, C.; Brinton, R.D.; Mosconi, L. Sex and Gender Driven Modifiers of Alzheimer's: The Role for Estrogenic Control Across Age, Race, Medical, and Lifestyle Risks. *Front. Aging Neurosci.* **2019**, *11*. [CrossRef]

280. McEwen, B.S.; Akama, K.T.; Spencer-Segal, J.L.; Milner, T.A.; Waters, E.M. Estrogen effects on the brain: Actions beyond the hypothalamus via novel mechanisms. *Behav. Neurosci.* **2012**, *126*, 4–16. [CrossRef] [PubMed]

281. Yao, J.; Brinton, R.D. Estrogen Regulation of Mitochondrial Bioenergetics: Implications for Prevention of Alzheimer's Disease. In *Advances in Pharmacolog*; Michaelis, E.K., Michaelis, M.L., Eds.; Academic Press: Cambridge, MA, USA, 2012; Volume 64, pp. 327–371.

282. Mattson, M.P.; Magnus, T. Ageing and neuronal vulnerability. *Nat. Rev. Neurosci.* **2006**, *7*, 278–294. [CrossRef] [PubMed]

283. Benz, V.; Kintscher, U.; Foryst-Ludwig, A. Sex-Specific Differences in Type 2 Diabetes Mellitus and Dyslipidemia Therapy: PPAR Agonists. In *Sex and Gender Differences in Pharmacology*; Regitz-Zagrosek, V., Ed.; Springer: Berlin/Heidelberg, Germany, 2012; pp. 387–410.

284. Wong, M.W.K.; Braidy, N.; Crawford, J.; Pickford, R.; Song, F.; Mather, K.A.; Attia, J.; Brodaty, H.; Sachdev, P.; Poljak, A. APOE Genotype Differentially Modulates Plasma Lipids in Healthy Older Individuals, with Relevance to Brain Health. *J. Alzheimer's Dis.* **2019**, *72*, 703–716. [CrossRef] [PubMed]

285. Díaz, M.; Fabelo, N.; Ferrer, I.; Marín, R. "Lipid raft aging" in the human frontal cortex during nonpathological aging: Gender influences and potential implications in Alzheimer's disease. *Neurobiol. Aging* **2018**, *67*, 42–52. [CrossRef]

286. Mielke, M.M.; Haughey, N.J.; Han, D.; An, Y.; Bandaru, V.V.R.; Lyketsos, C.G.; Ferrucci, L.; Resnick, S.M. The Association Between Plasma Ceramides and Sphingomyelins and Risk of Alzheimer's Disease Differs by Sex and APOE in the Baltimore Longitudinal Study of Aging. *J. Alzheimer's Dis.* **2017**, *60*, 819–828. [CrossRef]

287. Couttas, T.A.; Kain, N.; Tran, C.; Chatterton, Z.; Kwok, J.B.; Don, A.S. Age-Dependent Changes to Sphingolipid Balance in the Human Hippocampus are Gender-Specific and May Sensitize to Neurodegeneration. *J. Alzheimer's Dis.* **2018**, *63*, 503–514. [CrossRef]

Review

Therapeutic Potential of Peroxisome Proliferator-Activated Receptor (PPAR) Agonists in Substance Use Disorders: A Synthesis of Preclinical and Human Evidence

Justin Matheson [1,2,*] and **Bernard Le Foll** [1,3,4,5,6,7,8]

1. Department of Pharmacology and Toxicology, Faculty of Medicine, University of Toronto, 27 King's College Circle, Toronto, ON M5S 3H7, Canada; bernard.lefoll@camh.ca
2. Institute for Mental Health Policy Research, Centre for Addiction and Mental Health, 33 Russell Street, Toronto, ON M5S 2S1, Canada
3. Translational Addiction Research Laboratory, Centre for Addiction and Mental Health, 33 Russell Street, Toronto, ON M5S 2S1, Canada
4. Addictions Division, Centre for Addiction and Mental Health, 100 Stokes Street, Toronto, ON M6J 1H4, Canada
5. Campbell Family Mental Health Research Institute, Centre for Addiction and Mental Health, 250 College Street, Toronto, ON M5T 1R8, Canada
6. Department of Psychiatry, Faculty of Medicine, University of Toronto, 250 College Street, Toronto, ON M5T 1R8, Canada
7. Institute of Medical Sciences, University of Toronto, 1 King's College Circle, Room 2374, Toronto, ON M5S 1A8, Canada
8. Department of Family and Community Medicine, University of Toronto, 500 University Avenue, 5th Floor, Toronto, ON M5G 1V7, Canada
* Correspondence: justin.matheson@camh.ca; Tel.: +1-416-535-8501 (ext. 34727)

Received: 17 April 2020; Accepted: 8 May 2020; Published: 12 May 2020

Abstract: Targeting peroxisome proliferator-activated receptors (PPARs) has received increasing interest as a potential strategy to treat substance use disorders due to the localization of PPARs in addiction-related brain regions and the ability of PPAR ligands to modulate dopamine neurotransmission. Robust evidence from animal models suggests that agonists at both the PPAR-α and PPAR-γ isoforms can reduce both positive and negative reinforcing properties of ethanol, nicotine, opioids, and possibly psychostimulants. A reduction in the voluntary consumption of ethanol following treatment with PPAR agonists seems to be the most consistent finding. However, the human evidence is limited in scope and has so far been less promising. There have been no published human trials of PPAR agonists for treatment of alcohol use disorder, despite the compelling preclinical evidence. Two trials of PPAR-α agonists as potential smoking cessation drugs found no effect on nicotine-related outcomes. The PPAR-γ agonist pioglitazone showed some promise in reducing heroin, nicotine, and cocaine craving in two human laboratory studies and one pilot trial, yet other outcomes were unaffected. Potential explanations for the discordance between the animal and human evidence, such as the potency and selectivity of PPAR ligands and sex-related variability in PPAR physiology, are discussed.

Keywords: PPAR; nuclear receptors; addiction; alcohol; nicotine; opioids; psychostimulants; animal models; human studies

---

1. Introduction

Substance use disorders (SUDs) continue to represent a significant global public health burden. In 2017, of the estimated 271 million people aged 16–64 years worldwide who had used drugs in the past year, nearly 35 million (~13%) were estimated to suffer from an SUD [1]. An SUD is a diagnostic entity in the Diagnostic and Statistical Manual, 5th Edition (DSM-V) that refers to the repeated use of a substance that causes significant impairment, e.g., continued use despite physical and psychological harms and failure to meet expectations at work or school [2]. The term "addiction" is often used to refer to the severe stage of an SUD characterized by compulsive drug-seeking despite negative consequences [3,4] that runs a chronic, relapsing course with poor long-term durability of abstinence from drug-taking even with treatment [5].

Research into the neurobiology of addictions over the past few decades has substantively advanced our understanding of the key facets of compulsive drug-taking [6,7]. For example, while the focus of early addictions research was the acute, positively reinforcing properties of drugs of abuse, it is now recognized that negatively reinforcing states involving anhedonia, dysphoria, and anxiety become more important in maintaining drug-taking over time [7]. As a result, motivation to use the drug shifts from seeking pleasure to avoiding negative affect. Thus, pharmacotherapeutic strategies to treat addictions need to not only reduce the reinforcing or rewarding properties of drugs, but also target the negatively reinforcing states associated with chronic drug-taking that contribute to the significant risk of relapse [7]. Agonist substitution therapies have been successful in mitigating this negative reinforcement in some SUDs, e.g., methadone or buprenorphine for managing withdrawal and craving associated with opioid use disorder [8] and nicotine replacement therapy (NRT) for managing nicotine withdrawal [9]. Other medications, such as naltrexone or acamprosate for alcohol use disorder [10] and varenicline or bupropion for nicotine dependence [9], have demonstrated some efficacy in reducing positive and/or negative reinforcing aspects of drug use. Nevertheless, long-term abstinence rates remain low across SUDs, highlighting the need for novel pharmacological treatment approaches.

Peroxisome proliferator-activated receptors (PPARs) are a subfamily of nuclear receptors that dimerize with retinoid X receptors (RXRs) to regulate gene expression by binding to specific peroxisome proliferator response elements (PPREs) in enhancer sites of select genes [11]. Three isoforms of PPARs have been identified: α, γ, and β/δ. So far, the therapeutic potential of PPAR ligands had been in non-psychiatric fields. While PPARs were initially identified as lipid sensors [11], burgeoning evidence has demonstrated a role of these nuclear receptors in a wide range of physiological functions, including central nervous system (CNS) functions such as memory consolidation and modulation of pain perception [12]. PPAR agonists have been recently considered for their potential to treat neuropsychiatric disorders, largely due to their ability to target levels of neuroinflammation thought to be involved in the pathophysiology of these illnesses [13]. In particular, mounting evidence of an important relationship between neuroimmune function and addition-related processes has generated interest in investigating the role of PPARs in drug-related behaviors [14,15].

Converging lines of evidence have also suggested a more direct role of PPARs in addiction-relevant neurocircuitry. Initial evidence came from studies demonstrating that selective inhibition of fatty acid amide hydrolase (FAAH), an enzyme responsible for degradation of the endogenous cannabinoid anandamide and the endogenous PPAR ligands oleoylethanolamide (OEA) and palmitoylethanolamide (PEA), could suppress nicotine-induced activation of dopamine neurons in rats [16,17]. Importantly, this effect was mimicked by OEA and PEA, but not anandamide, suggesting the effect was due to PPAR activation specifically [16]. Exogenous PPAR agonists have also been demonstrated to attenuate nicotine-induced [18,19] and heroin-induced [20] excitation of dopamine neurons in the ventral tegmental area (VTA) and elevations of dopamine in the nucleus accumbens (NAc) shell in rats. Further confirmatory evidence comes from rodent studies demonstrating that PPAR isoforms are indeed localized in addiction-relevant brain regions such as the VTA [21,22], an important part of the mesocorticolimbic dopaminergic system that plays a central role in drug-related reward [7], and that PPAR-γ colocalizes with tyrosine-hydroxylase-positive cells in the VTA, suggesting direct expression in dopaminergic neurons [23]. A detailed presentation of the neurobiological substrates mediating the impact of PPAR agonists on addiction-related behaviors is beyond the topic of the present review (see [18,19] for some mechanistic studies).

The goal of the present review is to expand upon our previous review of the preclinical evidence for a role of PPARs in addiction [24] to incorporate novel preclinical findings as well as the current state of evidence from clinical and laboratory studies in humans.

2. Preclinical Behavioral Evidence

Evidence for the role of PPAR agonists in modifying addiction-like behaviors in animal models is broadly divided into two categories: drug consumption/motivation to use and withdrawal/relapse. A summary of key methodological details and relevant findings of the studies reviewed is provided in Table 1.

2.1. PPAR-α Agonists

2.1.1. Consumption/Motivation

A significant body of evidence has consistently demonstrated that PPAR-α agonists can attenuate voluntary consumption and operant self-administration of ethanol in rodents [25–32]. Using the two-bottle choice paradigm, studies have found a decrease in voluntary consumption of ethanol following administration of the clinically useful drugs gemfibrozil [25] and fenofibrate [26,27,29,30,32], the endogenous agonist OEA [28], the experimental agonist WY14643 [28], and the dual PPAR-α/γ agonist tesaglitazar [26,29,30]. In addition, operant self-administration of ethanol was attenuated following administration of OEA and WY14643 under a one-response fixed ratio (i.e., FR1) schedule [28] and fenofibrate under FR2 and progressive ratio (PR) schedules [31]. Importantly, the effects of the PPAR-α agonists on attenuation of voluntary consumption of ethanol were reversed when animals were pre-treated with the PPAR-α antagonists GW6471 [28] or MK886 [30]. Overall, these results strongly support a role of PPAR-α agonism in reducing willingness to consume ethanol and in reducing the reinforcing properties of ethanol.

Two studies have assessed the effects of fenofibrate on the development of ethanol conditioned place preference (CPP) as a measure of the rewarding effects of ethanol, with mixed results [32,33]. Blednov et al. (2016) found no effect of oral administration of 150 mg/kg fenofibrate or 1.5 mg/kg of the dual PPAR-α/γ agonist tesaglitazar on the development of ethanol CPP in male mice [33]. However, Rivera-Meza et al. (2017) found that oral administration of 50 mg/kg fenofibrate attenuated the development of ethanol CPP in male rats selectively bred for high ethanol intake (i.e., UChB rats) [32]. The inconsistency between these two studies is unclear but could be due to the different doses of fenofibrate used or differences in ethanol-related behaviors of the two different animal models.

More limited, but robust, evidence has supported a role of PPAR-α agonists in attenuating operant self-administration of nicotine in rodents and non-human primates [18,19]. Mascia et al. (2011) found that both WY14643 and methyl-OEA reduced nicotine self-administration under an FR5 schedule in rats and an FR10 schedule in monkeys, and that these effects were reversed by co-administration of the PPAR-α antagonist MK886 [18]. WY14643 had no effect on operant self-administration of cocaine in monkeys, demonstrating specificity to nicotine [18]. Panlilio et al. (2012) found further evidence that the clinically useful drug clofibrate prevented the acquisition of self-administration in naïve rats and decreased self-administration in experienced rats and monkeys, an effect that was reversed by treatment with MK886 [19]. Neither study found an effect of PPAR-α agonists on nicotine discrimination [18,19].

Table 1. Overview of methodological details and primary findings of the key studies providing behavioral evidence for a role of PPAR agonists in modulating addiction-related behaviors in animal models.

Reference	Species/Strain and Sex	Addiction Model and Task	PPAR Agonist, Dose, and Route of Administration	Treatment Regimen	Primary Findings
Maeda et al., 2007 [34]	Male ICR mice	Behavioral sensitization to methamphetamine	0.5–5 µg i.c.v. CIG and PIO (PPAR-γ)	Once daily administration either for 5 days concurrently with methamphetamine or for 6 days during the withdrawal period	No effect of CIG or PIO (5 µg) when administered concurrently with methamphetamine. When administered during the withdrawal period, both CIG and FIO (at 5 µg, but not 0.5 µg or 1.5 µg) attenuated behavioral sensitization, while 1.5 and 5 µg (but not 0.5 µg) GW9662 (PPAR-γ antagonist) augmented behavioral sensitization
Barson et al., 2009 [25]	Male Sprague-Dawley rats	Voluntary ethanol consumption (2BC paradigm)	50 mg/kg p.o. GEM (PPAR-α)	One gavage 2 h prior to 4-h access to ethanol	GEM reduced intake of 7% ethanol, with a significant effect at 1 h and 4 h (and reduced ethanol consumption during the first hour of access in a separate experiment)
Mascia et al., 2011 [18]	Male Sprague-Dawley rats; Male squirrel monkeys	Operant SA (FR5 schedule of i.v. nicotine) (rats); Nicotine seeking and relapse (nicotine/cue-induced reinstatement) (rats); Nicotine discrimination (rats); Operant SA (FR10 schedule of i.v. nicotine or cocaine) (monkeys); Nicotine seeking and relapse (nicotine/cue-induced reinstatement) (monkeys)	20 or 40 mg/kg i.p. WY14643 and 10 mg/kg i.p. methOEA (PPAR-α) (rats); 10, 20, or 40 mg/kg i.m. WY14643 and 10 mg/kg i.m. methOEA (PPAR-α) (monkeys)	Single injections of WY14643 20 min prior or methOEA 40 min prior to SA sessions (rats and monkeys); WY14643 20 min prior to reinstatement (rats and monkeys); WY14643 substituted for training dose of nicotine and co-administered with various doses of nicotine during discrimination sessions (rats)	Both WY14643 and methOEA (at all tested doses) reduced nicotine SA in rats and monkeys; co-administration with MK886 (PPAR-α antagonist) attenuated this effect in monkeys. WY14643 attenuated nicotine/cue-induced reinstatement at both doses tested in rats and monkeys; MK886 attenuated this effect in monkeys. WY14643 had no effect on cocaine SA in monkeys or nicotine discrimination in rats
Stopponi et al., 2011 [35]	Male msP (alcohol-preferring Marchigian Sardinian) rats	Voluntary ethanol consumption (2BC paradigm); Operant SA (FR1 schedule of oral ethanol); Ethanol seeking and relapse (stress- and cue-induced reinstatement); Ethanol withdrawal (ventromedial distal flexion response, tail stiffness/rigidity, and tremors)	10 or 30 mg/kg p.o PIO or ROSI (PPAR-γ)	Twice daily treatment (12 h and 1 h prior to dark period) for 7 consecutive days (2BC) or 3 consecutive days (2BC with antagonism treatment); Twice daily treatment every fourth day (SA)Single treatment 12 h and 1 h prior to reinstatement test and evaluation of withdrawal symptoms	PIO significantly reduced voluntary intake of 10% ethanol on all treatment days at 30 mg/kg, but only on treatment days 5 and 7 at 10 mg/kg; ROSI also significantly reduced intake at the 30 mg/kg dose on all treatment days except day 4, while only on days 1, 2, and 7 at the 10 mg/kg dose. The effect of PIO (30 mg/kg) on ethanol intake was attenuated by pre-treatment with 5 µg GW9662 (PPAR-γ antagonist) across all 3 treatment days. PIO (at 30 mg/kg, but not 10 mg/kg) significantly reduced operant SA of 10% ethanol. Pre-treatment with both doses of PIO significantly attenuated yohimbine-induced reinstatement of ethanol-seeking, but had no effect on cue-induced reinstatement. PIO (at both doses) significantly reduced total withdrawal signs

Table 1. *Cont.*

Reference	Species/Strain and Sex	Addiction Model and Task	PPAR Agonist, Dose, and Route of Administration	Treatment Regimen	Primary Findings
Panlilio et al., 2012 [19]	Male Sprague-Dawley rats Male squirrel monkeys	Operant SA (FR1 or FR5 schedule of i.v. nicotine) (rats) Nicotine discrimination (rats) Operant SA (FR10 schedule of i.v. nicotine) (monkeys) Nicotine seeking and relapse (nicotine- and cue-induced reinstatement) (monkeys)	100, 200, or 300 mg/kg i.p. CLO (PPAR-α) (rats) 25, 50, or 100 mg/kg i.m. CLO (PPAR-α) (monkeys)	Single injections once daily beginning two days prior to 18 days of testing (FR1, rats) Single injections once daily for 3 days (FR5, rats; FR10, monkeys) Single injection prior to priming injection of nicotine (reinstatement, monkeys) Single injection 100 min prior to discrimination sessions (rats)	CLO (300 mg/kg) prevented the acquisition of nicotine SA in naïve rats CLO decreased SA of nicotine in experienced rats (at all three doses) and monkeys (at 50 mg/kg and 100 mg/kg, but not 25 mg/kg); this effect was attenuated by pre-treatment with 3 mg/kg MK886 (PPAR-α antagonist) In monkeys, 100 mg/kg CLO attenuated both nicotine- and cue-induced reinstatement of nicotine-seeking; these effects were attenuated with MK886 pre-treatment CLO did not alter nicotine discrimination in rats
Bilbao et al., 2013 [36]	Male PPAR-α KO mice and WT counterparts	Behavioral sensitization to cocaine Cocaine CPP	1, 5, or 20 mg/kg i.p OEA (PPAR-α)	Single injection prior to tests (motor response and CPP) followed by injections every other day for 3 additional days (sensitization)	OEA (5 mg/kg and 20 mg/kg, but not 1 mg/kg) attenuated acute cocaine-induced motor activation and sensitization to the motor effects of cocaine OEA attenuated cocaine CPP at 1 and 5 mg/kg and completely abolished the development of CPP at 20 mg/kg The ability of OEA (20 mg/kg) to attenuate cocaine sensitization and CPP was intact in PPAR-α KO mice
Stopponi et al., 2013 [37]	Male msP rats	Voluntary ethanol consumption (2BC paradigm) Ethanol seeking and relapse (stress- and cue-induced reinstatement)	10 or 30 mg/kg p.o. PIO (PPAR-γ)	Two treatments (12 h and 1 h prior to dark period) prior to testing sessions	PIO (30 mg/kg, but not 10 mg/kg) reduced intake of 10% ethanol at 24 h (but not 2 or 8 h); 10 mg/kg PIO co-administered with 0.25 mg/kg naltrexone also significantly reduced intake at 8 and 24 h PIO (at both doses) and co-administration of 1 mg/kg naltrexone with either dose of PIO significantly attenuated yohimbine-induced reinstatement of ethanol-seeking PIO alone did not significantly alter cue-induced reinstatement of ethanol-seeking, but co-administration of 1 mg/kg naltrexone with either PIO dose did

Table 1. *Cont.*

Reference	Species/Strain and Sex	Addiction Model and Task	PPAR Agonist, Dose, and Route of Administration	Treatment Regimen	Primary Findings
De Guglielmo et al., 2014 [38]	Male C57 mice and conditional neuronal PPAR-γ KO mice and WT counterparts	Analgesic tolerance to morphine	10 or 30 mg/kg p.o. PIO (PPAR-γ)	Single gavage prior to morphine injections for 9 days (or only on days 8 and 9 for reversal of morphine tolerance experiments)	PIO (at both doses) attenuated the development of tolerance to the analgesic effects of morphine; this effect was blocked by pretreatment with 5 mg/kg GW9962 (PPAR-γ antagonist) and was absent in the PPAR-γ KO mice compared to their WT counterparts. GW9962 alone accelerated the development of morphine tolerance. PIO (at both doses) also reversed morphine tolerance when administered only on the last two days of treatment
Ferguson et al., 2014 [26]	Male C57BL/6J mice	Voluntary ethanol consumption (2BC paradigm)	150 mg/kg p.o. FEN (PPAR-α) 75 mg/kg p.o. BEZA (pan-PPAR) 1.5 mg/kg p.o. TESA (dual PPAR-α/γ)	Single treatment for 8 days (ethanol consumption measured on days 5 and 6)	FEN and TESA decreased voluntary consumption of and preference for 15% ethanol, while BEZA had no significant effect
Karahanian et al., 2014 [27]	Male UChB (selectively bred high-drinker) rats	Voluntary ethanol consumption (24-h 2BC and limited 2BC drinking in the dark paradigms)	50 mg/kg p.o. FEN (PPAR-α)	Single daily treatment for 14 consecutive days following 60 days of continuous free choice of ethanol or water	In the 24-h access paradigm, FEN reduced voluntary consumption of 10% ethanol, starting on day 4 of treatment and reaching a maximum reduction at day 12. In the drinking in the dark paradigm, FEN significantly reduced ethanol intake, starting on day 2 and reaching a maximum reduction at day 5
Blednov et al., 2015 [29]	Male C57BL/6J mice	Voluntary ethanol consumption (24-h 2BC and limited 2BC drinking in the dark paradigms)	10 or 30 mg/kg p.o. PIO (PPAR-γ) 50 or 150 mg/kg p.o. FEN (PPAR-α) 10 mg/kg p.o. GW0742 (PPAR-δ/β) 1.5 mg/kg p.o. TESA (dual PPAR-α/γ) 25 or 75 mg/kg p.o. BEZA (pan PPAR-α/γ/δ/β)	Once daily treatment for up to 10 days following 2 days of saline treatment	In the 24-h access paradigm, PIO (30 mg/kg), FEN (150 mg/kg), and TESA reduced intake of and preference for 15% ethanol; BEZA (75 mg/kg) reduced preference, but not intake; GW0742 had no effect. In the drinking in the dark paradigm, FEN (150 mg/kg), TESA (1.5 mg/kg), and BEZA (75 mg/kg) reduced intake and preference; PIO and GW0742 had no effect
De Guglielmo et al., 2015 [20]	Male Wistar rats	Operant SA (FR1 or PR schedule of i.v. heroin)	30 or 60 mg/kg p.o. PIO (PPAR-γ)	Twice-daily treatment (12 and 1 h prior to SA session) for 5 days	PIO significantly reduced heroin SA under an FR1 schedule (at 60 mg/kg, but not 30 mg/kg) and significantly decreased the breakpoint in the PR schedule (at 30 and 60 mg/kg); the reduction in responding under FR1 with 60 mg/kg PIO was blocked by pre-treatment with 5 mg/kg GW9662 (PPAR-γ antagonist)

Table 1. *Cont.*

Reference	Species/Strain and Sex	Addiction Model and Task	PPAR Agonist, Dose, and Route of Administration	Treatment Regimen	Primary Findings
Bilbao et al., 2016 [28]	Male Wistar rats	Voluntary ethanol consumption (2BC paradigm) Operant SA (FR1 schedule of oral ethanol) Ethanol seeking and relapse (cue-induced reinstatement) Ethanol withdrawal (vocalizations, head tremor and rigidity, tail tremor, and body tremor)	1, 5, or 20 mg/kg i.p OEA (PPAR-α) 5, 20, or 40 mg/kg i.p. WY14643 (PPAR-α)	Single injections 30 min prior to testing sessions	OEA (5 mg/kg) significantly decreased voluntary intake of 10% ethanol at all time points (2, 4, and 6 h); this effect was reversed by pre-treatment with 1 mg/kg GW6471 (PPAR-α antagonist) OEA (5 mg/kg and 20 mg/kg, but not 1 mg/kg) and WY14643 (20 mg/kg and 40 mg/kg, but not 5 mg/kg) significantly decreased SA of 10% ethanol OEA (5 mg/kg and 20 mg/kg, but not 1 mg/kg) and WY14643 (20 and 40 mg/kg) significantly attenuated cue-induced reinstatement of ethanol-seeking OEA (5 mg/kg, but not 1 mg/kg) and WY14643 (20 mg/kg) decreased ethanol SA following a deprivation periodOEA (5 mg/kg) significantly reduced ethanol withdrawal scores
Blednov et al., 2016 [30]	Male and female C57BL/6J and PPAR-α KO mice	Voluntary ethanol consumption (continuous and intermittent 2BC paradigm)	10, 50, 100, or 150 mg/kg p.o. FEN (PPAR-α) 1.5 mg/kg p.o. TESA (dual PPAR-α/γ)	Once daily treatment for up to 14 days after 2 days of saline treatment	In the continuous access paradigm, FEN reduced both intake of and preference for 15% ethanol (at 100 and 150 mg/kg, but not 10 mg/kg or 50 mg/kg) in male, but not female, mice; TESA reduced both intake and preference in both male and female mice In the intermittent (every other day) access paradigm, FEN (150 mg/kg, but not 100 mg/kg) reduced both intake and preference in male and female mice Pre-treatment with 5 mg/kg MK886 (PPAR-α antagonist), but not 5 mg/kg GW9662 (PPAR-γ antagonist), reduced the effect of FEN on ethanol intake; pre-treatment with GW9662 or MK886 did not block the effects of TESA on ethanol intake Both FEN and TESA had no effect on ethanol consumption in mice lacking PPAR-α
Blednov et al., 2016 [33]	Male and female C57BL/6J and B6 × 129S4 mice	Ethanol CPPEthanol withdrawal (handling-induced convulsions)	150 mg/kg p.o. FEN (PPAR-α) 1.5 mg/kg p.o. TESA (dual PPAR-α/γ)	Once daily treatment for the duration of each experiment after 2 days of saline treatment	No effect of either agonist on CPP in male B6x129S4 mice FEN increased withdrawal severity in male mice of both genotypes, while TESA increased withdrawal severity in only the B6x129S4 male mice; neither drug significantly altered withdrawal in female mice

Table 1. *Cont.*

Reference	Species/Strain and Sex	Addiction Model and Task	PPAR Agonist, Dose, and Route of Administration	Treatment Regimen	Primary Findings
De Guglielmo et al., 2017 [39]	Male Wistar rats and male CD1 mice	Morphine withdrawal (jumps, paw tremors, teeth chattering, and wet dog shakes) Heroin seeking and relapse (stress-, cue-, and heroin-induced reinstatement)	10, 30, or 60 mg/kg p.o. PIO (PPAR-γ)	Single treatment 1 h prior to morphine injection the evening of day 5 and morning of day 6 (withdrawal expression) Treatment twice daily (12 h and 1 h prior to tests) for 5 consecutive days, then again on the morning of day 6 1 h prior to final morphine injection (withdrawal development) Two treatments, 12 h and 1 h prior to reinstatement tests	In mice, PIO (10 and 30 mg/kg) attenuated the expression of morphine withdrawal and the development of morphine withdrawal (at 30 mg/kg); pre-treatment with 5 mg/kg GW9662 (PPAR-γ antagonist) reversed the effect of PIO on expression of withdrawal In rats, PIO significantly reduced yohimbine-induced reinstatement (at 30 mg/kg, but not 10 mg/kg) and heroin-induced reinstatement (at 30 mg/kg and 60 mg/kg, but not 10 mg/kg) of heroin-seeking, but had no effect on cue-induced reinstatement (at 10, 30, or 60 mg/kg)
Haile & Kosten, 2017 [31]	Wistar rats (sex not reported)	Operant SA of ethanol (FR2 and PR)	25, 50, or 100 mg/kg p.o. FEN (PPAR-α)	Single treatment 1 h prior to test sessions for 5 consecutive days (four days of FR2 schedule then one day of PR schedule)	Under the FR2 schedule, there was a significant difference between all doses tested, though the effect was dependent on day (by day 4, all three active doses of FEN significantly decreased active lever presses for 10% ethanol) Under the PR schedule, all three doses of FEN reduced active lever presses
Jackson et al., 2017 [40]	Male ICR mice	Nicotine (and cocaine) CPP Nicotine withdrawal (anxiety-like behavior, somatic withdrawal signs, and hyperalgesia)	0.3, 0.6, 1, and 5 mg/kg i.p. WY14643 (PPAR-α) 1, 9, 50, or 100 mg/kg i.p. FEN (PPAR-α)	For CPP experiments, WY14643 was administered 15 min prior to and FEN 1 h prior to nicotine Following 14 days of infusion with nicotine, mice were given a single treatment with WY14643 15 min prior to or FEN 1 h prior to precipitated withdrawal on day 15	WY14643 (at all three doses) significantly attenuated nicotine CPP WY14643 did not shift the potency of nicotine in the CPP paradigm WY14643 (1 mg/kg) did not attenuate cocaine CPP FEN attenuated nicotine CPP at 50 mg/kg (not 1, 9, or 100 mg/kg) WY14643 attenuated signs of nicotine withdrawal (anxiety-like behaviors and hyperalgesia attenuated at 5 mg/kg only; somatic withdrawal symptoms attenuated at 1 and 5 mg/kg; no effect of 0.3 mg/kg) FEN did not attenuate anxiety-like behaviors or hyperalgesia at either dose tested (50 or 100 mg/kg), but did attenuate somatic withdrawal symptoms at 100 mg/kg

Table 1. *Cont.*

Reference	Species/Strain and Sex	Addiction Model and Task	PPAR Agonist, Dose, and Route of Administration	Treatment Regimen	Primary Findings
Rivera-Meza et al., 2017 [32]	Male UChB rats	Voluntary ethanol consumption (2BC paradigm) Ethanol CPP	25, 50, or 100 mg/kg p.o. FEN (PPAR-α)	Following 60 days free choice between ethanol and water, rats were treated once daily for 14 days (in the CPP experiment, ethanol access was restricted during this period, and testing occurred on day 14 of FEN treatment) In a separate experiment, rats were deprived of ethanol on day 60 and treated once during two deprivation periods (days 61–74 and 103–116), voluntary consumption of ethanol was once again measured after each of these two periods	FEN (all three doses) significantly decreased voluntary consumption of 10% ethanol beginning on day 2 of treatment and continuing for the duration of treatment FEN (50 mg/kg) prevented the development of ethanol CPP FEN (50 mg/kg) significantly decreased voluntary consumption of ethanol following both periods of deprivation
Miller et al., 2018 [41]	Male Sprague-Dawley rats	Behavioral sensitization to cocaine Cocaine cue reactivity (lever-pressing for cocaine-associated cues during forced abstinence)	50 mg of PIO per kg of chow	PIO treatment initiated 4 days prior to behavioral sensitization protocol and immediately following final session of cocaine SA (continued during 30-day forced abstinence period)	PIO reduced both the development and expression of behavioral sensitization to cocaine PIO reduced cue reactivity following prolonged abstinence from cocaine; this effect was attenuated by pre-treatment with 1 mg/kg GW9662 (PPAR-γ antagonist)
Domi et al., 2019 [42]	Male Wistar rats and conditional neuronal PPAR-γ KO mice and WT counterparts	Nicotine withdrawal (somatic withdrawal signs and anxiety-like behaviors)	15 or 30 mg/kg p.o. PIO (PPAR-γ)	Two treatments, 12 h and 1 h prior to assessment of withdrawal	PIO (at both doses) reduced somatic signs of nicotine withdrawal and anxiety-like behaviors in rats and WT mice, but had no effect in conditional neuronal PPAR-γ KO mice; the effect of 30 mg/kg PIO on somatic and anxiety-like withdrawal signs was blocked by pre-treatment with GW9662 (PPAR-γ antagonist) in WT mice
Donvito et al., 2019 [43]	Male ICR mice	Nicotine withdrawal (anxiety-like behavior and somatic withdrawal signs) Nicotine (and morphine) CPP	10, 30, or 60 mg/kg i.p. OlGly (PPAR-α)	Single injection 15 min prior to nicotine injection in the CPP experiments or to precipitated withdrawal	OlGly (at 60 mg/kg, but not 10 mg/kg or 30 mg/kg) significantly attenuated anxiety-like and somatic nicotine withdrawal signs OlGly (at all three doses) attenuated the development of nicotine (but not morphine) CPP; this effect was blocked by pre-treatment with 2 mg/kg GW6471 (PPAR-α antagonist)

2BC, two-bottle choice; BEZA, bezafibrate; CIG, ciglitazone; CLO, clofibrate; CPP, conditioned place preference; FEN, fenofibrate; FR, fixed ratio; GEM, gemfibrozil; i.c.v., intracerebroventricular; i.m., intramuscular; i.p., intraperitoneal; i.v., intravenous; KO, knock-out; methOEA, methyl oleoylethanolamide; OEA, oleoylethanolamide; OlGly, N-Oleoyl-glycine; PIO, pioglitazone; p.o., per os (oral); ROSI, rosiglitazone; SA, self-administration; TESA, tesaglitazar; WT, wild-type.

Two additional studies have suggested a role of PPAR-α agonists in attenuating nicotine CPP [40,43]. Jackson et al. (2017) found that both WY14643 and fenofibrate significantly reduced nicotine preference in the CPP experiments, though fenofibrate was less potent [40]. Importantly, WY14643 did not shift the potency of nicotine in the CPP paradigm, and the effect of WY14643 was specific to nicotine as it had no effect on cocaine preference [40]. In support of these findings, Donvito et al. (2019) found that exogenous administration of the lipid transmitter N-Oleoyl-glycine (OlGly) prevented the development of nicotine, but not morphine, CPP, and that this effect was blocked by the PPAR-α antagonist GW6471 [43]. Taken together, the results of the operant self-administration and CPP experiments provide strong support for a role of PPAR-α agonism in reducing the reinforcing and rewarding properties of nicotine.

Finally, one study found that OEA reduced behavioral sensitization to cocaine and cocaine CPP, though this effect was intact in PPAR-α KO mice, suggesting this was due to a PPAR-independent mechanism [36].

2.1.2. Withdrawal/Relapse

Conflicting evidence exists regarding how PPAR-α agonists influence withdrawal from ethanol [28,33]. Bilbao et al. (2016) found that i.p. injection of 5 mg/kg of the endogenous PPAR-α agonist OEA significantly reduced total ethanol withdrawal scores in male rats, and furthermore decreased each of the individual withdrawal signs evaluated (vocalizations, head tremor and rigidity, tail tremor, and body tremor) [28]. Blednov et al. (2016) found that oral administration of 150 mg/kg fenofibrate or 1.5 mg/kg of the dual PPAR-α/γ agonist tesaglitazar actually increased withdrawal severity (handling-induced convulsions score) in male (but not female) mice [33]. The results of these two studies are difficult to compare given the different choices of PPAR-α agonist, dose, and route of administration, withdrawal signs evaluated, and animal models, but do suggest some role of PPAR-α in modulating ethanol withdrawal.

In the same experiments described above, Bilbao et al. (2016) found that both OEA and WY14643 were able to attenuate cue-induced reinstatement of ethanol-seeking after a period of deprivation [28], providing preliminary evidence that PPAR-α agonism may help to prevent alcohol relapse.

Two studies have suggested a role of PPAR-α agonists in reducing nicotine withdrawal signs. Jackson et al. (2017) assessed the impact of PPAR-α agonists on symptoms of precipitated nicotine withdrawal. They observed that WY14643 attenuated anxiety-like behaviors, hyperalgesia, and somatic withdrawal signs, while fenofibrate attenuated only somatic withdrawal signs [40]. Similarly, Donvito et al. (2019) found that exogenous administration of the lipid transmitter OlGly attenuated anxiety-like and somatic signs of nicotine withdrawal [43].

Finally, two studies have provided evidence that PPAR-α agonists can block reinstatement of nicotine-responding following a period of extinction [18,19]. Mascia et al. (2011) found that WY14643 attenuated reinstatement in both rats and monkeys using a procedure that combines both nicotine- and cue-induced reinstatement, and that this effect was reversed by co-administration of the PPAR-α antagonist MK886 [18]. Similarly, Panlilio et al. (2012) found that clofibrate attenuated both nicotine- and cue-induced reinstatement of nicotine responding in monkeys, and that these effects were reversed by pre-treatment with MK866 [19]. The reduction in withdrawal symptoms and the attenuation of both drug- and cue-induced reinstatement suggest that PPAR-α agonists may be useful in preventing relapse in nicotine-dependent smokers.

2.2. PPAR-γ Agonists

2.2.1. Consumption/Motivation

Similar to the evidence for PPAR-α agonists, the results of several studies support a role of PPAR-γ agonists in attenuating voluntary consumption and operant self-administration of ethanol [26,29,30,35,37]. In the two-bottle choice paradigm, voluntary ethanol consumption was found to be attenuated by treatment with rosiglitazone [35] and pioglitazone [29,35,37], as well as the dual PPAR-α/γ agonist tesaglitazar [26,29,30]. Stopponi et al. (2011) additionally observed that pioglitazone significantly reduced operant self-administration of ethanol under an FR1 schedule [35]. While one study found that pre-treatment with the PPAR-γ antagonist GW9662 reversed the effects of pioglitazone on voluntary ethanol consumption [35], another study found no effect of GW9662 pre-treatment on the ethanol-reducing effects of the dual PPAR-α/γ agonist tesaglitazar, suggesting that the PPAR-α isoform may be more important in modulating ethanol-related behaviors than the PPAR-γ isoform [30].

Limited evidence suggests that PPAR-γ agonists may not influence ethanol CPP. As described above, Blednov et al. (2016) found no effect of tesaglitazar on ethanol CPP [33].

One study found that pioglitazone reduced operant self-administration of heroin under an FR1 schedule and significantly decreased the breakpoint in a PR schedule [20]. Furthermore, the effects of pioglitazone on self-administration were reversed by pre-treatment with the PPAR-γ antagonist GW9662 [20]. This preliminary evidence suggests that PPAR-γ agonists may be useful in reducing the reinforcing effects of opioids such as heroin.

Two studies have suggested that PPAR-γ agonists can attenuate behavioral sensitization to stimulant drugs [34,41]. Maeda et al. (2007) found that treatment with ciglitazone or pioglitazone during a withdrawal period, but not concurrently with methamphetamine, significantly attenuated behavioral sensitization to methamphetamine, while the PPAR-γ antagonist GW9662 significant augmented behavioral sensitization [34]. Miller et al. (2018) found that treatment with pioglitazone 4 days prior to testing significantly attenuated both the development and expression of behavioral sensitization to cocaine and attenuated lever-pressing for cocaine-associated cues during a period of forced abstinence [41].

2.2.2. Withdrawal/Relapse

Similar to the results for PPAR-α agonists, the current evidence for an effect of PPAR-γ in modulating ethanol withdrawal signs is split [33,35]. As previously described, Blednov et al. (2016) found that the dual PPAR-α/γ agonist tesaglitazar increased withdrawal severity in mice [33]. In contrast, Stopponi et al. (2011) found that oral administration of both 10 and 30 mg/kg pioglitazone significantly reduced total withdrawal signs (composite score of ventromedial distal flexion responses, tail rigidity, and tremors) in rats [35]. While once again significant methodological differences prevent clear comparison of these results, it is important to note that in the same set of experiments, Blednov and colleagues did not find that the effects of tesaglitazar on ethanol-related behaviors were blocked by pre-treatment with the PPAR-γ antagonist GW9662 [30]. Thus, the ability of tesaglitazar to increase ethanol withdrawal severity in their experiment may not have been due to its actions at PPAR-γ.

Two studies have provided evidence for a role of PPAR-γ agonism in blocking reinstatement of ethanol-responding [35,37]. Both studies found that pioglitazone alone significantly attenuated stress-induced reinstatement (using yohimbine as a stressor), but not cue-induced reinstatement [35,37]. However, when pioglitazone was co-administered with naltrexone, there was an attenuation of cue-induced reinstatement [37]. These results suggest that PPAR-γ agonists may be useful in preventing alcohol relapse, possibly to a greater extent when administered concurrently with naltrexone, a non-selective opioid receptor antagonist that is already approved by the United States Food and Drug Administration (FDA) to treat alcohol use disorder [10].

One recent study found that PPAR-γ activation may play a role in nicotine withdrawal. Administration of pioglitazone prior to assessment of nicotine withdrawal attenuated somatic and anxiety-like signs of withdrawal in rats and in wild-type mice with intact PPAR-γ, but not in conditional neuronal PPAR-γ KO mice [42]. In addition, the effect of pioglitazone on both somatic and anxiety-like signs of nicotine withdrawal was blocked by pre-treatment with the PPAR-γ antagonist GW9662 in WT mice [42].

Finally, one study provided evidence that PPAR-γ agonists can reduce opioid withdrawal and relapse [39]. Treatment with pioglitazone significantly attenuated both the development and expression of morphine withdrawal in mice, and the PPAR-γ antagonist GW9662 blocked the ability of pioglitazone to attenuate the expression of morphine withdrawal [39]. Furthermore, pioglitazone significantly attenuated yohimbine- and heroin-induced reinstatement of heroin-responding in rats, while having no effect on cue-induced reinstatement [39]. Previously, the same group reported that pioglitazone significantly attenuated the development of analgesic tolerance to morphine [38], which provides additional evidence for a role of PPAR-γ in the effects of repeated morphine administration.

2.3. Summary of Preclinical Evidence

The majority of the preclinical behavioral evidence suggesting a role of PPAR agonists in addiction-like behaviors has focused on ethanol. Currently, the literature strongly supports a role of PPAR-α agonists (gemfibrozil, fenofibrate, OEA, and WY14643), and PPAR-γ agonists (rosiglitazone and pioglitazone) or a dual PPAR-α/γ agonist (tesaglitazar) to a lesser extent, in attenuating the voluntary consumption and reinforcing properties of ethanol in rodents. Limited evidence suggests that the PPAR-α agonist fenofibrate may additionally reduce the rewarding properties of ethanol, as assessed in the CPP paradigm. While agonists at both PPAR-α (OEA and fenofibrate) and PPAR-γ (pioglitazone) seem to have some role in modulating ethanol withdrawal signs, the nature of this role is unclear. However, the evidence does suggest that PPAR agonists may be useful in reducing the likelihood of alcohol relapse after a period of abstinence. PPAR-α agonists (OEA and WY14643) were shown to attenuate cue-induced reinstatement of ethanol-seeking, while a PPAR-γ agonist (pioglitazone) was shown to attenuate stress-induced reinstatement (and possibly also cue-induced reinstatement when co-administered with naltrexone).

Robust evidence from a limited number of studies strongly supports a role of PPAR-α (and possibly PPAR-γ) agonists in modulating nicotine-related behaviors in both rodents and non-human primates. The PPAR-α agonists methyl-OEA, WY14643, and clofibrate were found to reduce the reinforcing properties of nicotine. In addition, WY14643, fenofibrate, and OlGly were found to reduce the rewarding effects of nicotine in the CPP paradigm. WY14643 was shown to decrease behavioral and somatic signs of nicotine withdrawal, while both WY14643 and clofibrate reduced drug- and cue-induced reinstatement of nicotine-seeking. Finally, the PPAR-γ agonist pioglitazone reduced somatic and anxiety-like signs of nicotine withdrawal.

Preliminary evidence suggests that PPAR-γ agonists may have a role in modulating opioid-related behaviors. Studies found that pioglitazone was able to reduce the reinforcing effects of heroin in an operant self-administration paradigm, decrease both drug- and stress-induced reinstatement of heroin-seeking, and reduce the development and expression of morphine tolerance and withdrawal.

Finally, there seems to be a role of PPAR agonists in psychostimulant-related behaviors, yet the evidence is mixed. The PPAR-γ agonists ciglitazone and pioglitazone attenuated behavioral sensitization to methamphetamine, while pioglitazone attenuated behavioral sensitization to cocaine. Additionally, the endogenous PPAR-α agonist OEA attenuated behavioral sensitization to cocaine and cocaine CPP, but through a PPAR-α-independent mechanism. However, it is important to note that studies of nicotine-related outcomes found no effect of PPAR-α agonists on operant self-administration of cocaine or cocaine CPP.

3. Clinical or Human Laboratory Evidence

A summary of the methodological details and relevant findings of the human studies reviewed is provided in Table 2.

3.1. PPAR-α Agonists

Two published placebo-controlled studies have evaluated the potential of PPAR-α agonists in treatment of nicotine dependence [44,45]. Perkins et al. (2016) recruited nicotine-dependent smokers high in quit interest for a double-blind, counterbalanced, crossover trial with a target dose of 160 mg of fenofibrate administered once daily for 4 days following a 4-day dose run-up period [44]. There was no difference between fenofibrate and placebo on quit days, the primary outcome of the trial. In addition, there were no drug effects on any of the secondary outcomes, including pre-quit smoking reinforcement (i.e., number of puffs taken from participants' preferred brand of cigarettes and self-reported rewarding effects), craving responses during a cue reactivity task, and mean daily reductions in smoking [44]. In support of these negative findings, our lab found no effect of gemfibrozil (600 mg administered orally twice daily) on total self-reported days abstinent in a sample of nicotine-dependent smokers intent on quitting [45]. Similarly, we found no effects on secondary outcomes including a forced choice procedure (i.e., reinforcing effects) and both physiological and subjective measures of cue reactivity. In sum, despite the compelling preclinical evidence, the limited human evidence has not supported a role of PPAR-α agonists in treating nicotine dependence.

3.2. PPAR-γ Agonists

Three placebo-controlled studies have examined the potential for PPAR-γ agonists in modulating opioid-related outcomes [46–48]. In a sample of healthy, non-medical users of prescription opioids, there was no effect of 15 or 45 mg oral pioglitazone administered daily for 2–3 weeks on self-reported positive and negative subjective effects of oxycodone in a single-blind, within-subjects design [46]. In addition, pioglitazone had no impact on self-reported drug wanting (opioids, alcohol, cannabis, and tobacco) during the maintenance phases [46]. In a follow-up study, Jones and colleagues assessed the effects of 45 mg oral pioglitazone administered daily for 3 weeks in a sample of non-treatment-seeking adults with an opioid use disorder using a single-blind, randomized, between-subjects design [47]. Pioglitazone did not alter the reinforcing effects of heroin, its abuse liability, or cue reactivity, though self-reported ratings of "I want heroin" were significantly reduced [47]. Finally, Schroeder et al. (2018) assessed the potential for pioglitazone as an adjunct pharmacotherapy for patients with an opioid use disorder undergoing buprenorphine taper [48]. Pioglitazone treatment had no effect on withdrawal severity, and may actually have increased subjective withdrawal; yet, this trial was limited by very low recruitment numbers [48].

Two additional studies have investigated the role of pioglitazone in nicotine dependence and cocaine use disorder. In a single-blind, between-subjects laboratory study of nicotine-dependent smokers not interested in quitting, compared to placebo treatment, 45 mg oral pioglitazone administered daily for 3 weeks decreased self-reported measures of nicotine craving, though had minimal or no impact on reinforcing effects, self-reported positive or negative subjective effects, or cue reactivity [49]. In a pilot study to assess the potential of pioglitazone to target craving and white matter integrity in treatment-seeking adults with cocaine use disorder, daily administration of 45 mg oral pioglitazone for 12 weeks conferred benefit over placebo in reducing cocaine craving [50].

Taken together, the limited available human evidence suggests that the PPAR-γ agonist pioglitazone may be beneficial in reducing heroin, nicotine, and cocaine craving. However, it remains unclear how PPAR-γ agonists may impact more direct measures of treatment efficacy such as quit days or reductions in drug use.

Table 2. Overview of methodological details and primary findings of the key clinical and human laboratory studies of PPAR agonists in drug-related outcomes.

References	Study Sample	PPAR Agonist, Dose, and Route of Administration	Study Design	Primary Findings
Jones et al., 2016 [46]	Healthy non-medical users of prescription opioids, N = 17 (15 M, 2 F), 21–55 years old (mean 35 years)	15 or 45 mg p.o. PIO (PPAR-γ)	Single-blind, within-subjects, placebo-controlled design. Participants received PIO doses in ascending order and maintained on each dose for 2–3 weeks. Subjective, analgesic, and physiological effects of oral oxycodone examined at the end of each maintenance phase.	No effect of PIO on self-reported positive or negative subjective effects of oxycodone In addition, PIO did not affect drug wanting (opioids, alcohol, cannabis, or tobacco) during the maintenance phase
Perkins et al., 2016 [44]	Nicotine-dependent smokers high in quit interest, N = 38 (27 M, 11 F), 18–5 years old (mean 30.3 years)	160 mg p.o. FEN (PPAR-α)	Double-blind, within-subjects, counterbalanced, placebo-controlled design. Participants received FEN for 8 days (4-day dose run-up followed by 4-day quit period). A week of ad libitum smoking separated the two quit periods. Self-report of no smoking and expired-air CO < 5 ppm were assessed daily during quit periods. Secondary outcome measures included acute smoking reinforcement and cue reactivity (pre-quit) and amount of daily smoking exposure (post-quit).	FEN did not increase quit days compared to placebo Additionally, FEN had no impact on acute smoking reinforcement (SA paradigm), cue-induced craving, or mean daily smoking
Jones et al., 2017 [49]	Nicotine-dependent smokers not interested in quitting, N = 27 (14 active, 13 placebo; 25 M, 2 F), 21–55 years old (mean 44.9 years in active group, 41.6 years in placebo group)	45 mg p.o. PIO (PPAR-γ)	Single-blind, between-subjects, randomized, placebo-controlled design. Participants received PIO daily for 3 weeks. Laboratory testing (reinforcing effects, cue reactivity, subjective effects, and physiological effects) began after the first week of nicotine patch stabilization.	PIO did not alter the reinforcing effects of nicotine (verbal choice and progressive choice paradigms) or subjective/physiological reactivity to smoking cues PIO had minimal impact on positive subjective effects (increased one measure of nicotine "high") and no impact on negative subjective effects PIO decreased subjective ratings of "craving" and "desire"
Schmitz et al., 2017 [50]	Treatment-seeking adults with cocaine use disorder, N = 30 (15 active, 15 placebo; 22 M, 8 F), 18–60 years old (mean 48.3 in active group, 47.4 in placebo group)	Target dose of 45 mg p.o. PIO (PPAR-γ)	Double-blind, between-subjects, randomized placebo-controlled pilot trial design. Following a 1-week baseline period and a 2-week dose titration period, participants were maintained on 45 mg/day PIO for duration of study (12 weeks total). Periodic measures of craving and cocaine use.	High probability that PIO conferred benefit over placebo in reducing cocaine craving In addition, there was evidence that PIO decreased the odds of using cocaine during the treatment period

Table 2. *Cont.*

References	Study Sample	PPAR Agonist, Dose, and Route of Administration	Study Design	Primary Findings
Gendy et al., 2018 [45]	Nicotine-dependent smokers high in quit interest, N = 27 (17 M, 10 F), 19–65 years old (mean 43 years old)	2 × 600 mg p.o. GEM (PPAR-α)	Double-blind, within-subjects, counterbalanced, placebo-controlled design. Two 2-week phases separated by 1-week washout period. During the first week, participants smoked normally, and laboratory measures of cue-elicited craving and forced-choice paradigms were taken. During the second week, participants were instructed to stop smoking, and abstinence was assessed.	GEM did not increase number of days of self-reported abstinence compared to placebo GEM had no impact on subjective/physiological reaction to smoking cues or reinforcing effects of nicotine (forced choice paradigm)
Jones et al., 2018 [47]	Non-treatment-seeking adults with opioid dependence, N = 30 (14 active, 16 placebo; 28 M, 2 F), 21–55 years old (mean 42.4 years in active group, 44.5 years in placebo group)	45 mg p.o. PIO (PPAR-γ)	Single-blind, between-subjects, randomized placebo-controlled design. Participants received PIO daily for 3 weeks. Laboratory testing (reinforcing effects, cue reactivity, subjective effects, cognitive effects, and physiological effects) began after the first week of buprenorphine/naloxone stabilization.	PIO did not influence the reinforcing effects of heroin (verbal choice SA or progressive choice paradigms) or physiological/subjective reactivity to active drug cues PIO did not influence the positive subjective effects of heroin PIO did further attenuate self-report ratings of anxiety during heroin self-administration, but had no impact on any other negative subjective effects PIO reduced ratings of "I want heroin"
Schroeder et al., 2018 [48]	Opioid-dependent adults undergoing a buprenorphine taper, N = 21 randomized (8 active, 13 placebo; 15 M, 6 F), N = 17 received at least one dose (6 active, 11 placebo), 18–65 years old (mean 38.4 years of participants randomized to active, 39.5 years placebo)	15 or 45 mg p.o. PIO (PPAR-γ)	Randomized, between-subjects design. Initial outpatient design (12 weeks of PIO treatment following 1-week buprenorphine stabilization), then subsequent outpatient/inpatient combination (5 weeks of PIO treatment following buprenorphine stabilization). Measures of opiate withdrawal collected daily throughout the study.	PIO significantly increased scores on the SOWS during the taper and post-taper phases, and had no effect on COWS scores In addition, there was no effect of PIO on opioid-positive urine samples during the post-taper phase

COWS, Clinical Opiate Withdrawal Scale; FEN, fenofibrate; GEM, gemfibrozil; p.o., per os (oral); PIO, pioglitazone; SA, self-administration; SOWS, Subjective Opiate Withdrawal Scale.

4. Synthesis of the Preclinical and Human Evidence

Given the robust preclinical evidence that both PPAR-α and PPAR-γ play a role in addiction-related behaviors, the lack of significant findings from human studies is somewhat surprising. For example, multiple preclinical studies demonstrated that PPAR-α agonists were effective in reducing the reinforcing and rewarding properties of nicotine and reducing nicotine withdrawal and reinstatement of nicotine-seeking [18,19,40,43], yet two human trials found no effect of the PPAR-α agonists fenofibrate [44] or gemfibrozil [45] on smoking cessation outcomes. Potential explanations for the poor concordance between the animal and human evidence to data are discussed below.

Perhaps the most salient discordance between the animal and human literature is the complete lack of placebo-controlled trials of PPAR agonists for treatment of alcohol use disorder. One Phase II trial of pioglitazone for treatment of alcohol craving and other alcohol-related outcomes in adults with alcohol use disorder (ClinicalTrials.gov identifier: NCT01631630) was terminated due to feasibility problems. A similar Phase II trial of fenofibrate (ClinicalTrials.gov identifier: NCT02158273) has been completed, though the results are unpublished. The most consistently reported and robust addiction-related outcome associated with PPAR agonists in the preclinical literature is a reduction in voluntary consumption of ethanol. Yet, as of this writing, the potential for PPAR agonists in treatment of alcohol use disorders in human has not been reported in the published literature. Thus, this is an important priority for future research. Currently, most pharmacotherapies available for the treatment of substance use disorders are substance-specific (although some are able to affect different substance use disorders). Therefore, it would be important to study the role of PPAR agonists in various substances use disorders, as it is unlikely that a single drug would be able to cure all substance use disorders.

The choice of PPAR agonist and dose is likely an important source of the poor translation from animal to human studies. For example, Jones and colleagues noted that the pioglitazone dosing parameters they employed were based on clinical utility in treating type-II diabetes [46,49], which may not be sufficient to elicit an effect in attenuating the abuse liability or reinforcing effects of opioids or nicotine. Similarly, while the preclinical evidence for a role of fibrate drugs in attenuating nicotine-related behaviors came from a study administering clofibrate [19], Perkins et al. (2016) used fenofibrate instead, as clofibrate was removed from the U.S. market due to its adverse effects [44]. Fibrate drugs, in general, may be less effective in reducing the rewarding and reinforcing effects of nicotine compared to experimental compounds such as WY14643 [40]. This could be due to the poor blood-brain barrier penetrance of fibrates like fenofibrate [51,52] or the low potency and PPAR-α selectivity of fenofibrate [53]. It should be noted in general that the PPAR agonists available do not act with 100% selectivity on specific PPAR isoforms and therefore, action on multiple PPAR isoforms is a possibility that should be kept in mind while interpreting the research results. Thus, different dosing paradigms, or perhaps more potent and selective PPAR agonists, may be needed to elicit clinically meaningful outcomes.

Similarly, species differences in the distribution and signaling of PPARs could also play in a role in the negative human findings. For example, significant differences in the expression [54] and activity [55] of hepatic PPAR-α has been demonstrated in human and rat, in part due to differences in the PPREs of target genes [55]. In addition, species differences have been demonstrated in PPAR-α binding of and response to specific ligands (including clofibrate) [56]. While one recent study did suggest a similar brain distribution of PPARs in adult mice and humans [57], it is still possible that species differences in PPAR-ligand dynamics and in PPAR distribution and signaling could limit the translation of findings from animal models to human studies. The fact that there is poor inter-species comparability in the activity of PPAR agonists is not something unique for PPAR ligands. There have been multiple cases of drugs that appear to be effective in preclinical studies that have not been effective in clinical trials. For example, despite an extensive preclinical literature showing that corticotropin-releasing hormone (CRH) acting via its CRH1 receptor can affect alcohol-seeking behavior, the drug pexacerfont, a CRH1 brain-penetrant antagonist, had no clinical efficacy in a clinical trial in subjects with alcohol dependence [58]. Although it is yet too early to determine if PPAR agonists would similarly fail in humans, this remains a possibility.

Another possibility is simply that the published human studies were underpowered and too few in number to draw conclusions. Jones and colleagues note in two of their pioglitazone studies that they did not reach their recruitment goals [47,49]. Schroeder et al. (2018) noted significant difficulty in recruiting for their study of pioglitazone effects on opioid withdrawal during buprenorphine taper, reaching less than half of their target recruitment [48]. Schmitz et al. (2017), despite finding a potentially meaningful effect of pioglitazone on cocaine craving, note that their study was a pilot trial not specifically powered to detect a difference between drug conditions [50]. Appropriately powered randomized clinical trials are required to clarify the human evidence.

Finally, one possibility that has yet to be considered is the role of sex-related factors in the behavioral pharmacology of PPAR agonists. As seen in Table 1, the overwhelming majority of preclinical studies reviewed included only male animals in their experiments. In the two papers that did report sex differences, the PPAR-α agonist fenofibrate was shown to have more consistent and robust effects on ethanol-related outcomes (voluntary consumption and withdrawal severity) in male mice compared to female mice [30,33]. Furthermore, emerging evidence has found higher expression of PPAR-α mRNA and protein in immune cells of male rodents [59,60]; a role of PPAR-α in neuroprotection [61] and hippocampal synaptic plasticity [62] in male, but not female, rodents; and sex differences in the adverse effects and pharmacokinetics of PPAR-γ agonists such as pioglitazone in humans [63]. Given that all human studies reviewed included female participants (though consistently a small minority), sex differences in the effects of PPAR agonists on drug-related outcomes could have obscured overall drug effects.

5. Future Directions

Given the robust preclinical evidence for an effect of PPAR-α agonists in particular on ethanol-related outcomes, an important first step in moving forward with translating the animal evidence will be conducting human laboratory studies to determine if PPAR agonists (such as gemfibrozil or fenofibrate) modulate the subjective and reinforcing effects of alcohol. Subsequent to this, or in parallel, pilot RCTs to evaluate the efficacy and feasibility of administering PPAR agonists in alcohol use disorder will be necessary.

PPAR-α agonists showed promise for targeting nicotine-related behaviors in animal models, yet two adequately powered human trials found no benefit of fenofibrate or gemfibrozil on smoking cessation or other nicotine-related outcomes. It is possible that these agonists do not have sufficient pharmacological activity at PPAR-α to elicit clinically meaningful outcomes. Indeed, preclinical evidence has shown that more potent compounds like WY14643 confer benefit in attenuating nicotine-related behaviors over fibrates [40]. Selective PPAR modulators (SPPARMS), such as the highly potent and selective PPAR-α agonist K-877, have already shown some promise in treating dyslipidemias and insulin resistance with favorable adverse effect profiles compared to approved drugs such as fenofibrate [53]. If these compounds continue to show efficacy with limited adverse effects, it may be worth testing SPPARMS as smoking cessation drugs in RCTs.

It is possible that targeting PPAR isoforms alone may not be sufficient to treat addictions. For example, as discussed previously, pioglitazone was more effective in reducing reinstatement to ethanol-seeking when it was co-administered with naltrexone [37], an opioid receptor antagonist, suggesting some degree of synergy between PPAR activation and opioid receptor inhibition. Similarly, it has been proposed that simultaneous inhibition of FAAH and activation of PPARs may have an additive or even synergistic effect in treating cancers [64], and this approach may similarly hold promise in the context of addiction pharmacotherapy [65]. Future studies should consider possible synergistic effects that could be achieved by modulation of multiple signaling systems.

It will also be important to validate that the PPAR ligands that are used for SUD treatment are able to occupy/activate brain PPARs. Use of brain imaging approaches such as positron emission tomography could be useful for such target engagement validation. This is critical as some of the previous drug indications for PPAR ligands were likely mediated through PPAR action at the periphery [66].

The PPAR-β/δ isoform was not discussed in this review due to the lack of evidence implicating PPAR-β/δ agonists in addiction-related behaviors. However, it is important to note that PPAR-β/δ is present in the rodent brain at higher levels than the other two isoforms [67] and may play a role in regulating the expression and activity of PPAR-α and PPAR-γ [68]. Furthermore, limited evidence has suggested a role of PPAR-β/δ in neurodevelopmental and neurodegenerative disorders, possibly related to its anti-inflammatory properties [13]. Thus, future studies should investigate the role of PPAR-β/δ agonists in behavioral models of addiction.

A robust body of literature has demonstrated sex-related variability in the effects of common drugs of abuse and in addiction-related processes across animal and human studies [69–71], and emerging evidence suggests similar sex-related variability in the pharmacology of PPAR ligands and in PPAR signaling and function [59,61–63]. Considering sex as a biological variable in future animal studies of PPAR agonists and addiction-related behaviors will be another important next step.

Taken together, this review highlights the robust findings obtained in preclinical studies with agonists at both the PPAR-α and PPAR-γ isoforms that appear effective to reduce both positive and negative reinforcing properties of various drugs of abuse. However, the clinical findings are so far mixed and seem to indicate that the potential is much lower in human subjects. At this point, it is still important to perform small-scale appropriately powered proof of principle studies with PPAR drugs engaging brain PPARs to validate these findings in humans. Positive signals should then be followed by larger RCTs for further validation.

Author Contributions: Writing—original draft preparation, J.M. & B.L.F.; writing—review and editing, J.M. & B.L.F. All authors have read and agreed to the published version of the manuscript.

Funding: Le Foll has obtained funding in the past from Pfizer (GRAND Awards, including salary support and donation of product) for investigator-initiated projects, notably on the impact of gemfibrozil on smoking cessation. All other fundings from Le Foll are unrelated to this review topic. Le Foll is supported by a clinician-scientist award from Department of Family and Community Medicine from the University of Toronto and is supported by the Center for Addiction and Mental Health.

Conflicts of Interest: The authors declare no conflict of interest.

References

1. UNODC. World Drug Report. 2019. Available online: https://wdr.unodc.org/wdr2019/ (accessed on 14 April 2020).
2. American Psychiatric Association. *Diagnostic and Statistical Manual of Mental Disorders (DSM-5®)*; American Psychiatric Pub: Arlington, VA, USA, 2013.
3. Volkow, N.D.; Koob, G.F.; McLellan, A.T. Neurobiologic Advances from the Brain Disease Model of Addiction. *N. Engl. J. Med.* **2016**, *374*, 363–371. [CrossRef] [PubMed]
4. McLellan, A.T. Substance Misuse and Substance use Disorders: Why do they Matter in Healthcare? *Trans. Am. Clin. Climatol. Assoc.* **2017**, *128*, 112–130. [PubMed]
5. McLellan, A.T.; Lewis, D.C.; O'Brien, C.P.; Kleber, H.D. Drug dependence, a chronic medical illness: Implications for treatment, insurance, and outcomes evaluation. *JAMA* **2000**, *284*, 1689–1695. [CrossRef] [PubMed]
6. Koob, G.F.; Volkow, N.D. Neurobiology of addiction: A neurocircuitry analysis. *Lancet Psychiatry* **2016**, *3*, 760–773. [CrossRef]
7. Koob, G.F.; Volkow, N.D. Neurocircuitry of addiction. *Neuropsychopharmacology* **2010**, *35*, 217–238. [CrossRef]
8. Bell, J. Pharmacological maintenance treatments of opiate addiction. *Br. J. Clin. Pharmacol.* **2014**, *77*, 253–263. [CrossRef]
9. Cahill, K.; Stevens, S.; Perera, R.; Lancaster, T. Pharmacological interventions for smoking cessation: An overview and network meta-analysis. *Cochrane Database Syst. Rev.* **2013**. [CrossRef]

10. Kranzler, H.R.; Soyka, M. Diagnosis and Pharmacotherapy of Alcohol Use Disorder: A Review. *JAMA* **2018**, *320*, 815–824. [CrossRef]

11. Berger, J.; Moller, D.E. The mechanisms of action of PPARs. *Annu. Rev. Med.* **2002**, *53*, 409–435. [CrossRef]

12. Fidaleo, M.; Fanelli, F.; Ceru, M.P.; Moreno, S. Neuroprotective properties of peroxisome proliferator-activated receptor alpha (PPARalpha) and its lipid ligands. *Curr. Med. Chem.* **2014**, *21*, 2803–2821. [CrossRef]

13. Tufano, M.; Pinna, G. Is There a Future for PPARs in the Treatment of Neuropsychiatric Disorders? *Molecules* **2020**, *25*, 1062. [CrossRef] [PubMed]

14. Erickson, E.K.; Grantham, E.K.; Warden, A.S.; Harris, R.A. Neuroimmune signaling in alcohol use disorder. *Pharmacol. Biochem. Behav.* **2019**, *177*, 34–60. [CrossRef] [PubMed]

15. Kohno, M.; Link, J.; Dennis, L.E.; McCready, H.; Huckans, M.; Hoffman, W.F.; Loftis, J.M. Neuroinflammation in addiction: A review of neuroimaging studies and potential immunotherapies. *Pharmacol. Biochem. Behav.* **2019**, *179*, 34–42. [CrossRef] [PubMed]

16. Melis, M.; Pillolla, G.; Luchicchi, A.; Muntoni, A.L.; Yasar, S.; Goldberg, S.R.; Pistis, M. Endogenous fatty acid ethanolamides suppress nicotine-induced activation of mesolimbic dopamine neurons through nuclear receptors. *J. Neurosci.* **2008**, *28*, 13985–13994. [CrossRef]

17. Scherma, M.; Panlilio, L.V.; Fadda, P.; Fattore, L.; Gamaleddin, I.; Le Foll, B.; Justinova, Z.; Mikics, E.; Haller, J.; Medalie, J.; et al. Inhibition of anandamide hydrolysis by cyclohexyl carbamic acid 3′-carbamoyl-3-yl ester (URB597) reverses abuse-related behavioral and neurochemical effects of nicotine in rats. *J. Pharmacol. Exp. Ther.* **2008**, *327*, 482–490. [CrossRef]

18. Mascia, P.; Pistis, M.; Justinova, Z.; Panlilio, L.V.; Luchicchi, A.; Lecca, S.; Scherma, M.; Fratta, W.; Fadda, P.; Barnes, C.; et al. Blockade of nicotine reward and reinstatement by activation of alpha-type peroxisome proliferator-activated receptors. *Biol. Psychiatry* **2011**, *69*, 633–641. [CrossRef]

19. Panlilio, L.V.; Justinova, Z.; Mascia, P.; Pistis, M.; Luchicchi, A.; Lecca, S.; Barnes, C.; Redhi, G.H.; Adair, J.; Heishman, S.J.; et al. Novel use of a lipid-lowering fibrate medication to prevent nicotine reward and relapse: Preclinical findings. *Neuropsychopharmacology* **2012**, *37*, 1838–1847. [CrossRef]

20. De Guglielmo, G.; Melis, M.; De Luca, M.A.; Kallupi, M.; Li, H.W.; Niswender, K.; Giordano, A.; Senzacqua, M.; Somaini, L.; Cippitelli, A.; et al. PPARgamma activation attenuates opioid consumption and modulates mesolimbic dopamine transmission. *Neuropsychopharmacology* **2015**, *40*, 927–937. [CrossRef]

21. Melis, M.; Carta, S.; Fattore, L.; Tolu, S.; Yasar, S.; Goldberg, S.R.; Fratta, W.; Maskos, U.; Pistis, M. Peroxisome Proliferator-Activated Receptors-Alpha Modulate Dopamine Cell Activity Through Nicotinic Receptors. *Biol. Psychiatry* **2010**, *68*, 256–264. [CrossRef]

22. Moreno, S.; Farioli-Vecchioli, S.; Cerù, M.P. Immunolocalization of peroxisome proliferator-activated receptors and retinoid x receptors in the adult rat CNS. *Neuroscience* **2004**, *123*, 131–145. [CrossRef]

23. Sarruf, D.A.; Yu, F.; Nguyen, H.T.; Williams, D.L.; Printz, R.L.; Niswender, K.D.; Schwartz, M.W. Expression of peroxisome proliferator-activated receptor-gamma in key neuronal subsets regulating glucose metabolism and energy homeostasis. *Endocrinology* **2009**, *150*, 707–712. [CrossRef] [PubMed]

24. Le Foll, B.; Di Ciano, P.; Panlilio, L.V.; Goldberg, S.R.; Ciccocioppo, R. Peroxisome proliferator-activated receptor (PPAR) agonists as promising new medications for drug addiction: Preclinical evidence. *Curr. Drug Targets* **2013**, *14*, 768–776. [CrossRef] [PubMed]

25. Barson, J.R.; Karatayev, O.; Chang, G.Q.; Johnson, D.F.; Bocarsly, M.E.; Hoebel, B.G.; Leibowitz, S.F. Positive relationship between dietary fat, ethanol intake, triglycerides, and hypothalamic peptides: Counteraction by lipid-lowering drugs. *Alcohol* **2009**, *43*, 433–441. [CrossRef] [PubMed]

26. Ferguson, L.B.; Most, D.; Blednov, Y.A.; Harris, R.A. PPAR agonists regulate brain gene expression: Relationship to their effects on ethanol consumption. *Neuropharmacology* **2014**, *86*, 397–407. [CrossRef] [PubMed]

27. Karahanian, E.; Quintanilla, M.E.; Fernandez, K.; Israel, Y. Fenofibrate—A lipid-lowering drug–reduces voluntary alcohol drinking in rats. *Alcohol* **2014**, *48*, 665–670. [CrossRef] [PubMed]

28. Bilbao, A.; Serrano, A.; Cippitelli, A.; Pavon, F.J.; Giuffrida, A.; Suarez, J.; Garcia-Marchena, N.; Baixeras, E.; Gomez de Heras, R.; Orio, L.; et al. Role of the satiety factor oleoylethanolamide in alcoholism. *Addict. Biol.* **2016**, *21*, 859–872. [CrossRef]

29. Blednov, Y.A.; Benavidez, J.M.; Black, M.; Ferguson, L.B.; Schoenhard, G.L.; Goate, A.M.; Edenberg, H.J.; Wetherill, L.; Hesselbrock, V.; Foroud, T.; et al. Peroxisome proliferator-activated receptors alpha and gamma are linked with alcohol consumption in mice and withdrawal and dependence in humans. *Alcohol. Clin. Exp. Res.* **2015**, *39*, 136–145. [CrossRef]

30. Blednov, Y.A.; Black, M.; Benavidez, J.M.; Stamatakis, E.E.; Harris, R.A. PPAR Agonists: I. Role of Receptor Subunits in Alcohol Consumption in Male and Female Mice. *Alcohol. Clin. Exp. Res.* **2016**, *40*, 553–562. [CrossRef]

31. Haile, C.N.; Kosten, T.A. The peroxisome proliferator-activated receptor alpha agonist fenofibrate attenuates alcohol self-administration in rats. *Neuropharmacology* **2017**, *116*, 364–370. [CrossRef]

32. Rivera-Meza, M.; Munoz, D.; Jerez, E.; Quintanilla, M.E.; Salinas-Luypaert, C.; Fernandez, K.; Karahanian, E. Fenofibrate Administration Reduces Alcohol and Saccharin Intake in Rats: Possible Effects at Peripheral and Central Levels. *Front. Behav. Neurosci.* **2017**, *11*, 133. [CrossRef]

33. Blednov, Y.A.; Black, M.; Benavidez, J.M.; Stamatakis, E.E.; Harris, R.A. PPAR Agonists: II. Fenofibrate and Tesaglitazar Alter Behaviors Related to Voluntary Alcohol Consumption. *Alcohol. Clin. Exp. Res.* **2016**, *40*, 563–571. [CrossRef]

34. Maeda, T.; Kiguchi, N.; Fukazawa, Y.; Yamamoto, A.; Ozaki, M.; Kishioka, S. Peroxisome proliferator-activated receptor gamma activation relieves expression of behavioral sensitization to methamphetamine in mice. *Neuropsychopharmacology* **2007**, *32*, 1133–1140. [CrossRef] [PubMed]

35. Stopponi, S.; Somaini, L.; Cippitelli, A.; Cannella, N.; Braconi, S.; Kallupi, M.; Ruggeri, B.; Heilig, M.; Demopulos, G.; Gaitanaris, G.; et al. Activation of nuclear PPARgamma receptors by the antidiabetic agent pioglitazone suppresses alcohol drinking and relapse to alcohol seeking. *Biol. Psychiatry* **2011**, *69*, 642–649. [CrossRef] [PubMed]

36. Bilbao, A.; Blanco, E.; Luque-Rojas, M.J.; Suarez, J.; Palomino, A.; Vida, M.; Araos, P.; Bermudez-Silva, F.J.; Fernandez-Espejo, E.; Spanagel, R.; et al. Oleoylethanolamide dose-dependently attenuates cocaine-induced behaviours through a PPARalpha receptor-independent mechanism. *Addict. Biol.* **2013**, *18*, 78–87. [CrossRef] [PubMed]

37. Stopponi, S.; de Guglielmo, G.; Somaini, L.; Cippitelli, A.; Cannella, N.; Kallupi, M.; Ubaldi, M.; Heilig, M.; Demopulos, G.; Gaitanaris, G.; et al. Activation of PPARgamma by pioglitazone potentiates the effects of naltrexone on alcohol drinking and relapse in msP rats. *Alcohol. Clin. Exp. Res.* **2013**, *37*, 1351–1360. [CrossRef]

38. De Guglielmo, G.; Kallupi, M.; Scuppa, G.; Stopponi, S.; Demopulos, G.; Gaitanaris, G.; Ciccocioppo, R. Analgesic tolerance to morphine is regulated by PPARgamma. *Br. J. Pharmacol.* **2014**, *171*, 5407–5416. [CrossRef]

39. De Guglielmo, G.; Kallupi, M.; Scuppa, G.; Demopulos, G.; Gaitanaris, G.; Ciccocioppo, R. Pioglitazone attenuates the opioid withdrawal and vulnerability to relapse to heroin seeking in rodents. *Psychopharmacology* **2017**, *234*, 223–234. [CrossRef]

40. Jackson, A.; Bagdas, D.; Muldoon, P.P.; Lichtman, A.H.; Carroll, F.I.; Greenwald, M.; Miles, M.F.; Damaj, M.I. In vivo interactions between alpha7 nicotinic acetylcholine receptor and nuclear peroxisome proliferator-activated receptor-alpha: Implication for nicotine dependence. *Neuropharmacology* **2017**, *118*, 38–45. [CrossRef]

41. Miller, W.R.; Fox, R.G.; Stutz, S.J.; Lane, S.D.; Denner, L.; Cunningham, K.A.; Dineley, K.T. PPARgamma agonism attenuates cocaine cue reactivity. *Addict. Biol.* **2018**, *23*, 55–68. [CrossRef]

42. Domi, E.; Caputi, F.F.; Romualdi, P.; Domi, A.; Scuppa, G.; Candeletti, S.; Atkins, A.; Heilig, M.; Demopulos, G.; Gaitanaris, G.; et al. Activation of PPARgamma Attenuates the Expression of Physical and Affective Nicotine Withdrawal Symptoms through Mechanisms Involving Amygdala and Hippocampus Neurotransmission. *J. Neurosci.* **2019**, *39*, 9864–9875. [CrossRef]

43. Donvito, G.; Piscitelli, F.; Muldoon, P.; Jackson, A.; Vitale, R.M.; D'Aniello, E.; Giordano, C.; Ignatowska-Jankowska, B.M.; Mustafa, M.A.; Guida, F.; et al. N-Oleoyl-glycine reduces nicotine reward and withdrawal in mice. *Neuropharmacology* **2019**, *148*, 320–331. [CrossRef]

44. Perkins, K.A.; Karelitz, J.L.; Michael, V.C.; Fromuth, M.; Conklin, C.A.; Chengappa, K.N.; Hope, C.; Lerman, C. Initial Evaluation of Fenofibrate for Efficacy in Aiding Smoking Abstinence. *Nicotine Tob. Res.* **2016**, *18*, 74–78. [CrossRef] [PubMed]

45. Gendy, M.N.S.; Di Ciano, P.; Kowalczyk, W.J.; Barrett, S.P.; George, T.P.; Heishman, S.; Le Foll, B. Testing the PPAR hypothesis of tobacco use disorder in humans: A randomized trial of the impact of gemfibrozil (a partial PPARalpha agonist) in smokers. *PLoS ONE* **2018**, *13*, e0201512. [CrossRef] [PubMed]

46. Jones, J.D.; Sullivan, M.A.; Manubay, J.M.; Mogali, S.; Metz, V.E.; Ciccocioppo, R.; Comer, S.D. The effects of pioglitazone, a PPARgamma receptor agonist, on the abuse liability of oxycodone among nondependent opioid users. *Physiol. Behav.* **2016**, *159*, 33–39. [CrossRef] [PubMed]
47. Jones, J.D.; Bisaga, A.; Metz, V.E.; Manubay, J.M.; Mogali, S.; Ciccocioppo, R.; Madera, G.; Doernberg, M.; Comer, S.D. The PPARgamma Agonist Pioglitazone Fails to Alter the Abuse Potential of Heroin, But Does Reduce Heroin Craving and Anxiety. *J. Psychoact. Drugs* **2018**, *50*, 390–401. [CrossRef] [PubMed]
48. Schroeder, J.R.; Phillips, K.A.; Epstein, D.H.; Jobes, M.L.; Furnari, M.A.; Kennedy, A.P.; Heilig, M.; Preston, K.L. Assessment of pioglitazone and proinflammatory cytokines during buprenorphine taper in patients with opioid use disorder. *Psychopharmacology* **2018**, *235*, 2957–2966. [CrossRef]
49. Jones, J.D.; Comer, S.D.; Metz, V.E.; Manubay, J.M.; Mogali, S.; Ciccocioppo, R.; Martinez, S.; Mumtaz, M.; Bisaga, A. Pioglitazone, a PPARgamma agonist, reduces nicotine craving in humans, with marginal effects on abuse potential. *Pharmacol. Biochem. Behav.* **2017**, *163*, 90–100. [CrossRef]
50. Schmitz, J.M.; Green, C.E.; Hasan, K.M.; Vincent, J.; Suchting, R.; Weaver, M.F.; Moeller, F.G.; Narayana, P.A.; Cunningham, K.A.; Dineley, K.T.; et al. PPAR-gamma agonist pioglitazone modifies craving intensity and brain white matter integrity in patients with primary cocaine use disorder: A double-blind randomized controlled pilot trial. *Addiction* **2017**, *112*, 1861–1868. [CrossRef]
51. Grabacka, M.; Waligorski, P.; Zapata, A.; Blake, D.A.; Wyczechowska, D.; Wilk, A.; Rutkowska, M.; Vashistha, H.; Ayyala, R.; Ponnusamy, T.; et al. Fenofibrate subcellular distribution as a rationale for the intracranial delivery through biodegradable carrier. *J. Physiol. Pharmacol.* **2015**, *66*, 233–247.
52. Deplanque, D.; Gele, P.; Petrault, O.; Six, I.; Furman, C.; Bouly, M.; Nion, S.; Dupuis, B.; Leys, D.; Fruchart, J.C.; et al. Peroxisome proliferator-activated receptor-alpha activation as a mechanism of preventive neuroprotection induced by chronic fenofibrate treatment. *J. Neurosci.* **2003**, *23*, 6264–6271. [CrossRef]
53. Liu, Z.M.; Hu, M.; Chan, P.; Tomlinson, B. Early investigational drugs targeting PPAR-alpha for the treatment of metabolic disease. *Expert Opin. Investig. Drugs* **2015**, *24*, 611–621. [CrossRef] [PubMed]
54. Palmer, C.N.; Hsu, M.H.; Griffin, K.J.; Raucy, J.L.; Johnson, E.F. Peroxisome proliferator activated receptor-alpha expression in human liver. *Mol. Pharmacol.* **1998**, *53*, 14–22. [CrossRef] [PubMed]
55. Ammerschlaeger, M.; Beigel, J.; Klein, K.U.; Mueller, S.O. Characterization of the species-specificity of peroxisome proliferators in rat and human hepatocytes. *Toxicol. Sci.* **2004**, *78*, 229–240. [CrossRef] [PubMed]
56. Oswal, D.P.; Balanarasimha, M.; Loyer, J.K.; Bedi, S.; Soman, F.L.; Rider, S.D., Jr.; Hostetler, H.A. Divergence between human and murine peroxisome proliferator-activated receptor alpha ligand specificities. *J. Lipid Res.* **2013**, *54*, 2354–2365. [CrossRef] [PubMed]
57. Warden, A.; Truitt, J.; Merriman, M.; Ponomareva, O.; Jameson, K.; Ferguson, L.B.; Mayfield, R.D.; Harris, R.A. Localization of PPAR isotypes in the adult mouse and human brain. *Sci. Rep.* **2016**, *6*, 27618. [CrossRef]
58. Kwako, L.E.; Spagnolo, P.A.; Schwandt, M.L.; Thorsell, A.; George, D.T.; Momenan, R.; Rio, D.E.; Huestis, M.; Anizan, S.; Concheiro, M.; et al. The corticotropin releasing hormone-1 (CRH1) receptor antagonist pexacerfont in alcohol dependence: A randomized controlled experimental medicine study. *Neuropsychopharmacology* **2015**, *40*, 1053–1063. [CrossRef]
59. Dunn, S.E.; Ousman, S.S.; Sobel, R.A.; Zuniga, L.; Baranzini, S.E.; Youssef, S.; Crowell, A.; Loh, J.; Oksenberg, J.; Steinman, L. Peroxisome proliferator-activated receptor (PPAR)alpha expression in T cells mediates gender differences in development of T cell-mediated autoimmunity. *J. Exp. Med.* **2007**, *204*, 321–330. [CrossRef]
60. Jalouli, M.; Carlsson, L.; Ameen, C.; Linden, D.; Ljungberg, A.; Michalik, L.; Eden, S.; Wahli, W.; Oscarsson, J. Sex difference in hepatic peroxisome proliferator-activated receptor alpha expression: Influence of pituitary and gonadal hormones. *Endocrinology* **2003**, *144*, 101–109. [CrossRef]
61. Dotson, A.L.; Wang, J.; Chen, Y.; Manning, D.; Nguyen, H.; Saugstad, J.A.; Offner, H. Sex differences and the role of PPAR alpha in experimental stroke. *Metab. Brain Dis.* **2016**, *31*, 539–547. [CrossRef]
62. Pierrot, N.; Ris, L.; Stancu, I.C.; Doshina, A.; Ribeiro, F.; Tyteca, D.; Bauge, E.; Lalloyer, F.; Malong, L.; Schakman, O.; et al. Sex-regulated gene dosage effect of PPARalpha on synaptic plasticity. *Life Sci. Alliance* **2019**, *2*. [CrossRef]
63. Benz, V.; Kintscher, U.; Foryst-Ludwig, A. Sex-specific differences in Type 2 Diabetes Mellitus and dyslipidemia therapy: PPAR agonists. In *Handbook of Experimental Pharmacology*; Springer: Berlin/Heidelberg, Germany, 2012. [CrossRef]

64. Brunetti, L.; Loiodice, F.; Piemontese, L.; Tortorella, P.; Laghezza, A. New Approaches to Cancer Therapy: Combining Fatty Acid Amide Hydrolase (FAAH) Inhibition with Peroxisome Proliferator-Activated Receptors (PPARs) Activation. *J. Med. Chem.* **2019**, *62*, 10995–11003. [CrossRef] [PubMed]

65. Panlilio, L.V.; Justinova, Z.; Goldberg, S.R. Inhibition of FAAH and activation of PPAR: New approaches to the treatment of cognitive dysfunction and drug addiction. *Pharmacol. Ther.* **2013**, *138*, 84–102. [CrossRef] [PubMed]

66. Hong, F.; Xu, P.; Zhai, Y. The Opportunities and Challenges of Peroxisome Proliferator-Activated Receptors Ligands in Clinical Drug Discovery and Development. *Int. J. Mol. Sci.* **2018**, *19*, 2189. [CrossRef] [PubMed]

67. Braissant, O.; Foufelle, F.; Scotto, C.; Dauca, M.; Wahli, W. Differential expression of peroxisome proliferator-activated receptors (PPARs): Tissue distribution of PPAR-alpha, -beta, and -gamma in the adult rat. *Endocrinology* **1996**, *137*, 354–366. [CrossRef] [PubMed]

68. Aleshin, S.; Strokin, M.; Sergeeva, M.; Reiser, G. Peroxisome proliferator-activated receptor (PPAR)beta/delta, a possible nexus of PPARalpha- and PPARgamma-dependent molecular pathways in neurodegenerative diseases: Review and novel hypotheses. *Neurochem. Int.* **2013**, *63*, 322–330. [CrossRef] [PubMed]

69. Riley, A.L.; Hempel, B.J.; Clasen, M.M. Sex as a biological variable: Drug use and abuse. *Physiol. Behav.* **2018**, *187*, 79–96. [CrossRef]

70. Becker, J.B.; Chartoff, E. Sex differences in neural mechanisms mediating reward and addiction. *Neuropsychopharmacology* **2019**, *44*, 166–183. [CrossRef]

71. Becker, J.B.; Koob, G.F. Sex Differences in Animal Models: Focus on Addiction. *Pharmacol. Rev.* **2016**, *68*, 242–263. [CrossRef]

Review

PPAR-Mediated Toxicology and Applied Pharmacology

Yue Xi [1,2], Yunhui Zhang [1], Sirui Zhu [1], Yuping Luo [1], Pengfei Xu [2,*] and Zhiying Huang [1,*]

[1] School of Pharmaceutical Sciences, Sun Yat-sen University, Guangzhou 510006, China; xiyue3@mail2.sysu.edu.cn (Y.X.); zhangyh75@mail2.sysu.edu.cn (Y.Z.); zhusr3@mail2.sysu.edu.cn (S.Z.); luoyp26@mail2.sysu.edu.cn (Y.L.)

[2] Center for Pharmacogenetics and Department of Pharmaceutical Sciences, University of Pittsburgh, Pittsburgh, PA 15213, USA

* Correspondence: pex9@pitt.edu (P.X.); hzhiying@mail.sysu.edu.cn (Z.H.); Tel.: +1-412-708-4694 (P.X.); +86-20-39943092 (Z.H.)

Received: 5 January 2020; Accepted: 30 January 2020; Published: 3 February 2020

Abstract: Peroxisome proliferator-activated receptors (PPARs), members of the nuclear hormone receptor family, attract wide attention as promising therapeutic targets for the treatment of multiple diseases, and their target selective ligands were also intensively developed for pharmacological agents such as the approved drugs fibrates and thiazolidinediones (TZDs). Despite their potent pharmacological activities, PPARs are reported to be involved in agent- and pollutant-induced multiple organ toxicity or protective effects against toxicity. A better understanding of the protective and the detrimental role of PPARs will help to preserve efficacy of the PPAR modulators but diminish adverse effects. The present review summarizes and critiques current findings related to PPAR-mediated types of toxicity and protective effects against toxicity for a systematic understanding of PPARs in toxicology and applied pharmacology.

Keywords: PPARs; toxicology; pharmacology; ligand

---

1. Introduction

Peroxisome proliferator-activated receptors (PPARs), a group of nuclear hormone receptors, are composed of three isoforms which were identified as PPARα, PPARγ, and PPARβ. Each is encoded by distinct genes and has different targeting ligands, tissue distribution, and biological activities. PPAR family proteins, like other nuclear receptors, have three main functional segments, activation function 1 (AF1) and the conserved DNA-binding domain (DBD), the hinge region, and the ligand-binding domain (LBD) and AF2. The variable N-terminal regulatory AF1 domain binds co-regulators and the conserved DBD, which can bind to the peroxisome proliferator response elements (PPREs). The mobile hinge region links DBD and the conserved LBD in the middle. LBD and the variable C-terminal AF2 domain form a large ligand binding pocket [1,2]. Due to the large LBD pocket, PPARs have the capacity to bind various compounds, including endogenous or synthetic ligands and xenobiotic chemicals. In pharmacology, the ligands of PPARs are classified into full agonists, partial agonists, neutral antagonists, and inverse agonists. Recently, we summarized the 84 types of PPAR synthetic ligands for the treatment of various diseases in current clinical drug applications [3]. The LBD contains a C-terminal AF2 motif that is a ligand-dependent activation region [4]. Under physiological conditions, PPARs bind with co-repressors and form heterodimers with retinoid X receptor (RXR) [5]. In response to ligand activation, the protein conformation is changed and stabilized, which leads to dissociation of co-repressors and the recruitment of transcription co-activators and DNA-binding cofactors. This complex regulates transcription of target genes by binding specific DNA sequences, called peroxisome proliferator response elements (PPREs), on promoter regions of target genes [6,7].

PPAR activated genes play critical roles in fatty acid transportation and catabolism, glucose metabolism, adipogenesis, thermogenesis, cholesterol transportation and biosynthesis, and anti-inflammatory response [4,8]. Because of their broad-spectrum biological activities, PPARs arouse much attention, and they are studied intensively. Accumulated studies show that activation of PPARs has unique pharmacological effects on cardiovascular function, neurodegeneration, inflammation, cancer, fertility, and reproduction, and it is well established for managing dyslipidemia, diabetes, insulin resistance, and metabolic syndrome, which stimulates researchers to persistently develop more new drugs targeting PPARs [9,10]. Some PPAR agonists are approved as clinical agents such as thiazolidinediones, fibrates, and glitazars, for the treatment of diabetes, dyslipidemia, and diabetes-associated complications, respectively.

Despite the multiple biological activities of PPARs, several studies and clinical cases indicated that PPARs mediate various adverse effects of drugs, especially PPAR ligands or xenobiotic chemical-induced toxicity in different systems. Thiazolidinediones (TZDs), one class of PPARγ agonists, can cause fluid retention, heart failure, and hepatotoxicity [11,12]. Glitazones, another form of PPARγ ligand, were reported to cause peripheral edema, congestive heart failure, and body weight gain. Gemfibrozil, as a valuable agent to coronary heart disease, was shown to induce tumorigenesis, muscle weakness, and liver hypertrophy [13]. The detailed information and adverse reactions or toxicity of the 18 approved clinical agents that target PPARs are summarized in Table 1. Because of the high prevalence of tumorigenesis in PPAR activation by synthetic compounds, the Food and Drug Administration (FDA) requires that any PPAR agonists undergo a two-year rodent carcinogenicity study before being tested in clinical trials [13]. Moreover, PPARs were shown to be involved in pollutant-induced toxicity in the cardiovascular system, liver, reproductive and developmental system, gastrointestinal tract, muscle, and nervous system.

Based on the above information, this review is focused on the reports of PPAR activation-mediated toxicity and protective effects to date, aiming to provide an overview of studies evaluating the toxic role of PPARs in various systems and the molecular mechanisms of PPAR-elicited toxicity.

Table 1. The information and adverse reactions or toxicity of peroxisome proliferator-activated receptor (PPAR) targets related to 18 approved clinical drugs.

Generic Name (Brand Name)	Type of PPAR Agonist	Molecular Weight and Molecular Formula	Structure	Company	Indications	Adverse Reaction or Toxicity
Rosiglitazone maleate (Avandia)	PPARγ agonist	473.5 $C_{22}H_{23}N_3O_7S$	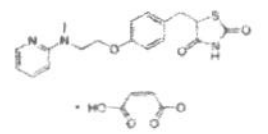	GlaxoSmithKline	Diabetes	Headache, cough, cold symptoms, and back pain
Pioglitazone hydrochloride (Actos)	PPARγ agonist	392.898 $C_{19}H_{21}ClN_2O_3S$	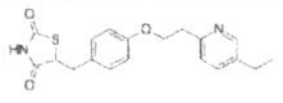	Takeda/Lilly	Diabetes	Cold or flu-like symptoms, headache, gradual weight gain, muscle pain, back pain, tooth problems, and mouth pain
Lobeglitazone sulfate (Duvie)	Dual PPARα/γ agonist	578.61 $C_{24}H_{26}N_4O_9S_2$	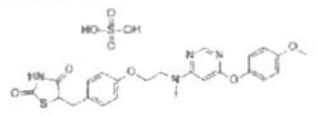	Chong Kun Dang	Diabetes	Edema and weight gain
Alogliptin benzoate/pioglitazone hydrochloride (Oseni)	Dipeptidyl peptidase IV inhibitor/PPARγ agonist	461.519 ($C_{25}H_{27}N_5O_4$)/ 392.898 ($C_{19}H_{21}ClN_2O_3S$)	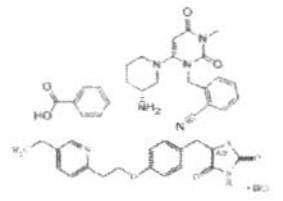	Takeda	Diabetes	Upper respiratory tract infection, bone fracture, headache, nasopharyngitis, and pharyngitis
Glimepiride/pioglitazone hydrochloride (Duetact)	Sulfonylurea receptor modulator/PPARγ agonist	490.62($C_{24}H_{34}N_4O_5S$)/ 392.898 ($C_{19}H_{21}ClN_2O_3S$)	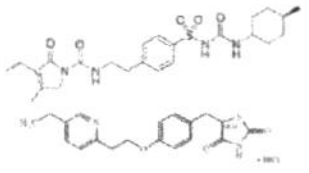	Takeda	Diabetes	Congestive heart failure, hypoglycemia, edema, fractures, and hemolytic anemia
Pioglitazone hydrochloride/metformin hydrochloride (Actoplus Met)	PPARγ agonist/adenosine monophosphate-activated protein kinase (AMPK) activator	392.898 ($C_{19}H_{21}ClN_2O_3S$)/ 165.6($C_4H_{12}ClN_5$)	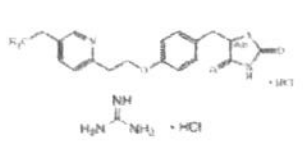	Takeda	Diabetes	Headache, nausea, vomiting, stomach upset, diarrhea, weakness, sore throat, muscle pain, weight gain, tooth problems, a metallic taste in the mouth, and sneezing, runny nose, cough, or other signs of a cold
Rosiglitazone maleate/metformin hydrochloride (Avandamet)	PPARγ agonist; AMPK activator	473.5($C_{22}H_{23}N_3O_7S$)/ 165.6($C_4H_{12}ClN_5$)	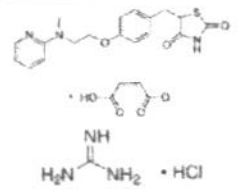	GlaxoSmithKline	Diabetes	Lactic acidosis, cardiac failure, adverse cardiovascular events, edema, weight gain, hepatic effects, macular edema, fractures, hematologic effects, and ovulation
Glimepiride/rosiglitazone maleate (Avandaryl)	Sulfonylurea receptor modulator/PPARγ agonist	490.62 ($C_{24}H_{34}N_4O_5S$)/ 473.5($C_{22}H_{23}N_3O_7S$)	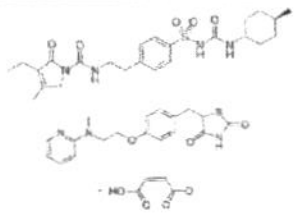	GlaxoSmithKline	Diabetes	Cardiac failure with rosiglitazone, major adverse cardiovascular events, hypoglycemia, edema, weight gain, hepatic effects, macular edema, fractures, hypersensitivity reactions, hematologic effects, hemolytic anemia, and increased risk of cardiovascular mortality for sulfonylurea drugs
Clofibrate (Atromid-S)	PPARα agonist	242.699 $C_{12}H_{15}ClO_3$	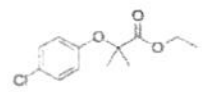	Pfizer	Hyperlipidemia Hypertriglyceridemia Hypercholesterolemia	Common: diarrhea, nausea Rare: abnormal heart rhythm, acute inflammation of the pancreas, anemia, angina, gallstones, kidney failure, and low levels of white blood cells

Table 1. *Cont.*

Generic Name (Brand Name)	Type of PPAR Agonist	Molecular Weight and Molecular Formula	Structure	Company	Indications	Adverse Reaction or Toxicity
Fenofibrate (Antara)	PPARα agonist	360.834 $C_{20}H_{21}ClO_4$	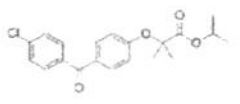	Abbvie	Hypercholesterolemia Hypertriglyceridemia	Common: abnormal liver tests including aspartate aminotransferase (AST) and alanine aminotransferase (ALT), and headache Rare: high blood pressure, dizziness, itching, nausea, upset stomach, constipation, diarrhea, urinary tract infections, muscle pain, kidney problems, and respiratory tract infections
Choline fenofibrate (Fenofibric Acid)	PPARα agonist	421.918 $C_{22}H_{28}ClNO_5$	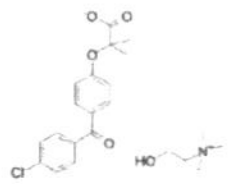	Abbvie	Hyperlipidemia	Diarrhea, dyspepsia, nasopharyngitis, sinusitis, upper respiratory tract infection, arthralgia, myalgia, pain in extremities, dizziness
Bezafibrate (Bezalip)	PPARα agonist	361.822 $C_{19}H_{20}ClNO_4$	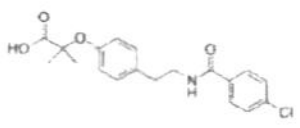	Roche Diagnostics	Hypertriglyceridemia hypercholesterolemia mixed hyperlipidemia	Stomach upset, stomach pain, gas, or nausea may occur in the first several days; itchy skin, redness, headache, and dizziness
Gemfibrozil (Lopid)	PPARα agonist	250.338 $C_{15}H_{22}O_3$	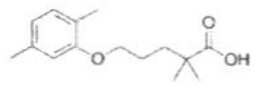	Pfizer	Hyperlipidemia Ischemic heart disorder	Stomach upset, stomach/abdominal pain, nausea, vomiting, diarrhea, constipation, rash, dizziness, headache, changes in the way things taste, muscle pain
Ciprofibrate (Lipanor)	PPARα agonist	289.152 $C_{13}H_{14}Cl_2O_3$	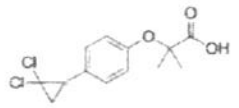	Sanofi-Aventis	Hyperlipidemia	Hair loss, balding, headache, balance problems, feeling dizzy, drowsiness or fatigue, feeling sick (nausea) or being sick (vomiting), diarrhea, indigestion or stomach pains, muscle pains
Pemafibrate (Parmodia)	PPARα agonist	490.556 $C_{28}H_{30}N_2O_6$	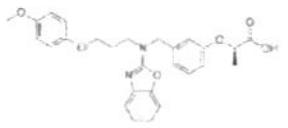	Kowa	Dyslipidemia	Cholelithiasis (upper abdominal pain, fever) and diabetes mellitus (dry mouth, excess intake of fluid, excessive urination, fatigue)
Pravastatin sodium/fenofibrate (Pravafenix)	3-hydroxy-3-methylglutaryl-CoA reductase (HMGCR) inhibitor/PPARα agonist	446.5($C_{23}H_{35}NaO_7$)/ 360.83($C_{20}H_{21}ClO_4$)	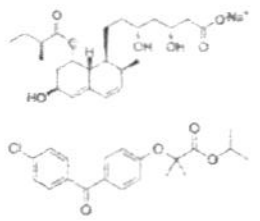	Laboratoires SMB	Mixed hyperlipidemia Coronary heart disease	Abdominal distension (bloating), abdominal pain (stomach ache), constipation, diarrhea, dry mouth, dyspepsia (heartburn), eructation (belching), flatulence (gas), nausea (feeling sick), abdominal discomfort, vomiting, and raised blood levels of liver enzymes

Table 1. *Cont.*

Generic Name (Brand Name)	Type of PPAR Agonist	Molecular Weight and Molecular Formula	Structure	Company	Indications	Adverse Reaction or Toxicity
Fenofibrate/simvastatin (Cholib)	PPARα agonist/ HMGCR inhibitor	360.834 $C_{20}H_{21}ClO_4$/ 418.57 ($C_{25}H_{38}O_5$)	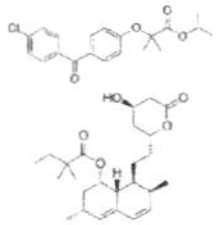	Mylan	Mixed hyperlipidemia	Raised blood creatinine levels, upper-respiratory-tract infection (colds), increased blood platelet counts, gastroenteritis (diarrhea and vomiting) and increased levels of alanine aminotransferase
Saroglitazar (Lipaglyn)	PPARα/γ agonist	439.57 $C_{25}H_{29}NO_4S$	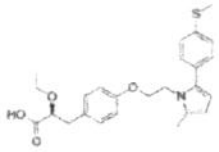	Zydus Cadila	Diabetic dyslipidemia	Asthenia, gastritis, chest discomfort, peripheral edema, dizziness, and tremors

2. PPARs in Cardiotoxicity

Considering the high expression level of PPARs in cardiac muscles and their strong implication in metabolic disorders and endocrine disruption, interference with PPARs can affect metabolic homeostasis and development of the cardiovascular system.

Cardiac edemas and the impairment of cardiac development were observed in marine medaka larvae fish exposed to perfluorooctane sulfonate (PFOS) by interfering PPARα and PPARβ [14,15]. Perfluorooctanoic acid (PFOA) exposure-induced right-ventricular wall thinning elevation in chicken embryos is also likely due to PPARα [16]. After exposure to di-ethyl-hexylphthalate (DEHP), changes in the metabolic profile via the PPARα pathway can be detected in rat cardiomyocytes [17]. Triclocarban (TCC) is a high-performance broad-spectrum fungicide, which can induce cardiac metabolic alterations in mice by suppression of PPARα messenger RNA (mRNA) expression and other enzymes involved in energy and lipid metabolism. A further study found TCC directly interacted with the active site of PPARα in both mice and human tissues [18]. Exposure to airborne particulate matter is positively correlated with cardiorespiratory mortality [19]. Some studies showed that heart abnormal energy metabolism caused by seasonal ambient fine particles (PM$_{2.5}$) was related to PPARα-regulated fatty-acid and glucose transporters. Unmanaged heart abnormal energy metabolism eventually leads to cardiac damage and heart failure [20]. Considerable research suggested that PPARs play pivotal roles in myocardial energy dysfunction. Energy substrate utilization showed a marked shift from fatty acid to glucose and lactate and cardiac hypertrophy in PPARα$^{-/-}$ hearts [21]. PPARα-null hearts with decreased contractile and metabolic remodeling were rescued by enhancing myocardial glucose transportation and utilization [21].

Furthermore, PPARγ inhibits cardiac growth and embryonic gene expression and decreases nuclear factor kappa B (NF-κB) activity in mice [22]. Cardiomyocyte-specific PPARγ knockout mice were more susceptible to cardiac hypertrophy with systolic cardiac function [22]. CKD-501, a new selective PPARγ agonist, induced heart toxicity in db/db mice by PPARγ-dependent mechanism [23]. Rosiglitazone leads to cardiac hypertrophy partially independent of cardiomyocyte PPARγ [22]. Another study indicated that rosiglitazone caused oxidative stress-induced mitochondrial dysfunction via PPARγ-independent pathways in mouse hearts [24]. Anna et al. reported that atorvastatin ameliorated cardiac hypertrophy by improving the protein expression of PPARα and PPARβ, which regulated the gene expression involved in fatty acid metabolism and avoided NF-κB activation by reducing the protein–protein interaction between PPARs and p65 [25]. Moreover, atorvastatin reduced the paraquat-induced cardiotoxicity via the PPARγ pathway [26]. Hesperidin, a flavanone glycoside and a known PPARγ ligand, improved cardiac hypertrophy by improving cardiac hemodynamics, as well as inhibiting oxidative stress and apoptosis through increasing PPARγ expression [27]. Piperine, a phenolic component of black pepper, attenuated cardiac fibrosis via PPARγ activation and the inhibition of protein kinase B (AKT)/glycogen synthase kinase 3 β (GSK3β) [28]. Interestingly, the regulation of PPARγ by pioglitazone suppressed cardiac hypertrophy as indicated by decreased heart/body weight ratio, wall thickness, and myocyte diameter [29], but the effect of pioglitazone on limiting myocardial infarct size was a PPARγ-independent event [30]. Epoxyeicosatrienoic acids (EET), a primary arachidonic acid metabolite, blocked tumor necrosis factor α (TNFα)-induced cardiotoxicity by reducing inflammation via upregulation of PPARγ expression [31].

Some dual PPARα/γ agonists such as tesaglitazar display an increased risk of cardiovascular events. Treatment with tesaglitazar in mice caused cardiac dysfunction associated with low mitochondrial abundance [32]. In addition, tesaglitazar increased acetylation of proliferator-activated receptor gamma coactivator 1α (PGC1α) and decreased the expression of sirtuin 1 (SIRT1), which was associated with competition between PPARα and PPARγ. LY510929, another dual PPARα/γ agonist, was shown to cause left-ventricular hypertrophy in rats [33]. However, aleglitazar inhibited hyperglycemia-induced cardiomyocyte apoptosis by activation of both PPARα and PPARγ [34,35]. Activation of PPARβ signaling mediated docosahexaenoic acid (DHA), and its metabolites elicited cytotoxicity in H9c2 cells via the de novo formation of ceramide [36]. Doxorubicin (DOX) caused a remarkable decrease in cardiac dP/dT and cardiac output by inhibition of PPARβ expression in rats [37].

PPARβ plays an important role in angiogenesis and cancers. Activation of PPARβ in blood vessels promotes tumor vascularization and the progression of different cancer cell types through direct activation of platelet-derived growth factor receptor beta (PDGFRβ), platelet-derived growth factor subunit B (PDGFB), and the c-Kit [38]. Figure 1 summarizes regulation of PPARs in cardiotoxicity.

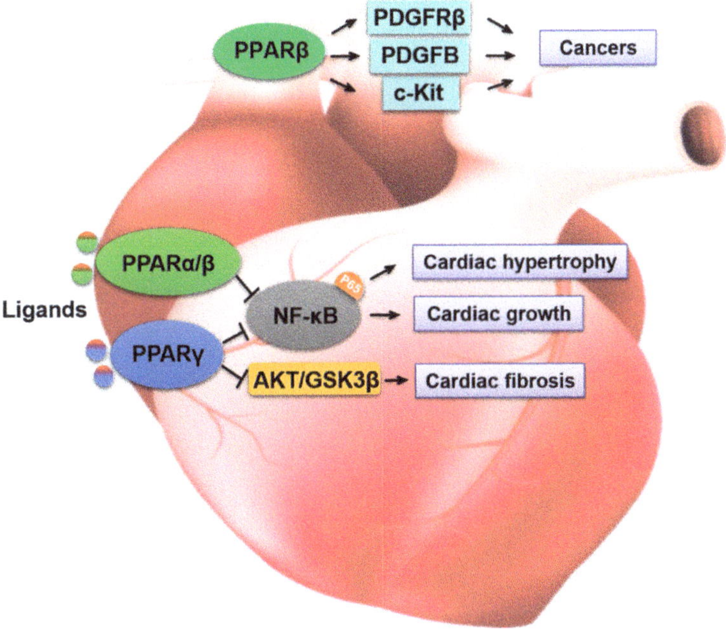

Figure 1. The regulation of PPARs in cardiotoxicity.

3. PPARs in Hepatotoxicity

In the liver, PPARs play indispensable roles in fatty-acid and glucose metabolism, and they supply energy to peripheral tissues. Numerous studies reported that xenobiotic chemicals and environmental contaminants disrupted the normal liver homeostasis by activating PPAR subtypes that are highly expressed in hepatocytes, especially PPARα. Indeed, PPARα was recognized as a target for pollutants, which could interact with the similar nuclear receptors and subsequently induce metabolic disorders.

Phthalates, common plasticizers in nearly all plastic consumer goods, are defined as PPAR modulators [6]. Accumulative studies showed that phthalates activated PPARα and other lipid-activated nuclear receptors in the liver, which induced metabolic disruption and endocrine disorders. The exposure concentration of phthalate metabolites such as DEHP and mono (2-ethylhexyl) phthalate (MEHP) positively correlated with insulin resistance and abdominal obesity in American male adults [39–41]. Di-*n*-butyl-di-(4-chlorobenzohydroxamato) tin (DBDCT), an organotin with high antitumor activity, was also demonstrated to induce notable toxicity in rat liver tissue via the PPAR signaling pathway [42]. DBDCT treatment aroused acute and focal necrosis and Kupffer cell hyperplasia in rat liver. The decreased expression levels of cluster of differentiation 36 (CD36), fatty acid binding protein 4 (FABP4), enoyl-CoA hydratase and 3-hydroxyacyl CoA dehydrogenase (EHHADH), acetyl-CoA acyltransferase 1 (ACAA1), phosphoenolpyruvate carboxykinase (PEPCK),

PPARα, and PPARγ in DBDCT-treated liver tissue were indicated by proteomics. Furthermore, the toxic effect was alleviated by PPARγ blocking agent T0070907 [42,43]. Additionally, organotins, the major components of agricultural fungicides and pesticides, were documented to exert similar functions as PPARγ and PPARβ ligands, which promote weight gain and increase fat storage by target gene induction in liver [44]. For example, tributyltin chloride (TBT) enhanced adipogenesis and adipocyte differentiation by directly stimulating downstream transcription of PPARγ in liver and adipose tissues. In mouse models, uterus exposure to TBT disrupted hepatic architecture and caused liver steatosis by increasing lipid accumulation and adipocyte maturation [40,45]. Thus, PPARs play an important role in contaminant-induced toxicity.

The hepatotoxicity of PPARα ligands was rarely documented. Few PPARα ligands are proven to be hepatotoxicants. Fenofibrate exerts only a minimal increase of alanine aminotransferase and aspartate aminotransferase [6,46]. In contrast to hepatotoxicity mediated by PPARs, PPAR ligands also display some protective effect against hepatotoxicity. PPARα ligand activation was proven to prevent acute liver toxicity induced by alcohol, carbon tetrachloride (CCl_4), acetaminophen, chloroform, thioacetamide, and bromobenzene due to the induction of fatty acid catabolism and anti-inflammatory properties [47–49]. PPARα agonists showed a reversal of fatty liver in mice even with continued ethanol consumption [50]. PPARγ agonist troglitazone and rosiglitazone are reported to induce mild liver toxicity in patients that might be PPARγ-independent because of the low expression level of PPARγ in the liver [12]. Despite the hepatotoxicity of PPARγ activation, PPARγ ligand treatment attenuated fibrogenesis by inhibiting the activation of hepatic stellate cells (HSCs) [51,52]. PPARγ ligands exhibited a suppressive effect on the expression of fibrogenic genes including collagen and α-smooth muscle actin. PPARβ activation by L-165041 enhanced the HSC proliferation and fibrogenic gene expression, and it exacerbated CCl_4-induced liver fibrotic progression [53]. PPARα, PPARγ, and PPARβ display different roles in hepatotoxicity. Activation of PPARα prevents acute liver toxicity. Activation of PPARγ induces mild liver toxicity but attenuates liver fibrogenesis. Activation of PPARβ promotes the progression of liver fibrosis.

Numerous studies reported that hepatocarcinogenesis was the major toxicity induced by PPARα activation [54,55]. Unmanaged peroxisomal proliferation and hepatomegaly observed in fibrate-treated livers can ultimately lead to hepatocellular carcinoma [56]. The hepatocarcinogenesis by PPARα activation was fully investigated over 30 years. The main target of PPARα is the liver, which induces pleiotropic impacts such as hypertrophy and hyperplasia [57,58]. These unmanaged responses cause hepatocellular carcinomas in rodents. The mechanisms remain elucidated. Some studies propose that PPARα-mediated DNA replication, proliferation, and suppressed apoptosis result in PPARα agonist-induced hepatocarcinogenesis [59]. Actually, the effect of PPARα on hepatocarcinogenesis varies among different species. In human, an increased risk of liver cancer of fibrates is not yet reported. This might be due to no significant peroxisome proliferation induced by hypolipidemic agents [60] and less expression of PPARα in patient livers compared to rodent liver. Although humans show resistance to the adverse effect of PPARα-induced hepatocarcinogenesis, vigilance is still required to develop new agents.

4. PPARs in Gastrointestinal Toxicity

As indicated by emerging evidence, PPARs and their ligands also play an important role in the regulation of immune and inflammatory reactions in the gastrointestinal (GI) system.

In view of modulation of several target genes involved in metabolic processes and immune response in the GI tract, PPARs and their ligands became a research hotspot in gastroenterology [61]. Accumulative evidence showed that inflammatory bowel diseases (IBDs) and colon cancer (CC), two important GI diseases, are related to PPARs and their ligands [62,63]. PPAR agonists might serve as a new effective pharmacotherapy for IBDs and CC. PPARα mediated the anti-inflammatory effect of glucocorticoid (GC) in a chemical-induced colitis mouse model [64]. More recently, it was shown that PPARα activation diminished the therapeutic effects of rSj16 in dextran sulfate sodium (DSS)-induced colitis mice, indicating that the PPARα signaling pathway plays a crucial role in DSS-induced colitis progression [65].

With the high expression in GI tract mucosa, especially in the intestine and colon [66–68], PPARγ is closely related to GI injury and inflammatory response. The inflammatory reaction is the common pathological process of many GI diseases and trauma. Once the homeostasis of GI is disrupted by exogenous factors or endogenous metabolites and shifts to the pro-inflammatory state, the pro-inflammatory cytokines such as TNF-α, interleukin 1β (IL-1β), IL-6 are liberated by the hyperactive immune cells. Transcription factor NF-κB is one of the most important regulatory mechanisms of immune and inflammatory responses mediated by PPARs and their ligands in the GI tract. In colon, PPARγ downregulated NF-κB and mitogen-activated protein kinase (MAPK) signaling pathways, which subsequently inhibited the mucosal production of inflammatory cytokines [69]. Furthermore, in intestinal cells, activation of PPARγ resulted in decreased expression of intercellular adhesion molecule 1 (ICAM-1) and TNF-α [70], which are downstream targets of NF-kB [71]. Treatment with troglitazone attenuated colitis induced by intrarectal administration of 2,4,6-trinitrobenzene sulfonic acid (TNBS) [69]. PPARγ could function as an endogenous anti-inflammatory pathway in a murine model of intestinal ischemia–reperfusion (I/R) injury. Activation of PPARγ by its agonist BRL-49653 had a protective effect on intestinal acute I/R injury [70]. However, the protective activity of BRL-49653 was abolished in PPARγ-deficient mice. In the investigation of the prevention and treatment of radiation-induced intestinal damage, accumulating evidence supported that the administration of PPARγ agonists alleviated radiation-induced intestinal toxicity. PPARγ agonists were shown to reverse radiation-induced apoptosis and inflammation and to exert radio-protective effects on healthy bowel upon irradiation [72,73]. Further research of acute intestinal injury reported that PPARγ agonist rosiglitazone reduced the expression of the fibrotic marker transforming growth factor β (TGFβ) and phosphorylation of the p65 subunit of NF-kB triggered by pro-inflammatory cytokine TNF-α [72]. In ulcerative colitis research, promoting the nuclear localization of PPARγ weakened the activity of NF-κB signaling in both rectal tissues from dextran sulfate sodium (DSS)-induced mice and lipopolysaccharide (LPS)-stimulated macrophages [74]. PPARs inhibited the expression of macrophage-related inflammatory mediators and macrophage infiltration in the acute irradiation intestinal damage [75]. Compared with wild-type mice, PPARγ-deficient mice showed significantly severe damage after an I/R injury procedure, which indicated the anti-inflammatory and protective role of PPARγ in GI damage [70]. These studies indicated the role of PPARγ in suppression of NF-kB activation and inflammatory response in intestinal tissues.

Additionally, several reports indicated that PPARs and their ligands could lead to carcinogenesis by affecting the metabolism of glucose and lipids. Intestinal PPARα exhibited a protective effect against colon carcinogenesis by inhibiting methylation of P21 and P27 [76]. Human colorectal tumors also show lower levels of PPARα compared to normal tissue [76]. PPARγ synthetic activator rosiglitazone has a radio-sensitizing effect on human bowel cancer cells [72]. PPARγ was reported to be associated with colorectal cancer via insulin and inflammatory mechanisms [77,78]. On the other hand, PPARγ was shown to be expressed in human colonic mucosa and cancer. The ability of PPARγ activation to decrease cyclooxygenase-2 (COX-2) expression and induce apoptosis suggests that the PPARγ pathway might be a tumor suppressor in humans [79]. Another study reports that 8% of primary colorectal tumors harbor function-dead mutations in one allele of the *PPARγ* gene and emphasizes the potential role of this receptor as a therapeutic target for cancer or in designing a mouse colon cancer model [80]. The treatment of colon cancer by suppressing the methylation of PPARγ promoter and enhancing PPARγ expression is also underway, because the hyper-methylation of promoter regions can induce PPARγ gene silence. Moreover, the risk of radiation-induced intestinal toxicity in methylated patients was also increased compared with un-methylated patients [81]. Furthermore, PPARβ is induced in intestinal stem cells and progenitor cells in high-fat diet-treated mice and enhances stemness and tumorigenicity of intestine [82]. Arachidonic acid derivative prostaglandin E2 (PGE2), which is a biologically active lipid, increases cell survival and improves intestinal adenoma formation by indirectly activating PPARβ via the phosphatidylinositol-3 kinase (PI3K)/Akt signaling pathway [83]. The activation of PPARβ also upregulates COX-2, which is a key activator for colon cancer cells [59].

We summarize the regulatory mechanism of PPARs in gastrointestinal toxicity in Figure 2. A better understanding of the role of PPARs in the GI system will help to develop novel pharmacotherapy against colon carcinogenesis and diminish intestinal toxicity.

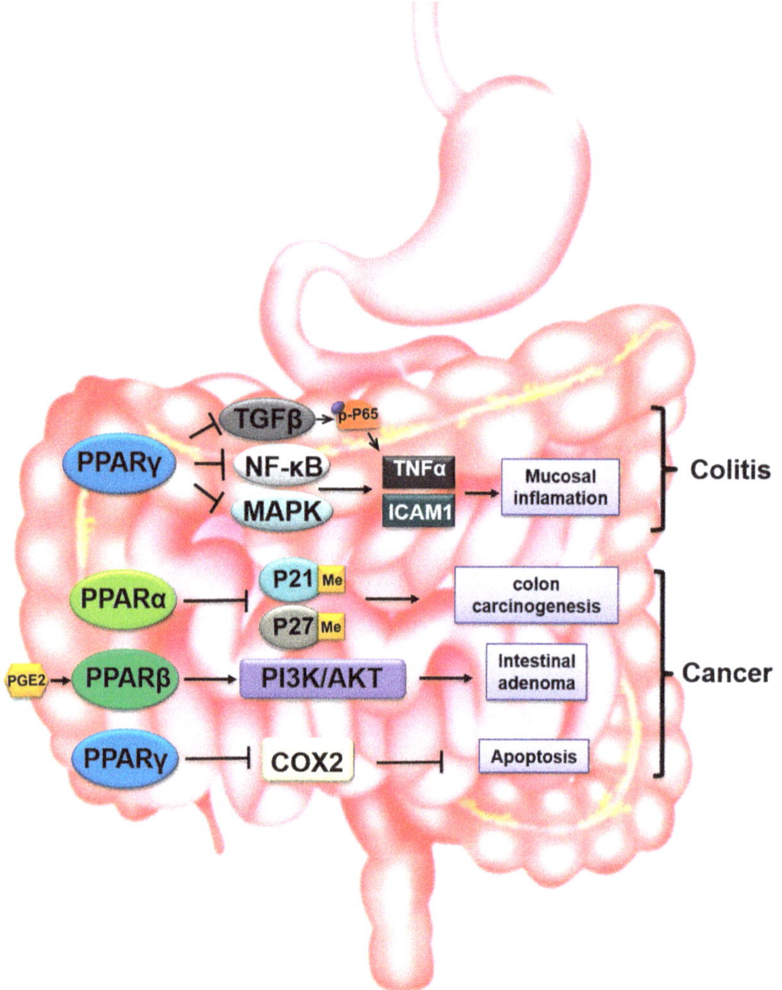

Figure 2. The regulation by PPARs in gastrointestinal toxicity.

5. PPARs in Reproductive and Developmental Toxicity

Three isoforms of PPARs were found in the reproductive system including the hypothalamus, pituitary, testis, ovary, uterus, adrenal, and mammary glands. Numerous studies showed that PPARs play a role in the normal reproductive and developmental functions, and abnormal regulation of PPARs by exposure to endogenous or exogenous compounds might lead to physiological dysfunction in reproductive system [84–86]. Thus, the research on reproductive and developmental disorders focuses on PPARs and their modulators.

Triptolide is a major active compound in Chinese herb *Tripterygium wilfordii* multiglycoside, and it is widely used for treatment of autoimmune diseases and nephrotic syndrome [87]. However, we previously reported that triptolide causes mitochondrial damage and dysregulates fatty-acid metabolism by upregulating expression and nuclear translocation of PPARα in mouse sertoli cells [88]. A metabolomics study revealed that triptolide caused impairment of spermatogenesis accompanied by abnormal lipid and energy metabolism in male mice through downregulation of PPARs [89]. Different concentrations and times of triptolide exposure led to the different behaviors of PPARs. These findings support that PPARs are key mediators in triptolide-induced reproduction toxicity.

Phthalates, which activate PPARs, have a remarkable effect on fertility rates, ovulation, development of the male reproductive tract, spermatogenesis, and teratogenesis [6]. Early exposure to phthalates influenced perinatal and postnatal cardiometabolic programming [90]. DEHP, a phthalate ester, is commonly used in industry as a plasticizer, which activates PPARα to regulate the expression of downstream target genes. DEHP treatment had no remarkable effect on body, liver, and ovary weight in female dams (F0) and offspring (F1) in either wild-type or PPARα-knockout mice. However, it suppressed the expression of ovarian estrogen receptor α, and the repression of ovarian estrogen receptor α expression by DEHP was lost in PPARα-knockout mice [91]. PPARα transcription is related to fertility impairment in female mice exposed to high doses of DEHP (500 mg/kg of body weight per day) [92]. Moreover, it was reported that MEHP, a principle active metabolite of DEHP, decreased the activity and production of aromatase, which converted testosterone to estradiol in ovarian granulosa cells by activating PPARα and PPARγ [93]. Benzo [a]pyrene (B [a]P) is a ubiquitous environmental contaminant, and the combination of B [a]P and DEHP induced ovotoxicity in female rats and suppressed sex hormone secretion via the PPAR-mediated signaling pathway [94].

Dehydroepiandrosterone (3c-hydroxy-5-androsten-17-one, DHEA) is a ligand of PPARα, and it also stimulates the production of PPARα. Some clinical studies showed that dietary supplementation of DHEA reversed the oocyte quality in mice and aged women [95,96]. Additionally, reduced DHEA and loss of function of PPARα result in the decreased follicle quality associated with the changes of fatty-acid metabolism, transport, and mitochondrial function. Perfluorooctanoic acid (PFOA), a synthetic perfluorinated compound (PFC) which is widely distributed, significantly inhibited mammary gland growth in mice through activation of PPARα, and this effect was reversed by supplementation with exogenous estrogen or progesterone [97]. Moreover, perfluorooctane sulfonate (PFOS) is a product of metabolic degradation of PFCs and has an estrogenic activity and endocrine-disruptive properties in the marine medaka embryos, partially through the regulation of PPARs.

Additionally, 15-deoxy-delta12,14-prostaglandin J2 (15dPGJ2), which is converted by arachidonic acid via successive dehydration and isomerization, acts as an endogenous ligand of PPARγ via direct covalent binding, and it plays a key role in lipid homeostasis [98,99]. Kurtz and colleagues found that 15dPGJ2 partially restored the mRNA expression of oxidizing enzymes including acyl-CoA 1 (ACO) and carnitine palmitoyltransferase 1 (CPT1) in the lungs of male fetuses from diabetic rats, but this effect was not observed in female fetuses [100]. Moreover, it was reported that 15dPGJ2 modulated lipid metabolism and nitric oxide production in diabetes-induced placental dysfunction partially through the PPAR pathway [101]. Trichloroethylene (TCE) reduced fertilizability of oocyte and its ability to bind sperm plasma membrane proteins in rats [102]. A systematic evaluation of TCE showed that TCE could cause cardiac defects in humans when the exposure is during a sensitive period of fetal development [103]. Tributyltin chloride (TBT) activates all three types of PPARs. TBT has effects on reproductive function and induces abnormal mammary gland fat accumulation by increasing PPARγ expression [104,105]. TZDs (e.g., pioglitazone, rosiglitazone, and troglitazone) activate PPARγ to regulate the transcription of genes responsible for glucose and lipid metabolism. TZDs clinically sensitize peripheral insulin in patients with type 2 diabetes by regulating glucose and lipid metabolism [106,107]. Oral administration of rosiglitazone 4 mg once a day for three months improves hyperandrogenemia, insulin resistance, lipidemia, C-reactive protein levels, ovarian volume, and follicle number in patients with polycystic ovary syndrome (PCOS) [108].

Rosiglitazone exhibited significant protective effects on metabolic, hormonal, and morphological features of PCOS. Significant changes were also observed in the isovaleryl carnitine levels and lipid oxidation rates after pioglitazone treatment [109]. Rosiglitazone significantly improved oocyte quality in diet-induced obesity (DIO) mice, indicating the positive effect of PPARγ on ovarian function [110]. Rosiglitazone affects steroidogenesis in porcine ovarian follicles by stimulating PPARγ [111,112]. In vivo experiments demonstrated that fenofibrate inhibited ovarian estrogen synthesis [113]. A review concluded that clofibrate and gemfibrozil caused atypical changes in maternal and fetal liver during pregnancy, but there was no direct evidence of developmental toxicity or teratogenicity of clofibrate and gemfibrozil [6]. Irbesartan (IRB) is one of the most widely used angiotensin type 1 (AT1) receptor blockers (ARBs) with PPARγ agonistic activity. Rats treated with IRB showed an increase in estradiol and follicle-stimulating hormone levels, which subsequently ameliorated ovarian dysfunction [114]. These studies indicate that activation of PPARγ signaling protects ovarian function.

Genistein (49,5,7-trihydroxyisoflavone, GEN), a kind of isoflavones derived from soybeans, was investigated for its antioxidant, anticancer, and anti-inflammatory activities [115]. It is a natural ligand of PPARs, and it can improve the development and metabolism of chick embryos through the activation of PPARs [116,117]. Prostacyclin (PGI2) activated its nuclear receptor PPARβ to accelerate blastocyst hatching in mice [118]. These studies suggest that the activation of PPARs is involved in toxicant-induced reproductive toxicity.

The anti-tumor effects of PPAR agonists were documented. Rosiglitazone and troglitazone, both PPARγ activators, showed inhibitory effects on pituitary adenoma cells in mice and human, and they were considered to be a new oral drug for the treatment of pituitary tumors [119]. Moreover, troglitazone treatment stabilized the prostate-specific antigen levels in patients with advanced prostate cancer clinically by upregulating E-cadherin and glutathione peroxidase 3 [120]. Rosiglitazone showed an inhibitory effect on proliferation of primary human prostate cancer cells [121]. However, the activation of PPARβ by selective agonist GW501516 was reported to stimulate proliferation of human breast and prostate cancer cells which are responsive to sexual hormones [122]. PPARβ activation by GW501516 increased cyclin-dependent kinase 2 (CDK2) and vascular endothelial growth factor α (VEGFα) expression, indicating the improved cell proliferation and angiogenesis. This study suggested the possibility of PPARβ antagonists in treating breast and prostate cancer.

6. Other Systemic Toxicity and Protective Effects Mediated by PPARs

Fibrates, PPARα synthetic ligands, were developed for treatment of hyperlipidemia in the clinic, such as fenofibrate, bezafibrate, ciprofibrate, and so on [123–125]. However, muscle weakness, muscle pain, and even rhabdomyolysis were observed during their application [6]. Different fibrates lead to different degrees of myopathy, and that might be due to different mechanisms. The underlying mechanism is still unclear. Some studies reported that PPARα activation in skeletal muscle transactivated the genes encoding muscle proteases, and the increased expression of skeletal muscle proteases led to severe myopathy [126,127]. The muscle toxicity might result from the blood concentration of the drug, because remarkably higher incidence occurs in patients with kidney failure or hypoalbuminemia [128]. Moreover, Motojma et al. proposed that the increase in pyruvate dehydrokinase isoenzyme4 (PDK4) and the decrease in serum triglyceride (TG) level mediated by PPARα in skeletal muscle caused the degradation of protein in muscle, ultimately resulting in myopathy and even rhabdomyolysis [129]. Due to the low incidence of rhabdomyolysis, no drug was withdrawn from the market because of the muscular toxicity.

In contrast to the adverse effect mediated by PPARs, PPARs also exert protective effects against nephrotoxicity and neuron injury.

Diabetic kidney disease is one complication of type 2 diabetes. PPARα and PPARγ are famous targets for treating diabetes, especially PPARγ. Increasing studies indicated that PPARs play important roles in kidney physiology and pathology. In most cases, PPARγ serves as a therapeutic target for treating nephrotoxicity. PPARγ-null mice showed spontaneous diabetic nephropathy. PPARγ knockout mice exhibited kidney hypertrophy accompanied by increased glucosuria, albuminuria, renal fibrosis, and mesangial expansion [130,131].

PPARs also play key roles in regulating brain self-repair. Central nervous system diseases, neuron injury, and cell death are closely related to neuroinflammation [132,133]. Lovastatin (LOV) can protect vulnerable oligodendrocytes in a mouse model of multiple sclerosis (MS) by inhibiting guanosine triphosphate (GTP)-binding proteins, small Rho GTPases, via a PPARα-dependent mechanism [134]. Healthy oligodendrocytes are essential for the synaptic survival of MS neurons. PPARα activation increases the seizure threshold and controls the seizure frequency [134]. The high expression of PPARα in the brain region also prevents nicotine-induced neuronal damage by regulating tyrosine kinases and phosphokinases in neuronal current. It decreases the frequency of seizures caused by the activation of nicotine receptors in vertebral neurons [135]. Animal model studies showed that fenofibrate prevented convulsions caused by dysregulation of neurotransmitters [136]. Substantia nigra has high-density microglia which show two polarization states, M1 and M2, which have pro-inflammatory or anti-inflammatory effects, respectively [132,137]. Therefore, inhibiting the activation of M1 microglia and promoting the activation of M2 microglia are beneficial to central system diseases. In the condition of inflammation, M1 microglia are activated and release pro-inflammatory factors and neurotoxic substances, such as cytokines, reactive oxygen species, prostaglandins, and complements, which aggravate inflammatory injury [138]. Recent studies showed that PPARs (mainly PPARγ) regulate microglia-mediated inflammation in Parkinson's disease (PD) and other neurodegenerative diseases [138–140]. Pioglitazone, a PPARγ ligand, was shown to inhibit the activation and secretion of glial cells by activating PPARγ [141]. Pioglitazone also inhibits the degeneration of dopamine neurons, which induces inflammation and promotes neuron death [141]. Rosiglitazone has a protective effect on neurotoxin 1-methy-4-phenyl-1,2,3,4,6-tetrahydropyidine (MPTP)-induced PD mouse model via upregulation of M2 phenotypic-related anti-inflammatory factors and the downregulation of M1 phenotypic-related pro-inflammatory factors [142]. A recent study found that PPARα/γ dual agonist MHY908 protects dopamine neurons from MPTP-induced loss in PD mice by reducing neuroinflammation and microglia activation [141]. Moreover, L-165041, a PPARβ agonist, can inhibit the radiation-induced inflammation in microglia by inhibition of the NF-kB signaling pathway [143]. At present, Alzheimer's disease is also considered to be a neuroinflammatory disease and is characterized by abnormal accumulation of β-amyloid (Aβ). Under the condition of Aβ accumulation, M1 microglia were activated, resulting in neuronal injury and apoptosis [144]. It was shown that adiponectin can activate M2 microglia and enhance the clearance of Aβ by activating the PPARγ signaling pathway.

7. Conclusions

Better understanding of the role of PPARs in toxicology and pharmacology and the underlying molecular basis is necessary for PPARs-related clinical drug discovery and development. Unfortunately, there are limited studies reviewing the integrated network of relationships in these aspects. Lots of PPAR ligands have beneficial effects on applied pharmacology, but they are also accompanied by various toxicities. Here, we mainly summarized the regulation of PPARs in toxicology and protection against toxicity in various systems, such as cardiotoxicity, hepatotoxicity, gastrointestinal toxicity, and reproductive and developmental toxicity (Figure 3). We hope that a comprehensive understanding of PPAR-mediated toxicology and applied pharmacology will contribute to the safety of PPAR-targeted therapies in the future.

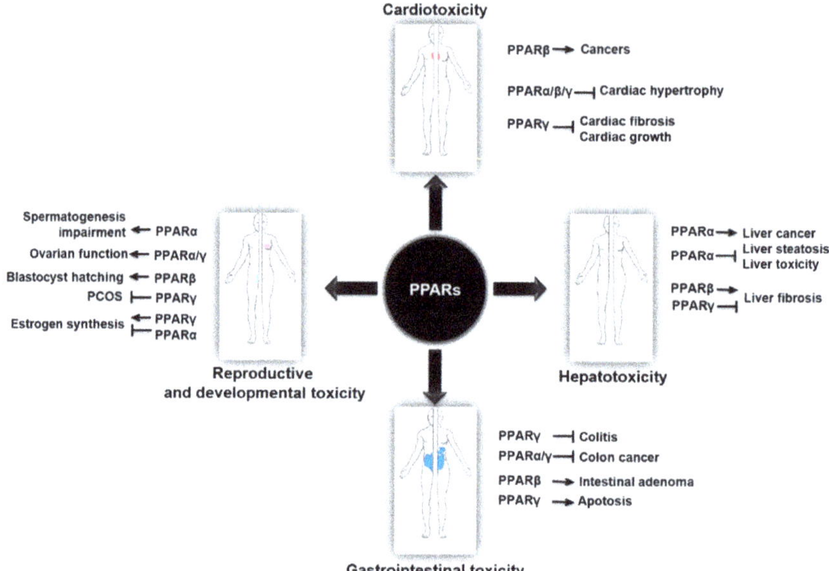

Figure 3. Concept map of the PPARs in various systemic toxicities.

Author Contributions: Conceptualization, P.X. and Y.X.; writing—original draft preparation, Y.X., P.X., Y.Z., S.Z. and Y.L.; writing—review and editing, Y.X., P.X. and Z.H.; supervision, P.X. and Z.H.; project administration, P.X. and Z.H.; funding acquisition, Z.H. All authors have read and agreed to the published version of the manuscript.

Funding: This work was supported by the National Natural Science Foundation of China (Grant No. 81773992), the General Projects in the Natural Science Foundation of Guangdong Province (Grant No. 2018A0303130170), and the National Engineering and Technology Research Center for New drug Druggability Evaluation (Seed Program of Guangdong Province, 2017B090903004). All support is gratefully acknowledged.

Conflicts of Interest: The authors declare no conflict of interest.

References

1. Grygiel-Gorniak, B. Peroxisome proliferator-activated receptors and their ligands: Nutritional and clinical implications—A review. *Nutr. J.* **2014**, *13*, 17. [CrossRef]
2. Lagana, A.S.; Vitale, S.G.; Nigro, A.; Sofo, V.; Salmeri, F.M.; Rossetti, P.; Rapisarda, A.M.; La Vignera, S.; Condorelli, R.A.; Rizzo, G.; et al. Pleiotropic Actions of Peroxisome Proliferator-Activated Receptors (PPARs) in Dysregulated Metabolic Homeostasis, Inflammation and Cancer: Current Evidence and Future Perspectives. *Int. J. Mol. Sci.* **2016**, *17*, 999. [CrossRef] [PubMed]
3. Hong, F.; Xu, P.; Zhai, Y. The Opportunities and Challenges of Peroxisome Proliferator-Activated Receptors Ligands in Clinical Drug Discovery and Development. *Int. J. Mol. Sci.* **2018**, *19*, 2189. [CrossRef] [PubMed]
4. Berger, J.; Wagner, J.A. Physiological and therapeutic roles of peroxisome proliferator-activated receptors. *Diabetes Technol. Ther.* **2002**, *4*, 163–174. [CrossRef] [PubMed]
5. Amber-Vitos, O.; Chaturvedi, N.; Nachliel, E.; Gutman, M.; Tsfadia, Y. The effect of regulating molecules on the structure of the PPAR-RXR complex. *Biochim. Biophys. Acta* **2016**, *1861*, 1852–1863. [CrossRef]
6. Peraza, M.A.; Burdick, A.D.; Marin, H.E.; Gonzalez, F.J.; Peters, J.M. The toxicology of ligands for peroxisome proliferator-activated receptors (PPAR). *Toxicol. Sci.* **2006**, *90*, 269–295. [CrossRef]
7. Corrales, P.; Vidal-Puig, A.; Medina-Gomez, G. PPARs and Metabolic Disorders Associated with Challenged Adipose Tissue Plasticity. *Int. J. Mol. Sci.* **2018**, *19*, 2124. [CrossRef]
8. Tyagi, S.; Gupta, P.; Saini, A.S.; Kaushal, C.; Sharma, S. The peroxisome proliferator-activated receptor: A family of nuclear receptors role in various diseases. *J. Adv. Pharm. Technol. Res.* **2011**, *2*, 236–240. [CrossRef]

9. Hong, F.; Pan, S.; Guo, Y.; Xu, P.; Zhai, Y. PPARs as Nuclear Receptors for Nutrient and Energy Metabolism. *Molecules* **2019**, *24*, 2545. [CrossRef]

10. Boeckmans, J.; Natale, A.; Rombaut, M.; Buyl, K.; Rogiers, V.; De Kock, J.; Vanhaecke, T.; Rodrigues, M.R. Anti-NASH Drug Development Hitches a Lift on PPAR Agonism. *Cells* **2019**, *9*, 37. [CrossRef]

11. Nakamura, T.; Funahashi, T.; Yamashita, S.; Nishida, M.; Nishida, Y.; Takahashi, M.; Hotta, K.; Kuriyama, H.; Kihara, S.; Ohuchi, N.; et al. Thiazolidinedione derivative improves fat distribution and multiple risk factors in subjects with visceral fat accumulation–double-blind placebo-controlled trial. *Diabetes Res. Clin. Pract.* **2001**, *54*, 181–190. [CrossRef]

12. Watkins, P.B.; Whitcomb, R.W. Hepatic dysfunction associated with troglitazone. *N. Engl. J. Med.* **1998**, *338*, 916–917. [CrossRef] [PubMed]

13. Rubenstrunk, A.; Hanf, R.; Hum, D.W.; Fruchart, J.C.; Staels, B. Safety issues and prospects for future generations of PPAR modulators. *Biochim. Biophys. Acta* **2007**, *1771*, 1065–1081. [CrossRef] [PubMed]

14. Fang, C.; Wu, X.; Huang, Q.; Liao, Y.; Liu, L.; Qiu, L.; Shen, H.; Dong, S. PFOS elicits transcriptional responses of the ER, AHR and PPAR pathways in Oryzias melastigma in a stage-specific manner. *Aquat. Toxicol.* **2012**, *106*, 9–19. [CrossRef]

15. Huang, Q.; Fang, C.; Wu, X.; Fan, J.; Dong, S. Perfluorooctane sulfonate impairs the cardiac development of a marine medaka (Oryzias melastigma). *Aquat. Toxicol.* **2011**, *105*, 71–77. [CrossRef]

16. Zhao, M.; Jiang, Q.; Geng, M.; Zhu, L.; Xia, Y.; Khanal, A.; Wang, C. The role of PPAR alpha in perfluorooctanoic acid induced developmental cardiotoxicity and l-carnitine mediated protection-Results of in ovo gene silencing. *Environ. Toxicol. Pharmacol.* **2017**, *56*, 136–144. [CrossRef]

17. Posnack, N.G.; Swift, L.M.; Kay, M.W.; Lee, N.H.; Sarvazyan, N. Phthalate exposure changes the metabolic profile of cardiac muscle cells. *Environ. Health Perspect.* **2012**, *120*, 1243–1251. [CrossRef]

18. Xie, W.; Zhang, W.; Ren, J.; Li, W.; Zhou, L.; Cui, Y.; Chen, H.; Yu, W.; Zhuang, X.; Zhang, Z.; et al. Metabonomics Indicates Inhibition of Fatty Acid Synthesis, beta-Oxidation, and Tricarboxylic Acid Cycle in Triclocarban-Induced Cardiac Metabolic Alterations in Male Mice. *J. Agric. Food Chem.* **2018**, *66*, 1533–1542. [CrossRef]

19. Ebenstein, A.; Fan, M.; Greenstone, M.; He, G.; Zhou, M. New evidence on the impact of sustained exposure to air pollution on life expectancy from China's Huai River Policy. *Proc. Natl. Acad. Sci. USA* **2017**, *114*, 10384–10389. [CrossRef]

20. Jin, X.; Xue, B.; Ahmed, R.Z.; Ding, G.; Li, Z. Fine particles cause the abnormality of cardiac ATP levels via PPARa-mediated utilization of fatty acid and glucose using in vivo and in vitro models. *Environ. Pollut.* **2019**, *249*, 286–294. [CrossRef]

21. Luptak, I.; Balschi, J.A.; Xing, Y.; Leone, T.C.; Kelly, D.P.; Tian, R. Decreased contractile and metabolic reserve in peroxisome proliferator-activated receptor-alpha-null hearts can be rescued by increasing glucose transport and utilization. *Circulation* **2005**, *112*, 2339–2346. [CrossRef] [PubMed]

22. Duan, S.Z.; Ivashchenko, C.Y.; Russell, M.W.; Milstone, D.S.; Mortensen, R.M. Cardiomyocyte-specific knockout and agonist of peroxisome proliferator-activated receptor-γ both induce cardiac hypertrophy in mice. *Circ. Res.* **2005**, *97*, 372–379. [CrossRef] [PubMed]

23. Yang, H.I.; Kim, W.S.; Kim, D.H.; Kang, J.S. Histopathological Evaluation of Heart Toxicity of a Novel Selective PPAR-γ Agonists CKD-501 in db/db Mice. *Biomol. Ther.* **2013**, *21*, 84–88. [CrossRef] [PubMed]

24. He, H.; Tao, H.; Xiong, H.; Duan, S.Z.; McGowan, F.X., Jr.; Mortensen, R.M.; Balschi, J.A. Rosiglitazone causes cardiotoxicity via peroxisome proliferator-activated receptor gamma-independent mitochondrial oxidative stress in mouse hearts. *Toxicol. Sci.* **2014**, *138*, 468–481. [CrossRef]

25. Planavila, A.; Laguna, J.C.; Vazquez-Carrera, M. Atorvastatin improves peroxisome proliferator-activated receptor signaling in cardiac hypertrophy by preventing nuclear factor-kappa B activation. *Biochim. Biophys. Acta* **2005**, *1687*, 76–83. [CrossRef]

26. Malekinejad, M.; Masoumi Verki, M.; Khoramjouy, M.; Alenabi, A.; Hallaj-Salahipour, M.; Malekinejad, H. Cardioprotective Effects of Atorvastatin Are Mediated Through PPARgamma in Paraquat-Exposed Rats. *J. Cardiovasc. Pharmacol.* **2019**, *74*, 400–408. [CrossRef]

27. Bhargava, P.; Verma, V.K.; Malik, S.; Khan, S.I.; Bhatia, J.; Arya, D.S. Hesperidin regresses cardiac hypertrophy by virtue of PPAR-γ agonistic, anti-inflammatory, antiapoptotic, and antioxidant properties. *J. Biochem. Mol. Toxicol.* **2019**, *33*, e22283. [CrossRef]

28. Ma, Z.G.; Yuan, Y.P.; Zhang, X.; Xu, S.C.; Wang, S.S.; Tang, Q.Z. Piperine Attenuates Pathological Cardiac Fibrosis via PPAR-γ/AKT Pathways. *EBioMedicine* **2017**, *18*, 179–187. [CrossRef]

29. Asakawa, M.; Takano, H.; Nagai, T.; Uozumi, H.; Hasegawa, H.; Kubota, N.; Saito, T.; Masuda, Y.; Kadowaki, T.; Komuro, I. Peroxisome proliferator-activated receptor gamma plays a critical role in inhibition of cardiac hypertrophy in vitro and in vivo. *Circulation* **2002**, *105*, 1240–1246. [CrossRef]

30. Birnbaum, Y.; Long, B.; Qian, J.; Perez-Polo, J.R.; Ye, Y. Pioglitazone limits myocardial infarct size, activates Akt, and upregulates cPLA2 and COX-2 in a PPAR-γ-independent manner. *Basic Res. Cardiol.* **2011**, *106*, 431–446. [CrossRef]

31. Zhao, G.; Wang, J.; Xu, X.; Jing, Y.; Tu, L.; Li, X.; Chen, C.; Cianflone, K.; Wang, P.; Dackor, R.T.; et al. Epoxyeicosatrienoic acids protect rat hearts against tumor necrosis factor-alpha-induced injury. *J. Lipid Res.* **2012**, *53*, 456–466. [CrossRef] [PubMed]

32. Kalliora, C.; Kyriazis, I.D.; Oka, S.I.; Lieu, M.J.; Yue, Y.; Area-Gomez, E.; Pol, C.J.; Tian, Y.; Mizushima, W.; Chin, A.; et al. Dual peroxisome-proliferator-activated-receptor-alpha/gamma activation inhibits SIRT1-PGC1alpha axis and causes cardiac dysfunction. *JCI Insight* **2019**, *5*, 129556. [CrossRef] [PubMed]

33. Engle, S.K.; Solter, P.F.; Credille, K.M.; Bull, C.M.; Adams, S.; Berna, M.J.; Schultze, A.E.; Rothstein, E.C.; Cockman, M.D.; Pritt, M.L.; et al. Detection of left ventricular hypertrophy in rats administered a peroxisome proliferator-activated receptor alpha/gamma dual agonist using natriuretic peptides and imaging. *Toxicol. Sci.* **2010**, *114*, 183–192. [CrossRef] [PubMed]

34. Chen, Y.; Chen, H.; Birnbaum, Y.; Nanhwan, M.K.; Bajaj, M.; Ye, Y.; Qian, J. Aleglitazar, a dual peroxisome proliferator-activated receptor-alpha and -gamma agonist, protects cardiomyocytes against the adverse effects of hyperglycaemia. *Diab. Vasc. Dis. Res.* **2017**, *14*, 152–162. [CrossRef] [PubMed]

35. Khuchua, Z.; Glukhov, A.I.; Strauss, A.W.; Javadov, S. Elucidating the Beneficial Role of PPAR Agonists in Cardiac Diseases. *Int. J. Mol. Sci.* **2018**, *19*, 3464. [CrossRef]

36. Samokhvalov, V.; Zlobine, I.; Jamieson, K.L.; Jurasz, P.; Chen, C.; Lee, K.S.; Hammock, B.D.; Seubert, J.M. PPARdelta signaling mediates the cytotoxicity of DHA in H9c2 cells. *Toxicol. Lett.* **2015**, *232*, 10–20. [CrossRef]

37. Chen, Z.C.; Chen, L.J.; Cheng, J.T. Doxorubicin-Induced Cardiac Toxicity Is Mediated by Lowering of Peroxisome Proliferator-Activated Receptor delta Expression in Rats. *PPAR Res.* **2013**, *2013*, 456042. [CrossRef]

38. Wagner, K.D.; Du, S.; Martin, L.; Leccia, N.; Michiels, J.F.; Wagner, N. Vascular PPARbeta/delta Promotes Tumor Angiogenesis and Progression. *Cells* **2019**, *8*, 1623. [CrossRef]

39. Ito, Y.; Nakajima, T. PPARalpha- and DEHP-Induced Cancers. *PPAR Res.* **2008**, *2008*, 759716. [CrossRef]

40. Casals-Casas, C.; Feige, J.N.; Desvergne, B. Interference of pollutants with PPARs: Endocrine disruption meets metabolism. *Int. J. Obes. (Lond.)* **2008**, *32*, S53–S61. [CrossRef]

41. Stahlhut, R.W.; van Wijngaarden, E.; Dye, T.D.; Cook, S.; Swan, S.H. Concentrations of urinary phthalate metabolites are associated with increased waist circumference and insuslin resistance in adult U.S. males. *Environ. Health Perspect.* **2007**, *115*, 876–882. [CrossRef] [PubMed]

42. Li, Y.; Liu, X.; Niu, L.; Li, Q. Proteomics Analysis Reveals an Important Role for the PPAR Signaling Pathway in DBDCT-Induced Hepatotoxicity Mechanisms. *Molecules* **2017**, *22*, 1113. [CrossRef] [PubMed]

43. Qiao, X.; Li, Y.; Mai, J.; Ji, X.; Li, Q. Effect of Dibutyltin Dilaurate on Triglyceride Metabolism through the Inhibition of the mTOR Pathway in Human HL7702 Liver Cells. *Molecules* **2018**, *23*, 1654. [CrossRef] [PubMed]

44. Upham, J.; Acott, P.D.; O'Regan, P.; Sinal, C.J.; Crocker, J.F.; Geldenhuys, L.; Murphy, M.G. The pesticide adjuvant, Toximul, alters hepatic metabolism through effects on downstream targets of PPARalpha. *Biochim. Biophys. Acta* **2007**, *1772*, 1057–1064. [CrossRef]

45. Grun, F.; Watanabe, H.; Zamanian, Z.; Maeda, L.; Arima, K.; Cubacha, R.; Gardiner, D.M.; Kanno, J.; Iguchi, T.; Blumberg, B. Endocrine-disrupting organotin compounds are potent inducers of adipogenesis in vertebrates. *Mol. Endocrinol.* **2006**, *20*, 2141–2155. [CrossRef]

46. Jeng, L.B.; Velmurugan, B.K.; Hsu, H.H.; Wen, S.Y.; Shen, C.Y.; Lin, C.H.; Lin, Y.M.; Chen, R.J.; Kuo, W.W.; Huang, C.Y. Fenofibrate induced PPAR alpha expression was attenuated by oestrogen receptor alpha overexpression in Hep3B cells. *Environ. Toxicol.* **2018**, *33*, 234–247. [CrossRef]

47. Nakajima, T.; Kamijo, Y.; Tanaka, N.; Sugiyama, E.; Tanaka, E.; Kiyosawa, K.; Fukushima, Y.; Peters, J.M.; Gonzalez, F.J.; Aoyama, T. Peroxisome proliferator-activated receptor alpha protects against alcohol-induced liver damage. *Hepatology* **2004**, *40*, 972–980. [CrossRef]

48. Anderson, S.P.; Howroyd, P.; Liu, J.; Qian, X.; Bahnemann, R.; Swanson, C.; Kwak, M.K.; Kensler, T.W.; Corton, J.C. The transcriptional response to a peroxisome proliferator-activated receptor alpha agonist includes increased expression of proteome maintenance genes. *J. Biol. Chem.* **2004**, *279*, 52390–52398. [CrossRef]

49. Shankar, K.; Vaidya, V.S.; Corton, J.C.; Bucci, T.J.; Liu, J.; Waalkes, M.P.; Mehendale, H.M. Activation of PPAR-alpha in streptozotocin-induced diabetes is essential for resistance against acetaminophen toxicity. *FASEB J.* **2003**, *17*, 1748–1750. [CrossRef]

50. You, M.; Crabb, D.W. Recent advances in alcoholic liver disease II. Minireview: Molecular mechanisms of alcoholic fatty liver. *Am. J. Physiol. Gastrointest. Liver Physiol.* **2004**, *287*, G1–G6. [CrossRef]

51. Kon, K.; Ikejima, K.; Hirose, M.; Yoshikawa, M.; Enomoto, N.; Kitamura, T.; Takei, Y.; Sato, N. Pioglitazone prevents early-phase hepatic fibrogenesis caused by carbon tetrachloride. *Biochem. Bioph. Res. Commun.* **2002**, *291*, 55–61. [CrossRef] [PubMed]

52. Choi, J.H.; Kim, S.M.; Lee, G.H.; Jin, S.W.; Lee, H.S.; Chung, Y.C.; Jeong, H.G. Platyconic Acid A, Platycodi Radix-Derived Saponin, Suppresses TGF-1-induced Activation of Hepatic Stellate Cells via Blocking SMAD and Activating the PPAR Signaling Pathway. *Cells* **2019**, *8*, 1544. [CrossRef] [PubMed]

53. Hellemans, K.; Michalik, L.; Dittie, A.; Knorr, A.; Rombouts, K.; De Jong, J.; Heirman, C.; Quartier, E.; Schuit, F.; Wahli, W.; et al. Peroxisome proliferator-activated receptor-beta signaling contributes to enhanced proliferation of hepatic stellate cells. *Gastroenterology* **2003**, *124*, 184–201. [CrossRef] [PubMed]

54. Hays, T.; Rusyn, I.; Burns, A.M.; Kennett, M.J.; Ward, J.M.; Gonzalez, F.J.; Peters, J.M. Role of peroxisome proliferator-activated receptor-alpha (PPARalpha) in bezafibrate-induced hepatocarcinogenesis and cholestasis. *Carcinogenesis* **2005**, *26*, 219–227. [CrossRef]

55. Corton, J.C.; Peters, J.M.; Klaunig, J.E. The PPARalpha-dependent rodent liver tumor response is not relevant to humans: Addressing misconceptions. *Arch. Toxicol.* **2018**, *92*, 83–119. [CrossRef]

56. Reddy, J.K.; Chu, R. Peroxisome proliferator-induced pleiotropic responses: Pursuit of a phenomenon. *Ann. N. Y. Acad. Sci.* **1996**, *804*, 176–201. [CrossRef]

57. Werle-Schneider, G.; Wolfelschneider, A.; von Brevern, M.C.; Scheel, J.; Storck, T.; Muller, D.; Glockner, R.; Bartsch, H.; Bartelmann, M. Gene expression profiles in rat liver slices exposed to hepatocarcinogenic enzyme inducers, peroxisome proliferators, and 17alpha-ethinylestradiol. *Int. J. Toxicol.* **2006**, *25*, 379–395. [CrossRef]

58. Corton, J.C.; Lapinskas, P.J.; Gonzalez, F.J. Central role of PPARalpha in the mechanism of action of hepatocarcinogenic peroxisome proliferators. *Mutat. Res.* **2000**, *448*, 139–151. [CrossRef]

59. Michalik, L.; Desvergne, B.; Wahli, W. Peroxisome-proliferator-activated receptors and cancers: Complex stories. *Nat. Rev. Cancer* **2004**, *4*, 61–70. [CrossRef]

60. Boitier, E.; Gautier, J.C.; Roberts, R. Advances in understanding the regulation of apoptosis and mitosis by peroxisome-proliferator activated receptors in pre-clinical models: Relevance for human health and disease. *Comp. Hepatol.* **2003**, *2*, 3. [CrossRef]

61. Clark, R.B.; Bishop-Bailey, D.; Estrada-Hernandez, T.; Hla, T.; Puddington, L.; Padula, S.J. The nuclear receptor PPAR gamma and immunoregulation: PPAR gamma mediates inhibition of helper T cell responses. *J. Immunol.* **2000**, *164*, 1364–1371. [CrossRef] [PubMed]

62. Mirza, A.Z.; Althagafi, I.I.; Shamshad, H. Role of PPAR receptor in different diseases and their ligands: Physiological importance and clinical implications. *Eur. J. Med. Chem.* **2019**, *166*, 502–513. [CrossRef] [PubMed]

63. Kong, R.; Luo, H.; Wang, N.; Li, J.; Xu, S.; Chen, K.; Feng, J.; Wu, L.; Li, S.; Liu, T.; et al. Portulaca Extract Attenuates Development of Dextran Sulfate Sodium Induced Colitis in Mice through Activation of PPARgamma. *PPAR Res.* **2018**, *2018*, 6079101. [CrossRef] [PubMed]

64. Riccardi, L.; Mazzon, E.; Bruscoli, S.; Esposito, E.; Crisafulli, C.; Di Paola, R.; Caminiti, R.; Riccardi, C.; Cuzzocrea, S. Peroxisome proliferator-activated receptor-alpha modulates the anti-inflammatory effect of glucocorticoids in a model of inflammatory bowel disease in mice. *Shock* **2009**, *31*, 308–316. [CrossRef] [PubMed]

65. Wang, L.; Xie, H.; Xu, L.; Liao, Q.; Wan, S.; Yu, Z.; Lin, D.; Zhang, B.; Lv, Z.; Wu, Z.; et al. rSj16 Protects against DSS-Induced Colitis by Inhibiting the PPAR-alpha Signaling Pathway. *Theranostics* **2017**, *7*, 3446–3460. [CrossRef] [PubMed]

66. Auboeuf, D.; Rieusset, J.; Fajas, L.; Vallier, P.; Frering, V.; Riou, J.P.; Staels, B.; Auwerx, J.; Laville, M.; Vidal, H. Tissue distribution and quantification of the expression of mRNAs of peroxisome proliferator-activated receptors and liver X receptor-alpha in humans: No alteration in adipose tissue of obese and NIDDM patients. *Diabetes* **1997**, *46*, 1319–1327. [CrossRef]
67. Zieleniak, A.; Wojcik, M.; Wozniak, L.A. Structure and physiological functions of the human peroxisome proliferator-activated receptor gamma. *Arch. Immunol. Ther. Exp.* **2008**, *56*, 331–345. [CrossRef]
68. Da Silva, S.; Keita, A.V.; Mohlin, S.; Pahlman, S.; Theodorou, V.; Pahlman, I.; Mattson, J.P.; Soderholm, J.D. A Novel Topical PPARgamma Agonist Induces PPARgamma Activity in Ulcerative Colitis Mucosa and Prevents and Reverses Inflammation in Induced Colitis Models. *Inflamm. Bowel Dis.* **2018**, *24*, 792–805. [CrossRef]
69. Desreumaux, P.; Dubuquoy, L.; Nutten, S.; Peuchmaur, M.; Englaro, W.; Schoonjans, K.; Derijard, B.; Desvergne, B.; Wahli, W.; Chambon, P.; et al. Attenuation of colon inflammation through activators of the retinoid X receptor (RXR)/peroxisome proliferator-activated receptor gamma (PPARgamma) heterodimer. A basis for new therapeutic strategies. *J. Exp. Med.* **2001**, *193*, 827–838. [CrossRef]
70. Nakajima, A.; Wada, K.; Miki, H.; Kubota, N.; Nakajima, N.; Terauchi, Y.; Ohnishi, S.; Saubermann, L.J.; Kadowaki, T.; Blumberg, R.S.; et al. Endogenous PPAR gamma mediates anti-inflammatory activity in murine ischemia-reperfusion injury. *Gastroenterology* **2001**, *120*, 460–469. [CrossRef]
71. Barnes, P.J.; Karin, M. Nuclear factor-kappaB: A pivotal transcription factor in chronic inflammatory diseases. *N. Engl. J. Med.* **1997**, *336*, 1066–1071. [CrossRef] [PubMed]
72. Mangoni, M.; Sottili, M.; Gerini, C.; Desideri, I.; Bastida, C.; Pallotta, S.; Castiglione, F.; Bonomo, P.; Meattini, I.; Greto, D.; et al. A PPAR-γ agonist protects from radiation-induced intestinal toxicity. *United Eur. Gastroenterol. J.* **2017**, *5*, 218–226. [CrossRef] [PubMed]
73. Linard, C.; Souidi, M. PPARs in Irradiation-Induced Gastrointestinal Toxicity. *PPAR Res.* **2010**, *2010*, 528327. [CrossRef] [PubMed]
74. Tian, X.; Peng, Z.; Luo, S.; Zhang, S.; Li, B.; Zhou, C.; Fan, H. Aesculin protects against DSS-Induced colitis though activating PPARgamma and inhibiting NF-small ka, CyrillicB pathway. *Eur. J. Pharmacol.* **2019**, *857*, 172453. [CrossRef] [PubMed]
75. Linard, C.; Gremy, O.; Benderitter, M. Reduction of peroxisome proliferation-activated receptor gamma expression by gamma-irradiation as a mechanism contributing to inflammatory response in rat colon: Modulation by the 5-aminosalicylic acid agonist. *J. Pharmacol. Exp. Ther.* **2008**, *324*, 911–920. [CrossRef] [PubMed]
76. Luo, Y.; Xie, C.; Brocker, C.N.; Fan, J.; Wu, X.; Feng, L.; Wang, Q.; Zhao, J.; Lu, D.; Tandon, M.; et al. Intestinal PPARalpha Protects Against Colon Carcinogenesis via Regulation of Methyltransferases DNMT1 and PRMT6. *Gastroenterology* **2019**, *157*, 744–759 e4. [CrossRef] [PubMed]
77. Slattery, M.L.; Curtin, K.; Wolff, R.; Ma, K.N.; Sweeney, C.; Murtaugh, M.; Potter, J.D.; Levin, T.R.; Samowitz, W. PPARgamma and colon and rectal cancer: Associations with specific tumor mutations, aspirin, ibuprofen and insulin-related genes (United States). *Cancer Causes Control.* **2006**, *17*, 239–249. [CrossRef]
78. Sabatino, L.; Ziccardi, P.; Cerchia, C.; Muccillo, L.; Piemontese, L.; Loiodice, F.; Colantuoni, V.; Lupo, A.; Lavecchia, A. Chiral phenoxyacetic acid analogues inhibit colon cancer cell proliferation acting as PPARgamma partial agonists. *Sci. Rep.* **2019**, *9*, 5434. [CrossRef]
79. Yang, W.L.; Frucht, H. Activation of the PPAR pathway induces apoptosis and COX-2 inhibition in HT-29 human colon cancer cells. *Carcinogenesis* **2001**, *22*, 1379–1383. [CrossRef]
80. Sarraf, P.; Mueller, E.; Jones, D.; King, F.J.; DeAngelo, D.J.; Partridge, J.B.; Holden, S.A.; Chen, L.B.; Singer, S.; Fletcher, C.; et al. Differentiation and reversal of malignant changes in colon cancer through PPARgamma. *Nat. Med.* **1998**, *4*, 1046–1052. [CrossRef]
81. Lian, B.; Ren, Y.; Zhang, H.; Lin, T.; Wang, Y. An adenosine derivative (IFC-305) reduced the risk of radiation-induced intestinal toxicity in the treatment of colon cancer by suppressing the methylation of PPAR-r promoter. *Biomed. Pharmacother.* **2019**, *118*, 109202. [CrossRef] [PubMed]
82. Beyaz, S.; Mana, M.D.; Roper, J.; Kedrin, D.; Saadatpour, A.; Hong, S.J.; Bauer-Rowe, K.E.; Xifaras, M.E.; Akkad, A.; Arias, E.; et al. High-fat diet enhances stemness and tumorigenicity of intestinal progenitors. *Nature* **2016**, *531*, 53–58. [CrossRef] [PubMed]
83. You, M.; Yuan, S.; Shi, J.; Hou, Y. PPARdelta signaling regulates colorectal cancer. *Curr. Pharm. Des.* **2015**, *21*, 2956–2959. [CrossRef] [PubMed]

84. Vitti, M.; Di Emidio, G.; Di Carlo, M.; Carta, G.; Antonosante, A.; Artini, P.G.; Cimini, A.; Tatone, C.; Benedetti, E. Peroxisome Proliferator-Activated Receptors in Female Reproduction and Fertility. *PPAR Res.* **2016**, *2016*, 4612306. [CrossRef]

85. Froment, P.; Gizard, F.; Defever, D.; Staels, B.; Dupont, J.; Monget, P. Peroxisome proliferator-activated receptors in reproductive tissues: From gametogenesis to parturition. *J. Endocrinol.* **2006**, *189*, 199–209. [CrossRef]

86. Huang, Q.; Chen, Q. Mediating Roles of PPARs in the Effects of Environmental Chemicals on Sex Steroids. *PPAR Res.* **2017**, *2017*, 3203161. [CrossRef]

87. Ziaei, S.; Halaby, R. Immunosuppressive, anti-inflammatory and anti-cancer properties of triptolide: A mini review. *Avicenna J. Phytomed.* **2016**, *6*, 149–164.

88. Cheng, Y.; Chen, G.; Wang, L.; Kong, J.; Pan, J.; Xi, Y.; Shen, F.; Huang, Z. Triptolide-induced mitochondrial damage dysregulates fatty acid metabolism in mouse sertoli cells. *Toxicol. Lett.* **2018**, *292*, 136–150. [CrossRef]

89. Ma, B.; Qi, H.; Li, J.; Xu, H.; Chi, B.; Zhu, J.; Yu, L.; An, G.; Zhang, Q. Triptolide disrupts fatty acids and peroxisome proliferator-activated receptor (PPAR) levels in male mice testes followed by testicular injury: A GC-MS based metabolomics study. *Toxicology* **2015**, *336*, 84–95. [CrossRef]

90. Philips, E.M.; Jaddoe, V.W.V.; Trasande, L. Effects of early exposure to phthalates and bisphenols on cardiometabolic outcomes in pregnancy and childhood. *Reprod. Toxicol.* **2017**, *68*, 105–118. [CrossRef]

91. Kawano, M.; Qin, X.Y.; Yoshida, M.; Fukuda, T.; Nansai, H.; Hayashi, Y.; Nakajima, T.; Sone, H. Peroxisome proliferator-activated receptor alpha mediates di-(2-ethylhexyl) phthalate transgenerational repression of ovarian Esr1 expression in female mice. *Toxicol. Lett.* **2014**, *228*, 235–240. [CrossRef] [PubMed]

92. Schmidt, J.S.; Schaedlich, K.; Fiandanese, N.; Pocar, P.; Fischer, B. Effects of di(2-ethylhexyl) phthalate (DEHP) on female fertility and adipogenesis in C3H/N mice. *Environ. Health Perspect.* **2012**, *120*, 1123–1129. [CrossRef] [PubMed]

93. Lovekamp-Swan, T.; Jetten, A.M.; Davis, B.J. Dual activation of PPARalpha and PPARgamma by mono-(2-ethylhexyl) phthalate in rat ovarian granulosa cells. *Mol. Cell Endocrinol.* **2003**, *201*, 133–141. [CrossRef]

94. Xu, C.; Chen, J.A.; Qiu, Z.; Zhao, Q.; Luo, J.; Yang, L.; Zeng, H.; Huang, Y.; Zhang, L.; Cao, J.; et al. Ovotoxicity and PPAR-mediated aromatase downregulation in female Sprague-Dawley rats following combined oral exposure to benzo [a]pyrene and di-(2-ethylhexyl) phthalate. *Toxicol. Lett.* **2010**, *199*, 323–332. [CrossRef]

95. Ford, J.H. Reduced quality and accelerated follicle loss with female reproductive aging—Does decline in theca dehydroepiandrosterone (DHEA) underlie the problem? *J. Biomed. Sci.* **2013**, *20*, 93. [CrossRef]

96. Eftekhari Moghadam, A.R.; Saki, G.; Hemadi, M.; Mazaheri, Z.; Khodadadi, A. Effect of dehydroepiandrosterone on meiotic spindle structure and oocyte quality in mice. *Iran. J. Basic Med. Sci.* **2018**, *21*, 1020–1025.

97. Zhao, Y.; Tan, Y.S.; Strynar, M.J.; Perez, G.; Haslam, S.Z.; Yang, C. Perfluorooctanoic acid effects on ovaries mediate its inhibition of peripubertal mammary gland development in Balb/c and C57Bl/6 mice. *Reprod. Toxicol.* **2012**, *33*, 563–576. [CrossRef]

98. Soares, A.F.; Nosjean, O.; Cozzone, D.; D'Orazio, D.; Becchi, M.; Guichardant, M.; Ferry, G.; Boutin, J.A.; Lagarde, M.; Geloen, A. Covalent binding of 15-deoxy-delta12,14-prostaglandin J2 to PPARgamma. *Biochem. Biophys. Res. Commun.* **2005**, *337*, 521–525. [CrossRef]

99. Li, J.; Guo, C.; Wu, J. 15-Deoxy-(12,14)-Prostaglandin J2 (15d-PGJ2), an Endogenous Ligand of PPAR-γ: Function and Mechanism. *PPAR Res.* **2019**, *2019*, 7242030. [CrossRef]

100. Kurtz, M.; Capobianco, E.; Careaga, V.; Martinez, N.; Mazzucco, M.B.; Maier, M.; Jawerbaum, A. Peroxisome proliferator-activated receptor ligands regulate lipid content, metabolism, and composition in fetal lungs of diabetic rats. *J. Endocrinol.* **2014**, *220*, 345–359. [CrossRef]

101. Kaplan, J.M.; Cook, J.A.; Hake, P.W.; O'Connor, M.; Burroughs, T.J.; Zingarelli, B. 15-Deoxy-delta(12,14)-prostaglandin J(2) (15D-PGJ(2)), a peroxisome proliferator activated receptor gamma ligand, reduces tissue leukosequestration and mortality in endotoxic shock. *Shock* **2005**, *24*, 59–65. [CrossRef] [PubMed]

102. Berger, T.; Horner, C.M. In vivo exposure of female rats to toxicants may affect oocyte quality. *Reprod. Toxicol.* **2003**, *17*, 273–281. [CrossRef]

103. Makris, S.L.; Scott, C.S.; Fox, J.; Knudsen, T.B.; Hotchkiss, A.K.; Arzuaga, X.; Euling, S.Y.; Powers, C.M.; Jinot, J.; Hogan, K.A.; et al. A systematic evaluation of the potential effects of trichloroethylene exposure on cardiac development. *Reprod. Toxicol.* **2016**, *65*, 321–358. [CrossRef] [PubMed]

104. da Costa, C.S.; Miranda-Alves, L.; La Merrill, M.A.; Silva, I.V.; Graceli, J.B. The tributyltin leads to obesogenic mammary gland abnormalities in adult female rats. *Toxicol. Lett.* **2019**, *307*, 59–71. [CrossRef]

105. Crawford, K.A.; Clark, B.W.; Heiger-Bernays, W.J.; Karchner, S.I.; Hahn, M.E.; Nacci, D.E.; Schlezinger, J.J. Tributyltin disrupts fin development in Fundulus heteroclitus from both PCB-sensitive and resistant populations: Investigations of potential interactions between AHR and PPARgamma. *Aquat. Toxicol.* **2020**, *218*, 105334. [CrossRef]

106. Hsueh, W.A.; Law, R. The central role of fat and effect of peroxisome proliferator-activated receptor-gamma on progression of insulin resistance and cardiovascular disease. *Am. J. Cardiol.* **2003**, *92*, 3J–9J. [CrossRef]

107. Dennis, J.M.; Henley, W.E.; Weedon, M.N.; Lonergan, M.; Rodgers, L.R.; Jones, A.G.; Hamilton, W.T.; Sattar, N.; Janmohamed, S.; Holman, R.R.; et al. Sex and BMI Alter the Benefits and Risks of Sulfonylureas and Thiazolidinediones in Type 2 Diabetes: A Framework for Evaluating Stratification Using Routine Clinical and Individual Trial Data. *Diabetes Care* **2018**, *41*, 1844–1853. [CrossRef]

108. Mohiyiddeen, L.; Watson, A.J.; Apostolopoulos, N.V.; Berry, R.; Alexandraki, K.I.; Jude, E.B. Effects of low-dose metformin and rosiglitazone on biochemical, clinical, metabolic and biophysical outcomes in polycystic ovary syndrome. *J. Obstet. Gynaecol.* **2013**, *33*, 165–170. [CrossRef]

109. Vigerust, N.F.; Bohov, P.; Bjorndal, B.; Seifert, R.; Nygard, O.; Svardal, A.; Glintborg, D.; Berge, R.K.; Gaster, M. Free carnitine and acylcarnitines in obese patients with polycystic ovary syndrome and effects of pioglitazone treatment. *Fertil. Steril.* **2012**, *98*, 1620–1626 e1. [CrossRef]

110. Minge, C.E.; Bennett, B.D.; Norman, R.J.; Robker, R.L. Peroxisome proliferator-activated receptor-gamma agonist rosiglitazone reverses the adverse effects of diet-induced obesity on oocyte quality. *Endocrinology* **2008**, *149*, 2646–2656. [CrossRef]

111. Rak-Mardyla, A.; Karpeta, A. Rosiglitazone stimulates peroxisome proliferator-activated receptor gamma expression and directly affects in vitro steroidogenesis in porcine ovarian follicles. *Theriogenology* **2014**, *82*, 1–9. [CrossRef] [PubMed]

112. Kurzynska, A.; Bogacki, M.; Chojnowska, K.; Bogacka, I. Peroxisome proliferator activated receptor ligands affect progesterone and 17beta-estradiol secretion by porcine corpus luteum during early pregnancy. *J. Physiol. Pharmacol.* **2014**, *65*, 709–717.

113. Toda, K.; Okada, T.; Miyaura, C.; Saibara, T. Fenofibrate, a ligand for PPARalpha, inhibits aromatase cytochrome P450 expression in the ovary of mouse. *J. Lipid Res.* **2003**, *44*, 265–270. [CrossRef] [PubMed]

114. Abdel-Raheem, I.T.; Omran, G.A.; Katary, M.A. Irbesartan, an angiotensin II receptor antagonist, with selective PPAR-γ-modulating activity improves function and structure of chemotherapy-damaged ovaries in rats. *Fundam. Clin. Pharmacol.* **2015**, *29*, 286–298. [CrossRef] [PubMed]

115. Polkowski, K.; Mazurek, A.P. Biological properties of genistein. A review of in vitro and in vivo data. *Acta Pol. Pharm.* **2000**, *57*, 135–155. [PubMed]

116. Lv, Z.; Fan, H.; Zhang, B.; Ning, C.; Xing, K.; Guo, Y. Dietary genistein supplementation in laying broiler breeder hens alters the development and metabolism of offspring embryos as revealed by hepatic transcriptome analysis. *FASEB J.* **2018**, *32*, 4214–4228. [CrossRef] [PubMed]

117. Wang, L.; Waltenberger, B.; Pferschy-Wenzig, E.M.; Blunder, M.; Liu, X.; Malainer, C.; Blazevic, T.; Schwaiger, S.; Rollinger, J.M.; Heiss, E.H.; et al. Natural product agonists of peroxisome proliferator-activated receptor gamma (PPARgamma): A review. *Biochem. Pharmacol.* **2014**, *92*, 73–89. [CrossRef]

118. Kang, H.J.; Hwang, S.J.; Yoon, J.A.; Jun, J.H.; Lim, H.J.; Yoon, T.K.; Song, H. Activation of peroxisome proliferators-activated receptor delta (PPARdelta) promotes blastocyst hatching in mice. *Mol. Hum. Reprod.* **2011**, *17*, 653–660. [CrossRef]

119. Heaney, A.P.; Fernando, M.; Melmed, S. PPAR-γ receptor ligands: Novel therapy for pituitary adenomas. *J. Clin. Investig.* **2003**, *111*, 1381–1388. [CrossRef]

120. Chang, S.N.; Lee, J.M.; Oh, H.; Kim, U.; Ryu, B.; Park, J.H. Troglitazone inhibits the migration and invasion of PC-3 human prostate cancer cells by upregulating E-cadherin and glutathione peroxidase 3. *Oncol. Lett.* **2018**, *16*, 5482–5488. [CrossRef]

121. Xu, Y.; Iyengar, S.; Roberts, R.L.; Shappell, S.B.; Peehl, D.M. Primary culture model of peroxisome proliferator-activated receptor gamma activity in prostate cancer cells. *J. Cell Physiol.* **2003**, *196*, 131–143. [CrossRef] [PubMed]

122. Stephen, R.L.; Gustafsson, M.C.; Jarvis, M.; Tatoud, R.; Marshall, B.R.; Knight, D.; Ehrenborg, E.; Harris, A.L.; Wolf, C.R.; Palmer, C.N. Activation of peroxisome proliferator-activated receptor delta stimulates the proliferation of human breast and prostate cancer cell lines. *Cancer Res.* **2004**, *64*, 3162–3170. [CrossRef] [PubMed]

123. Xu, P.; Zhai, Y.; Wang, J. The Role of PPAR and Its Cross-Talk with CAR and LXR in Obesity and Atherosclerosis. *Int. J. Mol. Sci.* **2018**, *19*, 1260. [CrossRef] [PubMed]

124. Jahansouz, C.; Xu, H.; Hertzel, A.V.; Kizy, S.; Steen, K.A.; Foncea, R.; Serrot, F.J.; Kvalheim, N.; Luthra, G.; Ewing, K.; et al. Partitioning of adipose lipid metabolism by altered expression and function of PPAR isoforms after bariatric surgery. *Int. J. Obes.* **2018**, *42*, 139–146. [CrossRef] [PubMed]

125. Wang, J.; Xu, P.; Xie, X.; Li, J.; Zhang, J.; Wang, J.; Hong, F.; Li, J.; Zhang, Y.; Song, Y.; et al. DBZ (Danshensu Bingpian Zhi), a Novel Natural Compound Derivative, Attenuates Atherosclerosis in Apolipoprotein E-Deficient Mice. *J. Am. Heart Assoc.* **2017**, *6*, e006297. [CrossRef] [PubMed]

126. Levak-Frank, S.; Radner, H.; Walsh, A.; Stollberger, R.; Knipping, G.; Hoefler, G.; Sattler, W.; Weinstock, P.H.; Breslow, J.L.; Zechner, R. Muscle-specific overexpression of lipoprotein lipase causes a severe myopathy characterized by proliferation of mitochondria and peroxisomes in transgenic mice. *J. Clin. Investig.* **1995**, *96*, 976–986. [CrossRef] [PubMed]

127. Hoefler, G.; Noehammer, C.; Levak-Frank, S.; el-Shabrawi, Y.; Schauer, S.; Zechner, R.; Radner, H. Muscle-specific overexpression of human lipoprotein lipase in mice causes increased intracellular free fatty acids and induction of peroxisomal enzymes. *Biochimie* **1997**, *79*, 163–168. [CrossRef]

128. Hodel, C. Myopathy and rhabdomyolysis with lipid-lowering drugs. *Toxicol. Lett.* **2002**, *128*, 159–168. [CrossRef]

129. Motojima, K.; Seto, K. Fibrates and statins rapidly and synergistically induce pyruvate dehydrogenase kinase 4 mRNA in the liver and muscles of mice. *Biol. Pharm. Bull.* **2003**, *26*, 954–958. [CrossRef]

130. Toffoli, B.; Gilardi, F.; Winkler, C.; Soderberg, M.; Kowalczuk, L.; Arsenijevic, Y.; Bamberg, K.; Bonny, O.; Desvergne, B. Nephropathy in Pparg-null mice highlights PPARgamma systemic activities in metabolism and in the immune system. *PLoS ONE* **2017**, *12*, e0171474. [CrossRef]

131. Corrales, P.; Izquierdo-Lahuerta, A.; Medina-Gomez, G. Maintenance of Kidney Metabolic Homeostasis by PPAR Gamma. *Int. J. Mol. Sci.* **2018**, *19*, 2063. [CrossRef] [PubMed]

132. Lee, Y.; Lee, S.; Chang, S.C.; Lee, J. Significant roles of neuroinflammation in Parkinson's disease: Therapeutic targets for PD prevention. *Arch. Pharm Res.* **2019**, *42*, 416–425. [CrossRef] [PubMed]

133. D'Angelo, M.; Antonosante, A.; Castelli, V.; Catanesi, M.; Moorthy, N.; Iannotta, D.; Cimini, A.; Benedetti, E. PPARs and Energy Metabolism Adaptation during Neurogenesis and Neuronal Maturation. *Int. J. Mol. Sci.* **2018**, *19*, 1869. [CrossRef] [PubMed]

134. Paintlia, A.S.; Paintlia, M.K.; Singh, A.K.; Singh, I. Modulation of Rho-Rock signaling pathway protects oligodendrocytes against cytokine toxicity via PPAR-alpha-dependent mechanism. *Glia* **2013**, *61*, 1500–1517. [CrossRef]

135. Panlilio, L.V.; Justinova, Z.; Mascia, P.; Pistis, M.; Luchicchi, A.; Lecca, S.; Barnes, C.; Redhi, G.H.; Adair, J.; Heishman, S.J.; et al. Novel use of a lipid-lowering fibrate medication to prevent nicotine reward and relapse: Preclinical findings. *Neuropsychopharmacology* **2012**, *37*, 1838–1847. [CrossRef]

136. Porta, N.; Vallee, L.; Lecointe, C.; Bouchaert, E.; Staels, B.; Bordet, R.; Auvin, S. Fenofibrate, a peroxisome proliferator-activated receptor-alpha agonist, exerts anticonvulsive properties. *Epilepsia* **2009**, *50*, 943–948. [CrossRef]

137. Wang, G.; Zhou, Y.; Wang, Y.; Li, D.; Liu, J.; Zhang, F. Age-Associated Dopaminergic Neuron Loss and Midbrain Glia Cell Phenotypic Polarization. *Neuroscience* **2019**, *415*, 89–96. [CrossRef]

138. Loane, D.J.; Byrnes, K.R. Role of microglia in neurotrauma. *Neurotherapeutics* **2010**, *7*, 366–377. [CrossRef]

139. Villapol, S. Roles of Peroxisome Proliferator-Activated Receptor Gamma on Brain and Peripheral Inflammation. *Cell Mol. Neurobiol.* **2018**, *38*, 121–132. [CrossRef]

140. Le Menn, G.; Neels, J.G. Regulation of Immune Cell Function by PPARs and the Connection with Metabolic and Neurodegenerative Diseases. *Int. J. Mol. Sci.* **2018**, *19*, 1575. [CrossRef]

141. Dehmer, T.; Heneka, M.T.; Sastre, M.; Dichgans, J.; Schulz, J.B. Protection by pioglitazone in the MPTP model of Parkinson's disease correlates with I kappa B alpha induction and block of NF kappa B and iNOS activation. *J. Neurochem.* **2004**, *88*, 494–501. [CrossRef] [PubMed]

142. Breidert, T.; Callebert, J.; Heneka, M.T.; Landreth, G.; Launay, J.M.; Hirsch, E.C. Protective action of the peroxisome proliferator-activated receptor-gamma agonist pioglitazone in a mouse model of Parkinson's disease. *J. Neurochem.* **2002**, *82*, 615–624. [CrossRef] [PubMed]

143. Schnegg, C.I.; Kooshki, M.; Hsu, F.C.; Sui, G.; Robbins, M.E. PPARdelta prevents radiation-induced proinflammatory responses in microglia via transrepression of NF-kappaB and inhibition of the PKCalpha/MEK1/2/ERK1/2/AP-1 pathway. *Free Radic. Biol. Med.* **2012**, *52*, 1734–1743. [CrossRef] [PubMed]

144. Song, J.; Choi, S.M.; Kim, B.C. Adiponectin Regulates the Polarization and Function of Microglia via PPAR-γ Signaling under Amyloid β Toxicity. *Front. Cell Neurosci.* **2017**, *11*, 64. [CrossRef] [PubMed]

Review

PPAR Beta/Delta and the Hallmarks of Cancer

Nicole Wagner * and **Kay-Dietrich Wagner**

Université Côte d'Azur, CNRS, INSERM, iBV, 06107 Nice, France; kwagner@unice.fr
* Correspondence: nwagner@unice.fr; Tel.: +33-493-377665

Received: 20 April 2020; Accepted: 1 May 2020; Published: 4 May 2020

Abstract: Peroxisome proliferator-activated receptors (PPARs) belong to the nuclear hormone receptor family. Three different isoforms, PPAR alpha, PPAR beta/delta and PPAR gamma have been identified. They all form heterodimers with retinoic X receptors to activate or repress downstream target genes dependent on the presence/absence of ligands and coactivators or corepressors. PPARs differ in their tissue expression profile, ligands and specific agonists and antagonists. PPARs attract attention as potential therapeutic targets for a variety of diseases. PPAR alpha and gamma agonists are in clinical use for the treatment of dyslipidemias and diabetes. For both receptors, several clinical trials as potential therapeutic targets for cancer are ongoing. In contrast, PPAR beta/delta has been suggested as a therapeutic target for metabolic syndrome. However, potential risks in the settings of cancer are less clear. A variety of studies have investigated PPAR beta/delta expression or activation/inhibition in different cancer cell models in vitro, but the relevance for cancer growth in vivo is less well documented and controversial. In this review, we summarize critically the knowledge of PPAR beta/delta functions for the different hallmarks of cancer biological capabilities, which interplay to determine cancer growth.

Keywords: peroxisome proliferator-activated receptor; angiogenesis; proliferation; metastasis; immortality; resistance to cell death; growth suppressors; immune system; cellular metabolism

---

1. Introduction

Peroxisome proliferator-activated receptors (PPARs) belong to the group of nuclear receptors. They exist in three different isoforms: PPARα (NR1C1), PPARβ/δ (NR1C2) and PPARγ (NR1C3). They heterodimerize with RXR; and upon ligand binding act mainly as transcriptional regulators of specific target genes. Dependent on the tissue distribution, cofactors and availability of ligands, PPARs exert multiple functions (reviewed in [1]). PPARα is mainly expressed in liver, heart, brown adipose tissue, kidney and intestine and regulates energy homeostasis by activation of fatty acid catabolism and stimulation of gluconeogenesis [2]. PPARβ/δ is more or less ubiquitously expressed with some species differences, while PPARγ is expressed in white and brown adipose tissue, the gut and immune cells [1]. Endogenous ligands for PPARs are fatty acids, triglycerides, prostacyclins, prostaglandins and probably retinoic acid. Although varies different binding sites for PPARs in target genes have been reported, they share in general as a response element a direct repeat of the sequence AGGTCA, spaced by a single nucleotide, which was originally identified for PPARα (reviewed in [1]). Thus, in case more than one of the receptors is expressed in a certain cell-type, one could expect cross talk in response to endogenous or pan-PPAR pharmacological agonists. Specific agonists for PPARα are used classically for the treatment of dyslipidemia and agonists for PPARγ are insulin sensitizers to treat patients with type 2 diabetes. Currently, no PPARβ/δ activators or antagonists are in official clinical use. A recent review summarized novel developments regarding patents for PPAR modulators and possible novel clinical indications [3]. Clinical evidence for the use of PPAR agonists and antagonists is reviewed in [4]. Toxicological aspects and side effects of PPAR modulators have been reviewed recently [5]. Increasing interest focuses on potential implications of PPARs in cancer. The major clinical trials database (https://clinicaltrials.gov) lists one clinical trial for a PPARα antagonist for treatment of multiple kinds of cancer, 24 trials for modulators of PPARγ for cancer treatment, but none for PPARβ/δ. The human protein atlas (https://www.proteinatlas.org/ENSG00000112033-PPARD/pathology) lists low cancer type specificity, but detection of PPARβ/δ in all cancer types. A current major limitation for the investigation of PPARβ/δ expression in human cancer samples compared to healthy tissues is the quality of commercially available antibodies. In agreement with this, large differences for PPARβ/δ RNA and protein levels in tumors are noted in the human protein atlas. The protein expression is globally described, but not annotated to certain cell types in the different tumors. Correlations of tumor PPARβ/δ expression with patients outcome have been reviewed recently [6].

Earlier experimental results concerning the role of PPARβ/δ activation for cancer growth were completely controversial with one study showing that pharmacological activation with GW501516 enhanced tumor growth in Apc(min) mice [7], while another study in the same year in the same journal showed enhanced tumor growth in Apc(min) mice crossed with PPARβ/δ knockout mice [8]. Many studies using different cell models have been published afterwards. Several aspects of PPARβ/δ function with relevance for cancer growth have been reviewed recently [1,5,6,9–11].

On a global view, tumor progression is determined by the interplay of cancer cell proliferation, angiogenesis, resisting cell death, evading growth suppressors, activating invasion and metastasis, enabling replicative immortality, deregulating cellular metabolism and avoiding immune destruction, which was defined by Hanahan and Weinberg as the didactic concept of the "hallmarks of cancer" [12,13]. We will follow here this concept and review the knowledge of PPARβ/δ function for the different hallmarks of cancer capabilities.

2. PPARβ/δ and Cell Proliferation

Most published papers focused on tumor growth-promoting or tumor-inhibiting actions of PPARβ/δ. Unfortunately, only few manuscripts distinguished between direct effects on cell proliferation and secondary effects, which might affect tumor growth. Thus, for simplification, we will summarize in this chapter the published results on cell proliferation as well as on general tumor growth. Table 1 summarizes published effects of PPARβ/δ on cell proliferation and tumor growth.

Table 1. Effects of PPARβ/δ on cell proliferation and tumor growth.

Model	Intervention	Outcome	Reference
Wild-type mice	GW0742 agonist	LLC1 tumor growth ↑, Metastasis ↑	[14]
Endothelial-specific PPARβ/δ overexpression	LLC1 tumor induction	Tumor growth ↑, Metastasis ↑	[14]
Nude Mice, SW480 cells	GW501560 GW501516 + Metformin	Tumor growth ↑ Tumor growth ↓	[15]
Apc(Min/+) mice PPARδ$^{-/-}$/Apc$^{Min/+}$ mice	GW501560 GW501560	Tumor growth ↑ Tumor growth ↓	[8,16]
Apc$^{\Delta 580}$ mice Apc$^{\Delta 580}$ mice	GW501560 GSK3787	Tumor growth ↑ Tumor growth ↓	[7,17]
Wild-type mice colitis model	GW501560	Epithelial cell proliferation ↑, Tumor growth ↑	[18]
Colitis-associated colon cancer mice	PPARβ/δ overexpression	Tumor growth ↑	[19,20]
Several mouse models	High Fat diet, GW501516	Intestinal stem and progenitor cell proliferation ↑, Tumor growth ↑	[21]
Azoxymethane-induced colon tumors	Colon-specific PPARβ/δ knockout	Tumor growth ↓	[22]
Azoxymethane-induced colon tumors	GW0742	Tumor growth ↓	[23]
Nude mice with KM12C colon cells	PPARβ/δ silencing	Tumor growth ↑	[24]
PPARβ/δ- floxed mice	Prostate-specific knockout	Cellularity ↑	[25]
Nude mice with DU145 prostate cancer cells	PPARβ/δ silencing	Cell number ↓, Tumor growth ↓	[26]
PC3M prostate cancer cells	GW0742 PPARβ/δ silencing	Cell number ↑, Cell number ↓	[27]
Daudi CLL cells	PPARβ/δ overexpression	Cell number ↑	[28]
Neuroblastoma cell lines	PPARβ/δ overexpression	Cell number ↓ in NGP, Cell number ≈ SK-N-BE(2) and IMR-32	[29]
Transgenic hepatitis B virus (HBV) mice	GW0742	Hepatic tumor foci ↓	[30]
Hepatocellular carcinoma cell lines	GW501560, PPARβ/δ RNAi	Proliferation ↑, Proliferation ↓	[31]
Melanoma cell lines	GW0742, GW501560	Proliferation ↓	[32]
UACC903 melanoma cells	GW0742, GW501560	Proliferation ↓	[33]
MCF-7, MDA-MB-231 breast cancer cell lines	GW501560	Proliferation ↓, Proliferation ≈	[34]

Table 1. *Cont.*

Model	Intervention	Outcome	Reference
NOD-SCID mice with MCF-7 breast cancer cells	PPARβ/δ overexpression	Tumor growth ↑, Lung metastasis ↑	[35]
MCF-7 cells with PPARβ/δ overexpression	DG172, NXT1511 antagonists	Cell number ↓	[35]
PPARβ/δ overexpression in the mammary gland	GW501516	Spontaneous mammary carcinomas after of 12 months, after 5 months with agonist treatment	[36]
Cox2 overexpressing mice	PPARβ/δ knockout	Proliferation ↓, Tumor growth ↓	[37]
PDK1 overexpression in the mammary gland	GW501516	Tumor growth ↑	[38]
MMTV-ErbB2/HER2 onco-mice	FABP5 knockout	Tumor growth ↓	[39]
Mouse mammary tumorigenesis	GW501516	Tumor growth ↑	[40]
Testicular embryonal carcinoma cell lines	PPARβ/δ overexpression, GW0742	Tumor growth ↓, Proliferation ↓	[41]
PPARβ/δ-null mice	Chronic UV exposure	Tumor growth ↓	[42]
Non-small cell lung cancer cell lines	PPARβ/δ agonists PPARβ/δ silencing	Proliferation ↑ Proliferation ↓	[43]
Lung cancer cell lines	PPARβ/δ agonists	Proliferation ≈	[44]
RAF-induced lung adenoma	PPARβ/δ knockout	Tumor growth ↑	[45]
Liposarcoma cell lines	PPARβ/δ agonists PPARβ/δ silencing	Proliferation ↑ Proliferation ↓	[46]
Primary human thyroid cells	PPARβ/δ overexpression, GW501516	Proliferation ↑	[47]
Epithelial ovarian cancer cell lines	Dominant negative PPARβ/δ	Proliferation ↓	[48]

↑ Increase; ↓ decrease; ≈ not significantly different; CLL: chronic lymphocytic leukemia; Apc: Familial Adenomatous Polyposis gene mutated; GW501516, GW0742—specific PPARβ/δ agonists; DG172, NXT1511, GSK3787— PPARβ/δ antagonists.

It has been shown that the PPARβ/δ agonist GW501560 increased VEGF expression in tumor cell lines [16] and promoted tumorigenesis in Apc(Min/+) mice [7,16] linking tumor cell growth and angiogenesis in the gut. Genetical disruption of PPARβ/δ in colon epithelial cells resulted in a lower incidence of azoxymethane-induced colon tumors and reduced VEGF expression [22]. Unfortunately, vessel formation has not been analyzed in detail in these models [16]. In the same azoxymethane-induced colon tumor model, also reduced tumor growth in response to GW0742, which was abolished by PPARβ/δ knockout [23] and a general reduced colon tumor growth in PPARβ/δ knockout mice [8] have been reported. This discrepancy remains unexplained. PPARγ and PPARβ/δ activation of VEGF and cyclooxygenase-2 (COX2) was confirmed in in the colorectal tumor cell lines SW480 and HT29 [49]. Curiously, that opposite regarding cancer growth and VEGF regulation was described using the KM12C colon cancer cell line with silencing of PPARβ/δ in mouse xenograft models [24]. Whether this reflects an unusual behavior of this specialized cell line remains to be determined. The tumor promoting action of GW501516 in Apc mice was confirmed recently and extended by the findings that the PPARβ/δ antagonist GSK3787 suppressed tumorigenesis. PPARβ/δ expression was significantly higher in human colorectal cancers compared to adenomatous polyps and normal mucosa [50] and also in the malignant cells—invasive front versus their paired tumor centers and adenomas—and proinvasive pathways (connexin 43, PDGFRb, AKT1, EIF4G1 and CDK1) were upregulated in response to PPARβ/δ stimulation [17]. Again, the opposite result has also been published for human and mouse tumor samples [51], while another important report using human colorectal cancer samples confirmed high expression of PPARβ/δ and COX2, which was correlated with the incidence of liver metastasis and identified as significant independent prognostic factor [52]. In colitis-associated colon cancer mouse models, PPARβ/δ overexpression promoted tumorigenesis in mice [20] and increased IL-6 expression and STAT3 phosphorylation, whereas concomitant 15-Lipoxygenase-1 expression in colonic epithelial cells suppressed these effects [19]. In an elegant study using different mouse lines, Beyaz et al. showed that high fat diet (HFD) via activation of PPARβ/δ augments the numbers and function of intestinal stem and progenitor cells. Pharmacological activation of PPARβ/δ using GW501516 recapitulated the effects of HFD on these cells. PPARβ/δ activation in the setting of a loss of the APC tumor suppressor gene allowed stem and progenitor cells to initiate tumorigenesis [21], which is in agreement with the studies mentioned above.

Regarding mammary neoplasia, Yuan et al. [36] showed in transgenic animals that activation of PPARβ/δ in the mammary epithelium resulted in progressive histopathologic changes that culminated in the appearance of estrogen receptor- and progesterone receptor-positive and ErbB2-negative infiltrating ductal carcinomas after 12 months in transgenic animals, while treatment with GW501516 shortened the interval until tumor appearance to 5 months. Histologically, Ki-67 expression was increased demonstrating enhanced proliferation of the epithelial cells, and several metabolic changes were observed (see below). Additionally, in animals with 3-phosphoinositide-dependent kinase-1 (PDK1) overexpression in mammary epithelium, GW501516 accelerated tumorigenesis, which was more pronounced in mice with PDK1 overexpression [38]. This is in agreement with many other reports as PDK1 overexpression resulted in an increase in PPARβ/δ expression and profound metabolic changes. Furthermore, GW501516 increased PPARβ/δ and PDK1 expression in mammary tumors [40]. In MMTV-ErbB2/HER2 onco-mice, knockout of FABP5, which shuttles ligands from the cytosol to nuclear PPARβ/δ was sufficient to reduce mammary tumorigenesis highlighting the importance of this molecule and endogenous PPARβ/δ ligands for cancer growth [39]. On the molecular level, epidermal growth factor receptor ligands signal through the ERK and the phophatidylinositol-3-kinase cascades to activate the transcription factor NF-kappaB. NF-kappaB increases via direct transcriptional activation the expression of FABP5 in MCF-7 breast cancer cells, which stimulates proliferation [53]. In Cox-2 overexpressing mice, mammary tumorigenesis was increased, which could be reverted by crossing them with PPARβ/δ knockout mice [37]. In severely immunocompromised mice, MCF-7 breast cancer cells with overexpression of PPARβ/δ produced bigger tumors and more metastasis compared to wild-type cells. Treatment of MCF-7 cells with PPARβ/δ antagonists in culture reduced significantly the number of these cells [35,54].

Martín-Martín et al. showed an opposite result for prostate carcinoma. PPARβ/δ mRNA was downregulated in prostate cancer specimens compared to benign prostate hyperplasia samples; and prostate epithelium-specific knockout of PPARβ/δ increased cellularity. Additional supporting evidence was obtained by the generation of different overexpression or silencing clones from different human prostate cancer cell lines. Mechanistically, PPARβ/δ exerted its activity in a DNA binding-dependent and ligand-independent manner, which involved regulation of the secretory trefoil factor family member 1 [25]. To which extend the stable cell clones and mice exposed during the entire lifespan to the Cre corresponding to tumor development in humans in vivo remains to be determined. In contrast, silencing of PPARβ/δ in prostate cancer cell lines inhibited tumor cell proliferation and tumor growth, which was attributed to activation of the ABCA1 cholesterol transporter-Caveolin1-TGFβ receptor signaling axis [26]. A similar observation in prostate cancer cells was published by Morgan et al., identifying fatty acid binding protein 5 (FABP5) as a direct target gene of PPARβ/δ [27].

Overexpression of PPARβ/δ decreased the cell number in neuroblastoma NGP, but not in SK-N-BE(2) and IMR-32 cell clones. In xenograft models, PPARβ/δ overexpression reduced tumor growth in NGP cell clones, but to a lesser extent in SK-N-BE(2) and IMR-32 cell clones [29]. As the level of overexpression of PPARβ/δ was highest in NGP cells, it is difficult to judge whether the different outcome is due to the different cell lines used or a response to the different levels of PPARβ/δ overexpression. Whether the results correspond to neuroblastoma pathogenesis in vivo remained an open question. A comparable observation was published by the same group when using testicular embryonal carcinoma cell clones with PPARβ/δ overexpression and the agonist GW0742 [41].

In transgenic hepatitis B virus (HBV) mice, long term treatment with the PPARβ/δ agonist GW0742 reduced the number of hepatic tumor foci. Based on reduced expression of cyclin D1 and c-Myc, a reduction in tumor cell proliferation has been proposed [30]. In human hepatocellular carcinoma cell lines, GW501516 increased proliferation, while RNAi against PPARβ/δ inhibited cell growth. PPARβ/δ activation up-regulates the expression of cyclooxygenase (COX)-2, a rate-limiting enzyme for prostaglandin synthesis and tumor growth in hepatocellular cancer lines [31].

Chronic exposure to ultraviolet light (UV) induced PPARβ/δ activity in the skin of mice. Increased PPARβ/δ activity directly stimulated Src expression, increased Src kinase activity and enhanced the EGFR/Erk1/2 signaling pathway, resulting in increased epithelial-to-mesenchymal transition (EMT) marker expression. PPARβ/δ-null mice developed fewer and smaller skin tumors. Furthermore, topical application of the PPARβ/δ antagonist GSK0660 prevented UV-dependent Src stimulation; and the expression of PPARβ/δ positively correlated with the expression of SRC and EMT markers in human skin squamous cell carcinoma (SCC) highlighting the clinical relevance of these findings [42]. Another report claimed that the agonist GW0742 delayed chemical induced skin carcinogenesis; combination of GW0742 and the COX2 inhibitor nimesulide resulted in a further decrease of tumor multiplicity in wild-type mice, but not in PPARβ/δ-null mice [55]. Given that the graphs in the different groups for tumor incidence, multiplicity and size look comparable and no statistical information is provided, it is difficult to follow this line of evidence, which is in sharp contrast to many other published papers. Even more surprising, the same authors reported earlier for a comparable model no effect of GW0742 on chemical induced skin carcinogenesis [56], or no combined effects for GW0742 and the COX2 inhibitor nimesulide in induced colon cancers [57] or independence of the COX2 inhibitor effects on PPARβ/δ [58].

High PPARβ/δ expression was detected in human melanoma compared to normal skin [32]. PPARβ/δ activation using GW0742 or GW501516 inhibited proliferation of different melanoma cell lines [32,33], which was due to direct transcriptional repression of the Wilms' tumor suppressor WT1 and its downstream target genes zyxin [59] and nestin [59–61].

In non-small cell lung cancer (NSCLC) cell lines, PPARβ/δ activation increased proliferation and survival, while PPARβ/δ knock-down reduced viability and increased apoptosis. As reported for colon cancer, PPARβ/δ agonists induced VEGF transcription in NSCLC cell lines. Furthermore, increased expression of PPARβ/δ and VEGF in human non-small cell lung cancer samples compared to normal lung tissues has been detected [43,62]. In contrast, a study using only two lung cancer cell lines in vitro, did not find any effects on cell proliferation in response to PPARβ/δ activation [44]. In a transgenic mouse model of RAF-induced lung adenoma, tumor growth in mice lacking one or both alleles of PPARβ/δ was reported to be increased [45]. However, the histological analysis performed in this model was superficial and statistical information lacking.

We showed in liposarcoma cell lines that PPARβ/δ activation increases proliferation, which is abolished by a PPARβ/δ–siRNA or a specific PPARβ/δ antagonist. These effects were mediated via direct transcriptional repression of leptin by PPARβ/δ. PPARβ/δ was highly expressed in liposarcoma compared to lipoma and correlated with increased proliferation in human tumor samples [46].

PPARβ/δ was increased in benign and malignant human thyroid tumors and correlated with the proliferation marker Ki67. Overexpression of PPARβ/δ in thyroid cells and treatment with GW501516 increased cell proliferation in a cyclin E1-dependent manner. Specificity of the findings was proven by reduction of cyclin E1 expression and cell proliferation in response to RNAi against PPARβ/δ [47].

Epithelial ovarian cancer cell lines expressed high levels of PPARβ/δ. Inhibition of PPARβ/δ reduced epithelial ovarian cancer cell proliferation and reduced tumor growth in vivo. Mechanistically, aspirin, a nonsteroidal anti-inflammatory drug that preferentially inhibits COX-1, compromised PPARβ/δ function and cell growth by inhibiting extracellular signal-regulated kinases 1/2 [48].

Although still some controversies exist, PPARβ/δ expression has been documented in a broad variety of different tumor samples and cancer cell lines. In the majority of published reports, PPARβ/δ activation or overexpression was associated with increased cancer cell and tumor growth, some opposite results may be explained by use of different clonal cell lines or different genetic backgrounds and models in mice.

3. PPARβ/δ and Angiogenesis

In contrast to PPARα and PPARγ, PPARβ/δ is a proangiogenic member of the PPAR family [63]. Vascular cell expression of PPARβ/δ has first been reported in the late 90s by Xin et al., 1999, using mRNA analysis [64] and Bishop-Bailey and Hla, 1999, employing Northern blot techniques [65]. In addition, PPARβ/δ expression in vascular smooth muscle cells had been observed by Bishop-Bailey in 2000 [66].

However, no specific functions of PPARβ/δ in the vasculature were discovered at that time, due to the lack of specific ligands. The first synthetic PPARβ/δ and PPARγ non-thiazolidinedione agonist

L-165041 was established in 1999 [67], followed later by the highly selective PPARβ/δ agonists GW0742 and GW501516 [68].

A first report shading light on the function of PPARβ/δ in vascular cells appeared in 2001. Hatae and colleagues observed that prostacyclins induce apoptosis via PPARβ/δ activation in HEK293 cells whereas endothelial cells, which express cytoplasmic prostacyclin receptors are protected from apoptosis. They concluded that prostacyclin-dependent receptor activation results in increased cAMP levels in endothelial cells, which protects from apoptosis while direct prostacyclin activation of PPARβ/δ in cells lacking cytoplasmic prostacyclin receptors is proapoptotic [69]. A second investigation focusing on endothelial cell apoptosis demonstrated a protective action of L-165041 as well as of carbaprostacyclin (cPGL2) upon H_2O_2 induced apoptosis. Both substances increased expression of PPARβ/δ; knockdown of PPARβ/δ abrogated the apoptosis diminishing effects of both agents. As the molecular mechanism of this apoptosis protective function of PPARβ/δ in endothelial cells, the authors proposed the direct transcriptional activation of 14-3-3alpha protein, a cytosolic protein involved in apoptosis protection, by PPARβ/δ [70]. A later study further added activation of endothelial 14-3-3epsilon protein by PPARβ/δ agonists to the antiapoptotic role [71]. Non-steroidal anti-inflammatory drugs (NSAIDs) can induce endothelial cell apoptosis by disconcerting these transcriptional pathways [72].

PPARβ/δ agonists became of particular interest in vascular biology as they were shown to potently inhibit vascular inflammation and reduce atherosclerosis [73]. They inhibit tumor necrosis factor alpha (TNFα) mediated endothelial inflammation, evidenced by decreased expression of vascular cell adhesion molecule-1 (VCAM-1), monocyte chemotactic protein-1 (MCP-1) expression and inhibition of monocyte binding of TNFα stimulated endothelial cells treated with the PPARβ/δ agonist L-165041 [74]. It has been proposed that PPARβ/δ further controls inflammation via a ligand-dependent interaction with the transcriptional repressor BCL-6. In the absence of other ligands, PPARβ/δ binds BCL-6. When activated with a PPARβ/δ ligand, BCL-6 is released and can suppress proinflammatory pathways [65,75]. Later reports confirmed the anti-inflammatory effect of PPARβ/δ in endothelium [76,77]. PPARβ/δ also inhibits vascular smooth muscle inflammation by transcriptional activation of transforming growth factor (TGF)β1. The decreased MCP-1 expression induced by PPARβ/δ was shown to be mediated by the effector of TGF-β1, Smad3 [78].

Activation of PPARβ/δ has also been reported to prevent endothelial dysfunction by reducing oxidative stress [79]. In diabetic mice, PPARβ/δ activation mediated through phosphatidylinositol 3-kinase (PI3K) and Akt an increase of endothelial nitric oxide synthase (eNOS) activity and nitric oxide (NO) production and improved endothelium-dependent relaxation parameters [80]. In high glucose induced impairment of insulin signaling, PPARβ/δ activation restores endothelial function in part through pyruvate dehydrogenase kinase (PDK) 4 activation, thus preserving the insulin-Akt-eNOS pathway impaired by high glucose [81].

Despite its anti-inflammatory and anti-atherosclerotic functions in the vasculature, PPARβ/δ is a major factor for acute vascular hyperpermeability and vasodilatation, key features of allergic reactions, which can lead to lethal systemic anaphylaxis. The group of Michalik recently demonstrated that selective vessel-specific deletion of PPARβ/δ is sufficient to inhibit VEGF or IgE- induced acute vascular hyperpermeability and vasodilatation, most likely due to activity modulation of kinase pathways and destabilization of cell-to-cell adherens junctions. Inhibition of PPARβ/δ should be considered as a therapeutic approach in acute allergic and inflammatory diseases with disturbed endothelial integrity [82].

The first detailed report about the proangiogenic function of PPARβ/δ appeared in 2007. The selective PPARβ/δ ligand GW501516 was tested at this time in phase II clinical trials for the treatment of dyslipidemia. Using a variety of in vitro and ex vivo approaches, the authors clearly demonstrated that PPARβ/δ induces endothelial cell migration, proliferation and tube formation. They further described an increase of vascular endothelial growth factor (VEGF) expression upon activation of PPARβ/δ and already cautioned against possible negative side effects of agonist treatment in patients susceptible for "angiogenic diseases", such as elderly persons prone to cancer incidence or diabetic individuals with retinopathies [83].

In vivo studies further showed that pharmacological activation with GW0742 as well as muscle specific transgenic overexpression of PPARβ/δ resulted in a rapid increase of capillary density and oxidative fiber numbers in skeletal muscle, resembling the muscular phenotype induced by regular physical training. It had been proposed that the observed effects were the calcineurin-nuclear factor of activated T cells (NFAT) pathway dependent, as inhibition of calcineurin by cyclosporine A (CsA) totally abolished the observed effects of pharmacological activation of PPARβ/δ [84]. Our group further demonstrated the function of PPARβ/δ in physiological vascularization. Treatment of mice with the agonist GW0742 resulted in rapid cardiac growth and vascularization without functional impairment as reflected by normal echocardiographic parameters. The cardiac hypertrophy accompanied by intensive vascularization resembled the cardiac phenotype obtained by long-term voluntary exercise. As the underlying molecular mechanism of this PPARβ/δ action, we identified the calcineurin-nuclear factor of activated T cells (NFAT) pathway [85]. However, it was unclear if the observed increased vascularization was a secondary effect of the myocardial hypertrophy or if the induction of cardiac growth was due to the increased angiogenesis. We therefore generated conditional mice with inducible vessel specific overexpression of PPARβ/δ and observed that vascular overexpression of PPARβ/δ was sufficient to induce a rapid cardiac hypertrophy. Nevertheless, the increased angiogenesis did not ameliorate cardiac function after myocardial infarction [86]. Similar observations were made using pharmacological activation of PPARβ/δ after myocardial infarction; also in this setting the increase in angiogenesis did not ameliorate the clinical outcome [87]. The proangiogenic function of PPARβ/δ was also exploited in other therapeutic approaches in ischemic cardiovascular diseases. Bone marrow derived endothelial progenitor cells (EPCs) represent an interesting path in the therapy of ischemic diseases, but due to their low number their clinical use is limited. Han and colleagues investigated the effects of PPARβ/δ agonists GW501516 or L-165041 on EPCs and found an increase of angiogenic EPC properties including increased migration, proliferation and tube formation in response to activation of PPARβ/δ. These effects were phosphatidylinositol 3-kinase/Akt pathway dependent. Systemic administration of PPARβ/δ agonists led to an increase of hematopoietic stem cells in bone marrow and blood as well as to an enhanced vascularization in ischemic hindlimb models and corneal neovascularization in vivo [88].

The therapeutic potential of PPARβ/δ modulation on aspects of ocular neovascularization, a common feature of premature or diabetic retinopathy, as well as age-related macular degeneration, the leading causes of irreversible blindness, was studied using human retinal microvascular endothelial cells (HRMEC) and in vivo models of oxygen-induced retinopathy (OIR). The authors demonstrated a stimulation of ocular vascularization with PPARβ/δ activation. Furthermore, using the selective PPARβ/δ antagonist GSK0660 [89], the potential therapeutic utility of PPARβ/δ inhibition was proven. GSK0660 decreased HRMEC migration, proliferation, and tube formation and neovascularization in OIR [90].

The effects of PPARβ/δ on tumor angiogenesis were first investigated in 2007. Employing B16 melanoma and LLC1 (Lewis lung carcinoma) tumor cell inoculated in PPARβ/δ$^{-/-}$ mice, Müller-Brüsselbach and colleagues demonstrated cancer vascularization defects and diminished tumor blood flow, resulting in reduced tumor growth in animals lacking PPARβ/δ. In contrast to the report from Piqueras and colleagues [83], the authors observed a hyperproliferative state of endothelial cells, leading to the formation of immature and dysfunctional microvessels upon deletion of PPARβ/δ. On a molecular level, decreased expression of the antiproliferative cyclin-dependent kinase inhibitor 1C (Cdkn1c) was observed in PPARβ/δ$^{-/-}$ cells isolated from Matrigel plugs, which might explain the proliferative immature state of PPARβ/δ$^{-/-}$ endothelium [91].

However, in a second study from this group, diminished expression of chloride intracellular channel protein 4 (Clic4) and increased expression of cellular retinol binding protein 1 (Crbp1) were observed in PPARβ/δ$^{-/-}$ fibroblasts and endothelial cells as compared to wildtype cells [92]. Clic4 promotes endothelial cell proliferation, capillary network and lumen formation [93], whereas Crbp1 binding retinoids in contrast favors growth arrest and differentiation [94]. This is in discrepancy to the observed hyperproliferative state of PPARβ/δ$^{-/-}$ endothelial cells observed by the group of Müller-Brüsselbach [91] and fits to the conclusions made by Piqueras and colleagues that PPARβ/δ stimulates endothelial cell proliferation [83].

An important study further confirmed the strong implication of PPARβ/δ in proangiogenic stimulation favoring tumor progression. Abdollahi and coworkers aimed to identify genes involved in the "angiogenic switch", the shift of an angiogenic balance to a proangiogenic state, one hallmark of cancer progression. Human microvascular cells were submitted to proangiogenic stimuli and subsequent cDNA arrays performed to identify differentially expressed genes upon proangiogenic stimulation. Further selection of genes based on their involvement in the angiogenic network identified PPARβ/δ as a "hubnode" in the "angiogenic switch". The authors confirmed their findings in vivo using B16 melanoma and LLC1 Lewis lung carcinoma inoculated in PPARβ/δ$^{-/-}$ mice, in which they observed dramatically reduced tumor angiogenesis and growth. PPARβ/δ expression levels in human cancer samples further correlated with advanced stages of tumor progression and metastasis [95,96].

Recently, PPARβ/δ activators (L165041 and GW501516) were shown to induce interleukin 8 (Il-8) expression in endothelial cells by transcriptional and posttranscriptional mechanisms [97], and enhanced production of IL-8 due to PPARβ/δ activation caused not only elevated tumor angiogenesis, but also metastasis formation in vivo [98].

Our group further confirmed the general tumor-angiogenesis and cancer growth promoting effect of PPARβ/δ [14]. Although we observed a decrease of LLC1 cancer cell proliferation in vitro upon treatment with GW0742, tumor growth and metastases formation in LLC1 cancer bearing animals was enhanced upon administration of the PPARβ/δ agonist. Tumor vascularization was strongly increased, which supports the hypothesis that enhancement of angiogenesis by PPARβ/δ dominates the eventually growth-inhibiting function on cancer cells. To further determine the functional relevance of PPARβ/δ for tumor vascularization and identify angiogenic signaling pathways, we made use of mice with conditional inducible vascular overexpression of PPARβ/δ subcutaneously implanted with LLC1 cells. Vessel-specific overexpression of PPARβ/δ was sufficient to increase cancer growth, progression and metastases formation. Tumor-sorted endothelial cells were submitted to RNA-sequencing; 283 genes were found to be differentially expressed and cluster analysis revealed mostly up-regulation of genes upon overexpression of PPARβ/δ in endothelial cells. This argues for an angiogenesis boosting effect of PPARβ/δ rather than a repression of antiangiogenic molecules to enhance angiogenesis. We identified six potential target genes of PPARβ/δ, all of them known to be involved in tumor angiogenesis, by combining the top ten network analysis with a search for PPAR responsive elements: Vegf receptors 1 (Flt1), 2 (Kdr) and 3 (Flt4), [99,100], and platelet-derived growth factor receptor beta (Pdgfrβ) [101], platelet-derived growth factor subunit B (Pdgfb) [102] and the tyrosinkinase KIT c-kit [103,104]. Finally, we confirmed that PPARβ/δ directly transcriptional activates Pdgfrβ, Pdgfb, and c-Kit. PPARβ/δ tumor-angiogenesis promoting effects are mediated via activation of the PDGF/PDGFR pathway, c-Kit and probably the VEGF/VEGFR pathway [14].

Despite their beneficial effects on vascular inflammation and atherosclerosis, the therapeutic use of PPARβ/δ agonists could be critical in cancer patients and should therefore in general not be considered as a therapeutic option.

4. PPARβ/δ and Cell Death

The first study demonstrating an inhibitory function of PPARβ/δ in cancer cell death appeared in 1999. He and colleagues revealed that the adenomatous polyposis coli (APC) tumor suppressor represses PPARβ/δ expression through inhibition of β-catenin/Tcf-4 regulated transcription (CRT). APC/β-catenin mutations can therefore lead to increased PPARβ/δ activity. Nonsteroidal anti-inflammatory drugs (NSAIDs) Sulindac and Indomethacin promoted apoptosis of colorectal cancer cells, which could be inhibited by overexpression of PPARβ/δ. The authors demonstrated that NSAIDs suppressed activity of PPARβ/δ through the direct inhibition of DNA binding activity. As fatty acids and eicosanoids are ligands and modifiers of PPAR activity, NSAID-dependent changes in eicosanoid metabolism could also contribute to inhibition of PPARβ/δ activity. NSAIDs were therefore considered as an important therapeutic approach in colorectal carcinoma as they inhibited apoptosis-preventing PPARβ/δ activity also in the context of frequently occurring APC/β-catenin mutations [105]. Other groups demonstrated that cyclooxygenase-derived prostaglandin E2 (PGE2) inhibits colon cancer cell apoptosis through the indirect transactivation of PPARβ/δ. Of note, the authors showed that PGE2 specifically regulates PPARβ/δ, not the other PPARs. The apoptosis inhibiting effects of PGE2 are mediated through indirect mediation of PPARβ/δ by activation of the PI3K/Akt signaling pathway [106]. Gupta et al. confirmed the antiapoptotic effect of PPARβ/δ activation in wildtype and PPARβ/δ-deficient HCT116 colon carcinoma cells. Pretreatment of wildtype HCT116 cells with GW501516 reduced serum withdrawal induced apoptosis, which was not the case in PPARβ/δ-deficient HCT116 cells, suggesting a specific effect of PPARβ/δ activation [7]. Nonsteroidal anti-inflammatory drugs (NSAIDs) were also shown to induce colorectal cancer cell apoptosis through other PPARβ/δ mediated mechanisms. NSAIDs inhibited 14-3-3ε protein expression, leading to apoptosis, accompanied by a decrease of cytosolic and an increase of mitochondrial Bad [107]. The authors had already shown that PPARβ/δ transcriptionally activates 14-3-3ε [70], and further confirmed their hypothesis in this study by overexpression of PPARβ/δ, which rescued colorectal cancer cells from NSAID induced apoptosis and upregulated 14-3-3ε protein levels. This additionally implicates the PPARβ/δ 14-3-3ε pathway in colon cancer cell survival [107]. Again, in the setting of colorectal cancer, it has been shown that PPARβ/δ overexpression or activation antagonizes PPARγ-induced apoptosis of cancer cells. PPARγ agonists induce apoptosis in these cancer cells through reduction of survivin, which in turn leads to apoptosis through increased caspase-3 activity. PPARβ/δ agonists inhibit induction of this apoptotic pathway by increasing survivin expression levels [108]. The apoptosis inducing effects of NSAIDs in colon cancer were also linked to 15-lipoxygenase-1 (15-LOX-1) upregulation. 13-S-hydroxyoctadecadienoic acid (13-S-HODE), the primary product of 15-LOX-1 metabolism of linoleic acid, was found to decrease activity and downregulate expression of PPARβ/δ in colon cancer cells, thereby inducing apoptosis [109]. An interesting study of Cutler and colleagues showed that fibroblasts isolated from the mucosa of hereditary non polyposis colorectal cancer (HNPCC) patients produced 50-fold more PGE2 than normal fibroblasts [110]. PGE2 inhibits apoptosis of colonic carcinoma cells through the activation of PPARβ/δ [106]. As HNPCC patients are more susceptible to develop colorectal cancer (CRC), the authors hypothesized that the overproduction of PGI2 from the stroma of HNPCC patients prevents apoptosis of neoplastic lesions through activation of PPARβ/δ and therefore facilitates progression into a malignant state of CRC [110]. In contrast to all these studies, indicating an antiapoptotic function of PPARβ/δ in colon cancer cells, one report suggested a proapoptotic function of PPARβ/δ in the setting of colon carcinoma. In a model of chemically induced colon carcinogenesis using wildtype and PPARβ/δ knockout mice, treatment of mice with the agonist GW0742 resulted in higher colonic cell apoptosis in wildtype animals as assessed by TUNEL staining and subsequent quantification of cell counts from colon sections, which does not really assure cancer cell specificity. No changes in apoptotic cell counts were observed in colons from PPARβ/δ knockout mice upon agonistic activation of PPARβ/δ [23].

Maggiora et al. investigated the effects of linoleic (LA) and conjugated-linoleic acids (CLA) on the growth of several human tumor cell lines, comprising prostate, bladder, liver, glioblastoma and breast cancer cells. In contrast to Las, CLAs had a strong growth inhibitory effect in the cancer cell lines tested and were able to induce apoptosis in the more deviated cells. PPARβ/δ levels decreased strongly in apoptotic cancer cells upon CLA treatment, but not in cell lines where only an inhibition of cell proliferation without subsequent cell death could be observed [111].

Our group investigated the effects of PPARβ/δ activation on human and mouse melanoma cells. Although we could observe a reduction of melanoma cell proliferation upon PPARβ/δ activation either with GW0742 or with GW501516 at nanomolar concentrations, we did not observe changes in melanoma cell apoptosis [32].

In one lung cancer cell line, PPARβ/δ activation with the agonist L165041 or treatment with the NSAID Indomethacin alone had no effect on apoptosis, however, a combination of these molecules induced apoptosis in this cancer cell line [112]. In contrast, activation of PPARβ/δ with the more specific agonist GW501516 has been demonstrated to inhibit cisplatin-induced apoptosis in different lung cancer cell lines [62]. In line with this latter finding, Genini et al. reported enhanced apoptosis in different human non-small cell lung cancer (NSCLC) lines upon knockdown of PPARβ/δ [43]. We investigated the effects of PPARβ/δ activation or antagonism on mouse Lewis lung carcinoma cells and observed no differences in apoptosis for neither modulation of PPARβ/δ activity [14].

In contrast to the studies mentioned above, which mostly describe an antiapoptotic function of PPARβ/δ in cancer cells, Foreman and colleagues postulated a proapoptotic action of PPARβ/δ in a mouse mammary gland cell line. Treatment with very high concentrations of the PPARβ/δ agonist GW501516 (10 μmolar) for 24 h increased early apoptosis in this cell line, as analyzed by annexin V staining. However, prolonged treatment for 48 h at the same concentrations had no effect on apoptosis, which could raise some doubts concerning the conclusions given in this study [113]. A study from the same group could neither confirm the observation that NSAIDs decrease PPAR activation and expression in colon cancer cells, nor that PPARβ/δ exerts an antiapoptotic function in the setting of colon cancer. Using different human colon cancer cell lines treated with hydrogen peroxide to induce apoptosis and NSAIDs and different concentrations of the PPARβ/δ agonist GW0742, the authors did not observe a decrease of early (evidenced by annexin V labeling) or late (analyzed by PARP cleavage) apoptosis upon PPARβ/δ activation [51]. Bell and colleagues demonstrated that inhibition of PPARβ/δ using siRNA mediated knockdown or the antagonist GSK0660 sensitized neuroblastoma cells to all-trans retinoic acid induced cell death [114]. In line with this proapoptotic function of PPARβ/δ, Péchery and colleagues reported enhanced apoptosis in tumor cells derived from high-grade bladder tumor upon activation with the PPARβ/δ agonist GW501516 [115]. Using only one prostate cancer cell line it has been postulated that the inhibition of PPARβ/δ with the antagonist GSK0660 partially inhibited ginsenoside Rh2 induced apoptosis [116]. In line with this study, another group recently reported a proapoptotic role of PPARβ/δ in prostate cancer cells. Treatment of one prostate cancer cell line with Telmisartan, an angiotensin receptor blocker, induced apoptosis, which could be partially inhibited by pharmacological or genetic down-regulation of PPARβ/δ activity or expression [117]. Additionally, in a nasopharyngeal carcinoma cell line, proapoptotic functions of PPARβ/δ could be demonstrated. Using in vitro and in vivo xenograft assays, high concentrations of GW501516 (10 or 30 μmolar) induced apoptosis of the nasopharyngeal cancer cells. The authors proposed as underlying mechanisms the activation of adenosine monophosphate-activated protein kinase (AMPKα) and downregulation of integrin-linked kinase (ILK), as the AMPK inhibitor compound C was able to inhibit the reduction of ILK expression induced by GW501516 [118]. Employing the same cell line, the authors further implicated the microRNA miR-206 in the apoptosis promoting effects of PPARβ/δ activation, as they observed an induction of miR-206 upon GW501516 mediated PPARβ/δ activation, which could be antagonized by the PPARβ/δ antagonist GSK3787 or the AMPK antagonist dorsomorphin [119].

In conclusion, it is not perfectly clear if PPARβ/δ prevents or stimulates cancer cell death. Although the majority of studies suggest that PPARβ/δ has an antiapoptotic function in cancer cells, some reports evoke the contrary and others do not observe implication of PPARβ/δ in apoptotic cancer cell death at all. This might be due to cancer cell type specific differences, but also to discrepancies in experimental set ups.

5. PPARβ/δ and Tumor Suppressors

In addition to positive regulation of growth-promoting signals, cancer development also requires inhibition of negative growth regulators, i.e., escaping the action of tumor suppressor genes [12]. Although a large number of publications described the overall effects of PPARβ/δ modulation on tumor growth, knowledge on PPARβ/δ and tumor suppressor genes is relatively limited. Mice with mutations in the adenomatous polyposis coli (APC) tumor suppressor are frequently used as a tool for PPAR research in colon cancer, but also a direct function of the APC tumor suppressor on PPARβ/δ expression has been described. APC represses PPARβ/δ expression through inhibition of β-catenin/Tcf-4 regulated transcription in colon cancer cells [105]. Besides colon cancer cells, inactivating mutations in APC or the Axin tumor suppressor proteins or activating mutations in β-catenin resulting in positive effects on T-cell factor (TCF)-regulated transcription have been described in several cancer types. Zhai et al. reported mutations leading to β-catenin deregulation in half of ovarian endometrioid adenocarcinomas. They found elevated expression of the MMP-7, CCND1 (Cyclin D1), CX43 (Connexin 43), ITF2 and also PPARβ/δ genes in ovarian endometrioid adenocarcinomas with deregulated β-catenin [120]. Transformation of intestinal epithelial cells with the K-Ras oncogene led to increased expression and activity of PPARβ/δ. Mechanistically, PPARβ/δ up-regulation was due to increased mitogen-activated protein kinase activity; and PPARβ/δ activation required the endogenous production of prostacyclins via the cyclooxygenase-2 pathway [121]. An initial important report from mice with inactivation of the APC tumor suppressor showed that treatment with the PPARβ/δ agonist GW501516 resulted in a significant increase in the number and size of intestinal polyps [7]. In contrast to the reports mentioned above, another study confirmed APC/beta-catenin-dependent expression of Cyclin D1, while expression of PPARβ/δ was not different in colon or intestinal polyps from wild-type or Apc(min) heterozygous mice or in human colon cancer cell lines with mutations in APC or beta-catenin [122]. This study based exclusively on the use of a polyclonal antibody in Western blots. The quality of the available PPARβ/δ antibodies is still a matter of concern.

Regarding the Wilms' tumor suppressor WT1, we showed that PPARβ/δ activation in melanoma cells inhibits its expression via direct transcriptional repression [32]. WT1 was originally identified as a tumor suppressor based on its mutational inactivation in nephroblastoma [123,124], but later studies provided evidence that WT1 might act as an oncogene [60,101,104,125,126]. Wt1 was up-regulated instead of downregulated in endothelial cells with PPARβ/δ overexpression [14], which suggests cell-type dependent differential regulation of Wt1 by PPARβ/δ. Whether PPARβ/δ is a direct activator of WT1 in endothelial cells and other cell-types remains to be determined.

Epidermal growth factor receptor (EGFR) signaling promotes breast cancer cell proliferation and tumorigenesis. It has been shown that EGFR ligands signal through the ERK and the phophatidylinositol-3-kinase cascades, resulting in activation of the transcription factor NF-kappaB. The NF-kappaB transcription factor directly activates the promoter of fatty-acid binding protein 5 (FABP5) resulting in increased FABP5 protein expression, which in turn shuttles endogenous ligands to PPARβ/δ [53]. In contrast, Krüppel-like factor KLF2 inhibits FABP5 protein expression and subsequent PPARβ/δ activation and thus, might act as a tumor suppressor in breast cancer cells [53].

Transducer of ErbB-2.1 (Tob1) is another tumor suppressor protein, which is inactivated in different cancer types including gastrointestinal cancers. Overexpression of Tob1 in gastric cancer cell lines induced the expression of Smad4 and p15. Tob1 decreased the phosphorylation of Akt and glycogen synthase kinase-3β (GSK3β), resulting in reduced expression and the transcriptional activity of β-catenin, which in turn decreased the expression of PPARβ/δ, cyclin D1, cyclin-dependent kinase-4 (CDK4) and urokinase plasminogen activator receptor (uPAR) in gastric cancer cells [127]. These data are in agreement with the general regulation of PPARβ/δ by β-catenin and provide an additional complex signaling pathway for stimulation of PPARβ/δ activity in cancer progression.

In neuroblastoma cell lines, all-trans-retinoic acid reduced expression of the stem cell factor Sox2 in cell lines with low expression of the tumor suppressor p53, while this was not the case in cells with wild type p53. However, PPARβ/δ activation with GW0742 reduced SOX2 expression independent on the p53 status of the cells. The authors concluded that activating PPARβ/δ induces cell differentiation through p53- and SOX2-dependent signaling pathways in neuroblastoma cells and tumors [29]. However, the exact interaction between retinoic acid and PPARβ/δ signaling on SOX2 expression and the possible role of p53 therein remains to be determined.

In smooth muscle cells, the PPARβ/δ agonist L-165041 inhibited dose-dependently proliferation by blocking G(1) to S phase progression and repressing the phosphorylation of retinoblastoma protein (Rb). In a carotid artery injury model in vivo, L-165041 inhibited neointima formation [128]. To our knowledge, this is the only report linking retinoblastoma protein and PPARβ/δ activation. Whether these findings are relevant for cancer cell proliferation or tumor angiogenesis remains to be determined.

6. PPARβ/δ and Invasion and Metastasis

Abdollahi et al. were the first to correlate PPARβ/δ expression levels with advanced pathological tumor stage and increased risk for distant metastasis. Statistical analyses of PPARβ/δ expression in published large-scale microarray data from cancer patients with prostate, breast, and endometrial adenocarcinoma revealed significantly increased PPARβ/δ expression levels in cases of higher malignant grade and distant metastasis formation [95]. Similar observations were made by Yoshinaga and colleagues who found an increased risk for colorectal cancer patients with high expression of PPARβ/δ and cyclooxygenase (COX)2 in the primary tumor to develop distant liver metastasis, consequently leading to a poor prognostic outcome [96]. In contrast to these studies, one group reported decreased invasion capacity of pancreatic cancer cells in vitro upon PPARβ/δ activation with GW501516 as well as downregulated prometastatic Matrix metalloproteinase-9 (MMP9) expression [129]. A similar study implying in vitro approaches using breast cancer cell lines demonstrated decreased migration and invasion upon PPARβ/δ activation with GW501516. PPARβ/δ mediated inhibition of breast cancer cell migration and invasion was proposed to be regulated via thrombospondin-1 (TSP-1) and its degrading protease, a disintegrin and metalloprotease domains with thrombospondin motifs 1 (ADAMTS1), as knockdown of ADAMTS1 reduced the effects of PPARβ/δ activation; and ADAMTS1 promoter activity was increased by GW501516 [130].

Interestingly, yeast-two hybrid screening identified the metastasis suppressor NDP Kinase alpha (NM23-H2) as a binding protein of PPARβ/δ [131]. NM23 genes have been shown to suppress metastasis development [132]. Overexpression of NM23-H2 in cholangiocarcinoma cells downregulated PPARβ/δ expression, impedes PPARβ/δ promoter activity and diminishes GW501516 induced cholangiocarcinoma cell proliferation. Reactivation of NM23-H2 was suggested as a therapeutic approach in cholangiocarcinoma metastasis [131].

Zuo and collaborators further demonstrated the importance of PPARβ/δ in metastatic cancer. Using an experimental mouse model of metastasis formation by tail vein injection of syngenic tumor cells (B16 melanoma and LLC1 Lewis lung carcinoma cells), the authors showed that PPARβ/δ knockdown in the respective cancer cells inhibited metastasis formation. Additionally, the potential of colon cancer cells (HCT116) to form metastasis in vivo was abolished completely upon genetic deletion of PPARβ/δ. Treatment of mice with the PPARβ/δ agonist GW0742 enhanced metastasis formation. The metastatic potential of PPARβ/δ in cancer cells was confirmed in orthotopic tumor models, confirming that also spontaneous metastasis formation was dramatically reduced upon knockdown of PPARβ/δ. Using heterozygous PPARβ/δ mice for syngenic tumor cell vein injection the authors further demonstrated that high expression of PPARβ/δ in cancer cells is the most important factor for metastasis formation as heterozygous PPARβ/δ mice developed fewer metastasis than their wildtype littermates, but exhibited the most important reduction of metastasis formation when injected with PPARβ/δ knockdown cancer cells. Transcriptome profiling of HCT116 wildtype and PPARβ/δ knockout cells identified gap junction protein alpha 1 (GJA1), vimentin (VIM), secreted protein acidic rich in cysteine (SPARC), neuregulin-1 (NRG1), CXCL8 (IL-8), stanniocalcin-1 (STC1), and synuclein gamma (breast cancer-specific protein 1; SNCG) as pro-metastatic PPARβ/δ targets. Finally, the authors further confirmed the correlation of high PPARβ/δ expression and significantly reduced metastasis-free survival in various cancer patient (colorectal, lung, breast) cohorts, including the largest reported cohort of 1609 breast cancer patients [98].

In profound contrast to the extensive in vivo study of Zuo, Lim and coworkers reported increased melanoma cell migration and invasion upon treatment with the PPARβ/δ antagonist 10 h as well as increased metastasis formation in PPARβ/δ knockout mice [133]. This antagonist had so far not been used in other studies and results were not confirmed employing well established antagonists as GSK0660 or GSK3787. Conversely, Ham and colleagues demonstrated that activation of PPARβ/δ in highly metastatic melanoma cell lines provoked an upregulation of Snail, a decrease of E-cadherin, and a stimulation of migration and invasion, which could be reversed by knockdown of PPARβ/δ. PPARβ/δ therefore seems to promote the high metastatic potential of aggressive melanoma [134].

Our group confirmed pro-metastatic effects of PPARβ/δ activation. Syngenic subcutaneous LLC1 tumor cell implantation resulted in significantly increased lung and liver metastasis when animals received the PPARβ/δ agonist GW0742. Interestingly, we also observed increased spontaneous metastatic spreading in a model with inducible conditional vascular-specific overexpression of PPARβ/δ, indicating that the proangiogenic function of PPARβ/δ importantly contributes to metastatic tumor progression [14].

Recently, an elegant study demonstrated the implication of PPARβ/δ in the pro-metastatic effects of dietary fats in colorectal cancer. Activation of PPARβ/δ with GW501516 induces cancer stem-like cell (CSC) expansion and accelerates liver metastasis in vivo. Analysis of promoters of self-renewal regulatory factors such as Oct4, Nanog, Sox2, and KLF4 identified a PPAR responsive element in the Nanog promoter. Activation of PPARβ/δ with GW501516 increased whereas knockout of PPARβ/δ decreased Nanog expression. Colonic CSC expansion was shown to be induced by PPARβ/δ through direct induction of Nanog expression via binding to its promoter. Furthermore, knockdown of Nanog abolished PPARβ/δ stimulation of hepatic metastasis formation. Similar to the exposure to GW501516, a high fat diet induced expression of Nanog, accelerated tumor growth and liver metastasis formation and knockout of PPARβ/δ completely inhibited these effects. This identifies a novel PPARβ/δ-mediated mechanism responsible for the contribution of dietary fat to colorectal cancer initiation and metastasis [135].

In conclusion, overwhelming evidence suggests that PPARβ/δ promotes metastasis.

7. PPARβ/δ and Replicative Immortality

Activation of PPARβ/δ with GW501516 was shown to inhibit angiotensin (Ang) II induced premature senescence of human vascular smooth muscle cells (hVSMCs). Ang II treatment of hVSMCs provoked an increase of senescence associated beta galactosidase activity (SA β-gal), which was inhibited by GW501515, an effect that could be reversed by hVSMCs knockdown. A significant reduction of SA β-gal activity was also observed upon pretreatment with N-acetyl-l-cysteine (NAC), a thiol antioxidant, suggesting that reactive oxygen species (ROS) mediate Ang II-induced premature senescence of hVSMCs. Activation of hVSMCs significantly reduced ROS accumulation as well as DNA damage in hVSMCs treated with Ang II. PPARβ/δ mediated transcriptional up-regulation of antioxidant genes (glutathione peroxidase (GPx)-1, manganese superoxide dismutase (Mn-SOD), heme oxygenase (HO)-1, and Thioredoxin (Trx)-1) had been identified as the major mechanism in the inhibition of premature senescence of hVSMCs [136]. In a following study, the authors identified upregulation of phosphatase and tensin homolog deleted on chromosome 10 (PTEN), leading to suppression of phosphatidylinositol 3-kinase (PI3K)/Akt pathway, by PPARβ/δ as a second mechanism of senescence inhibition in hVSMCs [137]. Increase of PTEN and suppression of PI3K/Akt by PPARβ/δ activation was also the main pathway identified for senescence inhibition of UV-induced keratinocytes by the agonist GW501516 [138]. In human coronary artery endothelial cells, inhibition of Ang II induced senescence by PPARβ/δ was found to be dependent of transcriptional activation of Sirtuin (SIRT) 1. Downregulation or inhibition of SIRT1 abolished the effects of PPARβ/δ on Ang II induced ROS production and premature senescence, and resveratrol, a SIRT1 activator, mimicked PPARβ/δ agonist effects [139]. PPARβ/δ activation has also been shown to prevent doxorubicin induced cardiomyocyte senescence. The PPARβ/δ agonist L165041 prevented telomeric repeat factor (TRF) 2 downregulation, partially rescued cell proliferation blockage, significantly attenuated cytoskeletal remodeling and the early loss of plasma membrane integrity and significantly reduced SA-β-gal activity. Senescence inhibition was in this case shown to be dependent of B-cell lymphoma 6 protein (Bcl6) as a potent inhibitor of senescence, rendering cells unresponsive to antiproliferative signals from the p19ARF–p53 pathway. L1650141 increased the expression of Bcl6, which upon ligand binding, was released from PPARβ/δ and repressed its target genes, involved in DNA damage sensing and proliferation of checkpoint control [140]. In this context, it might be interesting to mention that our group observed an increase of TRF2, a protein that has a key role in the protective activity of telomeres [141], in tumor sorted endothelial cells from mice with vascular specific overexpression of PPARβ/δ (Wagner et al., unpublished results).

In contrast to these studies reporting senescence inhibition upon PPARβ/δ activation, Zhu and coworkers observed stimulation of Harvey sarcoma ras virus gene (Hras)-induced senescence by PPARβ/δ. 7,12-dimethylbenz[a]anthracene (DMBA)-initiation led to a higher percentage of malignant squamous cell carcinomas and a lower percentage of benign papillomas in PPARβ/δ knockout compared to wildtype animals. In vitro, Hras expressing PPARβ/δ knockout keratinocytes displayed less senescence as investigated by SA β-gal staining. The authors identified as the molecular mechanisms of this senescence induction by PPARβ/δ a potentiation of the RAF/MEK/ERK pathway and an inhibition of the PI3K/AKT pathway [142]. In a very similar study appearing in the same year, the authors showed that increased endoplasmatic reticulum (ER) stress attenuated senescence in part by up-regulating phosphorylated protein kinase B (p-AKT) and decreasing phosphorylated extracellular signal-regulated kinase (p-ERK), which was repressed by PPARβ/δ [143].

Cellular senescence has been linked to the development of endothelial cell dysfunction in atherosclerosis. Especially oxidative stress induced by ROS from lipid loaded macrophage foam cells has been linked to premature senescence of the vasculature. Riahi and coworkers exposed endothelial cells to the secretome of such foam cells and observed an increase of endothelial SA β-gal activity, p16 and p21 expression as well as a decrease of phosphorylated retinoblastoma protein. They found that senescence was induced by 4-hydroxnonenal (4-HNE) through stimulation of pro-oxidant thioredoxin-interacting protein (TXNIP). The lipid peroxidation product 4-HNE activated PPARβ/δ promoter activity. The PPARβ/δ agonist GW501516 enhanced TXNIP expression, whereas the antagonist GSK0660 reduced TXNIP promoter activity and inhibited 4-HNE induced senescence [144].

In contrast to the prosenescent effects of PPARβ/δ in endothelial cells, Bernal and colleagues reported that PPARβ/δ maintains the proliferative undifferentiated phenotype of adult neuronal precursor cells, probably through activation of SOX2, one self-renewal regulatory factor [145]. This is in line with the findings from Wang and colleagues showing that colonic cancer stem cell expansion was induced by PPARβ/δ through direct transcriptional activation of Nanog [135].

It has been described that PPARβ/δ amplifies Wnt signaling activity through direct interaction with β-catenin and direct transcriptional activation of the Wnt coreceptor low-density lipoprotein receptor-related protein (LRP) 5 [146]. Senescence associated reprogramming has been shown to upregulate an adult tissue stem-cell signature in lymphoma cells, activate Wnt signaling and distinct stem-cell markers. Former senescent lymphoma cells had a higher in vivo tumor initiation potential than their non-senescent counterparts [147]. Given these highly interesting findings, it will be extremely exciting to further clarify the role of PPARβ/δ in cancer related senescence, replicative immortality and cancer stemness.

8. PPARβ/δ and Metabolism

It is well established that the high tumor cell growth rate due to proliferation is connected to profound metabolic changes [12]. As early as 1927, Otto Warburg described an anomaly in cancer cell metabolism compared to normal cells—cancer cells largely depend on aerobic glycolysis for energy production [148–150]. Cancer metabolism is not only linked to proliferation, but also to tumor angiogenesis as rapidly growing tumor cells will turn on the "angiogenic switch" for increased oxygen supply in the tissue. Lack of oxygen results in hypoxia in the tissue, which results in stabilization of hypoxia- induced factors (Hif) [151–153] and subsequent activation/inhibition of downstream target genes, e.g., VEGF [154], erythropoietin [155], WT1 [156], PPARα [157], glucose transporters (Glut-1 and Glut-3) and many other target genes involved in cancer metabolism (for a recent review see [158]). In contrast to PPARα, PPARβ/δ seems not to be directly regulated by Hif-1; but Hif-1 expression is stimulated by calcineurin A [159] and PPARβ/δ activates calcineurin [85]. Consequently, we observed an increase in calcineurin and Hif-1 expression in the hearts of mice treated with the PPARβ/δ agonists GW0742 and GW501516 [85]. Whether this signaling cascade is relevant for PPARβ/δ-dependent cancer progression remains to be established. Hypoxic stress has been shown to induce transcriptional activation of PPARβ/δ in HCT116 colon cancer cells. PPARβ/δ associated with p300 upon hypoxic stress in these cells. The p300 and the PI3K/Akt pathways seem to play a role in the regulation of PPARβ/δ transactivation as PI3K inhibitors or siRNA knockdown of Akt suppressed the PPARβ/δ transactivation in response to hypoxia [160]. Interestingly, hypoxia-induced IL-8 and VEGF expression was significantly attenuated in PPARβ/δ-deficient colon cancer cells linking expression of PPARβ/δ in cancer cells to tumor angiogenesis and immune response [160]. The in vivo relevance of these findings for tumor growth remains to be determined. In addition, prostacyclin synthase, which catalyzes the conversion of prostaglandin H2 (PGH2) to prostaglandin I2 (PGI2) is upregulated in fibroblasts and cancer cells in response to hypoxia. PGI2 in turn stimulates PPARβ/δ and subsequent VEGF expression [161], which provides an additional link between hypoxia, metabolism, PPARβ/δ in the tumor stroma and angiogenesis. PPARβ/δ also protects chronic lymphocytic leukemia and breast cancer cells from harsh environmental conditions, i.e., hypoxia and low glucose concentrations, which was related to increased antioxidant expression, substrate utilization and mitochondrial performance providing additional evidence for PPARβ/δ as a positive regulator of cancer growth [28,35].

Long chain fatty acids (LCFA) represent energy sources, components of cell membranes and are further processed into signaling molecules. Dietary fatty acids are linked to cancer risk especially colon cancer. Saturated fatty acids were positively associated with colon cancer risk, while polyunsaturated fatty acids showed inverse associations [162]. Experimental studies, however suggested that saturated long chain fatty acids (SLCFA) inhibit while unsaturated long chain fatty acids (ULCFA) might increase proliferation of different cancer cell lines [163,164]. A recent report provided novel mechanistic insights into this problem linking long chain fatty acid metabolism and cancer [165]. Saturated fatty acids bind to fatty acid binding protein 5 (FABP5) and displace endogenous ligands and retinoic acid (RA) from this transport protein. Thus, these ligands are not delivered to PPARβ/δ and its transcriptional activity is reduced while RA is diverted to the retinoic acid receptor (RAR), which becomes activated. In contrast, binding of unsaturated long-chain fatty acids to FABP5 has similar consequences for the displacement of RA and its subsequent binding to RARs, but results in nuclear import of the ULCFA/FABP5 complex and subsequent activation of PPARβ/δ, which in turn results in increased cancer cell proliferation [165]. Although these results identify a central role for FABP5 for cancer cell proliferation and might explain the differences observed regarding PPARβ/δ and cancer cell proliferation dependent on the presence/absence of FABP5 and amounts of RA and endogenous PPAR ligands, the situation for in vivo experimental and clinical studies might be even more complex due to the interplay of the different hallmark capabilities.

PPARβ/δ is, however, not only activated by fatty acids presented by FABP5 in tumorigenesis. In mammary epithelium, overexpression of PDK1 resulted in increased phosphorylation of Akt and GSK3β and augmented expression of PPARβ/δ protein. Treatment with GW501516 increased the number of mammary tumors and reduced survival, which was even more pronounced in animals with PDK1 overexpression. This dramatic effect correlated with an increase in a specific metabolic gene signature indicative of glycolysis and greater levels of fatty acid and phospholipid metabolites in PDK1 overexpressing mice treated with GW501516 compared to treated wild-type control mice [38]. As these metabolic changes are common also in human tumors [166] and enable high tumor cell proliferation [167], it is possible that this mechanism plays a common role in tumor types with PPARβ/δ overexpression. In addition, GW501516 increases expression of glucose transporter 1 (Glut-1) and solute carrier family 1 member 5 (SLC1A5), which results in an increased influx of glucose and glutamine in different cancer cell types and subsequently augments cancer cell proliferation [15]. Furthermore, animals with direct specific overexpression of PPARβ/δ in the mammary epithelium were prone to the development of mammary tumors [36]. Infiltrating mammary ductal carcinomas developed after a latency of 12 months; GW501516 reduced tumor latency to 5 months. Histologically, PPARβ/δ overexpression was confirmed in the mammary epithelium. In agreement with the study by Pollock et al. [38], increased Akt phosphorylation was detected, but also mTOR was activated. Inhibition of mTOR by everolimus reduced cell proliferation and the malignant phenotype indicating the importance of this signaling pathway for PPARβ/δ-dependent mammary tumorigenesis. Microarray and metabolomic analyses revealed a marked increase in the levels of phosphatidylcholine metabolites, lysophosphatidylcholine, lysophosphatidic acid and arachidonic acid metabolites, which correspond to PPARβ/δ-dependent gene regulation involved in prostaglandin biosynthesis. Lysophosphatidic acid stimulated mTOR activation through Akt, and phosphatidic acid directly mediates activation of mTOR [36]. These results provided robust evidence for PPARβ/δ induced metabolic changes resulting in mTOR activation in mammary tumorigenesis. Taken together, several metabolites increase PPARβ/δ activity and PPARβ/δ stimulation induces complex metabolic alterations, which are mostly protumorigenic.

9. PPARβ/δ and Immune Function

PPARβ/δ agonists have been reported to inhibit the tumor necrosis factor (TNF) α induced up-regulation of monocyte chemoattractant protein (MCP)-1 and vascular cell adhesion protein (VCAM)-1 in endothelial cells, to inhibit cytokine induced nuclear translocation of NF-kappaB and to reduce monocyte binding to activated vascular cells [74]. They modulate acute inflammation by targeting the neutrophil-endothelial cell interaction and reducing tumor necrosis factor alpha induced endothelial chemokine ligand (CXCL) 1 release and VCAM-1, E-selectin and ICAM-1 expression [77]. Another study described potent inhibitory effects of the PPARβ/δ agonist GW0742 on lipopolysaccharide target genes as cyclooxygenase (COX)-2 and inducible nitric oxide synthase (iNOS) in macrophages. Lipopolysaccharide (LPS) is the most abundant component within the cell wall of Gram-negative bacteria. It can stimulate the release of inflammatory cytokines in various cell types, leading to an acute inflammatory response towards pathogens. It has been suggested that PPARβ/δ functions in modulating the program of macrophages during inflammatory responses [168]. PPARβ/δ modulation has been proposed to attenuate inflammation in atherosclerosis. A comparison between wildtype and PPARβ/δ knockout macrophages revealed that proinflammatory genes such as MMP9, IL-1β and MCP-1 were down-regulated in PPARβ/δ knockout macrophages. However, activation of PPARβ/δ with GW501516 suppressed the expression of MCP-1 and IL-1β, indicating that activation of PPARβ/δ is anti-inflammatory. As an explanation for this seemingly discrepancy of PPARβ/δ function in inflammation, a ligand-dependent interaction of PPARβ/δ with the anti-inflammatory transcriptional repressor BCL-6 had been suggested. Without ligand, PPARβ/δ binds BCL-6. When activated with a PPARβ/δ ligand, BCL-6 is released and suppresses proinflammatory pathways [75]. Monocytes can be differentiated in either a proinflammatory (M1 or classically activated macrophage, induced by TNFα, bacterial LPS or interferon gamma) or an anti-inflammatory (M2 or alternatively activated macrophage, induced by interleukins). PPARβ/δ has an important role in the development of the M2 phenotype, as PPARβ/δ knockout cells were unable to acquire this alternatively activated macrophage phenotype upon interleukin-4 or-10 stimulation [169]. In contrast, Thulin and colleagues demonstrated that PPARβ/δ is regulated by the microRNA miR-9 in monocytes and that activation of PPARβ/δ might be of importance in M1 proinflammatory, but not in M2 anti-inflammatory macrophages, as the PPARβ/δ agonist GW501516 induced expression of PPARβ/δ target genes in proinflammatory M1, but not in M2 macrophages [170]. Further studies confirmed the implication of PPARβ/δ in the modification of macrophage functions and the reprogramming of their activation status. Treatment of macrophages with modified low-density lipoproteins (LDLs) induced arginase I expression, which was abolished by the PPARβ/δ antagonist GW9662. In contrast, the PPARβ/δ agonist GW0742 strongly induced arginase I expression. PPARβ/δ activity in macrophages therefore impacts the balance of Th1/Th2 responses through specific induction of arginase I expression and activity [171]. Myelin-derived phosphatidylserine was found to mediate PPARβ/δ activation in macrophages after myelin uptake, a pathway leading to suppression of the production of inflammatory mediators, ameliorating experimental autoimmune encephalomyelitis (EAE), an animal model of multiple sclerosis [172]. Mukundan et al. identified PPARβ/δ as a transcriptional sensor of apoptotic cells in macrophages. Apoptotic cell feeding stimulated PPARβ/δ expression in macrophages, which then induced expression of opsonins, enhanced apoptotic cell clearance by macrophages and increased anti-inflammatory cytokine production [173]. As another mechanism of PPARβ/δ function in macrophages, induction of the immunoreceptor CD300a has been postulated. The PPARβ/δ agonist GW501516 activated CD300a expression in macrophages. Mice lacking CD300a showed chronic intestinal inflammation upon high fat diet and an increase in proinflammatory cytokines, specific for the M1 macrophage type. The PPARβ/δ/CD300a pathway could therefore contribute to the anti-inflammatory action in macrophages [174]. Adhikary and colleagues investigated the global PPARβ/δ-regulated signaling network in human monocyte-derived macrophages. They found a robust induction of PPARβ/δ expression upon monocyte to macrophage differentiation. Using PPARβ/δ agonists and inverse agonists, they identified two mechanisms by which PPARβ/δ regulates immune-modulatory genes: 1) canonical regulation through DNA binding at PPARβ/δ–RXR sites (PPREs), induced by agonists and repressed by inverse agonists, and 2) repression by agonists in the absence of PPARβ/δ DNA binding (inverse regulation). Inverse regulation concerned NF-kappaB and the signal transducer and activator of transcription (STAT)1 target genes, resulting in the inhibition of multiple proinflammatory mediators consistent with anti-inflammatory effects of PPARβ/δ activation. Interestingly, they could also demonstrate specific immune stimulatory effects induced by PPARβ/δ agonists, a pro-survival effect on macrophages and inhibition of CD32B surface expression and stimulation of T cell activation. This confirms the strong anti-inflammatory function of PPARβ/δ, but also indicates context-dependent specific immune-stimulatory actions of PPARβ/δ activation [175]. The same group aimed at elucidating the role of PPARβ/δ in the pro-tumorigenic polarization of tumor associated macrophages (TAMs) in ovarian cancer. In vitro, PPARβ/δ target genes such as pyruvate dehydrogenase kinase (PDK) 4 and angiopoietin-like protein (ANGPTL) 4 were robustly induced in monocyte derived macrophages, but the ligand response in TAMs was impaired and most PPARβ/δ target genes were refractory to synthetic agonists. Next, the authors compared freshly isolated ascites-associated TAMs from ovarian cancer patients with monocyte-derived macrophages from healthy donors. Many PPARβ/δ target genes as PDK4, ANGPTL4, and carnitine palmitoyl transferase (CPT) 1A were found to be up-regulated in TAMs and were refractory to stimulation with the PPARβ/δ agonist L-165041. The deregulation and unresponsiveness of target genes in TAMs was found to be due to the presence of endogenous activators in malignancy associated ascites, as ascites caused an equal deregulation in normal macrophages. Lipidomic analysis of ascites samples revealed high levels of polyunsaturated fatty acids (PUFA) [176], known as PPARβ/δ activators [177]. The deregulation of PPARβ/δ target genes by PUFA ligands stimulates the pro-tumorigenic conversion of host-derived monocytic cells and might contribute to tumor progression [176]. Very little is known about the PPARβ/δ function in other key immune cell types except macrophages. In 2008, protein expression of PPARβ/δ in activated human T-cells was described. It has been shown that PPARβ/δ is a transcriptional target of human type I interferon (IFN), stimulates T-cell proliferation and inhibits IFN induced apoptosis, which is partially mediated through enhanced extracellular signal-regulated kinases (ERK) 1/2 signaling [178]. More recently, it has been demonstrated that PPARβ/δ overexpression/activation in vivo inhibits thymic T-cell development by decreasing proliferation of $CD4^-CD8^-$ double-negative stage 4 (DN4) thymocytes [179]. PPARβ/δ has further been reported to drive maturation of monocyte–derived dendritic cells towards an atypical phenotype with reduced stimulatory effects on T-cells [180]. An interesting in vivo study using murine models of septic shock induction confirmed general anti-inflammatory effects of PPARβ/δ activation. PPARβ/δ deletion had detrimental effects on cardiac and renal function, liver injury, lung inflammation and survival, which could not be attenuated by administration of the specific agonist PPARβ/δ GW0742. In wildtype animals, selective activation of PPARβ/δ attenuated the multiple organ injury and dysfunction and improved survival when administered acutely in rodent models of endotoxemia and polymicrobial sepsis. PPARβ/δ activation was proposed as an anti-inflammatory therapeutic approach for the treatment of conditions involving local and systemic inflammation [181]. Using an experimental model for multiple sclerosis, it has been shown that PPARβ/δ limits the expansion of pathogenic T helper cells and production of Interleukin 12 and Interferon gamma, thereby limiting autoinflammation in the central nervous system [182]. Similar, in acute cerulein and taurocholate induced pancreatitis mouse models, treatment with the PPARβ/δ agonist GW0742 reduced expression of proinflammatory enzymes and cytokines, neutrophil invasion and tissue and inflammatory deterioration in the pancreas [183]. In contrast, PPARβ/δ has been shown to be a negative regulator of mesenchymal stem cell (MSC) immunosuppressive function, as PPARβ/δ inhibition or genetic deletion enhanced the immunosuppressive properties of MSCs, involving an increased NF-kappaB, ICAM-1 and VCAM-1 activity [184]. Interestingly, also in natural killer (NK) cells, inhibition of PPARβ/δ was beneficial to restore cytotoxic anti-tumor activity. Obesity induced a PPAR driven lipid accumulation in NK cells causing inhibition of their cellular metabolism and inhibiting their function. PPARβ/δ agonists mimicked obesity effects and inhibited trafficking of the cytotoxic machinery to the NK cell-tumor junction, disenabling NK cells to reduce tumor growth in obesity in vivo. Inhibition of PPARβ/δ restored NK cell cytotoxicity [185]. Finally, it may be concluded that most studies identified PPARβ/δ function as anti-inflammatory, mainly in the setting of atherosclerosis. However, only few cancer related investigations exist. In this context, PPARβ/δ has pro-inflammatory and pro-tumorigenic functions by converting host monocytes in macrophages favoring tumor progression [176,185] or impairing antitumor cytotoxicity of NK cells. Surely, more cancer related studies addressing the question how PPARβ/δ acts in different immune regulatory cells, tissues and conditions, are needed.

10. Conclusions and Outlook

PPARβ/δ functions have been studied extensively. We summarized here known PPARβ/δ effects on cell proliferation, induction of angiogenesis, cell death, function of tumor suppressors, replicative immortality and senescence, invasion and metastasis, tumor metabolism and immune function and mentioned underlying molecular mechanisms. Although not all cited manuscripts were directly related to cancer, one has to keep in mind that the different hallmark capabilities interplay during tumor progression [12,13]. Some controversies regarding the effects of PPARβ/δ activation for cancer progression still exist, which might relate to the different cellular or animal models used. The majority of reports, however, suggest that activation of PPARβ/δ might result in modifications of the hallmark capabilities in favor of a pro-tumorigenic profile. Thus, in contrast to the earlier notion of the therapeutic potential of PPARβ/δ agonists as "exercise mimetics" and potential treatments for metabolic syndrome [186–188], extreme caution should be applied when considering PPARβ/δ agonists for therapeutic purposes given their pro-tumorigenic properties.

For future approaches using PPARβ/δ modulation for potential cancer therapy, collaborations between different laboratories and pathologists are urgently needed to define exact expression patterns of PPARβ/δ in different types, stages and grades of cancer. Currently, already the antibody validation is a limiting factor. Reproducible immunostaining protocols established between different laboratories and precise annotation of cell types would be required to define, which patients might benefit from PPARβ/δ modulation according to expression pattern in cells of the different hallmark capabilities.

Author Contributions: Conceptualization, N.W. and K.-D.W.; writing—original draft preparation, N.W. and K.-D.W.; writing—review and editing, N.W. and K.-D.W.; funding acquisition, N.W. and K.-D.W. Both authors have read and agreed to the published version of the manuscript.

Funding: This research was funded by "Fondation ARC pour la recherche sur le cancer", grant number n_PJA 20161204650 (N.W.), Gemluc (N.W.), and Plan Cancer INSERM, Fondation pour la Recherche Médicale (K.-D.W.).

Conflicts of Interest: The authors declare no conflict of interest.

References

1. Wagner, K.D.; Wagner, N. Peroxisome proliferator-activated receptor beta/delta (PPARbeta/delta) acts as regulator of metabolism linked to multiple cellular functions. *Pharmacol. Ther.* **2010**, *125*, 423–435. [CrossRef]
2. Lefebvre, P.; Chinetti, G.; Fruchart, J.C.; Staels, B. Sorting out the roles of PPAR alpha in energy metabolism and vascular homeostasis. *J. Clin. Investig.* **2006**, *116*, 571–580. [CrossRef]
3. Takada, I.; Makishima, M. Peroxisome proliferator-activated receptor agonists and antagonists: A patent review (2014-present). *Expert Opin. Ther. Pat.* **2020**, *30*, 1–13. [CrossRef] [PubMed]
4. Cheng, H.S.; Tan, W.R.; Low, Z.S.; Marvalim, C.; Lee, J.Y.H.; Tan, N.S. Exploration and Development of PPAR Modulators in Health and Disease: An Update of Clinical Evidence. *Int. J. Mol. Sci.* **2019**, *20*, 5055. [CrossRef] [PubMed]
5. Xi, Y.; Zhang, Y.; Zhu, S.; Luo, Y.; Xu, P.; Huang, Z. PPAR-Mediated Toxicology and Applied Pharmacology. *Cells* **2020**, *9*, 352. [CrossRef] [PubMed]
6. Müller, R. PPARβ/δ in human cancer. *Biochimie* **2017**, *136*, 90–99. [CrossRef]
7. Gupta, R.A.; Wang, D.; Katkuri, S.; Wang, H.; Dey, S.K.; DuBois, R.N. Activation of nuclear hormone receptor peroxisome proliferator-activated receptor-delta accelerates intestinal adenoma growth. *Nat. Med.* **2004**, *10*, 245–247. [CrossRef]
8. Harman, F.S.; Nicol, C.J.; Marin, H.E.; Ward, J.M.; Gonzalez, F.J.; Peters, J.M. Peroxisome proliferator-activated receptor-delta attenuates colon carcinogenesis. *Nat. Med.* **2004**, *10*, 481–483. [CrossRef]
9. Liu, Y.; Colby, J.K.; Zuo, X.; Jaoude, J.; Wei, D.; Shureiqi, I. The Role of PPAR-δ in Metabolism, Inflammation, and Cancer: Many Characters of a Critical Transcription Factor. *Int. J. Mol. Sci.* **2018**, *19*, 3339. [CrossRef]
10. Laganà, A.S.; Vitale, S.G.; Nigro, A.; Sofo, V.; Salmeri, F.M.; Rossetti, P.; Rapisarda, A.M.; La Vignera, S.; Condorelli, R.A.; Rizzo, G.; et al. Pleiotropic Actions of Peroxisome Proliferator-Activated Receptors (PPARs) in Dysregulated Metabolic Homeostasis, Inflammation and Cancer: Current Evidence and Future Perspectives. *Int. J. Mol. Sci.* **2016**, *17*, 999. [CrossRef]
11. Montagner, A.; Wahli, W.; Tan, N.S. Nuclear receptor peroxisome proliferator activated receptor (PPAR) β/δ in skin wound healing and cancer. *Eur. J. Dermatol.* **2015**, *25* (Suppl. 1), 4–11. [CrossRef]
12. Hanahan, D.; Weinberg, R.A. Hallmarks of cancer: The next generation. *Cell* **2011**, *144*, 646–674. [CrossRef] [PubMed]
13. Hanahan, D.; Weinberg, R.A. The hallmarks of cancer. *Cell* **2000**, *100*, 57–70. [CrossRef]
14. Wagner, K.D.; Du, S.; Martin, L.; Leccia, N.; Michiels, J.F.; Wagner, N. Vascular PPARβ/δ Promotes Tumor Angiogenesis and Progression. *Cells* **2019**, *8*, 1623. [CrossRef] [PubMed]
15. Ding, J.; Gou, Q.; Jin, J.; Shi, J.; Liu, Q.; Hou, Y. Metformin inhibits PPARδ agonist-mediated tumor growth by reducing Glut1 and SLC1A5 expressions of cancer cells. *Eur. J. Pharmacol.* **2019**, *857*, 172425. [CrossRef] [PubMed]
16. Wang, D.; Wang, H.; Guo, Y.; Ning, W.; Katkuri, S.; Wahli, W.; Desvergne, B.; Dey, S.K.; DuBois, R.N. Crosstalk between peroxisome proliferator-activated receptor delta and VEGF stimulates cancer progression. *Proc. Natl. Acad. Sci. USA* **2006**, *103*, 19069–19074. [CrossRef] [PubMed]
17. Liu, Y.; Deguchi, Y.; Tian, R.; Wei, D.; Wu, L.; Chen, W.; Xu, W.; Xu, M.; Liu, F.; Gao, S.; et al. Pleiotropic Effects of PPARD Accelerate Colorectal Tumorigenesis, Progression, and Invasion. *Cancer Res.* **2019**, *79*, 954–969. [CrossRef]
18. Zhou, D.; Jin, J.; Liu, Q.; Shi, J.; Hou, Y. PPARδ agonist enhances colitis-associated colorectal cancer. *Eur. J. Pharmacol.* **2019**, *842*, 248–254. [CrossRef]
19. Mao, F.; Xu, M.; Zuo, X.; Yu, J.; Xu, W.; Moussalli, M.J.; Elias, E.; Li, H.S.; Watowich, S.S.; Shureiqi, I. 15-Lipoxygenase-1 suppression of colitis-associated colon cancer through inhibition of the IL-6/STAT3 signaling pathway. *FASEB J.* **2015**, *29*, 2359–2370. [CrossRef]
20. Zuo, X.; Xu, M.; Yu, J.; Wu, Y.; Moussalli, M.J.; Manyam, G.C.; Lee, S.I.; Liang, S.; Gagea, M.; Morris, J.S.; et al. Potentiation of colon cancer susceptibility in mice by colonic epithelial PPAR-δ/β overexpression. *J. Natl. Cancer Inst.* **2014**, *106*, dju052. [CrossRef]

21. Beyaz, S.; Mana, M.D.; Roper, J.; Kedrin, D.; Saadatpour, A.; Hong, S.J.; Bauer-Rowe, K.E.; Xifaras, M.E.; Akkad, A.; Arias, E.; et al. High-fat diet enhances stemness and tumorigenicity of intestinal progenitors. *Nature* **2016**, *531*, 53–58. [CrossRef] [PubMed]

22. Zuo, X.; Peng, Z.; Moussalli, M.J.; Morris, J.S.; Broaddus, R.R.; Fischer, S.M.; Shureiqi, I. Targeted genetic disruption of peroxisome proliferator-activated receptor-delta and colonic tumorigenesis. *J. Natl. Cancer Inst.* **2009**, *101*, 762–767. [CrossRef] [PubMed]

23. Marin, H.E.; Peraza, M.A.; Billin, A.N.; Willson, T.M.; Ward, J.M.; Kennett, M.J.; Gonzalez, F.J.; Peters, J.M. Ligand activation of peroxisome proliferator-activated receptor beta inhibits colon carcinogenesis. *Cancer Res.* **2006**, *66*, 4394–4401. [CrossRef] [PubMed]

24. Yang, L.; Zhou, J.; Ma, Q.; Wang, C.; Chen, K.; Meng, W.; Yu, Y.; Zhou, Z.; Sun, X. Knockdown of PPAR δ gene promotes the growth of colon cancer and reduces the sensitivity to bevacizumab in nude mice model. *PLoS ONE* **2013**, *8*, e60715. [CrossRef] [PubMed]

25. Martín-Martín, N.; Zabala-Letona, A.; Fernández-Ruiz, S.; Arreal, L.; Camacho, L.; Castillo-Martin, M.; Cortazar, A.R.; Torrano, V.; Astobiza, I.; Zúñiga-García, P.; et al. PPARδ Elicits Ligand-Independent Repression of Trefoil Factor Family to Limit Prostate Cancer Growth. *Cancer Res.* **2018**, *78*, 399–409. [CrossRef] [PubMed]

26. Her, N.G.; Jeong, S.I.; Cho, K.; Ha, T.K.; Han, J.; Ko, K.P.; Park, S.K.; Lee, J.H.; Lee, M.G.; Ryu, B.K.; et al. PPARδ promotes oncogenic redirection of TGF-β1 signaling through the activation of the ABCA1-Cav1 pathway. *Cell Cycle* **2013**, *12*, 1521–1535. [CrossRef] [PubMed]

27. Morgan, E.; Kannan-Thulasiraman, P.; Noy, N. Involvement of Fatty Acid Binding Protein 5 and PPARβ/δ in Prostate Cancer Cell Growth. *PPAR Res.* **2010**, *2010*. [CrossRef]

28. Li, Y.J.; Sun, L.; Shi, Y.; Wang, G.; Wang, X.; Dunn, S.E.; Iorio, C.; Screaton, R.A.; Spaner, D.E. PPAR-delta promotes survival of chronic lymphocytic leukemia cells in energetically unfavorable conditions. *Leukemia* **2017**, *31*, 1905–1914. [CrossRef]

29. Yao, P.L.; Chen, L.; Dobrzański, T.P.; Zhu, B.; Kang, B.H.; Müller, R.; Gonzalez, F.J.; Peters, J.M. Peroxisome proliferator-activated receptor-β/δ inhibits human neuroblastoma cell tumorigenesis by inducing p53- and SOX2-mediated cell differentiation. *Mol. Carcinog.* **2017**, *56*, 1472–1483. [CrossRef]

30. Balandaram, G.; Kramer, L.R.; Kang, B.H.; Murray, I.A.; Perdew, G.H.; Gonzalez, F.J.; Peters, J.M. Ligand activation of peroxisome proliferator-activated receptor-β/δ suppresses liver tumorigenesis in hepatitis B transgenic mice. *Toxicology* **2016**, *363-364*, 1–9. [CrossRef]

31. Xu, L.; Han, C.; Lim, K.; Wu, T. Cross-talk between peroxisome proliferator-activated receptor delta and cytosolic phospholipase A(2)alpha/cyclooxygenase-2/prostaglandin E(2) signaling pathways in human hepatocellular carcinoma cells. *Cancer Res.* **2006**, *66*, 11859–11868. [CrossRef] [PubMed]

32. Michiels, J.F.; Perrin, C.; Leccia, N.; Massi, D.; Grimaldi, P.; Wagner, N. PPARbeta activation inhibits melanoma cell proliferation involving repression of the Wilms' tumour suppressor WT1. *Pflugers Arch.* **2010**, *459*, 689–703. [CrossRef] [PubMed]

33. Girroir, E.E.; Hollingshead, H.E.; Billin, A.N.; Willson, T.M.; Robertson, G.P.; Sharma, A.K.; Amin, S.; Gonzalez, F.J.; Peters, J.M. Peroxisome proliferator-activated receptor-beta/delta (PPARbeta/delta) ligands inhibit growth of UACC903 and MCF7 human cancer cell lines. *Toxicology* **2008**, *243*, 236–243. [CrossRef] [PubMed]

34. Ham, S.A.; Kim, E.; Yoo, T.; Lee, W.J.; Youn, J.H.; Choi, M.J.; Han, S.G.; Lee, C.H.; Paek, K.S.; Hwang, J.S.; et al. Ligand-activated interaction of PPARδ with c-Myc governs the tumorigenicity of breast cancer. *Int. J. Cancer* **2018**, *143*, 2985–2996. [CrossRef]

35. Wang, X.; Wang, G.; Shi, Y.; Sun, L.; Gorczynski, R.; Li, Y.J.; Xu, Z.; Spaner, D.E. PPAR-delta promotes survival of breast cancer cells in harsh metabolic conditions. *Oncogenesis* **2016**, *5*, e232. [CrossRef]

36. Yuan, H.; Lu, J.; Xiao, J.; Upadhyay, G.; Umans, R.; Kallakury, B.; Yin, Y.; Fant, M.E.; Kopelovich, L.; Glazer, R.I. PPARδ induces estrogen receptor-positive mammary neoplasia through an inflammatory and metabolic phenotype linked to mTOR activation. *Cancer Res.* **2013**, *73*, 4349–4361. [CrossRef]

37. Ghosh, M.; Ai, Y.; Narko, K.; Wang, Z.; Peters, J.M.; Hla, T. PPARdelta is pro-tumorigenic in a mouse model of COX-2-induced mammary cancer. *Prostaglandins Other Lipid Mediat* **2009**, *88*, 97–100. [CrossRef]

38. Pollock, C.B.; Yin, Y.; Yuan, H.; Zeng, X.; King, S.; Li, X.; Kopelovich, L.; Albanese, C.; Glazer, R.I. PPARδ activation acts cooperatively with 3-phosphoinositide-dependent protein kinase-1 to enhance mammary tumorigenesis. *PLoS ONE* **2011**, *6*, e16215. [CrossRef]

39. Levi, L.; Lobo, G.; Doud, M.K.; Von Lintig, J.; Seachrist, D.; Tochtrop, G.P.; Noy, N. Genetic ablation of the fatty acid-binding protein FABP5 suppresses HER2-induced mammary tumorigenesis. *Cancer Res.* **2013**, *73*, 4770–4780. [CrossRef]

40. Yin, Y.; Russell, R.G.; Dettin, L.E.; Bai, R.; Wei, Z.L.; Kozikowski, A.P.; Kopelovich, L.; Kopleovich, L.; Glazer, R.I. Peroxisome proliferator-activated receptor delta and gamma agonists differentially alter tumor differentiation and progression during mammary carcinogenesis. *Cancer Res.* **2005**, *65*, 3950–3957. [CrossRef]

41. Yao, P.L.; Chen, L.P.; Dobrzański, T.P.; Phillips, D.A.; Zhu, B.; Kang, B.H.; Gonzalez, F.J.; Peters, J.M. Inhibition of testicular embryonal carcinoma cell tumorigenicity by peroxisome proliferator-activated receptor-β/δ- and retinoic acid receptor-dependent mechanisms. *Oncotarget* **2015**, *6*, 36319–36337. [CrossRef] [PubMed]

42. Montagner, A.; Delgado, M.B.; Tallichet-Blanc, C.; Chan, J.S.; Sng, M.K.; Mottaz, H.; Degueurce, G.; Lippi, Y.; Moret, C.; Baruchet, M.; et al. Src is activated by the nuclear receptor peroxisome proliferator-activated receptor β/δ in ultraviolet radiation-induced skin cancer. *EMBO Mol. Med.* **2014**, *6*, 80–98. [CrossRef] [PubMed]

43. Genini, D.; Garcia-Escudero, R.; Carbone, G.M.; Catapano, C.V. Transcriptional and Non-Transcriptional Functions of PPARβ/δ in Non-Small Cell Lung Cancer. *PLoS ONE* **2012**, *7*, e46009. [CrossRef] [PubMed]

44. He, P.; Borland, M.G.; Zhu, B.; Sharma, A.K.; Amin, S.; El-Bayoumy, K.; Gonzalez, F.J.; Peters, J.M. Effect of ligand activation of peroxisome proliferator-activated receptor-beta/delta (PPARbeta/delta) in human lung cancer cell lines. *Toxicology* **2008**, *254*, 112–117. [CrossRef]

45. Müller-Brüsselbach, S.; Ebrahimsade, S.; Jäkel, J.; Eckhardt, J.; Rapp, U.R.; Peters, J.M.; Moll, R.; Müller, R. Growth of transgenic RAF-induced lung adenomas is increased in mice with a disrupted PPARbeta/delta gene. *Int. J. Oncol.* **2007**, *31*, 607–611.

46. Wagner, K.D.; Benchetrit, M.; Bianchini, L.; Michiels, J.F.; Wagner, N. Peroxisome proliferator-activated receptor β/δ (PPARβ/δ) is highly expressed in liposarcoma and promotes migration and proliferation. *J. Pathol.* **2011**, *224*, 575–588. [CrossRef]

47. Zeng, L.; Geng, Y.; Tretiakova, M.; Yu, X.; Sicinski, P.; Kroll, T.G. Peroxisome proliferator-activated receptor-delta induces cell proliferation by a cyclin E1-dependent mechanism and is up-regulated in thyroid tumors. *Cancer Res.* **2008**, *68*, 6578–6586. [CrossRef]

48. Daikoku, T.; Tranguch, S.; Chakrabarty, A.; Wang, D.; Khabele, D.; Orsulic, S.; Morrow, J.D.; Dubois, R.N.; Dey, S.K. Extracellular signal-regulated kinase is a target of cyclooxygenase-1-peroxisome proliferator-activated receptor-delta signaling in epithelial ovarian cancer. *Cancer Res.* **2007**, *67*, 5285–5292. [CrossRef]

49. Röhrl, C.; Kaindl, U.; Koneczny, I.; Hudec, X.; Baron, D.M.; König, J.S.; Marian, B. Peroxisome-proliferator-activated receptors γ and β/δ mediate vascular endothelial growth factor production in colorectal tumor cells. *J. Cancer Res. Clin. Oncol.* **2011**, *137*, 29–39. [CrossRef]

50. Takayama, O.; Yamamoto, H.; Damdinsuren, B.; Sugita, Y.; Ngan, C.Y.; Xu, X.; Tsujino, T.; Takemasa, I.; Ikeda, M.; Sekimoto, M.; et al. Expression of PPARdelta in multistage carcinogenesis of the colorectum: Implications of malignant cancer morphology. *Br. J. Cancer* **2006**, *95*, 889–895. [CrossRef]

51. Foreman, J.E.; Chang, W.C.; Palkar, P.S.; Zhu, B.; Borland, M.G.; Williams, J.L.; Kramer, L.R.; Clapper, M.L.; Gonzalez, F.J.; Peters, J.M. Functional characterization of peroxisome proliferator-activated receptor-β/δ expression in colon cancer. *Mol. Carcinog.* **2011**, *50*, 884–900. [CrossRef]

52. Yoshinaga, M.; Taki, K.; Somada, S.; Sakiyama, Y.; Kubo, N.; Kaku, T.; Tsuruta, S.; Kusumoto, T.; Sakai, H.; Nakamura, K.; et al. The expression of both peroxisome proliferator-activated receptor delta and cyclooxygenase-2 in tissues is associated with poor prognosis in colorectal cancer patients. *Dig. Dis. Sci.* **2011**, *56*, 1194–1200. [CrossRef]

53. Kannan-Thulasiraman, P.; Seachrist, D.D.; Mahabeleshwar, G.H.; Jain, M.K.; Noy, N. Fatty acid-binding protein 5 and PPARbeta/delta are critical mediators of epidermal growth factor receptor-induced carcinoma cell growth. *J. Biol. Chem.* **2010**, *285*, 19106–19115. [CrossRef] [PubMed]

54. Yao, P.L.; Morales, J.L.; Zhu, B.; Kang, B.H.; Gonzalez, F.J.; Peters, J.M. Activation of peroxisome proliferator-activated receptor-β/δ (PPAR-β/δ) inhibits human breast cancer cell line tumorigenicity. *Mol. Cancer Ther.* **2014**, *13*, 1008–1017. [CrossRef] [PubMed]

55. Zhu, B.; Bai, R.; Kennett, M.J.; Kang, B.H.; Gonzalez, F.J.; Peters, J.M. Chemoprevention of chemically induced skin tumorigenesis by ligand activation of peroxisome proliferator-activated receptor-beta/delta and inhibition of cyclooxygenase 2. *Mol. Cancer Ther.* **2010**, *9*, 3267–3277. [CrossRef] [PubMed]

56. Bility, M.T.; Zhu, B.; Kang, B.H.; Gonzalez, F.J.; Peters, J.M. Ligand activation of peroxisome proliferator-activated receptor-beta/delta and inhibition of cyclooxygenase-2 enhances inhibition of skin tumorigenesis. *Toxicol. Sci.* **2010**, *113*, 27–36. [CrossRef]

57. Hollingshead, H.E.; Borland, M.G.; Billin, A.N.; Willson, T.M.; Gonzalez, F.J.; Peters, J.M. Ligand activation of peroxisome proliferator-activated receptor-beta/delta (PPARbeta/delta) and inhibition of cyclooxygenase 2 (COX2) attenuate colon carcinogenesis through independent signaling mechanisms. *Carcinogenesis* **2008**, *29*, 169–176. [CrossRef]

58. Kim, D.J.; Prabhu, K.S.; Gonzalez, F.J.; Peters, J.M. Inhibition of chemically induced skin carcinogenesis by sulindac is independent of peroxisome proliferator-activated receptor-beta/delta (PPARbeta/delta). *Carcinogenesis* **2006**, *27*, 1105–1112. [CrossRef]

59. Wagner, N.; Panelos, J.; Massi, D.; Wagner, K.D. The Wilms' tumor suppressor WT1 is associated with melanoma proliferation. *Pflugers Arch.* **2008**, *455*, 839–847. [CrossRef]

60. Wagner, N.; Michiels, J.F.; Schedl, A.; Wagner, K.D. The Wilms' tumour suppressor WT1 is involved in endothelial cell proliferation and migration: Expression in tumour vessels in vivo. *Oncogene* **2008**, *27*, 3662–3672. [CrossRef]

61. Wagner, N.; Wagner, K.D.; Scholz, H.; Kirschner, K.M.; Schedl, A. Intermediate filament protein nestin is expressed in developing kidney and heart and might be regulated by the Wilms' tumor suppressor Wt1. *Am. J. Physiol. Regul. Integr. Comp. Physiol.* **2006**, *291*, R779–R787. [CrossRef] [PubMed]

62. Pedchenko, T.V.; Gonzalez, A.L.; Wang, D.; DuBois, R.N.; Massion, P.P. Peroxisome proliferator-activated receptor beta/delta expression and activation in lung cancer. *Am. J. Respir Cell Mol. Biol.* **2008**, *39*, 689–696. [CrossRef] [PubMed]

63. Bishop-Bailey, D. PPARs and angiogenesis. *Biochem. Soc. Trans.* **2011**, *39*, 1601–1605. [CrossRef] [PubMed]

64. Xin, X.; Yang, S.; Kowalski, J.; Gerritsen, M.E. Peroxisome proliferator-activated receptor gamma ligands are potent inhibitors of angiogenesis in vitro and in vivo. *J. Biol. Chem.* **1999**, *274*, 9116–9121. [CrossRef]

65. Bishop-Bailey, D.; Hla, T. Endothelial cell apoptosis induced by the peroxisome proliferator-activated receptor (PPAR) ligand 15-deoxy-Delta12, 14-prostaglandin J2. *J. Biol. Chem.* **1999**, *274*, 17042–17048. [CrossRef]

66. Bishop-Bailey, D. Peroxisome proliferator-activated receptors in the cardiovascular system. *Br. J. Pharmacol.* **2000**, *129*, 823–834. [CrossRef]

67. Berger, J.; Leibowitz, M.D.; Doebber, T.W.; Elbrecht, A.; Zhang, B.; Zhou, G.; Biswas, C.; Cullinan, C.A.; Hayes, N.S.; Li, Y.; et al. Novel peroxisome proliferator-activated receptor (PPAR) gamma and PPARdelta ligands produce distinct biological effects. *J. Biol. Chem.* **1999**, *274*, 6718–6725. [CrossRef]

68. Sznaidman, M.L.; Haffner, C.D.; Maloney, P.R.; Fivush, A.; Chao, E.; Goreham, D.; Sierra, M.L.; LeGrumelec, C.; Xu, H.E.; Montana, V.G.; et al. Novel selective small molecule agonists for peroxisome proliferator-activated receptor delta (PPARdelta)–synthesis and biological activity. *Bioorg. Med. Chem. Lett.* **2003**, *13*, 1517–1521. [CrossRef]

69. Hatae, T.; Wada, M.; Yokoyama, C.; Shimonishi, M.; Tanabe, T. Prostacyclin-dependent apoptosis mediated by PPAR delta. *J. Biol. Chem.* **2001**, *276*, 46260–46267. [CrossRef]

70. Liou, J.Y.; Lee, S.; Ghelani, D.; Matijevic-Aleksic, N.; Wu, K.K. Protection of endothelial survival by peroxisome proliferator-activated receptor-delta mediated 14-3-3 upregulation. *Arterioscler. Thromb. Vasc. Biol.* **2006**, *26*, 1481–1487. [CrossRef]

71. Brunelli, L.; Cieslik, K.A.; Alcorn, J.L.; Vatta, M.; Baldini, A. Peroxisome proliferator-activated receptor-delta upregulates 14-3-3 epsilon in human endothelial cells via CCAAT/enhancer binding protein-beta. *Circ. Res.* **2007**, *100*, e59–e71. [CrossRef] [PubMed]

72. Liou, J.Y.; Wu, C.C.; Chen, B.R.; Yen, L.B.; Wu, K.K. Nonsteroidal anti-inflammatory drugs induced endothelial apoptosis by perturbing peroxisome proliferator-activated receptor-delta transcriptional pathway. *Mol. Pharmacol.* **2008**, *74*, 1399–1406. [CrossRef] [PubMed]

73. Bishop-Bailey, D. Peroxisome proliferator-activated receptor beta/delta goes vascular. *Circ. Res.* **2008**, *102*, 146–147. [CrossRef] [PubMed]

74. Rival, Y.; Bénéteau, N.; Taillandier, T.; Pezet, M.; Dupont-Passelaigue, E.; Patoiseau, J.F.; Junquéro, D.; Colpaert, F.C.; Delhon, A. PPARalpha and PPARdelta activators inhibit cytokine-induced nuclear translocation of NF-kappaB and expression of VCAM-1 in EAhy926 endothelial cells. *Eur. J. Pharmacol.* **2002**, *435*, 143–151. [CrossRef]

75. Lee, C.H.; Chawla, A.; Urbiztondo, N.; Liao, D.; Boisvert, W.A.; Evans, R.M.; Curtiss, L.K. Transcriptional repression of atherogenic inflammation: Modulation by PPARdelta. *Science* **2003**, *302*, 453–457. [CrossRef]

76. Fan, Y.; Wang, Y.; Tang, Z.; Zhang, H.; Qin, X.; Zhu, Y.; Guan, Y.; Wang, X.; Staels, B.; Chien, S.; et al. Suppression of pro-inflammatory adhesion molecules by PPAR-delta in human vascular endothelial cells. *Arterioscler. Thromb. Vasc. Biol.* **2008**, *28*, 315–321. [CrossRef]

77. Piqueras, L.; Sanz, M.J.; Perretti, M.; Morcillo, E.; Norling, L.; Mitchell, J.A.; Li, Y.; Bishop-Bailey, D. Activation of PPARbeta/delta inhibits leukocyte recruitment, cell adhesion molecule expression, and chemokine release. *J. Leukoc. Biol.* **2009**, *86*, 115–122. [CrossRef]

78. Kim, H.J.; Ham, S.A.; Kim, S.U.; Hwang, J.Y.; Kim, J.H.; Chang, K.C.; Yabe-Nishimura, C.; Seo, H.G. Transforming growth factor-beta1 is a molecular target for the peroxisome proliferator-activated receptor delta. *Circ. Res.* **2008**, *102*, 193–200. [CrossRef]

79. D'Uscio, L.V.; Das, P.; Santhanam, A.V.; He, T.; Younkin, S.G.; Katusic, Z.S. Activation of PPARδ prevents endothelial dysfunction induced by overexpression of amyloid-β precursor protein. *Cardiovasc. Res.* **2012**, *96*, 504–512. [CrossRef]

80. Tian, X.Y.; Wong, W.T.; Wang, N.; Lu, Y.; Cheang, W.S.; Liu, J.; Liu, L.; Liu, Y.; Lee, S.S.; Chen, Z.Y.; et al. PPARδ activation protects endothelial function in diabetic mice. *Diabetes* **2012**, *61*, 3285–3293. [CrossRef]

81. Quintela, A.M.; Jiménez, R.; Piqueras, L.; Gómez-Guzmán, M.; Haro, J.; Zarzuelo, M.J.; Cogolludo, A.; Sanz, M.J.; Toral, M.; Romero, M.; et al. PPARβ activation restores the high glucose-induced impairment of insulin signalling in endothelial cells. *Br. J. Pharmacol.* **2014**, *171*, 3089–3102. [CrossRef] [PubMed]

82. Wawrzyniak, M.; Pich, C.; Gross, B.; Schütz, F.; Fleury, S.; Quemener, S.; Sgandurra, M.; Bouchaert, E.; Moret, C.; Mury, L.; et al. Endothelial, but not smooth muscle, peroxisome proliferator-activated receptor β/δ regulates vascular permeability and anaphylaxis. *J. Allergy Clin. Immunol.* **2015**, *135*, 1625–1635.e1625. [CrossRef] [PubMed]

83. Piqueras, L.; Reynolds, A.R.; Hodivala-Dilke, K.M.; Alfranca, A.; Redondo, J.M.; Hatae, T.; Tanabe, T.; Warner, T.D.; Bishop-Bailey, D. Activation of PPARbeta/delta induces endothelial cell proliferation and angiogenesis. *Arterioscler. Thromb. Vasc. Biol.* **2007**, *27*, 63–69. [CrossRef] [PubMed]

84. Gaudel, C.; Schwartz, C.; Giordano, C.; Abumrad, N.A.; Grimaldi, P.A. Pharmacological activation of PPARbeta promotes rapid and calcineurin-dependent fiber remodeling and angiogenesis in mouse skeletal muscle. *Am. J. Physiol. Endocrinol. Metab.* **2008**, *295*, E297–E304. [CrossRef] [PubMed]

85. Wagner, N.; Jehl-Piétri, C.; Lopez, P.; Murdaca, J.; Giordano, C.; Schwartz, C.; Gounon, P.; Hatem, S.N.; Grimaldi, P.; Wagner, K.D. Peroxisome proliferator-activated receptor beta stimulation induces rapid cardiac growth and angiogenesis via direct activation of calcineurin. *Cardiovasc. Res.* **2009**, *83*, 61–71. [CrossRef] [PubMed]

86. Wagner, K.D.; Vukolic, A.; Baudouy, D.; Michiels, J.F.; Wagner, N. Inducible Conditional Vascular-Specific Overexpression of Peroxisome Proliferator-Activated Receptor Beta/Delta Leads to Rapid Cardiac Hypertrophy. *PPAR Res.* **2016**, *2016*, 7631085. [CrossRef] [PubMed]

87. Park, J.R.; Ahn, J.H.; Jung, M.H.; Koh, J.S.; Park, Y.; Hwang, S.J.; Jeong, Y.H.; Kwak, C.H.; Lee, Y.S.; Seo, H.G.; et al. Effects of Peroxisome Proliferator-Activated Receptor-δ Agonist on Cardiac Healing after Myocardial Infarction. *PLoS ONE* **2016**, *11*, e0148510. [CrossRef]

88. Han, J.K.; Lee, H.S.; Yang, H.M.; Hur, J.; Jun, S.I.; Kim, J.Y.; Cho, C.H.; Koh, G.Y.; Peters, J.M.; Park, K.W.; et al. Peroxisome proliferator-activated receptor-delta agonist enhances vasculogenesis by regulating endothelial progenitor cells through genomic and nongenomic activations of the phosphatidylinositol 3-kinase/Akt pathway. *Circulation* **2008**, *118*, 1021–1033. [CrossRef]

89. Shearer, B.G.; Steger, D.J.; Way, J.M.; Stanley, T.B.; Lobe, D.C.; Grillot, D.A.; Iannone, M.A.; Lazar, M.A.; Willson, T.M.; Billin, A.N. Identification and characterization of a selective peroxisome proliferator-activated receptor beta/delta (NR1C2) antagonist. *Mol. Endocrinol.* **2008**, *22*, 523–529. [CrossRef]

90. Capozzi, M.E.; McCollum, G.W.; Savage, S.R.; Penn, J.S. Peroxisome proliferator-activated receptor-β/δ regulates angiogenic cell behaviors and oxygen-induced retinopathy. *Invest. Ophthalmol. Vis. Sci.* **2013**, *54*, 4197–4207. [CrossRef]

91. Müller-Brüsselbach, S.; Kömhoff, M.; Rieck, M.; Meissner, W.; Kaddatz, K.; Adamkiewicz, J.; Keil, B.; Klose, K.J.; Moll, R.; Burdick, A.D.; et al. Deregulation of tumor angiogenesis and blockade of tumor growth in PPARbeta-deficient mice. *EMBO J.* **2007**, *26*, 3686–3698. [CrossRef] [PubMed]

92. Adamkiewicz, J.; Kaddatz, K.; Rieck, M.; Wilke, B.; Müller-Brüsselbach, S.; Müller, R. Proteomic profile of mouse fibroblasts with a targeted disruption of the peroxisome proliferator activated receptor-beta/delta gene. *Proteomics* **2007**, *7*, 1208–1216. [CrossRef] [PubMed]

93. Tung, J.J.; Hobert, O.; Berryman, M.; Kitajewski, J. Chloride intracellular channel 4 is involved in endothelial proliferation and morphogenesis in vitro. *Angiogenesis* **2009**, *12*, 209–220. [CrossRef] [PubMed]

94. Liu, T.; Bohlken, A.; Kuljaca, S.; Lee, M.; Nguyen, T.; Smith, S.; Cheung, B.; Norris, M.D.; Haber, M.; Holloway, A.J.; et al. The retinoid anticancer signal: Mechanisms of target gene regulation. *Br. J. Cancer* **2005**, *93*, 310–318. [CrossRef]

95. Abdollahi, A.; Schwager, C.; Kleeff, J.; Esposito, I.; Domhan, S.; Peschke, P.; Hauser, K.; Hahnfeldt, P.; Hlatky, L.; Debus, J.; et al. Transcriptional network governing the angiogenic switch in human pancreatic cancer. *Proc. Natl. Acad. Sci. USA* **2007**, *104*, 12890–12895. [CrossRef]

96. Yoshinaga, M.; Kitamura, Y.; Chaen, T.; Yamashita, S.; Tsuruta, S.; Hisano, T.; Ikeda, Y.; Sakai, H.; Nakamura, K.; Takayanagi, R.; et al. The simultaneous expression of peroxisome proliferator-activated receptor delta and cyclooxygenase-2 may enhance angiogenesis and tumor venous invasion in tissues of colorectal cancers. *Dig. Dis. Sci.* **2009**, *54*, 1108–1114. [CrossRef]

97. Meissner, M.; Hrgovic, I.; Doll, M.; Naidenow, J.; Reichenbach, G.; Hailemariam-Jahn, T.; Michailidou, D.; Gille, J.; Kaufmann, R. Peroxisome proliferator-activated receptor {delta} activators induce IL-8 expression in nonstimulated endothelial cells in a transcriptional and posttranscriptional manner. *J. Biol. Chem.* **2010**, *285*, 33797–33804. [CrossRef]

98. Zuo, X.; Xu, W.; Xu, M.; Tian, R.; Moussalli, M.J.; Mao, F.; Zheng, X.; Wang, J.; Morris, J.S.; Gagea, M.; et al. Metastasis regulation by PPARD expression in cancer cells. *JCI Insight* **2017**, *2*, e91419. [CrossRef]

99. Sadremomtaz, A.; Mansouri, K.; Alemzadeh, G.; Safa, M.; Rastaghi, A.E.; Asghari, S.M. Dual blockade of VEGFR1 and VEGFR2 by a novel peptide abrogates VEGF-driven angiogenesis, tumor growth, and metastasis through PI3K/AKT and MAPK/ERK1/2 pathway. *Biochim. Biophys. Acta Gen. Subj.* **2018**, *1862*, 2688–2700. [CrossRef]

100. Tammela, T.; Zarkada, G.; Wallgard, E.; Murtomäki, A.; Suchting, S.; Wirzenius, M.; Waltari, M.; Hellström, M.; Schomber, T.; Peltonen, R.; et al. Blocking VEGFR-3 suppresses angiogenic sprouting and vascular network formation. *Nature* **2008**, *454*, 656–660. [CrossRef]

101. El Maï, M.; Wagner, K.D.; Michiels, J.F.; Ambrosetti, D.; Borderie, A.; Destree, S.; Renault, V.; Djerbi, N.; Giraud-Panis, M.J.; Gilson, E.; et al. The Telomeric Protein TRF2 Regulates Angiogenesis by Binding and Activating the PDGFRβ Promoter. *Cell Rep.* **2014**, *9*, 1047–1060. [CrossRef] [PubMed]

102. Battegay, E.J.; Rupp, J.; Iruela-Arispe, L.; Sage, E.H.; Pech, M. PDGF-BB modulates endothelial proliferation and angiogenesis in vitro via PDGF beta-receptors. *J. Cell Biol.* **1994**, *125*, 917–928. [CrossRef] [PubMed]

103. Sihto, H.; Tynninen, O.; Bützow, R.; Saarialho-Kere, U.; Joensuu, H. Endothelial cell KIT expression in human tumours. *J. Pathol.* **2007**, *211*, 481–488. [CrossRef] [PubMed]

104. Wagner, K.D.; Cherfils-Vicini, J.; Hosen, N.; Hohenstein, P.; Gilson, E.; Hastie, N.D.; Michiels, J.F.; Wagner, N. The Wilms' tumour suppressor Wt1 is a major regulator of tumour angiogenesis and progression. *Nat. Commun.* **2014**, *5*, 5852. [CrossRef] [PubMed]

105. He, T.C.; Chan, T.A.; Vogelstein, B.; Kinzler, K.W. PPARdelta is an APC-regulated target of nonsteroidal anti-inflammatory drugs. *Cell* **1999**, *99*, 335–345. [CrossRef]

106. Wang, D.; Wang, H.; Shi, Q.; Katkuri, S.; Walhi, W.; Desvergne, B.; Das, S.K.; Dey, S.K.; DuBois, R.N. Prostaglandin E(2) promotes colorectal adenoma growth via transactivation of the nuclear peroxisome proliferator-activated receptor delta. *Cancer Cell* **2004**, *6*, 285–295. [CrossRef]

107. Liou, J.Y.; Ghelani, D.; Yeh, S.; Wu, K.K. Nonsteroidal anti-inflammatory drugs induce colorectal cancer cell apoptosis by suppressing 14-3-3epsilon. *Cancer Res.* **2007**, *67*, 3185–3191. [CrossRef]

108. Wang, D.; Ning, W.; Xie, D.; Guo, L.; DuBois, R.N. Peroxisome proliferator-activated receptor δ confers resistance to peroxisome proliferator-activated receptor γ-induced apoptosis in colorectal cancer cells. *Oncogene* **2012**, *31*, 1013–1023. [CrossRef]

109. Shureiqi, I.; Jiang, W.; Zuo, X.; Wu, Y.; Stimmel, J.B.; Leesnitzer, L.M.; Morris, J.S.; Fan, H.Z.; Fischer, S.M.; Lippman, S.M. The 15-lipoxygenase-1 product 13-S-hydroxyoctadecadienoic acid down-regulates PPAR-delta to induce apoptosis in colorectal cancer cells. *Proc. Natl. Acad. Sci. USA* **2003**, *100*, 9968–9973. [CrossRef]

110. Cutler, N.S.; Graves-Deal, R.; LaFleur, B.J.; Gao, Z.; Boman, B.M.; Whitehead, R.H.; Terry, E.; Morrow, J.D.; Coffey, R.J. Stromal production of prostacyclin confers an antiapoptotic effect to colonic epithelial cells. *Cancer Res.* **2003**, *63*, 1748–1751.

111. Maggiora, M.; Bologna, M.; Cerù, M.P.; Possati, L.; Angelucci, A.; Cimini, A.; Miglietta, A.; Bozzo, F.; Margiotta, C.; Muzio, G.; et al. An overview of the effect of linoleic and conjugated-linoleic acids on the growth of several human tumor cell lines. *Int. J. Cancer* **2004**, *112*, 909–919. [CrossRef] [PubMed]

112. Fukumoto, K.; Yano, Y.; Virgona, N.; Hagiwara, H.; Sato, H.; Senba, H.; Suzuki, K.; Asano, R.; Yamada, K.; Yano, T. Peroxisome proliferator-activated receptor delta as a molecular target to regulate lung cancer cell growth. *FEBS Lett.* **2005**, *579*, 3829–3836. [CrossRef] [PubMed]

113. Foreman, J.E.; Sharma, A.K.; Amin, S.; Gonzalez, F.J.; Peters, J.M. Ligand activation of peroxisome proliferator-activated receptor-beta/delta (PPARbeta/delta) inhibits cell growth in a mouse mammary gland cancer cell line. *Cancer Lett.* **2010**, *288*, 219–225. [CrossRef] [PubMed]

114. Bell, E.; Ponthan, F.; Whitworth, C.; Westermann, F.; Thomas, H.; Redfern, C.P. Cell survival signalling through PPARδ and arachidonic acid metabolites in neuroblastoma. *PLoS ONE* **2013**, *8*, e68859. [CrossRef] [PubMed]

115. Péchery, A.; Fauconnet, S.; Bittard, H.; Lascombe, I. Apoptotic effect of the selective PPARβ/δ agonist GW501516 in invasive bladder cancer cells. *Tumour Biol.* **2016**, *37*, 14789–14802. [CrossRef]

116. Tong-Lin Wu, T.; Tong, Y.C.; Chen, I.H.; Niu, H.S.; Li, Y.; Cheng, J.T. Induction of apoptosis in prostate cancer by ginsenoside Rh2. *Oncotarget* **2018**, *9*, 11109–11118. [CrossRef]

117. Wu, T.T.; Niu, H.S.; Chen, L.J.; Cheng, J.T.; Tong, Y.C. Increase of human prostate cancer cell (DU145) apoptosis by telmisartan through PPAR-delta pathway. *Eur. J. Pharmacol.* **2016**, *775*, 35–42. [CrossRef]

118. Ji, Y.; Li, H.; Wang, F.; Gu, L. PPARβ/δ Agonist GW501516 Inhibits Tumorigenicity of Undifferentiated Nasopharyngeal Carcinoma in C666-1 Cells by Promoting Apoptosis. *Front. Pharmacol.* **2018**, *9*, 648. [CrossRef]

119. Gu, L.; Shi, Y.; Xu, W.; Ji, Y. PPARβ/δ Agonist GW501516 Inhibits Tumorigenesis and Promotes Apoptosis of the Undifferentiated Nasopharyngeal Carcinoma C666-1 Cells by Regulating miR-206. *Oncol. Res.* **2019**, *27*, 923–933. [CrossRef]

120. Zhai, Y.; Wu, R.; Schwartz, D.R.; Darrah, D.; Reed, H.; Kolligs, F.T.; Nieman, M.T.; Fearon, E.R.; Cho, K.R. Role of beta-catenin/T-cell factor-regulated genes in ovarian endometrioid adenocarcinomas. *Am. J. Pathol* **2002**, *160*, 1229–1238. [CrossRef]

121. Shao, J.; Sheng, H.; DuBois, R.N. Peroxisome proliferator-activated receptors modulate K-Ras-mediated transformation of intestinal epithelial cells. *Cancer Res.* **2002**, *62*, 3282–3288. [PubMed]

122. Foreman, J.E.; Sorg, J.M.; McGinnis, K.S.; Rigas, B.; Williams, J.L.; Clapper, M.L.; Gonzalez, F.J.; Peters, J.M. Regulation of peroxisome proliferator-activated receptor-beta/delta by the APC/beta-CATENIN pathway and nonsteroidal antiinflammatory drugs. *Mol. Carcinog* **2009**, *48*, 942–952. [CrossRef] [PubMed]

123. Huff, V.; Miwa, H.; Haber, D.A.; Call, K.M.; Housman, D.; Strong, L.C.; Saunders, G.F. Evidence for WT1 as a Wilms tumor (WT) gene: Intragenic germinal deletion in bilateral WT. *Am. J. Hum. Genet.* **1991**, *48*, 997–1003. [PubMed]

124. Haber, D.A.; Buckler, A.J.; Glaser, T.; Call, K.M.; Pelletier, J.; Sohn, R.L.; Douglass, E.C.; Housman, D.E. An internal deletion within an 11p13 zinc finger gene contributes to the development of Wilms' tumor. *Cell* **1990**, *61*, 1257–1269. [CrossRef]

125. Wagner, K.D.; El Maï, M.; Ladomery, M.; Belali, T.; Leccia, N.; Michiels, J.F.; Wagner, N. Altered VEGF Splicing Isoform Balance in Tumor Endothelium Involves Activation of Splicing Factors Srpk1 and Srsf1 by the Wilms' Tumor Suppressor Wt1. *Cells* **2019**, *8*, 41. [CrossRef]

126. Sugiyama, H. WT1 (Wilms' tumor gene 1): Biology and cancer immunotherapy. *Jpn. J. Clin. Oncol.* **2010**, *40*, 377–387. [CrossRef]

127. Kundu, J.; Wahab, S.M.; Kundu, J.K.; Choi, Y.L.; Erkin, O.C.; Lee, H.S.; Park, S.G.; Shin, Y.K. Tob1 induces apoptosis and inhibits proliferation, migration and invasion of gastric cancer cells by activating Smad4 and inhibiting β-catenin signaling. *Int. J. Oncol.* **2012**, *41*, 839–848. [CrossRef]

128. Lim, H.J.; Lee, S.; Park, J.H.; Lee, K.S.; Choi, H.E.; Chung, K.S.; Lee, H.H.; Park, H.Y. PPAR delta agonist L-165041 inhibits rat vascular smooth muscle cell proliferation and migration via inhibition of cell cycle. *Atherosclerosis* **2009**, *202*, 446–454. [CrossRef]

129. Coleman, J.D.; Thompson, J.T.; Smith, R.W.; Prokopczyk, B.; Vanden Heuvel, J.P. Role of Peroxisome Proliferator-Activated Receptor β/δ and B-Cell Lymphoma-6 in Regulation of Genes Involved in Metastasis and Migration in Pancreatic Cancer Cells. *PPAR Res.* **2013**, *2013*, 121956. [CrossRef]
130. Ham, S.A.; Yoo, T.; Lee, W.J.; Hwang, J.S.; Hur, J.; Paek, K.S.; Lim, D.S.; Han, S.G.; Lee, C.H.; Seo, H.G. ADAMTS1-mediated targeting of TSP-1 by PPARδ suppresses migration and invasion of breast cancer cells. *Oncotarget* **2017**, *8*, 94091–94103. [CrossRef]
131. He, F.; York, J.P.; Burroughs, S.G.; Qin, L.; Xia, J.; Chen, D.; Quigley, E.M.; Webb, P.; LeSage, G.D.; Xia, X. Recruited metastasis suppressor NM23-H2 attenuates expression and activity of peroxisome proliferator-activated receptor δ (PPARδ) in human cholangiocarcinoma. *Dig. Liver Dis.* **2015**, *47*, 62–67. [CrossRef] [PubMed]
132. Steeg, P.S.; Palmieri, D.; Ouatas, T.; Salerno, M. Histidine kinases and histidine phosphorylated proteins in mammalian cell biology, signal transduction and cancer. *Cancer Lett.* **2003**, *190*, 1–12. [CrossRef]
133. Lim, J.C.W.; Kwan, Y.P.; Tan, M.S.; Teo, M.H.Y.; Chiba, S.; Wahli, W.; Wang, X. The Role of PPARβ/δ in Melanoma Metastasis. *Int. J. Mol. Sci.* **2018**, *19*, 2860. [CrossRef]
134. Ham, S.A.; Yoo, T.; Hwang, J.S.; Kang, E.S.; Lee, W.J.; Paek, K.S.; Park, C.; Kim, J.H.; Do, J.T.; Lim, D.S.; et al. Ligand-activated PPARδ modulates the migration and invasion of melanoma cells by regulating Snail expression. *Am. J. Cancer Res.* **2014**, *4*, 674–682. [PubMed]
135. Wang, D.; Fu, L.; Wei, J.; Xiong, Y.; DuBois, R.N. PPARδ Mediates the Effect of Dietary Fat in Promoting Colorectal Cancer Metastasis. *Cancer Res.* **2019**, *79*, 4480–4490. [CrossRef] [PubMed]
136. Kim, H.J.; Ham, S.A.; Paek, K.S.; Hwang, J.S.; Jung, S.Y.; Kim, M.Y.; Jin, H.; Kang, E.S.; Woo, I.S.; Lee, J.H.; et al. Transcriptional up-regulation of antioxidant genes by PPARδ inhibits angiotensin II-induced premature senescence in vascular smooth muscle cells. *Biochem. Biophys. Res. Commun.* **2011**, *406*, 564–569. [CrossRef] [PubMed]
137. Kim, H.J.; Ham, S.A.; Kim, M.Y.; Hwang, J.S.; Lee, H.; Kang, E.S.; Yoo, T.; Woo, I.S.; Yabe-Nishimura, C.; Paek, K.S.; et al. PPARδ coordinates angiotensin II-induced senescence in vascular smooth muscle cells through PTEN-mediated inhibition of superoxide generation. *J. Biol. Chem.* **2011**, *286*, 44585–44593. [CrossRef]
138. Ham, S.A.; Hwang, J.S.; Yoo, T.; Lee, H.; Kang, E.S.; Park, C.; Oh, J.W.; Lee, H.T.; Min, G.; Kim, J.H.; et al. Ligand-activated PPARδ inhibits UVB-induced senescence of human keratinocytes via PTEN-mediated inhibition of superoxide production. *Biochem. J.* **2012**, *444*, 27–38. [CrossRef]
139. Kim, M.Y.; Kang, E.S.; Ham, S.A.; Hwang, J.S.; Yoo, T.S.; Lee, H.; Paek, K.S.; Park, C.; Lee, H.T.; Kim, J.H.; et al. The PPARδ-mediated inhibition of angiotensin II-induced premature senescence in human endothelial cells is SIRT1-dependent. *Biochem. Pharmacol.* **2012**, *84*, 1627–1634. [CrossRef]
140. Altieri, P.; Spallarossa, P.; Barisione, C.; Garibaldi, S.; Garuti, A.; Fabbi, P.; Ghigliotti, G.; Brunelli, C. Inhibition of doxorubicin-induced senescence by PPARδ activation agonists in cardiac muscle cells: Cooperation between PPARδ and Bcl6. *PLoS ONE* **2012**, *7*, e46126. [CrossRef]
141. Van Steensel, B.; Smogorzewska, A.; De Lange, T. TRF2 protects human telomeres from end-to-end fusions. *Cell* **1998**, *92*, 401–413. [CrossRef]
142. Zhu, B.; Ferry, C.H.; Blazanin, N.; Bility, M.T.; Khozoie, C.; Kang, B.H.; Glick, A.B.; Gonzalez, F.J.; Peters, J.M. PPARβ/δ promotes HRAS-induced senescence and tumor suppression by potentiating p-ERK and repressing p-AKT signaling. *Oncogene* **2014**, *33*, 5348–5359. [CrossRef] [PubMed]
143. Zhu, B.; Ferry, C.H.; Markell, L.K.; Blazanin, N.; Glick, A.B.; Gonzalez, F.J.; Peters, J.M. The nuclear receptor peroxisome proliferator-activated receptor-β/δ (PPARβ/δ) promotes oncogene-induced cellular senescence through repression of endoplasmic reticulum stress. *J. Biol. Chem.* **2014**, *289*, 20102–20119. [CrossRef] [PubMed]
144. Riahi, Y.; Kaiser, N.; Cohen, G.; Abd-Elrahman, I.; Blum, G.; Shapira, O.M.; Koler, T.; Simionescu, M.; Sima, A.V.; Zarkovic, N.; et al. Foam cell-derived 4-hydroxynonenal induces endothelial cell senescence in a TXNIP-dependent manner. *J. Cell Mol. Med.* **2015**, *19*, 1887–1899. [CrossRef] [PubMed]
145. Bernal, C.; Araya, C.; Palma, V.; Bronfman, M. PPARβ/δ and PPARγ maintain undifferentiated phenotypes of mouse adult neural precursor cells from the subventricular zone. *Front. Cell Neurosci.* **2015**, *9*, 78. [CrossRef]
146. Scholtysek, C.; Katzenbeisser, J.; Fu, H.; Uderhardt, S.; Ipseiz, N.; Stoll, C.; Zaiss, M.M.; Stock, M.; Donhauser, L.; Böhm, C.; et al. PPARβ/δ governs Wnt signaling and bone turnover. *Nat. Med.* **2013**, *19*, 608–613. [CrossRef]

147. Milanovic, M.; Fan, D.N.Y.; Belenki, D.; Däbritz, J.H.M.; Zhao, Z.; Yu, Y.; Dörr, J.R.; Dimitrova, L.; Lenze, D.; Monteiro Barbosa, I.A.; et al. Senescence-associated reprogramming promotes cancer stemness. *Nature* **2018**, *553*, 96–100. [CrossRef]

148. WARBURG, O. On respiratory impairment in cancer cells. *Science* **1956**, *124*, 269–270.

149. WARBURG, O. On the origin of cancer cells. *Science* **1956**, *123*, 309–314. [CrossRef]

150. Warburg, O.; Wind, F.; Negelein, E. THE METABOLISM OF TUMORS IN THE BODY. *J. Gen. Physiol.* **1927**, *8*, 519–530. [CrossRef]

151. Semenza, G.L. Expression of hypoxia-inducible factor 1: Mechanisms and consequences. *Biochem. Pharmacol.* **2000**, *59*, 47–53. [CrossRef]

152. Maxwell, P.H.; Dachs, G.U.; Gleadle, J.M.; Nicholls, L.G.; Harris, A.L.; Stratford, I.J.; Hankinson, O.; Pugh, C.W.; Ratcliffe, P.J. Hypoxia-inducible factor-1 modulates gene expression in solid tumors and influences both angiogenesis and tumor growth. *Proc. Natl. Acad. Sci. USA* **1997**, *94*, 8104–8109. [CrossRef] [PubMed]

153. Richard, D.E.; Berra, E.; Pouysségur, J. Angiogenesis: How a tumor adapts to hypoxia. *Biochem. Biophys Res. Commun.* **1999**, *266*, 718–722. [CrossRef]

154. Forsythe, J.A.; Jiang, B.H.; Iyer, N.V.; Agani, F.; Leung, S.W.; Koos, R.D.; Semenza, G.L. Activation of vascular endothelial growth factor gene transcription by hypoxia-inducible factor 1. *Mol. Cell Biol.* **1996**, *16*, 4604–4613. [CrossRef] [PubMed]

155. Wang, G.L.; Semenza, G.L. General involvement of hypoxia-inducible factor 1 in transcriptional response to hypoxia. *Proc. Natl. Acad. Sci. USA* **1993**, *90*, 4304–4308. [CrossRef] [PubMed]

156. Wagner, K.D.; Wagner, N.; Wellmann, S.; Schley, G.; Bondke, A.; Theres, H.; Scholz, H. Oxygen-regulated expression of the Wilms' tumor suppressor Wt1 involves hypoxia-inducible factor-1 (HIF-1). *FASEB J.* **2003**, *17*, 1364–1366. [CrossRef] [PubMed]

157. Narravula, S.; Colgan, S.P. Hypoxia-inducible factor 1-mediated inhibition of peroxisome proliferator-activated receptor alpha expression during hypoxia. *J. Immunol* **2001**, *166*, 7543–7548. [CrossRef]

158. Balamurugan, K. HIF-1 at the crossroads of hypoxia, inflammation, and cancer. *Int. J. Cancer* **2016**, *138*, 1058–1066. [CrossRef]

159. Liu, Y.V.; Hubbi, M.E.; Pan, F.; McDonald, K.R.; Mansharamani, M.; Cole, R.N.; Liu, J.O.; Semenza, G.L. Calcineurin promotes hypoxia-inducible factor 1alpha expression by dephosphorylating RACK1 and blocking RACK1 dimerization. *J. Biol. Chem.* **2007**, *282*, 37064–37073. [CrossRef]

160. Jeong, E.; Koo, J.E.; Yeon, S.H.; Kwak, M.K.; Hwang, D.H.; Lee, J.Y. PPARδ deficiency disrupts hypoxia-mediated tumorigenic potential of colon cancer cells. *Mol. Carcinog.* **2014**, *53*, 926–937. [CrossRef]

161. Wang, J.; Ikeda, R.; Che, X.F.; Ooyama, A.; Yamamoto, M.; Furukawa, T.; Hasui, K.; Zheng, C.L.; Tajitsu, Y.; Oka, T.; et al. VEGF expression is augmented by hypoxia-induced PGIS in human fibroblasts. *Int. J. Oncol.* **2013**, *43*, 746–754. [CrossRef] [PubMed]

162. Hodge, A.M.; Williamson, E.J.; Bassett, J.K.; MacInnis, R.J.; Giles, G.G.; English, D.R. Dietary and biomarker estimates of fatty acids and risk of colorectal cancer. *Int. J. Cancer* **2015**, *137*, 1224–1234. [CrossRef] [PubMed]

163. Hardy, S.; Langelier, Y.; Prentki, M. Oleate activates phosphatidylinositol 3-kinase and promotes proliferation and reduces apoptosis of MDA-MB-231 breast cancer cells, whereas palmitate has opposite effects. *Cancer Res.* **2000**, *60*, 6353–6358. [PubMed]

164. Li, C.; Zhao, X.; Toline, E.C.; Siegal, G.P.; Evans, L.M.; Ibrahim-Hashim, A.; Desmond, R.A.; Hardy, R.W. Prevention of carcinogenesis and inhibition of breast cancer tumor burden by dietary stearate. *Carcinogenesis* **2011**, *32*, 1251–1258. [CrossRef]

165. Levi, L.; Wang, Z.; Doud, M.K.; Hazen, S.L.; Noy, N. Saturated fatty acids regulate retinoic acid signalling and suppress tumorigenesis by targeting fatty acid-binding protein 5. *Nat. Commun.* **2015**, *6*, 8794. [CrossRef]

166. Altenberg, B.; Greulich, K.O. Genes of glycolysis are ubiquitously overexpressed in 24 cancer classes. *Genomics* **2004**, *84*, 1014–1020. [CrossRef]

167. DeBerardinis, R.J.; Lum, J.J.; Hatzivassiliou, G.; Thompson, C.B. The biology of cancer: Metabolic reprogramming fuels cell growth and proliferation. *Cell Metab.* **2008**, *7*, 11–20. [CrossRef]

168. Welch, J.S.; Ricote, M.; Akiyama, T.E.; Gonzalez, F.J.; Glass, C.K. PPARgamma and PPARdelta negatively regulate specific subsets of lipopolysaccharide and IFN-gamma target genes in macrophages. *Proc. Natl. Acad. Sci. USA* **2003**, *100*, 6712–6717. [CrossRef]

169. Kang, K.; Reilly, S.M.; Karabacak, V.; Gangl, M.R.; Fitzgerald, K.; Hatano, B.; Lee, C.H. Adipocyte-derived Th2 cytokines and myeloid PPARdelta regulate macrophage polarization and insulin sensitivity. *Cell Metab.* **2008**, *7*, 485–495. [CrossRef]

170. Thulin, P.; Wei, T.; Werngren, O.; Cheung, L.; Fisher, R.M.; Grandér, D.; Corcoran, M.; Ehrenborg, E. MicroRNA-9 regulates the expression of peroxisome proliferator-activated receptor δ in human monocytes during the inflammatory response. *Int. J. Mol. Med.* **2013**, *31*, 1003–1010. [CrossRef]

171. Gallardo-Soler, A.; Gómez-Nieto, C.; Campo, M.L.; Marathe, C.; Tontonoz, P.; Castrillo, A.; Corraliza, I. Arginase I induction by modified lipoproteins in macrophages: A peroxisome proliferator-activated receptor-gamma/delta-mediated effect that links lipid metabolism and immunity. *Mol. Endocrinol.* **2008**, *22*, 1394–1402. [CrossRef] [PubMed]

172. Bogie, J.F.; Jorissen, W.; Mailleux, J.; Nijland, P.G.; Zelcer, N.; Vanmierlo, T.; Van Horssen, J.; Stinissen, P.; Hellings, N.; Hendriks, J.J. Myelin alters the inflammatory phenotype of macrophages by activating PPARs. *Acta Neuropathol. Commun.* **2013**, *1*, 43. [CrossRef] [PubMed]

173. Mukundan, L.; Odegaard, J.I.; Morel, C.R.; Heredia, J.E.; Mwangi, J.W.; Ricardo-Gonzalez, R.R.; Goh, Y.P.; Eagle, A.R.; Dunn, S.E.; Awakuni, J.U.; et al. PPAR-delta senses and orchestrates clearance of apoptotic cells to promote tolerance. *Nat. Med.* **2009**, *15*, 1266–1272. [CrossRef]

174. Tanaka, T.; Tahara-Hanaoka, S.; Nabekura, T.; Ikeda, K.; Jiang, S.; Tsutsumi, S.; Inagaki, T.; Magoori, K.; Higurashi, T.; Takahashi, H.; et al. PPARβ/δ activation of CD300a controls intestinal immunity. *Sci. Rep.* **2014**, *4*, 5412. [CrossRef] [PubMed]

175. Adhikary, T.; Wortmann, A.; Schumann, T.; Finkernagel, F.; Lieber, S.; Roth, K.; Toth, P.M.; Diederich, W.E.; Nist, A.; Stiewe, T.; et al. The transcriptional PPARβ/δ network in human macrophages defines a unique agonist-induced activation state. *Nucleic Acids Res.* **2015**, *43*, 5033–5051. [CrossRef] [PubMed]

176. Schumann, T.; Adhikary, T.; Wortmann, A.; Finkernagel, F.; Lieber, S.; Schnitzer, E.; Legrand, N.; Schober, Y.; Nockher, W.A.; Toth, P.M.; et al. Deregulation of PPARβ/δ target genes in tumor-associated macrophages by fatty acid ligands in the ovarian cancer microenvironment. *Oncotarget* **2015**, *6*, 13416–13433. [CrossRef]

177. Xu, H.E.; Lambert, M.H.; Montana, V.G.; Parks, D.J.; Blanchard, S.G.; Brown, P.J.; Sternbach, D.D.; Lehmann, J.M.; Wisely, G.B.; Willson, T.M.; et al. Molecular recognition of fatty acids by peroxisome proliferator-activated receptors. *Mol. Cell* **1999**, *3*, 397–403. [CrossRef]

178. Al Yacoub, N.; Romanowska, M.; Krauss, S.; Schweiger, S.; Foerster, J. PPARdelta is a type 1 IFN target gene and inhibits apoptosis in T cells. *J. Invest. Dermatol.* **2008**, *128*, 1940–1949. [CrossRef]

179. Mothe-Satney, I.; Murdaca, J.; Sibille, B.; Rousseau, A.S.; Squillace, R.; Le Menn, G.; Rekima, A.; Larbret, F.; Pelé, J.; Verhasselt, V.; et al. A role for Peroxisome Proliferator-Activated Receptor Beta in T cell development. *Sci. Rep.* **2016**, *6*, 34317. [CrossRef]

180. Jakobsen, M.A.; Petersen, R.K.; Kristiansen, K.; Lange, M.; Lillevang, S.T. Peroxisome proliferator-activated receptor alpha, delta, gamma1 and gamma2 expressions are present in human monocyte-derived dendritic cells and modulate dendritic cell maturation by addition of subtype-specific ligands. *Scand. J. Immunol.* **2006**, *63*, 330–337. [CrossRef]

181. Kapoor, A.; Shintani, Y.; Collino, M.; Osuchowski, M.F.; Busch, D.; Patel, N.S.; Sepodes, B.; Castiglia, S.; Fantozzi, R.; Bishop-Bailey, D.; et al. Protective role of peroxisome proliferator-activated receptor-β/δ in septic shock. *Am. J. Respir. Crit. Care Med.* **2010**, *182*, 1506–1515. [CrossRef] [PubMed]

182. Dunn, S.E.; Bhat, R.; Straus, D.S.; Sobel, R.A.; Axtell, R.; Johnson, A.; Nguyen, K.; Mukundan, L.; Moshkova, M.; Dugas, J.C.; et al. Peroxisome proliferator-activated receptor delta limits the expansion of pathogenic Th cells during central nervous system autoimmunity. *J. Exp. Med.* **2010**, *207*, 1599–1608. [CrossRef] [PubMed]

183. Paterniti, I.; Mazzon, E.; Riccardi, L.; Galuppo, M.; Impellizzeri, D.; Esposito, E.; Bramanti, P.; Cappellani, A.; Cuzzocrea, S. Peroxisome proliferator-activated receptor β/δ agonist GW0742 ameliorates cerulein- and taurocholate-induced acute pancreatitis in mice. *Surgery* **2012**, *152*, 90–106. [CrossRef]

184. Djouad, F.; Ipseiz, N.; Luz-Crawford, P.; Scholtysek, C.; Krönke, G.; Jorgensen, C. PPARβ/δ: A master regulator of mesenchymal stem cell functions. *Biochimie* **2017**, *136*, 55–58. [CrossRef] [PubMed]

185. Michelet, X.; Dyck, L.; Hogan, A.; Loftus, R.M.; Duquette, D.; Wei, K.; Beyaz, S.; Tavakkoli, A.; Foley, C.; Donnelly, R.; et al. Metabolic reprogramming of natural killer cells in obesity limits antitumor responses. *Nat. Immunol.* **2018**, *19*, 1330–1340. [CrossRef] [PubMed]

186. Fan, W.; Waizenegger, W.; Lin, C.S.; Sorrentino, V.; He, M.X.; Wall, C.E.; Li, H.; Liddle, C.; Yu, R.T.; Atkins, A.R.; et al. PPARδ Promotes Running Endurance by Preserving Glucose. *Cell Metab.* **2017**, *25*, 1186–1193.e1184. [CrossRef]

187. Narkar, V.A.; Downes, M.; Yu, R.T.; Embler, E.; Wang, Y.X.; Banayo, E.; Mihaylova, M.M.; Nelson, M.C.; Zou, Y.; Juguilon, H.; et al. AMPK and PPARdelta agonists are exercise mimetics. *Cell* **2008**, *134*, 405–415. [CrossRef]

188. Wang, Y.X.; Lee, C.H.; Tiep, S.; Yu, R.T.; Ham, J.; Kang, H.; Evans, R.M. Peroxisome-proliferator-activated receptor delta activates fat metabolism to prevent obesity. *Cell* **2003**, *113*, 159–170. [CrossRef]

Article

In Vitro-Generated Hypertrophic-Like Adipocytes Displaying *PPARG* Isoforms Unbalance Recapitulate Adipocyte Dysfunctions In Vivo

Marianna Aprile [1,*,†,‡], Simona Cataldi [1,†], Caterina Perfetto [1], Maria Rosaria Ambrosio [2], Paola Italiani [3], Rosarita Tatè [1], Matthias Blüher [4], Alfredo Ciccodicola [1,5] and Valerio Costa [1,*,‡]

1   Institute of Genetics and Biophysics "Adriano Buzzati-Traverso," CNR, Via P. Castellino 111, 80131 Naples, Italy; simona.cataldi@igb.cnr.it (S.C.); caterina.perfetto@igb.cnr.it (C.P.); rosarita.tate@igb.cnr.it (R.T.); alfredo.ciccodicola@igb.cnr.it (A.C.)
2   Department of Translational Medicine, University of Naples "Federico II" & URT "Genomic of Diabetes," Institute of Experimental Endocrinology and Oncology "G. Salvatore," CNR, Via Pansini 5, 80131 Naples, Italy; mariarosaria.ambrosio@unina.it
3   Institute of Biochemistry and Cell Biology CNR, Via P. Castellino 111, 80131 Naples, Italy; p.italiani@ibp.cnr.it
4   Department of Medicine, University of Leipzig, 4289 Leipzig, Germany; matthias.blueher@medizin.uni-leipzig.de
5   Department of Science and Technology, University of Naples "Parthenope," 80131 Naples, Italy
*   Correspondence: marianna.aprile@igb.cnr.it (M.A.); valerio.costa@igb.cnr.it (V.C.)
†   Co-first authors.
‡   Lead Contacts.

Received: 30 April 2020; Accepted: 19 May 2020; Published: 21 May 2020

Abstract: Reduced neo-adipogenesis and dysfunctional lipid-overloaded adipocytes are hallmarks of hypertrophic obesity linked to insulin resistance. Identifying molecular features of hypertrophic adipocytes requires appropriate in vitro models. We describe the generation of a model of human hypertrophic-like adipocytes directly comparable to normal adipose cells and the pathologic evolution toward hypertrophic state. We generate in vitro hypertrophic cells from mature adipocytes, differentiated from human mesenchymal stem cells. Combining optical, confocal, and transmission electron microscopy with mRNA/protein quantification, we characterize this cellular model, confirming specific alterations also in subcutaneous adipose tissue. Specifically, we report the generation and morphological/molecular characterization of human normal and hypertrophic-like adipocytes. The latter displays altered morphology and unbalance between canonical and dominant negative (PPARGΔ5) transcripts of *PPARG*, paralleled by reduced expression of PPARγ targets, including *GLUT4*. Furthermore, the unbalance of PPARγ isoforms associates with *GLUT4* down-regulation in subcutaneous adipose tissue of individuals with overweight/obesity or impaired glucose tolerance/type 2 diabetes, but not with normal weight or glucose tolerance. In conclusion, the hypertrophic-like cells described herein are an innovative tool for studying molecular dysfunctions in hypertrophic obesity and the unbalance between PPARγ isoforms associates with down-regulation of *GLUT4* and other PPARγ targets, representing a new hallmark of hypertrophic adipocytes.

Keywords: hypertrophic adipocytes; *PPARG* isoforms; *PPARG* splicing; dominant-negative isoform; in vitro adipocytes; adipogenesis; hypertrophic obesity; insulin-resistance

1. Introduction

The individual obesity-related risk for metabolic complications associates with storage capability of adipose tissue (AT). Energy buffering in the AT can occur either by tissue hyperplasia (i.e., de novo formation of new lipid-storing adipose cells) or hypertrophy of pre-existing adipocytes. According to the "overflow hypothesis", exceeding the storage capability of adipose tissue leads to ectopic lipid accumulation, insulin resistance (IR), and type 2 diabetes (T2D) [1,2]. Consequently, similar metabolic consequences occur in conditions of deficiency and the excess of body fat, i.e., in lipodystrophies and obesity, respectively [3,4]. Particularly, hypertrophic obesity is associated with the reduced capacity to recruit and differentiate precursor cells into mature adipocytes [5–8]. Therefore, limited AT expandability, along with the balance between hyperplasia and hypertrophy, are key factors to clarify why not all obese individuals develop metabolic complications.

However, identifying the determinants accounting for the pathologic shift toward AT hypertrophy requires appropriate in vitro models able to recapitulate both the physiological processes governing adipocyte differentiation and the pathological causes of cells' hypertrophy. In this regard, murine pre-adipocytes (i.e., 3T3-L1) have been widely used to study adipogenesis [9] as well as to generate hypertrophic cells in vitro [10]. Nevertheless, obvious differences between human and murine metabolism and physiology indicate the need to use more appropriate human models. Indeed, human primary pre-adipocytes [11–13] and adult mesenchymal stem cells—isolated from bone marrow, AT, umbilical cord and other tissues—represent the most reliable sources of cells able to differentiate toward the adipogenic lineage. The former cell type displays a proliferation/differentiation capacity that is strictly donor- and depot-related, showing unpredictable variability [11,14]. The latter displays low variability and high expansion/propagation capacity—especially for AT-derived cells—and are particularly useful for exploring early stages of differentiation, including the adipogenic commitment [15].

In this regard, we recently used a commercially available *hTERT*-immortalized cell line, i.e., AT-derived human mesenchymal stem cells (hMSCs), as model of human adipogenesis, in which we determined the negative impact of PPARγΔ5 isoform on PPARγ transcriptional activity and on adipocyte differentiation [16]. Together with the finding that PPARγΔ5 positively correlates with BMI and T2D [16], our results prompted us to evaluate whether the alteration of *PPARG* splicing is a feature of hypertrophic obesity.

Corroborating this hypothesis, our work reveals significant correlations between the expression of the different *PPARG* isoforms, subcutaneous adipocytes' size and the inducible glucose transporter Glut4 (i.e., *SLC2A4* gene) in human subcutaneous adipose tissue (SAT). However, the intrinsic inter-individual variability and methodological issues related to adipocyte diameter calculation [17] represent sources of bias threatening the reliability and reproducibility of the results. Indeed, according to our previous study revealing highly variable PPARGΔ5 expression in human SAT, and considering the presence of complex feedback mechanisms regulating different *PPARG* isoforms [16,18,19], unpredictable genetic/environmental factors may affect *PPARG* expression and splicing in vivo. Therefore, it is glaring the need of a cellular model offering a direct comparison between normal and hypertrophic adipocytes and able to avoid—or at least reduce—any masking effect due to multiple unpredictable factors.

Thus, to recapitulate in vitro in a unique and highly reproducible model all the main molecular hallmarks of human hypertrophic AT, we setup a protocol for generating (for the first time, to the best of our knowledge) human hypertrophic-like adipocytes (HAs) that can be directly compared to mature cells (MAs) without confounding variables. Hence, in this work we report an accurate morphological, ultrastructural and transcriptional analysis of hMSCs differentiating into mature adipocytes, providing also evidence that the hypertrophic state associates with marked alterations in cell morphology, gene expression and *PPARG* splicing. This cellular model represents a versatile tool for studying structural remodeling and altered functionality of adipose cells during their pathologic evolution toward the hypertrophic state, as well as to test short- and long-term pharmacological treatments. Remarkably, analyzing this cellular model we confirmed that—similarly to large SAT adipocytes in vivo—hypertrophic-like cells display higher PPARGΔ5/cPPARG ratio and that such unbalance associates with marked deregulation in the network of *PPARG*-regulated genes, including those responsible of glucose transport and metabolism, insulin signaling and lipid droplet remodeling.

2. Materials and Methods

2.1. Human Samples

RNAs from subcutaneous adipose tissue biopsies were available in our laboratory from a previous study [16]. Samples were obtained from a clinically well-characterized German cohort of individuals (n = 94; mean age = 55.5 ± 16.5 y.o.; mean BMI = 35.4 ± 11.8) [20,21] undergoing bariatric surgery. The study was carried out in accordance with the Declaration of Helsinki, the Bioethics Convention (Oviedo), and EU Directive on Clinical Trials (Directive 2001/20/ EC) and approved by the University of Leipzig (approval numbers: 159-12-21052012 and 017-12-23012012). Random selection of samples, as well as exclusion criteria and classifications of individuals were applied as described in Aprile et al. (2018) [16]. Clinical and biochemical parameters were provided by Prof. Blüher's unit, including visceral and subcutaneous mean and maximum diameter analyzed by Multisizer (Table 1).

Table 1. Characteristics of the study participants.

	Women	Men
n	51	35
Age (years)	53.4 ± 16.6	58.4 ± 16.3
Body weight (kg)	99.5 ± 34.4	109 ± 44.9
BMI (kg/m^2)	36.5 ± 12	34.1 ± 12.3
Body fat (%)	35 ± 11.7	30.7 ± 9.5
Visceral fat area (cm^2)	165.5 ± 119.6	172.9 ± 134
Subcutaneous fat area (cm^2)	471 ± 492.5	418.5 ± 333.3
Waist circumference (cm)	114 ± 33.1	120.3 ± 24
FPG (mmol/L)	5.9 ± 1.2	5.7 ± 0.9
FPI (pmol/L)	127.5 ± 133.1	81.4 ± 89.8
HbA1c (%)	5.9 ± 0.8	5.9 ± 0.62
Clamp GIR (µmol/kg/min)	79 ± 35	75.2 ± 32
Cholesterol (mmol/L)	5.1 ± 0.75	4.9 ± 1.02
HDL-Cholesterol (mmol/L)	1.2 ± 0.3	1.1 ± 0.3
LDL-Cholesterol (mmol/L)	3.6 ± 1.2	3.5 ± 1.2
Triglycerides (mmol/L)	1.42 ± 0.36	1.9 ± 1.6
Free fatty acids (mmol/L)	0.44 ± 0.38	0.47 ± 0.4
hsCRP (mg/L)	12.3 + 14.8	11.5 ± 14.1
IL-6 (pg/mL)	4.2 ± 4.1	3.3 ± 4.4
ALAT (µkat/L)	0.8 ± 1.1	0.67 ± 0.7
ASAT (µkat/L)	0.7 ± 0.85	0.63 ± 0.5
GGT (µkat/L)	1.9 ± 3.5	1.5 ± 2.5
Adiponectin (µg/mL)	9.1 ± 6.6	5 ± 3**
Leptin (pg/mL)	40 ± 20	18.5 ± 11.5**
Mean subcutaneous adipocyte diameter (µm)	110.6 ± 11.6	107.6 ± 9.3
Mean visceral adipocyte diameter (µm)	100 ± 7.4	97.6 ± 5.7

Data are means ± SD. ** p < 0.01 for gender differences. Abbreviations: ALAT—Alanine-Aminotransferase; ASAT—Aspartate-Aminotransferase; BMI—body mass index; FPG—fasting plasma glucose; FPI—fasting plasma insulin; GGT—gamma- glutamyl transpeptidase; HbA1c—glycated haemoglobin; HDL—high density lipoproteins; hsCRP—high sensitivity C-reactive protein; IL-6—Interleukin 6; LDL—low density lipoproteins.

2.2. Cell Lines and Cultures

hTERT-immortalized adipose derived mesenchymal stem cells (hMSCs) were purchased from American Type Culture Collection (ATCC SCRC-4000; Virginia, USA). Cells were cultured in DMEM-F12 (1:1) supplemented with 10% South American Fetal Bovine Serum (FBS), 2 mM glutamine, 30 units/mL penicillin, 30 mg/mL streptomycin, and maintained in humidified atmosphere of 5% CO_2 at 37 °C. Media, sera, and antibiotics for cell culture were from Thermo Fisher Scientific (Waltham, MA, USA).

2.3. In Vitro Differentiation of Mature and Hypertrophic-Like Adipocytes

hMSCs were pulsed to differentiate in mature adipocytes as previously reported [16]. Briefly, cells between 5 and 12 passages have been plated at density of 3–4 × 10^3/cm^2 and induced toward adipocyte differentiation after reaching maximum confluence (48–72 h after plating). Cells at confluence were treated with Adipogenic Induction Mix (AIM; constituted by 850 nM insulin, 10 mM dexamethasone, 0.5 mM 3-isobutyl-1- methylXanthine, 33 mM biotin, 17 mM pantothenate, 1 mM rosiglitazone), and adipogenic maintaining mix (AMM; consisting of 850 nM insulin and 1 mM rosiglitazone). AIM and AMM were alternatively used for three days until 19–21 day, considered the terminal point of the process. Additionally, alternative differentiation protocols were tested adding the Bone Morphogenic Protein 4 (BMP4) bioactive protein (10, 20, and 50 ng/mL) to the AIM, or by using different FBS formulations (i.e., FBS qualified Australia and South American origin, Thermo Fisher Scientific, Waltham, MA, USA). Afterward, hypertrophic-like cells were obtained by treating mature adipocytes for 12 days with AMM mix supplemented with fatty acids i.e., palmitate, oleate, or both (350 µM). Such concentration reflects the pathological levels of fatty acids (200–375 µM) adipose cells of obese individuals are exposed to. Mixes were added to the cells every 3 days, and hypertrophic-like adipocytes were obtained within 32 days from adipogenesis induction.

2.4. Immunofluorescence Microscopy

For immunofluorescence analysis, hMSCs at different time points of adipocyte differentiation, grown on coverslips, were fixed with 4% formaldehyde for 15 min and washed in PBS. After washing, the cells were incubated with WGA 632/647 (red; 5 µg/mL) as membrane marker following manufacturer's instructions. Afterward, cells were permeabilized with PBS/10% FBS/0.1% Triton X-100 for 5 min. Lipid droplets were marked with Bodipy 493/503 (green; 5 µL/mL) and cell nuclei were counterstained with DAPI (blue; 1 mg/mL). Reagents were purchased from Thermo Fisher Scientific (Waltham, MA, USA). Cells on coverslips were mounted with fluorescent mounting medium, and fluorescent labeling was examined using an A1 Resonance Plus confocal microscope (Nikon, Melville, NY, USA) and inverted (Leica DMI6000B) microscopy. Z-Stack imaging was performed by confocal microscopy for reconstruction of 3D images. Nikon Imaging Software (NIS) Elements Advanced Research software (version 4.50.00) was used for images acquisition/elaboration. All images were captured using a 20× Plan Apo lambda objective (1024 × 1024 pixels), numerical aperture 0.75, pinhole 1.2 AU, and exposure 6.2 s per pixel dwell. Detector sensitivity (gain) and laser power settings were kept the same for all collected images to allow for comparisons between images.

2.5. Cell Count and Oil Red O Staining for Quantifying Adipocyte Differentiation

Confocal microscopy images were processed by Image J [22] and analyzed for determining the percentage of differentiated cells. For each field, total cell number was determined by segmentation of nuclei stained with DAPI. Differentiated cells were identified and counted basing on the presence of lipid droplets stained with Bodipy 493/503. The percentage of cells that underwent adipogenic differentiation was calculated as cells positively stained with Bodipy ÷ the total number of cells (nuclei) × 100. Differentiation rate was calculated in five independent experiments and a total of ~6000 cells were analyzed. Additionally, adipocyte differentiation was estimated measuring lipid accumulation by Oil Red O staining [19,23]. Optical density determination at 510 nm was assessed by VICTOR Multilabel Plate Reader (Perkin Elmer, Massachusetts, USA), and the corrected subtracting background signal was determined by not specifically staining the undifferentiated cells.

2.6. Analysis of Adipocyte Size and Lipid Droplets

Cellular size and lipid droplet were analyzed in mature and hypertrophic-like adipocytes. Confocal Z-stack images were processed by open-source program Image J [22]. Adipocyte area was automatically measured after manual definition of cell perimeters. Lipid droplet number and size were analyzed by Image J macro "*MRI_Lipid_Droplets_Tool.ijm*," which applies a Gaussian filter to the input images and an automatic threshold (percentile method) to remove artifacts from the mask of the droplets image, finally separating the touching droplets by a binary watershed transform. Therefore, starting from Z-stack images at focal plane with higher number of maximum diameters, lipid droplets marked with Bodipy were automatically selected and segmented by size threshold setting to 2 pixel/microns (expected size of smaller droplets). Multiple Z-stack images were used for reconstructing 3D projections, and for each cell/lipid droplet the focal plane with maximum size/diameter was considered. For lipid droplets analysis, individual parameters for accurate 3D surface selection were manually adjusted, increasing the accuracy of geometrical setting of touching droplets and background removal. Maximum diameter, related area and optical density were measured for each lipid droplet. Similarly, nuclei segmentation was applied for cell number counting, setting the size threshold to ≈50 pixel/microns. Such analysis was performed on both mature and hypertrophic-like adipocytes, calculating the average of total lipid area and the number of lipid droplets for cell and the mean area of a droplet.

2.7. Transmission Electron Microscopy

The hMSCs at different time points of adipocyte differentiation and upon the adipogenic induction were fixed in 2% glutaraldehyde, post-fixed in 1% osmium tetraoxide, dehydrated by being passed through a graded ethanol series, and embedded in Poly/Bed 812 resin (Polyscience, Warrington, PA, USA). The embedded samples were cut using a Leica ultracut UCT ultramicrotome (Leica Microsystems, Wetzlar, Germany) into ultrathin sections (50 nm thickness) and, to increase the contrast of the samples, an additional staining with uranyl acetate was performed. Finally, the samples attached to Formvar/carbon copper grids were observed under a model JEM-1011 (JEOL, Tokyo, Japan) transmission electron microscope using an accelerating voltage of 100 kV. Low- and high-magnification images were captured by iTEM software (Olympus Soft Imaging System, Münster, Germany). At least 10 different microscopic fields on multiple thin sections from different independent samples were observed and captured to obtain good confidence.

2.8. RNA Extraction, RT-PCR and qPCR

Total RNA was isolated using TRIzol Reagent (Life Technologies, Carlsband, CA, USA) according to manufacturer's instructions. Quantification and purity of RNA was evaluated by NanoDrop spectrophotometer (Life Technologies, Carlsband, CA, USA). The synthesis of cDNA was performed with a High Capacity cDNA Reverse Transcription kit (Invitrogen, Carlsband, CA, USA) according to the manufacturer's instructions. Expression analyses were performed by RT-PCR and qPCR techniques. Gene specific primers were designed using Oligo 4.0 program and listed in Table S1. RT-PCR products were amplified using MyTaq DNA Polymerase (Bioline, Memphis, Tennesse, USA) and analyzed by electrophoresis on agarose gel. PowerUp Sybr Green Master Mix (Thermo Fisher Scientific, Waltham, Massachusetts, USA) was used for qPCR expression analysis on CFX Connect Detection System (Bio-Rad, Hercules, CA, USA) according to manufacturer's instructions. Relative mRNA expression was measured by $2^{-\Delta\Delta Ct}$ method. *PPIA* and *RPS23* were selected as housekeeping genes in hMSCs and SAT biopsies, respectively, evaluating the expression stability of at least three candidate housekeeping genes among *ACTB, HPRT, GAPDH, PPIA*, and *RPS23*. All reactions were performed in duplicates in at least three independent experiments.

2.9. Western Blot

Whole-cell lysates were obtained using RIPA lysis buffer supplemented with Halt Protease and Phosphatase Inhibitor Cocktail (Thermo Fisher Scientific, Waltham, Massachusetts, USA) and quantified by Bradford Assay Reagent (Bio-Rad, Hercules, CA, USA). For each sample, 40-60 mg of proteins were used for western blot analysis. According to manufacturer's instructions, primary antibodies were used to different dilutions: anti-PPARγ (1:1000, Cell Signaling Technology, Danvers, Massachusetts, USA), anti-Glut4, anti-Adiponectin, anti-aP2, anti-Irs2 (1:500, Elabscience, Houston, Texas). Anti-Hsp90 (1:5000; Origene, Rockville, Maryland, USA) was used as a loading control antibody. Secondary anti-IgG (goat, mouse, and rabbit) antibodies were used at dilution 1:5000 (Bio-Rad, Hercules, CA, USA). Pierce ECL Western Blotting Substrate (Thermo Fisher Scientific, Waltham, Massachusetts, USA) was used for detection of immunoreactive bands. Quantification of protein levels (pixel density) was performed by *GelQuant.NET* software (www.biochemlabsolutions.com). Intensity values were normalized on Hsp90 expression and reference sample (i.e., the first time point having detectable levels).

2.10. Flow Cytometry Analysis

The expression of mesenchymal markers was analyzed by flow cytometry in hMSCs (T = 0 h), hMSCs-derived adipocytes (T = 20 day), and hypertrophic-like adipocytes (T = 32 day). The cells were incubated with PE-conjugated anti-CD73 antibody and FITC-conjugated anti-CD90 antibody, as well as with dye/isotype-matched antibodies (all from BD Biosciences, USA). The incubation was carried for 30 min at 4 °C in a dark environment. Afterward, unbound antibodies were washed out and the samples were processed by a BD FACS canto II (BD Biosciences, San Jose, CA) and analyzed using BD FACSDiva software. For each sample, 10^4 events were acquired. Cells were counted and compared with the signals of the corresponding antibody isotype controls.

2.11. ELISA

Levels of the inflammatory cytokine IL-6 were determined by ELISA (R&D Systems, Minneapolis, USA) in cell-free supernatants according to the manufacturer's instructions. Absorbance of assay wavelength was measured at 450 nm using a Cytation 3 imaging reader (BioTek, Winooski, VT, USA).

2.12. Quantification and Statistical Analysis

Student's t test (one sample or two samples test; two tailed) was used for assessing statistical significance of differences in lipid accumulation (Oil Red O staining), cellular and lipid droplet area (3D analysis) between mature and hypertrophic-like adipocytes, as well as in gene expression assays (qPCR) for hMSCs at different stages of adipocyte differentiation. All assays were performed at least in triplicate. For each assay, the number of replicates, SD or SEM and statistical significance are reported in figure legends. The Kolmogorv–Smirnov test was used to analyze gene expression differences (qPCR) in SAT of patients. p value (p) ≤ 0.05 was considered significant. Statistical analysis of flow cytometry data was performed by BD FACSDiva software according to manufacturer's instructions. Linear models were fit by *lm* function in R using the equation "result = lm(feature ~ cond, data, na.action = na.omit). For each specific analysis, *feature* was the response variable, *cond* was the regressor, and *data* was the dataframe containing expression data and clinical parameters. Missing fields (*na*) were omitted from regression. Residuals, Coefficients, Residual standard errors, Multiple R-squared, Adjusted R-squared, as well as F-statistic and p-values (ANOVA) were analyzed by the *summary* function. Pairwise correlations between couple of variables were carried out by the *cor* function in R language and Pearson's (r) coefficient calculated as default parameter. Custom scripts in R language (using ggplot2) were used to generate the correlation, scatter, violin, and box plots.

Detailed information about all reagents and resources are provided in Table S2.

3. Results and Discussion

3.1. Unbalance of PPARG Isoforms in Patients with Hypertrophic Obesity

The subcutaneous adipose tissues of obese individuals and patients with T2D display reduced PPARγ activity [24] and increased relative amount of canonical and dominant negative transcripts (i.e., higher PPARGΔ5/cPPARG ratio) [16]. PPARGΔ5 mRNA levels positively correlate with BMI, and high levels of this dominant negative isoform reduce PPARγ transactivation ability and the adipogenic capacity of precursor cells [16]. Hence, considering that defective adipogenesis in adult AT and functional impairment of PPARγ are hallmarks of hypertrophic obesity, we decided to explore whether the pattern of *PPARG* splicing is modified in overweight and obese individuals, also considering the presence of related metabolic complications—i.e., the presence of impaired glucose tolerance (IGT) or T2D—and the adipocyte size.

In this regard, we previously reported that PPARGΔ5 mRNA levels in the SAT are higher in overweight/obese than in individuals with normal weight [16]. Here, we observe that cPPARG and PPARGΔ5 mRNA levels are positively correlated between them only in the SAT of normal weight individuals ($n = 20$) and not in overweight/obese patients ($n = 74$, Figure 1A), suggesting that physiological balance between *PPARG* isoforms occurs exclusively in "healthy" AT. Indeed, our previous results indicated that *PPARG* splicing is modified in presence of metabolic disorders as obesity and T2D [16]. To address whether the unbalance between canonical and dominant negative *PPARG* isoforms associates with other relevant clinical parameters, we used PPARGΔ5 and cPPARG relative mRNA abundance from a previously analyzed German cohort [16] and compared their expression with respect to the size of adipocytes in the SAT ($n = 86$). Interestingly, cPPARG expression inversely correlates with the size of subcutaneous adipocytes (Figure 1B), whereas the ratio between the transcripts has a positive trend (Figure 1C, Figure S1A), compatible with the impaired metabolic profile of large adipocytes. Although it is not possible to define unhealthy hypertrophic adipose cells by means of their size, we used the mean diameter of subcutaneous adipocytes (115μm) to stratify patients in two groups—"Low Mean Diameter" (LMD; $n = 63$) and "High Mean Diameter" (HMD; $n = 23$)—in line with the work of Stenkula and Erlanson-Albertsson (2018) [17]. As shown in Figure 1D, PPARGΔ5/cPPARG ratio is higher in the SAT of HMD (vs. LMD), mostly because of a pronounced drop in cPPARG expression in this group (Figure S1B). These data provide evidence of *PPARG* isoforms unbalance in large adipocytes compared to small/medium-sized ones, suggesting a contribution of both *PPARG* expression and alternative splicing in compromising the metabolic homeostasis of these cells. Finally, extending the analysis to other clinical and biochemical parameters (Figure S1C), we disclosed that PPARGΔ5 negatively correlates with low-density lipoprotein (LDL) cholesterol levels, whereas cPPARG does not (Figure 1E). Conversely, only cPPARG negatively correlates with leptin serum levels (Figure 1E). Although these data support the unbalance of *PPARG* isoforms in the SAT of individuals with enlarged adipocytes, inter-individual variability, adipocyte heterogeneity in AT, or technical drawbacks (e.g., measurement of adipocyte size on histological specimen) could affect results' consistency.

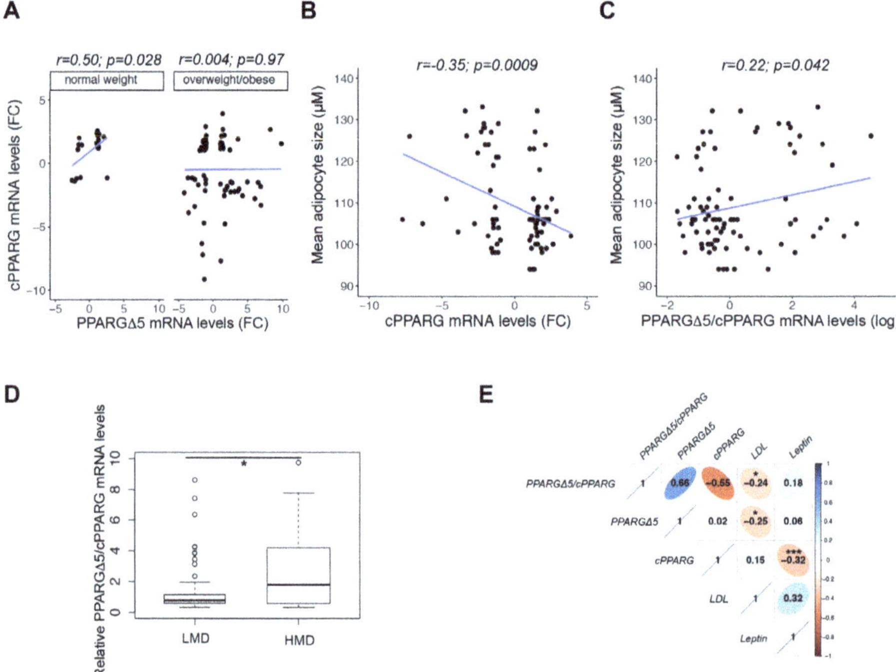

Figure 1. PPARGΔ5/cPPARG ratio correlates with adipocyte size, leptin and low density lipoprotein (LDL) cholesterol serum levels. Expression values of PPARGΔ5 and cPPARG have been previously measured in subcutaneous adipose tissue (SAT) of patients from a German cohort [16]. (**A**) Scatterplot reporting the correlation (linear regression analysis) between PPARGΔ5 and cPPARG expression levels in normal weight ($n = 20$) and overweight/obese individuals ($n = 74$). Pearson correlation coefficient (r) and p values (p) are shown. (**B–C**) Scatterplot resulting from regression analysis and indicating that cPPARG (B) and PPARGΔ5/cPPARG (C) levels oppositely correlate with mean diameter of subcutaneous adipocyte size ($n = 86$). Pearson correlation coefficient (r) and p values (p) are shown. (**D**) Boxplot showing PPARGΔ5/cPPARG levels in two subgroups, defined according to the mean diameter of subcutaneous adipocytes as "Low Mean Diameter" (LMD; mean diameter < 115μm, $n = 63$) and "High Mean Diameter" (HMD; mean diameter >115μm, $n = 23$) group. *$p < 0.05$. (**E**) Correlation plot indicating Pearson's correlation coefficient and p value (* $p < 0.05$, *** $p < 0.0001$) among cPPARG, PPARGΔ5, PPARGΔ5/cPPARG, leptin ($n = 61$), and LDL serum levels ($n = 59$).

3.2. From Human Mesenchymal Stem Cells to Mature Adipocytes

To study *PPARG* splicing alteration in large dysfunctional adipose cells, in a more controlled and unbiased system, we set up a new model of human hypertrophic-like adipocytes, directly comparable to starting mature cells. Hence, we first assessed different experimental conditions for optimizing the differentiation protocols previously used [16,25]. In particular, after plating increased number of cells (Figure S2A) we confirmed that high densities favor adipocyte differentiation [26], whereas low cell densities, or high passage numbers, reduce the differentiation rate. As we previously described [16], using a modified version of the protocol reported by Janderová et al. (2003) [25], adipocyte differentiation of hMSCs is completely reached in 19–21 days, alternating two different mixes. Additionally, we did not measure significant increase in the differentiation rate neither supplementing cells with recombinant Bone Morphogenic Protein 4 (BMP4; Figure S2B) (capable of triggering commitment of MSCs into pre-adipocytes [27]) nor using different FBS formulations (data not shown). Up to day 2 after the adipogenic induction, hMSCs show the typical fibroblast-like shape (Figure 2A).

Up to day 4 the cells maintain spindle-shape, even though cell circularity increases and immature droplets begin to be visible in the cytosol (Figure 2A). Therefore, according to the lipid droplets (LDs) formation model [28,29], this in vitro model requires about 4 days to complete triglyceride synthesis and to form oil lens within the endoplasmic reticulum and budding of lipid droplets into the cytosol. From day 4 to day 10, LDs number progressively increases and cells accentuate their characteristic whirlpool-like morphology. Afterward, both LDs number and volume markedly increase (Figure 2A,B). Particularly, on day 12, few clusters of "bunch of grapes"–like LDs became visible in a certain number of cells. Circularity is slightly emphasized, although cells maintain an elongated shape, similarly to mouse embryonic fibroblasts (MEFs) [30]. Around day 20, hMSCs appear evenly and appreciably differentiated into mature adipocytes, with an increased number of larger LDs and a marked whirling pattern with abundant clusters of "bunch of grapes"–like LDs (Figure 2A–C). In Figure 2B, hMSCs on day 12 and day 20—observed in dark field microscopy—show an increasing number of LDs and a typical swirled growth. Staining of neutral lipids combined to cell count by confocal microscopy ($n >$ 6000) indicates that about 80% of hTERT-MSCs are terminally differentiated ~20 days upon induction (Figure 2D). Then, FACS analysis revealed a significant decrease of hMSCs surface markers CD73 and CD90 in mature adipose cells compared to undifferentiated precursors (Figure S3). These results, verified by several independent experiments, confirm the very low variability in the differentiation rate for this in vitro model.

3.3. From Mature to Hypertrophic Adipocytes

Adipocyte hypertrophy is a feature of dysfunctional AT and tightly associates with IR and T2D onset. Molecular mechanisms causing adipocytes' dysfunctions in hypertrophic AT have not been completely clarified, and the cons and limitations of currently available in vitro models largely impede this task. Therefore, taking advantage of peculiar characteristics, such as the long propagation capacity, the good expansion and high population homogeneity of hMSCs, we set up a new protocol for generating in vitro human hypertrophic-like adipocytes—directly comparable to starting mature cells—that is useful to study the effects of adipocyte hypertrophy without confounding variables.

To this aim, hMSC-derived mature adipocytes were cultured with media containing saturated and/or monounsaturated fatty acids (MUFA) specifically selected for their high presence in human diet (i.e., palmitic and/or oleic acids), insulin and rosiglitazone for additional 12 days. Then, on day 32 (upon the adipogenic induction), cells supplemented with palmitate, oleate or their mixture were completely full of large LDs but no significant variation in lipid accumulation was measured among the different formulations of fatty acids (Figure 3A,B). Hence, to define which treatment induces the transcriptional alterations recapitulating at best the characteristics of hypertrophic adipocytes, we measured the expression of some selected genes. As shown in Figure 3C, HAs cultured in presence of palmitate (vs. fully differentiated MAs) revealed the most pronounced down-regulation of key adipogenic markers regulating AT homeostasis (*PPARG*, *ADIPOQ* and *FABP4*), LDs biogenesis (*PLIN1* and *PLIN2*) and insulin signaling (*SLC2A4* and *IRS2*). Overall, these results indicate prolonged palmitate treatment as the most reliable way to induce hypertrophic-like features in MAs differentiated from hMSCs. Therefore, the above-mentioned protocol was chosen for in vitro generation of HAs. Additionally, FACS analysis revealed, as expected, a marked decrease of CD73 and CD90 in hypertrophic-like cells (Figure S3). As evident by confocal microscopy, treating hMSCs-derived MAs for additional 12 days (up to day 32) with insulin, PPARγ agonist (rosiglitazone) and palmitic acid induces visible LDs enlargement and the progressive formation of giant LDs (Figure 3D). Cell circularity substantially increases on day 32, and lipid-overburden leads to evident reduction of the cytoplasmic layer surrounding LDs with compressive effects on cell nuclei (Figure 3D, Figure S6A). Both hMSC-derived MAs and HAs can be further propagated and show a very weak susceptibility to dedifferentiate in vitro. Indeed, HAs (T = 32 day) cultured in standard growth medium show only a modest reduction of lipid content and mild morphological variations even when cultured for additional 30 days (i.e., 62 days after adipogenesis' induction; Figure S4).

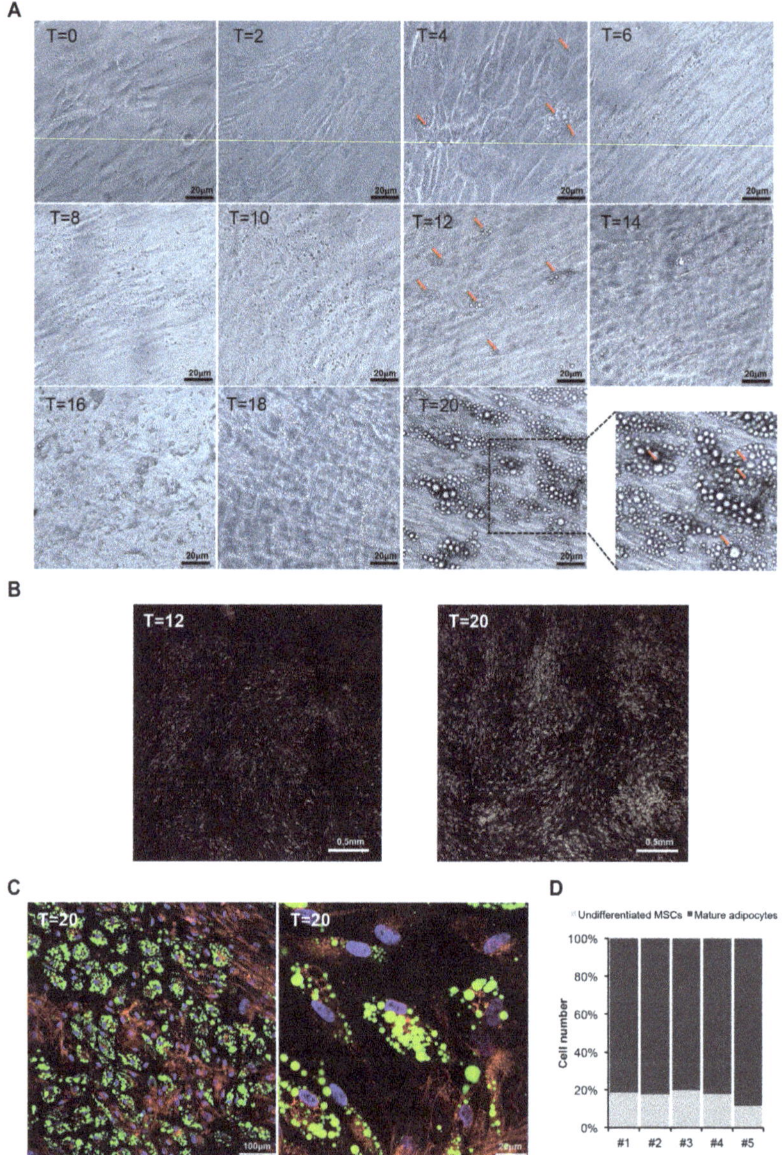

Figure 2. Human mesenchymal stem cells (hMSCs) are a reliable in vitro model of human adipocyte differentiation. (**A**) Representative phase-contrast images showing morphological changes of hMSCs along adipocyte differentiation, i.e., at starting point (T = 0 h), at different time points upon adipogenesis induction (T = 2, 4, 6, 8, 10, 12, 14, 16, 18 days) and at terminal differentiation (T = 20 days). Red arrows indicate some lipid droplets visible by optical microscopy during adipocyte differentiation (scale bar, 20 μm). (**B**) Representative images of hMSCs at 12 days and 20 days upon induction fixed with osmium tetraoxide and observed in dark field microscopy. The LDs are showed as white dots (scale bar, 1 mm). (**C**) Representative confocal microscopy images of hMSCs differentiated in mature adipocytes (T = 20 days):

nuclei in blue (DAPI), lipid droplets in green (Bodipy 495/503) and cell membranes in red (WGA 632/647). Clusters of "bunch of grapes"–like droplets are evident (scale bar, 100 µm left panel; scale bar, 20 µm right panel). (**D**) Bar graph indicating the percentage of undifferentiated and differentiated cells in five independent experiments measured analyzing confocal images of hMSCs at terminal adipogenic differentiation (T = 20 days). Total number of cells was calculated counting nuclei stained with DAPI, differentiated cells were identified by positive staining of lipid droplets (Bodipy 495/503), and the number of undifferentiated hMSCs was calculated as the difference between total and differentiated cells. A total of ~6000 cells were analyzed.

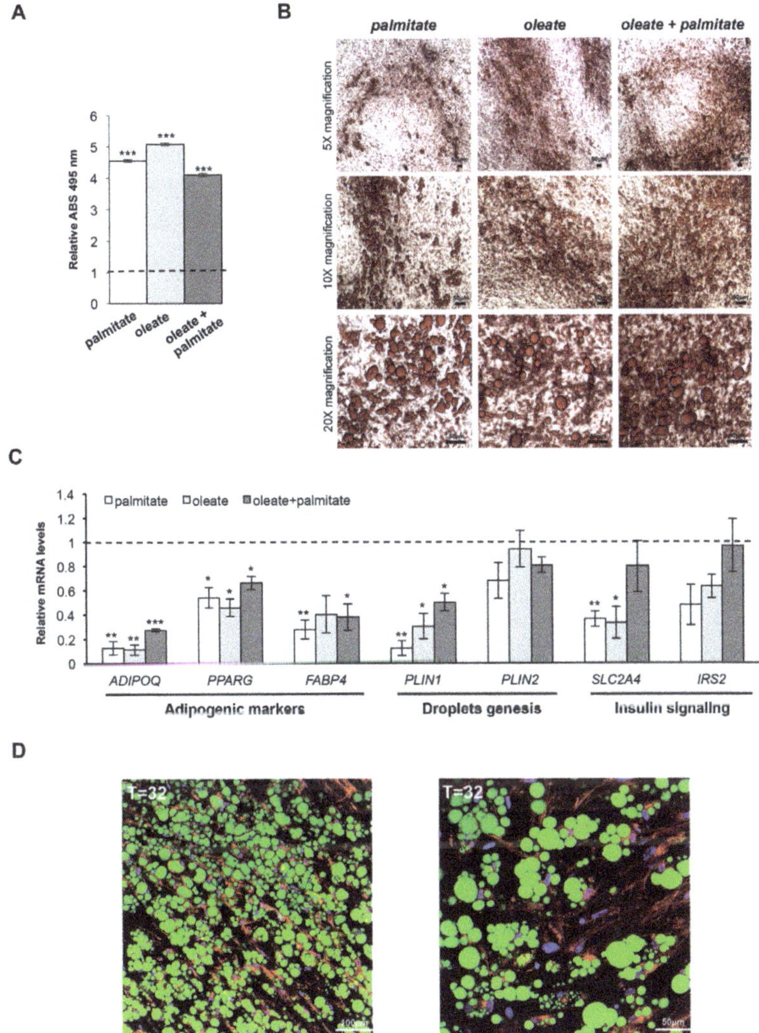

Figure 3. Hypertrophic-like cells generated from hMSCs-derived adipocytes. (**A**) Optical determination of lipid accumulation (Oil Red O staining) in hypertrophic-like adipose cells (HAs) generated from mature adipocytes (MAs)—differentiated in vitro by hMSCs—by supplementation of three different fatty acids mixes. Data are shown as mean ±SEM compared to mature adipocytes from three independent

experiments. ***p val ≤ 0.001. (**B**) Representative bright-field images of HAs—generated by three different fatty acids mixes—after lipid droplets staining by Oil Red O (scale bar, 50 μm). (**C**) Relative mRNA quantification (qPCR) of *PPARG* and key target genes in HAs generated by three different treatments. *PPIA* was used as reference gene. Data are reported as mean ±SEM vs. mature adipocytes (dotted line) from three independent experiments. * p val ≤ 0.05, ** p val ≤ 0.01 and *** p val ≤ 0.001. (**D**) Representative confocal microscopy images of HAs (T = 32d) generated by palmitate-containing mix. Nuclei were stained by DAPI (blue), lipid droplets, and cell membranes by Bodipy 495/503 (green) and WGA 632/647 (red), respectively (scale bar, 100 μm left panel; scale bar, 50 μm right panel).

Therefore, the long propagation, culturing time, and durable cell attachment of MAs and HAs make hMSCs an advantageous human cellular model for studying in vitro physiologic adipogenesis—from very early to late differentiation stages—as well as pathologic conditions such as the hypertrophic state. Our results also indicate the possibility to expose cells to different stimuli and experimental conditions even for long time, without the risk of dedifferentiation and/or cell detachment. To the best of our knowledge, we describe the first in vitro-generated model of human hypertrophic-like adipocytes, directly comparable to control mature adipocytes. It is a suitable in vitro model for studying the molecular mechanisms causing the functional defects observed in the adipocytes of hypertrophic obese patients.

3.4. Stepwise Expression from hMSCs to Hypertrophic-Like Adipocytes

The transcription factor PPARγ—master regulator of adipogenesis—belongs to the "second wave" of adipogenic factors, and therefore, it is expressed later than other factors in the adipogenic process [31]. Unexpectedly, undifferentiated cells show moderate basal expression of all *PPARG* transcripts, including the dominant negative PPARGΔ5 (Figure 4A, Figure S5A). Similar to our previous finding in human primary adipocyte precursor cells [19], also hMSCs display higher PPARG1 than PPARG2 mRNA basal levels (Figure S5A). However, *PPARG* expression progressively increases, reaching its highest levels 12 days after adipogenesis induction (Figure 4B), according to our previous analysis [16]. In line with PPARγ levels, its direct targets *FABP4*, *LPL*, and *ADIPOQ* [32,33] have almost undetectable levels in hMSCs (Figure 4A) and are markedly induced from day 6 up to terminal differentiation both at mRNA and protein level (Figure 4B–D).

Genes encoding perilipins—proteins involved in lipid storage and droplet biogenesis—show a peculiar expression pattern. As shown in Figure 4A and Figure S5B, *PLIN1* mRNA cannot be detected in undifferentiated hMSCs, whereas *PLIN2* gene (encoding the adipose differentiation-related protein, ADRP) is highly expressed, in line with the notion that ADRP locates on small LDs even in early differentiating cells and in 3T3-L1 murine pre-adipocytes [34,35]. Moreover, in line with previous analyses indicating that perilipin 1 is transcriptionally regulated by PPARγ and has AT-specific expression [34–37], *PLIN1* is strongly and rapidly induced upon stimulation with the adipogenic mix and shows the same trend of expression of other PPARγ target genes (Figure 4B, Figure S5B–D). *PLIN2* mRNA is modestly induced along hMSCs adipogenic differentiation and, starting from day 6, a progressive switch in the expression of perilipins occurs (Figure 4B, Figure S5B–D), as also reported for mouse embryonic fibroblasts (MEFs) and stromal vascular cells [38]. This finding suggests that perilipin 1 fully replaces ADRP on the surface of mature LDs, according to the hypothesized role of *PLIN1* in the formation of large LDs [39].

As previously described, adipocyte differentiation associates with profound changes of cell morphology [40]. In mouse pre-adipocytes, the interaction of monomeric G-actin with the transcriptional co-activator myocardin-related transcription factor A (MRTFA) prevents its nuclear translocation, inducing *Pparg* expression [41]. Interestingly, we observed an opposite trend of *MRTFA* gene expression compared to *PPARG* up to terminal differentiation (Figure S5E) suggesting that—also in the context of human neo-adipogenesis—MRTFA and PPARγ act in a mutually antagonistic manner.

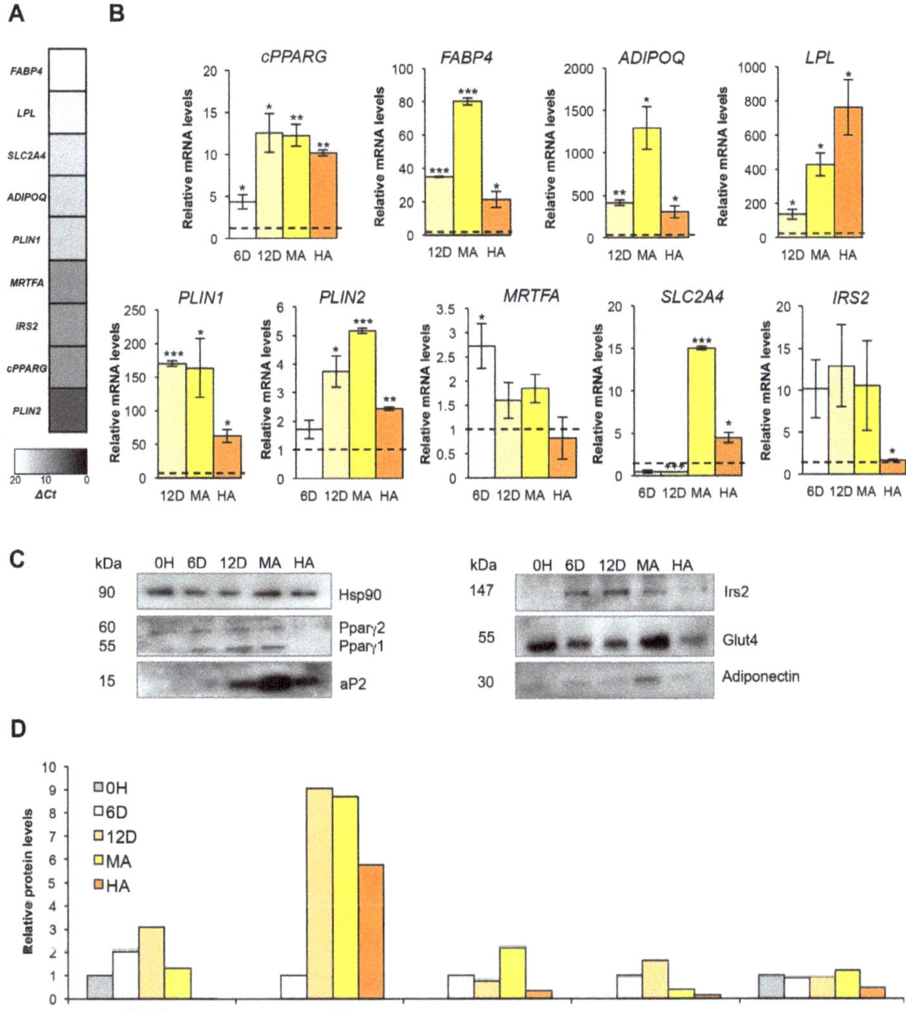

Figure 4. Expression analysis along differentiation of hMSCs in mature and hypertrophic-like adipocytes. (**A**) Gray-scale heatmap of normalized mRNA expression values (ΔCt = Ct gene – Ct *PPIA*) determined by qPCR in hMSCs (T = 0 h). *PPIA* was used as reference gene. (**B**) Relative mRNA expression analysis (qPCR) at different time points upon adipogenic induction (T = 0 h, T = 6 and T = 12 days) and in mature and hypertrophic-like adipocytes (MAs and HAs, respectively). For each gene, the first time point showing detectable levels was used as reference (dotted line; i.e., T = 0 h for *PPARG, PLIN2, MRTFA, SLC2A4, IRS2* genes; T = 6 days for *FABP4, ADIPOQ, LPL* and *PLIN1* genes). *PPIA* was used as reference gene in all assays. Data are reported as mean ±SEM from three independent experiments. *p val ≤ 0.05, **p val ≤ 0.01 and ***p val ≤ 0.001. (**C**) Western blots on lysates of hMSCs at different time points from adipogenic induction (T = 0 h, T = 6, and T = 12 days) and in MAs and HAs. Hsp90 was used as loading control. Representative autoradiographs are shown. (**D**) Bar graph reporting protein quantification (pixel density analysis of western blots). Values are normalized on Hsp90 (loading control) and—for each analyzed protein—the first time point having detectable levels (by the specific Ab) was used as reference (relative protein levels = 1).

Insulin resistance is hallmark of hypertrophic obesity [42–44], and PPARγ activation in adipose cells is sufficient to improve insulin sensitivity [45]. Among the genes involved in glucose uptake and insulin signaling, we measured *SLC2A4* (alias *GLUT4*) and *IRS2* expression since they are validated PPARγ targets [45,46]. As shown in Figure 4B, *SLC2A4* levels are higher in terminally differentiated adipocytes, whereas *IRS2* expression peaks on day 12, as also corroborated by protein quantification analysis performed along the entire adipogenic process (Figure 4C,D).

3.5. Hypertrophic-Like Adipocytes Display Morphological Features Resembling Adipose Tissue Hypertrophy

Our in vitro model of hypertrophic-like adipocytes allows a direct comparison with mature cells. Hence, we used optical (Figure 5A) and electron (Figure 5B) microscopy for qualitative morphological and ultrastructural comparison between the two conditions. Furthermore, using confocal microscopy we evaluated cellular size as well as the number, size and distribution of LDs within adipose cells (Figure 6A–D, Figure S6B,C).

As evident in Figure 5A, HAs show a massive and diffuse staining by Oil Red O corresponding to a larger amount of neutral lipids stored in giant LDs compared with smaller LDs present in mature adipocytes. The marked enlargement of LDs in hypertrophic cells induces a substantial reduction of the cytoplasmic layer increasing the pressure on cell nucleus (Figure 3D, Figure S6A). Ultrastructure analysis reveals a different electron density of LDs between MAs and HAs suggesting a shift in the lipid content not measurable by Oil Red O staining (Figure 5B). In mature adipocytes, magnifications show contacts between LDs and mitochondria, well-structured ER nearby LDs and LD–LD contacts indicating fusion sites between coalescent LDs (Figure 5B, upper panels). In hypertrophic-like cells, large LDs are surrounded by several mitochondria and an extensive network of intermediate filaments (IFs; Figure 5B, lower panels). Particularly, in line with previous studies on differentiating 3T3-L1 [47] and mouse macrophage foam cells [48], IFs form cage-like structures around LDs (Figure 5B, lower panel). These structures may serve both to prevent contacts with other droplets or organelles [49] and to regulate lipids influxes and effluxes [47,50], especially in hypertrophic cells with large LDs.

Estimation of cell size by confocal microscopy revealed a marked increase of cellular area (FC~1.7) in hypertrophic-like cells compared to mature ones (Figure 6A, Figure S6B). A direct comparison of lipid accumulation in mature and hypertrophic-like cells revealed a global increase (~35%) in lipid accumulation in HAs vs. MAs (Figure 5A). Interestingly, LDs analysis by confocal microscopy revealed a marked increase in the mean area (FC~5), and a doubling of lipid content, with a reduction (~41%) of total LD number (Figure 6B–D, Figure S6C). Overall, this analysis supports the formation of giant LDs by both increased lipid accumulation and coalescence of small droplets [51,52], suggesting that these processes govern the transformation of LDs during the transition of human adipocytes to hypertrophic state. Additionally, in line with the work of Skurk and colleagues (2007) [53], in vitro-generated HAs secrete higher levels of IL-6 compared to mature adipocytes (Figure 6E). Among the well-characterized cytokines released by hypertrophic adipocytes, IL-6 is the most studied due to its role in defective adipogenesis and IR onset [54–57].

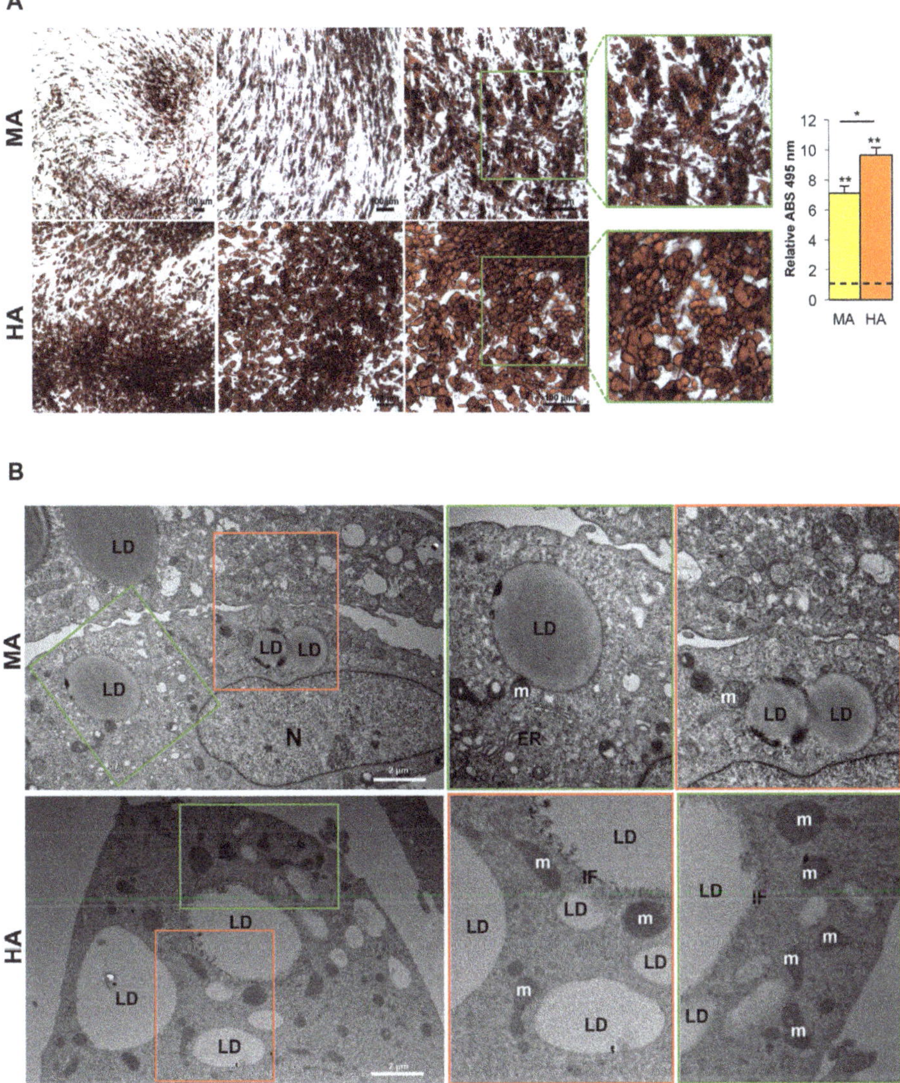

Figure 5. Mature and hypertrophic-like adipocytes display morphological differences. (**A**) Representative images (left panel) of mature and hypertrophic-like adipocytes (MAs and HAs, respectively) stained with Oil Red O (scale bars, 100 μm). Measurement of lipid accumulation (optical density) in hMSCs (T = 0 h used as control, CTR), in MAs and in HAs. Data are shown as mean ± SEM vs. CTR (dotted line). *p val ≤ 0.05 and **p val ≤ 0.01. (**B**) Representative micrographs of MAs and HAs by transmission electron microscopy (scale bars, 2 μm). Green and red squares in each left panel are observed at higher magnification in right panels. n = nucleus; m = mitochondria, LD = lipid droplets, ER = endoplasmic reticulum, IF = intermediate filaments.

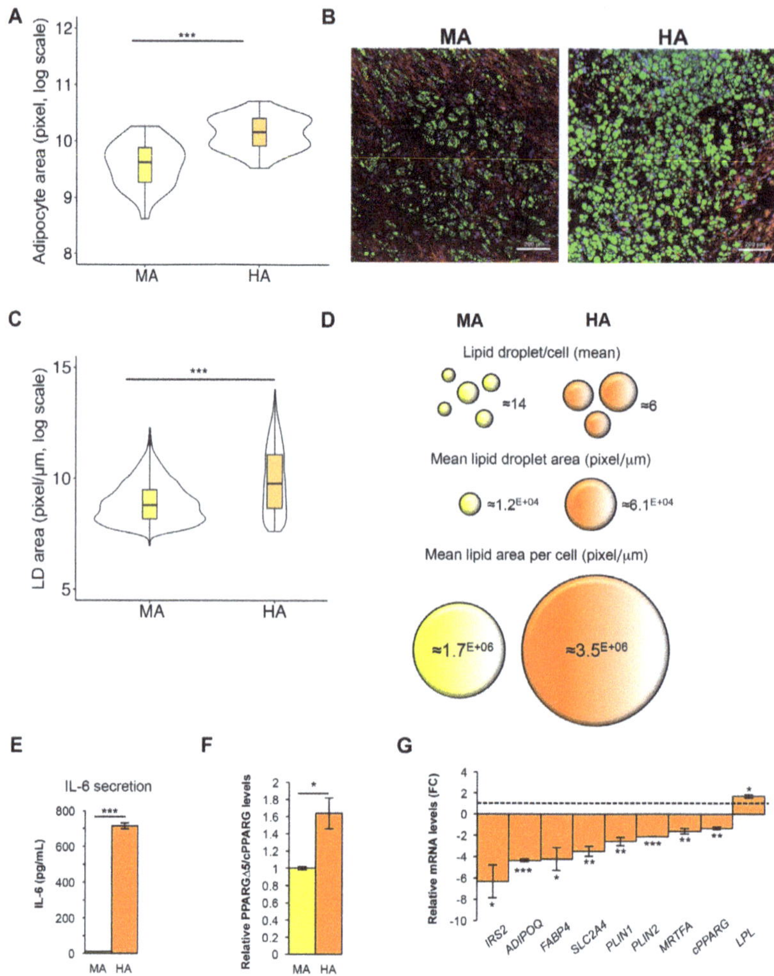

Figure 6. Differences between mature and hypertrophic-like adipocytes occur in cell size, lipid droplets, IL-6 secretion, and gene expression. (**A**) Violin plot showing cell area of mature (MAs) and hypertrophic-like (HAs) adipocytes ($n = 60$). *** p val ≤ 0.001. (**B**) Representative confocal microscopy images of hMSCs differentiated in MAs and HAs stained with DAPI (nuclei, blue), Bodipy 495/503 (lipid droplets, green) and WGA 632/647 (cell membranes, red; scale bars, 200 μm). (**C**) Violin plot showing lipid area/cell measured by 3D analysis on 2973 LDs (from 214 MAs) and on 1168 LDs (from 206 HAs). *** p val ≤ 0.001. (**D**) Schematic representation of 3D analysis results from LDs in MAs and HAs (i.e., mean of number of LD/cell, LD area, and total LD area/cell), as described in B. (**E**) Bar graph reporting IL-6 concentration (pg/mL) determined by ELISA on culture supernatant of MAs and HAs. Data are reported as mean ±SEM from three independent experiments. *** p val ≤ 0.001. (**F**) Relative PPARGΔ5/cPPARG mRNA levels (qPCR) in MAs and HAs. MAs were used as reference sample and *PPIA* as reference gene. Data are reported as mean ±SEM from three independent experiments. *** p val ≤ 0.001. (**G**) Relative mRNA quantifications (signed fold-changes) in HAs vs. MAs (dotted line). *PPIA* was used as reference gene. Data are reported as mean ±SEM from three independent experiments. * p val ≤ 0.05, ** p val ≤ 0.01 and *** p val ≤ 0.001.

Noteworthy, mirroring what observed in the SAT of hypertrophic obese patients, HAs display reduced levels of cPPARG mRNA and increased PPARGΔ5/cPPARG ratio (Figure 6F). Despite cPPARG expression being only modestly reduced in HAs (vs. MAs), PPARγ target genes are highly reduced, compatibly with the impaired metabolic activity in hypertrophic AT (Figure 6G). Hence, in light of the dominant negative activity of PPARGΔ5 [16], we speculate that increased PPARGΔ5/cPPARG ratio in hypertrophic adipose cells may contribute—at least in part—to inhibit the transcription of direct PPARγ target genes. Differently, despite *LPL* being a PPARγ target, its expression in HAs is increased compared to mature cells. However, this is in line with high *LPL* levels in SAT of obese patients [58] and the elevated enzymatic activity in hypertrophic AT [59,60]. Furthermore, insulin and rosiglitazone (contained in the AMM) are likely to promote *LPL* increase during the in vitro transition from mature to hypertrophic-like adipose cells [61]. Genes encoding proteins involved in insulin signaling and associated with insulin sensitivity, such as *IRS2* and *SLC2A4*, were among the most down-regulated PPARγ targets in hypertrophic cells, in line with the notion that IR is a hallmark of hypertrophic obesity. Since perlipin 1 can restrain the pro-inflammatory response—reducing futile lipolysis—and promote insulin sensitivity [62], *PLIN1* gene expression is strongly down-regulated in hypertrophic-like cells compared to matureones. The pathologic transition of adipocytes toward the hypertrophic state also induces a marked decrease of *MRTFA* expression (Figure 6G, Figure S5E), whereas cPPARG—albeit reduced (Figure 6G)—is still highly expressed in these cells (Figure 4B, Figure S5E), in line with its antagonist activity toward MRTFA in adipose cells.

3.6. GLUT4 Negatively Correlates with PPARGΔ5/cPPARG Ratio Only in Pathologic Conditions

PPARγΔ5 has a proven dominant negative activity on canonical PPARγ and is highly expressed in human SAT [16]. In light of this evidence, it is reasonable to consider that the global PPARγ activity in this tissue is tightly dependent on the relative amount between its canonical and dominant negative isoforms. As evidenced by the comparison between mature and hypertrophic-like adipose cells, we observed the unbalance between *PPARG* isoforms (increased PPARGΔ5/cPPARG ratio) and a concomitant pronounced alteration—in terms both of mRNA and protein expression—of PPARγ target genes involved in insulin signaling, and particularly of *SLC2A4* encoding the inducible glucose transporter 4 (Figure 4B–C). Accordingly, even in vivo large adipocytes display unbalanced PPARGΔ5/cPPARG ratio (Figures 1D and 6E), whose increase negatively impacts PPARγ transactivation ability in vitro [16].

Hence, we tested the hypothesis of a correlation between PPARGΔ5/cPPARG ratio and *SLC2A4* expression in vivo, measuring mRNA levels in the SAT of a subset of individuals from our German cohort (*n* = 56). As shown in Figure 7A, *SLC2A4* negatively correlates with PPARGΔ5/cPPARG ratio in the entire cohort, whereas cPPARG expression shows an opposite trend (Figure S7A). Interestingly, we disclosed significant correlation only in obese/overweight individuals (vs. individuals with normal weight; Figure 7B and Figure S7B), as well as in patients with altered glucose metabolism, i.e., IGT and T2D (vs. individuals with normal glucose tolerance, NGT; Figure 7C; Supplementary Figure S7C). PPARγ-mediated induction of *GLUT4* is a primary mechanism to establish insulin sensitivity of adipose tissue, liver and skeletal muscle [45,46,63]. Therefore, these data suggest that the unbalance of *PPARG* isoforms in the SAT—and particularly high PPARGΔ5/cPPARG ratio—can further contribute to IR onset in the adipose tissue of patients with hypertrophic obesity.

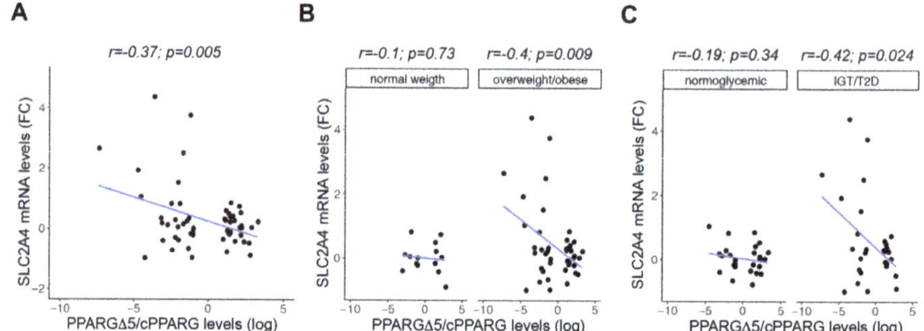

Figure 7. PPARGΔ5/cPPARG ratio correlates with *SLC2A4* levels. PPARGΔ5 and cPPARG expression was previously measured in Aprile et al. (2018). (**A–C**) Scatterplot reporting the correlations by linear regression analysis between *SLC2A4* and PPARGΔ5/cPPARG levels (qPCR) in the SAT of a subset of individuals (*n* = 56), stratified in subgroups according to BMI in normal weight (*n* = 14) and overweight/obese (*n* = 42), or to glucose-metabolizing capacity in NGT (*n* = 27), IGT and T2D (*n* = 29). *RPS23* was used as reference gene. Pearson's correlation coefficient (r) and *p* values (*p*) are shown.

4. Conclusions

Inappropriate expansion of adipocytes in the SAT is a characteristic of hypertrophic obesity, associated with a reduced adipogenesis and impaired insulin sensitivity. These primary events contribute to establish local inflammation and reduced insulin sensitivity in the AT, leading to ectopic fat deposition and systemic IR [5–8]. Hence, adipocyte size has been proposed as a predictor of IR and T2D onset. Nevertheless, technical drawbacks, the dynamical distribution of adipose cells in AT, and inter-individual variability make difficult accurately determining adipocyte size and establishing a value (or a range) that is indicative of metabolically defective cells. Then, adequate cellular models recapitulating the physiological aspects of neo-adipogenesis and the pathological features of hypertrophic metabolically unhealthy adipocytes are needed for addressing factors responsible of the AT shift toward the hypertrophic state.

Our recent work established PPARγΔ5 - a dominant negative isoform of PPARγ - as a potential contributor to the functional PPARγ impairment in the SAT of obese patients [16]. Our previous data suggest that the unbalanced ratio between dominant negative and canonical isoforms in the SAT can contribute to the transcriptional repression of metabolic genes and to the impairment of neo-adipogenesis. Both these pathologic features are hallmarks of hypertrophic obesity and are strictly related to IR and T2D onset [1,42,44]. It prompted us exploring whether *PPARG* splicing is affected in the context of hypertrophic obesity. Our in vivo finding that the ratio between PPARGΔ5 and canonical *PPARG* transcripts is significantly higher in SAT enriched of large adipocytes corroborates the finding that PPARγ activity is impaired in adipose tissue when reduced insulin sensitivity and defective neo-adipogenesis are in place. Of note, the observation that canonical and dominant negative *PPARG* transcripts have opposite correlation, not only with BMI and body fat [16], but also with LDL-cholesterol and leptin serum levels, further highlights the differential role of *PPARG* isoforms in the SAT.

However, our interest in studying in a more controlled and unbiased system, whether *PPARG* expression and splicing are affected in hypertrophic adipocytes guided us to set-up a new cellular model of human adipocyte hypertrophy through the generation of hypertrophic-like cells directly comparable to mature ones. Low variability in the differentiation rate, together with a weak susceptibility to dedifferentiate and detach in culture, represent only some advantages of this model. Indeed, by a detailed morphological, ultrastructural and transcriptional analysis, we provide a qualitative and quantitative estimation of the differentiation process, ranging from hMSCs to mature adipocytes as well as of the pathological shift toward the hypertrophic state. Indeed, the transition from terminally differentiated to hypertrophic-like cells mimics the dynamics of hypertrophic AT, showing the genesis of giant LDs—surrounded by an extensive network of IFs and mitochondria - which induce progressive cell engulfment, thickening of the cytoplasmic layer and increased pressure on cell nuclei. An accurate analysis of lipid droplets provided a quantitative estimation of lipid accumulation in hypertrophic-like cells, supporting the model in which giant LDs originate either by progressive lipid storage in single LDs and by coalescence of smaller droplets. Beyond the morphological changes, we also observed that hypertrophic-like adipocytes secrete high amount of IL-6, a pro-inflammatory cytokine typically observed in the microenvironment of hypertrophic AT. Notably, higher IL-6 levels are secreted by hypertrophic-like adipocytes compared to normal cells, further supporting the bona fide of this new cellular model. Along with the progressive lipid accumulation and the following morphological changes, we also systematically explored how these cells modulate gene/protein expression. The finding that throughout the differentiation of hMSCs several PPARγ target genes involved in LDs formation, insulin signaling and lipid metabolism display peculiar expression patterns compatible with PPARγ levels, further supports the use of hMSCs as model of human adipogenesis. Additionally, the progressive switch in the expression of perilipin genes—compatible with the replacement of ADRP with perilipin 1 in LDs formation—and the opposite trend of expression for *MRTFA* and *PPARG*, consistent with their supposed mutual antagonistic activity, are in line with previous studies in murine cells [38,41]. Mimicking the pathologic state of human AT, hypertrophic-like cells display a marked increase of PPARGΔ5/cPPARG ratio and a substantial reduction of PPARγ target genes involved in LDs biogenesis and insulin signaling, especially of Glut4. We found a significant negative correlation between PPARGΔ5/cPPARG ratio and *SLC2A4* expression only in patients with overweight/obesity as well as in those having altered glucose metabolism (i.e., with impaired glucose tolerance or T2D), but not in individuals with normal BMI or with normal glucose tolerance. These in vitro and in vivo data support the hypothesis that the unbalance between *PPARG* canonical and dominant negative isoforms is a characteristic of hypertrophic adipocytes, and that it associates with a marked perturbation of the PPARγ-dependent gene network, with a pronounced down-regulation of factors involved in glucose and lipid metabolism, such as Glut4. Furthermore, beyond the in vivo investigation of *PPARG* isoforms in hypertrophic AT, to the best of our knowledge, here we describe the first in vitro model of human hypertrophic-like adipocytes. This cellular model can be instrumental for dissecting—in the absence of confounding effects—the molecular mechanisms underlying the functional defects of the adipocytes in hypertrophic AT. Indeed, the use of a unique in vitro model—able to recapitulate each differentiation step from hMSCs to mature adipocyte and further toward hypertrophic state—is a powerful tool to decipher in a stepwise manner the pathological determinants of AT dysfunction in obesity and in its related comorbidities.

Supplementary Materials: The following are available online at http://www.mdpi.com/2073-4409/9/5/1284/s1, Figure S1. Correlation of PPARGΔ5 and cPPARG levels with clinical and biochemical parameters; Figure S2. Adipocyte differentiation of hMSCs is affected by densities of cell plating; Figure S3. Stemness markers expression in terminally differentiated hMSCs and in hypertrophic-like cells; Figure S4. Hypertrophic-like adipocytes are less prone to dedifferentiation in vitro; Figure S5. Expression trends of PPARγ and its target genes; Figure S6. Morphometric characteristics of hypertrophic-like adipocytes; Figure S7. The expression of canonical *PPARG* transcripts correlates with *SLC2A4* levels; Table S1. Sequences of oligonucleotides used in RT-PCR and qPCR assays; Table S2. Reagents and resources information.

Author Contributions: V.C., M.A., A.C., and S.C. designed the experiments; M.A., S.C., C.P., and M.R.A., performed experiments and analyzed data; P.I. carried out ELISA, and R.T. carried out the ultrastructural and morphologic analysis by microscopy; V.C., M.A., and S.C. processed data and wrote the paper; V.C., M.A., and A.C. supervised all experiments; M.B. provided patient biopsy specimens, clinical/biochemical parameters, and was involved in interpretation of ex vivo data. All authors contributed to data interpretation and discussion, also edited and approved the final manuscript. All authors have read and agreed to the published version of the manuscript.

Funding: This research and the APC was funded by PON Ricerca e Innovazione 2014–2020, PON Ars01_01270—2"Innovative Device For SHAping the Risk of Diabetes (IDF SHARID)" to V.C.

Acknowledgments: We express our gratitude to the FACS Facility of the IGB-CNR for the technical support and to Flavia Scognamiglio for useful comments.

Conflicts of Interest: The authors declare no conflict of interest.

References

1. Danforth, E. Failure of adipocyte differentiation causes type II diabetes mellitus? *Nat. Genet.* **2000**, *26*, 13. [CrossRef] [PubMed]
2. Shulman, G.I. Cellular mechanisms of insulin resistance. *J. Clin. Investig.* **2000**, *106*, 171–176. [CrossRef] [PubMed]
3. Savage, D.B.; Petersen, K.F.; Shulman, G.I. Disordered lipid metabolism and the pathogenesis of insulin resistance. *Physiol. Rev.* **2007**, *87*, 507–520. [CrossRef] [PubMed]
4. Savage, D.B. Mouse models of inherited lipodystrophy. *Dis. Model. Mech.* **2009**, *2*, 554–562. [CrossRef] [PubMed]
5. Gustafson, B.; Hammarstedt, A.; Hedjazifar, S.; Smith, U. Restricted Adipogenesis in Hypertrophic Obesity. *Diabetes* **2013**, *62*, 2997–3004. [CrossRef]
6. Muir, L.; Neeley, C.K.; Meyer, K.A.; Baker, N.A.; Brosius, A.M.; Washabaugh, A.R.; Varban, O.A.; Finks, J.F.; Zamarron, B.F.; Flesher, C.G.; et al. Adipose tissue fibrosis, hypertrophy, and hyperplasia: Correlations with diabetes in human obesity. *Obesity* **2016**, *24*, 597–605. [CrossRef]
7. Klöting, N.; Blüher, M. Adipocyte dysfunction, inflammation and metabolic syndrome. *Rev. Endocr. Metab. Disord.* **2014**, *15*, 277–287. [CrossRef]
8. Savage, D.B. PPAR[gamma] as a metabolic regulator: Insights from genomics and pharmacology. *Expert Rev. Mol. Med.* **2005**, *7*. [CrossRef]
9. Sadowski, H.B.; Wheeler, T.T.; Young, D.A. Gene expression during 3T3-L1 adipocyte differentiation. Characterization of initial responses to the inducing agents and changes during commitment to differentiation. *J. Boil. Chem.* **1992**, *267*, 4722–4731.
10. Kim, J.I.; Huh, J.Y.; Sohn, J.H.; Choe, S.S.; Lee, Y.S.; Lim, C.Y.; Jo, A.; Park, S.B.; Han, W.; Kim, J.B. Lipid-Overloaded Enlarged Adipocytes Provoke Insulin Resistance Independent of Inflammation. *Mol. Cell. Boil.* **2015**, *35*, 1686–1699. [CrossRef]
11. Cawthorn, W.P.; Scheller, E.L.; MacDougald, O. Adipose tissue stem cells meet preadipocyte commitment: Going back to the future[S]. *J. Lipid Res.* **2011**, *53*, 227–246. [CrossRef] [PubMed]
12. Huang, X.; Ordemann, J.; Müller, J.M.; Dubiel, W. The COP9 signalosome, cullin 3 and Keap1 super complex regulates CHOP stability and adipogenesis. *Biol. Open* **2012**, *1*, 705–710. [CrossRef] [PubMed]
13. Bunnell, B.A.; Flaat, M.; Gagliardi, C.; Patel, B.; Ripoll, C. Adipose-derived stem cells: Isolation, expansion and differentiation☆. *Methods* **2008**, *45*, 115–120. [CrossRef] [PubMed]
14. Ruiz-Ojeda, F.J.; Iris-Rupérez, A.; Gómez-Llorente, C.; Gil, A.; Aguilera, C.M. Cell Models and Their Application for Studying Adipogenic Differentiation in Relation to Obesity: A Review. *Int. J. Mol. Sci.* **2016**, *17*, 1040. [CrossRef]
15. Mohamed-Ahmed, S.; Fristad, I.; Lie, S.A.; Suliman, S.; Mustafa, K.; Vindenes, H.; Idris, S.B. Adipose-derived and bone marrow mesenchymal stem cells: A donor-matched comparison. *Stem Cell Res. Ther.* **2018**, *9*, 168. [CrossRef]
16. Aprile, M.; Cataldi, S.; Ambrosio, M.R.; D'Esposito, V.; Lim, K.; Dietrich, A.; Blüher, M.; Savage, D.B.; Formisano, P.; Ciccodicola, A.; et al. PPARγΔ5, a Naturally Occurring Dominant-Negative Splice Isoform, Impairs PPARγ Function and Adipocyte Differentiation. *Cell Rep.* **2018**, *25*, 1577–1592. [CrossRef]
17. Stenkula, K.G.; Erlanson-Albertsson, C. Adipose cell size: Importance in health and disease. *Am. J. Physiol. Integr. Comp. Physiol.* **2018**, *315*, R284–R295. [CrossRef]
18. Costa, V.; Gallo, M.A.; Letizia, F.; Aprile, M.; Casamassimi, A.; Ciccodicola, A. PPARG: Gene Expression Regulation and Next-Generation Sequencing for Unsolved Issues. *PPAR Res.* **2010**, *2010*, 1–17. [CrossRef]
19. Aprile, M.; Ambrosio, M.R.; D'Esposito, V.; Beguinot, F.; Formisano, P.; Costa, V.; Ciccodicola, A. PPARG in human adipogenesis: Differential contribution of canonical transcripts and dominant negative isoforms. *PPAR Res.* **2014**, *2014*, 18–20. [CrossRef]
20. Keller, M.; Hopp, L.; Liu, X.; Wohland, T.; Rohde, K.; Cancello, R.; Klös, M.; Bacos, K.; Kern, M.; Eichelmann, F.; et al. Genome-wide DNA promoter methylation and transcriptome analysis in human adipose tissue unravels novel candidate genes for obesity. *Mol. Metab.* **2016**, *6*, 86–100. [CrossRef]

21. Guiu-Jurado, E.; Unthan, M.; Böhler, N.; Kern, M.; Landgraf, K.; Dietrich, A.; Schleinitz, R.; Ruschke, K.; Klöting, N.; Faßhauer, M.; et al. Bone morphogenetic protein 2 (BMP2) may contribute to partition of energy storage into visceral and subcutaneous fat depots. *Obesity* **2016**, *24*, 2092–2100. [CrossRef] [PubMed]

22. Schneider, C.A.; Rasband, W.S.; Eliceiri, K.W. NIH Image to ImageJ: 25 years of image analysis. *Nat. Methods* **2012**, *9*, 671–675. [CrossRef] [PubMed]

23. Isakson, P.; Hammarstedt, A.; Gustafson, B.; Smith, U. Impaired Preadipocyte Differentiation in Human Abdominal Obesity. *Diabetes* **2009**, *58*, 1550–1557. [CrossRef] [PubMed]

24. Armoni, M.; Harel, C.; Karnieli, E. Transcriptional regulation of the GLUT4 gene: From PPAR-γ and FOXO1 to FFA and inflammation. *Trends Endocrinol. Metab.* **2007**, *18*, 100–107. [CrossRef] [PubMed]

25. Janderová, L.; McNeil, M.; Murrell, A.; Mynatt, R.L.; Smith, S.R.; Rossmeislová, L. Human Mesenchymal Stem Cells as an in Vitro Model for Human Adipogenesis. *Obes. Res.* **2003**, *11*, 65–74. [CrossRef] [PubMed]

26. McBeath, R.; Pirone, D.M.; Nelson, C.M.; Bhadriraju, K.; Chen, C.S. Cell Shape, Cytoskeletal Tension, and RhoA Regulate Stem Cell Lineage Commitment. *Dev. Cell* **2004**, *6*, 483–495. [CrossRef]

27. Modica, S.; Wolfrum, C. The dual role of BMP4 in adipogenesis and metabolism. *Adipocyte* **2017**, *6*, 141–146. [CrossRef]

28. Walther, T.C.; Chung, J.; Farese, R.V. Lipid Droplet Biogenesis. *Annu. Rev. Cell Dev. Boil.* **2017**, *33*, 491–510. [CrossRef]

29. Heid, H.; Rickelt, S.; Zimbelmann, R.; Winter, S.; Schumacher, H.; Dörflinger, Y.; Kuhn, C.; Franke, W. On the Formation of Lipid Droplets in Human Adipocytes: The Organization of the Perilipin–Vimentin Cortex. *PLoS ONE* **2014**, *9*, e90386. [CrossRef]

30. Khatau, S.B.; Hale, C.; Stewart-Hutchinson, P.J.; Patel, M.S.; Stewart, C.L.; Searson, P.; Hodzic, D.; Wirtz, D. A perinuclear actin cap regulates nuclear shape. *Proc. Natl. Acad. Sci. USA* **2009**, *106*, 19017–19022. [CrossRef]

31. Spiegelman, B.M.; Hu, E.; Kim, J.B.; Brun, R. PPAR gamma and the control of adipogenesis. *Biochimie* **1997**, *79*, 111–112.

32. Costa, V.; Ciccodicola, A. Is PPARG the key gene in diabetic retinopathy? *Br. J. Pharmacol.* **2012**, *165*, 1–3. [CrossRef] [PubMed]

33. Tontonoz, P.; Spiegelman, B.M. Fat and Beyond: The Diverse Biology of PPARγ. *Annu. Rev. Biochem.* **2008**, *77*, 289–312. [CrossRef]

34. Prokesch, A.; Smorlesi, A.; Perugini, J.; Manieri, M.; Ciarmela, P.; Mondini, E.; Trajanoski, Z.; Kristiansen, K.; Giordano, A.; Bogner-Strauss, J.G.; et al. Molecular aspects of adipoepithelial transdifferentiation in mouse mammary gland. *STEM CELLS* **2014**, *32*, 2756–2766. [CrossRef] [PubMed]

35. Wolins, N.E.; Quaynor, B.K.; Skinner, J.R.; Schoenfish, M.J.; Tzekov, A.; Bickel, P.E. S3-12, Adipophilin, and TIP47 Package Lipid in Adipocytes. *J. Boil. Chem.* **2005**, *280*, 19146–19155. [CrossRef] [PubMed]

36. Dalen, K.T.; Schoonjans, K.; Ulven, S.M.; Weedon-Fekjaer, M.S.; Bentzen, T.G.; Koutnikova, H.; Auwerx, J.; Nebb, H.I. Adipose tissue expression of the lipid droplet-associating proteins S3-12 and perilipin is controlled by peroxisome proliferator-activated receptor-gamma. *Diabetes* **2004**, *53*. [CrossRef]

37. Arimura, N.; Horiba, T.; Imagawa, M.; Shimizu, M.; Sato, R. The Peroxisome Proliferator-activated Receptor γ Regulates Expression of the Perilipin Gene in Adipocytes. *J. Boil. Chem.* **2004**, *279*, 10070–10076. [CrossRef]

38. Takahashi, Y.; Shinoda, A.; Kamada, H.; Shimizu, M.; Inoue, J.; Sato, R. Perilipin2 plays a positive role in adipocytes during lipolysis by escaping proteasomal degradation. *Sci. Rep.* **2016**, *6*, 20975. [CrossRef]

39. Itabe, H.; Yamaguchi, T.; Nimura, S.; Sasabe, N. Perilipins: A diversity of intracellular lipid droplet proteins. *Lipids Heal. Dis.* **2017**, *16*, 83. [CrossRef]

40. Kanzaki, M.; Pessin, J.E. Insulin-stimulated GLUT4 Translocation in Adipocytes Is Dependent upon Cortical Actin Remodeling. *J. Boil. Chem.* **2001**, *276*, 42436–42444. [CrossRef]

41. Nobusue, H.; Onishi, N.; Shimizu, T.; Sugihara, E.; Oki, Y.; Sumikawa, Y.; Chiyoda, T.; Akashi, K.; Saya, H.; Kano, K. Regulation of MKL1 via actin cytoskeleton dynamics drives adipocyte differentiation. *Nat. Commun.* **2014**, *5*, 3368. [CrossRef] [PubMed]

42. Gustafson, B.; Gogg, S.; Hedjazifar, S.; Jenndahl, L.; Hammarstedt, A.; Smith, U. Inflammation and impaired adipogenesis in hypertrophic obesity in man. *Am. J. Physiol. Metab.* **2009**, *297*, E999–E1003. [CrossRef]

43. Gustafson, B.; Hedjazifar, S.; Gogg, S.; Hammarstedt, A.; Smith, U. Insulin resistance and impaired adipogenesis. *Trends Endocrinol. Metab.* **2015**, *26*, 193–200. [CrossRef] [PubMed]

44. Hajer, G.R.; Van Haeften, T.W.; Visseren, F.L. Adipose tissue dysfunction in obesity, diabetes, and vascular diseases. *Eur. Hear. J.* **2008**, *29*, 2959–2971. [CrossRef] [PubMed]

45. Sugii, S.; Olson, P.; Sears, R.D.; Saberi, M.; Atkins, A.R.; Barish, G.D.; Hong, S.-H.; Castro, G.L.; Yin, Y.-Q.; Nelson, M.C.; et al. PPARgamma activation in adipocytes is sufficient for systemic insulin sensitization. *Proc. Natl. Acad. Sci. USA* **2009**, *106*, 22504–22509. [CrossRef] [PubMed]
46. Smith, U.; Gogg, S.; Johansson, A.; Olausson, T.; Rotter, V.; Svalstedt, B. Thiazolidinediones (PPARγ agonists) but not PPAR α agonists increase IRS-2 gene expression in 3T3-L1 and human adipocytes 1. *FASEB J.* **2001**, *15*, 215–220. [CrossRef]
47. Lieber, J.G.; Evans, R.M. Disruption of the vimentin intermediate filament system during adipose conversion of 3T3-L1 cells inhibits lipid droplet accumulation. *J. Cell Sci.* **1996**, *109*, 109.
48. McGookey, D.J.; Anderson, R.G. Morphological characterization of the cholesteryl ester cycle in cultured mouse macrophage foam cells. *J. Cell Boil.* **1983**, *97*, 1156–1168. [CrossRef]
49. Murphy, D.J.; Vance, J. Mechanisms of lipid-body formation. *Trends Biochem. Sci.* **1999**, *24*, 109–115. [CrossRef]
50. Van Meer, G. Caveolin, Cholesterol, and Lipid Droplets? *J. Cell Boil.* **2001**, *152*, F29–F34. [CrossRef]
51. Thiam, A.R.; Farese, R.V., Jr.; Walther, T.C. The biophysics and cell biology of lipid droplets. *Nat. Rev. Mol. Cell Boil.* **2013**, *14*, 775–786. [CrossRef] [PubMed]
52. Fei, W.; Shui, G.; Zhang, Y.; Krahmer, N.; Ferguson, C.; Kapterian, T.S.; Lin, R.C.Y.; Dawes, I.W.; Brown, A.J.; Li, P.; et al. A Role for Phosphatidic Acid in the Formation of "Supersized" Lipid Droplets. *PLoS Genet.* **2011**, *7*, e1002201. [CrossRef] [PubMed]
53. Skurk, T.; Alberti-Huber, C.; Herder, C.; Hauner, H. Relationship between Adipocyte Size and Adipokine Expression and Secretion. *J. Clin. Endocrinol. Metab.* **2007**, *92*, 1023–1033. [CrossRef] [PubMed]
54. Almuraikhy, S.; Kafienah, W.; Bashah, M.; Diboun, I.; Jaganjac, M.; Al-Khelaifi, F.; Abdesselem, H.; Mazloum, N.A.; Alsayrafi, M.; Mohamed-Ali, V.; et al. Interleukin-6 induces impairment in human subcutaneous adipogenesis in obesity-associated insulin resistance. *Diabetologia* **2016**, *59*, 2406–2416. [CrossRef]
55. Rotter, V.; Nagaev, I.; Smith, U. Interleukin-6 (IL-6) Induces Insulin Resistance in 3T3-L1 Adipocytes and Is, Like IL-8 and Tumor Necrosis Factor-α, Overexpressed in Human Fat Cells from Insulin-resistant Subjects. *J. Boil. Chem.* **2003**, *278*, 45777–45784. [CrossRef]
56. Bastard, J.-P.; Maachi, M.; van Nhieu, J.T.; Jardel, C.; Bruckert, E.; Grimaldi, A.; Robert, J.-J.; Capeau, J.; Hainque, B. Adipose Tissue IL-6 Content Correlates with Resistance to Insulin Activation of Glucose Uptake both in Vivo and in Vitro. *J. Clin. Endocrinol. Metab.* **2002**, *87*, 2084–2089. [CrossRef]
57. Kern, P.A.; Ranganathan, S.; Li, C.; Wood, L.; Ranganathan, G. Adipose tissue tumor necrosis factor and interleukin-6 expression in human obesity and insulin resistance. *Am. J. Physiol. Metab.* **2001**, *280*, E745–E751. [CrossRef]
58. Sadur, U.N.; Yost, T.J.; Eckel, R.H. Insulin Responsiveness of Adipose Tissue Lipoprotein Lipase Is Delayed but Preserved in Obesity*. *J. Clin. Endocrinol. Metab.* **1984**, *59*, 1176–1182. [CrossRef]
59. Serra, M.C.; Ryan, A.S.; Sorkin, J.D.; Favor, K.H.; Goldberg, A.P.; Favors, K.H. High adipose LPL activity and adipocyte hypertrophy reduce visceral fat and metabolic risk in obese, older women. *Obesity* **2015**, *23*, 602–607. [CrossRef]
60. Farnier, C.; Krief, S.; Blache, M.; Diot-Dupuy, F.; Mory, G.; Ferre, P.; Bazin, R. Adipocyte functions are modulated by cell size change: Potential involvement of an integrin/ERK signalling pathway. *Int. J. Obes.* **2003**, *27*, 1178–1186. [CrossRef]
61. McTernan, P.G.; Harte, A.L.; Anderson, L.A.; Green, A.; Smith, S.A.; Holder, J.C.; Barnett, A.H.; Eggo, M.C.; Kumar, S. Insulin and rosiglitazone regulation of lipolysis and lipogenesis in human adipose tissue in vitro. *Diabetes* **2002**, *51*, 1493–1498. [CrossRef] [PubMed]
62. Sohn, J.H.; Lee, Y.K.; Han, J.S.; Jeon, Y.G.; Kim, J.I.; Choe, S.S.; Kim, S.J.; Yoo, H.J.; Kim, J.B. Perilipin 1 (Plin1) deficiency promotes inflammatory responses in lean adipose tissue through lipid dysregulation. *J. Boil. Chem.* **2018**, *293*, 13974–13988. [CrossRef] [PubMed]
63. Czech, M.P. Insulin action and resistance in obesity and type 2 diabetes. *Nat. Med.* **2017**, *23*, 804–814. [CrossRef] [PubMed]

 *cells*

Article

Vascular Lipidomic Profiling of Potential Endogenous Fatty Acid PPAR Ligands Reveals the Coronary Artery as Major Producer of CYP450-Derived Epoxy Fatty Acids

Matthew L. Edin [1], Fred B. Lih [1], Bruce D. Hammock [2], Scott Thomson [3], Darryl C. Zeldin [1] and David Bishop-Bailey [4,*]

[1] Division of Intramural Research, NIEHS/NIH, Research Triangle Park, NC 27709, USA;
    edinm@niehs.nih.gov (M.L.E.); lih@niehs.nih.gov (F.B.L.); zeldin@niehs.nih.gov (D.C.Z.)
[2] Department of Entomology and Comprehensive Cancer Center, University of California, Davies,
    CA 95616-8584, USA; bdhammock@ucdavis.edu
[3] Royal Veterinary College, University of London, London N1 0TU, UK; scott.thomson@kcl.ac.uk
[4] North Cornwall Research Institute, Bude, Cornwall EX23 9EE, UK
*   Correspondence: bishopbaileyd@gmail.com; Tel.: +44-207-652-4513

Received: 13 March 2020; Accepted: 27 April 2020; Published: 29 April 2020

Abstract: A number of oxylipins have been described as endogenous PPAR ligands. The very short biological half-lives of oxylipins suggest roles as autocrine or paracrine signaling molecules. While coronary arterial atherosclerosis is the root of myocardial infarction, aortic atherosclerotic plaque formation is a common readout of in vivo atherosclerosis studies in mice. Improved understanding of the compartmentalized sources of oxylipin PPAR ligands will increase our knowledge of the roles of PPAR signaling in diverse vascular tissues. Here, we performed a targeted lipidomic analysis of ex vivo-generated oxylipins from porcine aorta, coronary artery, pulmonary artery and perivascular adipose. Cyclooxygenase (COX)-derived prostanoids were the most abundant detectable oxylipin from all tissues. By contrast, the coronary artery produced significantly higher levels of oxylipins from CYP450 pathways than other tissues. The TLR4 ligand LPS induced prostanoid formation in all vascular tissue tested. The 11-HETE, 15-HETE, and 9-HODE were also induced by LPS from the aorta and pulmonary artery but not coronary artery. Epoxy fatty acid (EpFA) formation was largely unaffected by LPS. The pig CYP2J homologue CYP2J34 was expressed in porcine vascular tissue and primary coronary artery smooth muscle cells (pCASMCs) in culture. Treatment of pCASMCs with LPS induced a robust profile of pro-inflammatory target genes: *TNFα, ICAM-1, VCAM-1, MCP-1* and *CD40L*. The soluble epoxide hydrolase inhibitor TPPU, which prevents the breakdown of endogenous CYP-derived EpFAs, significantly suppressed LPS-induced inflammatory target genes. In conclusion, PPAR-activating oxylipins are produced and regulated in a vascular site-specific manner. The CYP450 pathway is highly active in the coronary artery and capable of providing anti-inflammatory oxylipins that prevent processes of inflammatory vascular disease progression.

Keywords: PPARs; vascular; coronary artery; lipidomics; eicosanoids; inflammation; CYP450

---

1. Introduction

Peroxisome proliferator-activated receptors (PPARs) can be activated by a diverse group of endogenous fatty acid mediators including those produced from cyclooxygenase (COX), lipoxygenase and CYP450 enzymatic pathways [1]. These COX, lipoxygenase and CYP450 enzymes metabolize arachidonic acid and related polyunsaturated fatty acids, linoleic acid (LA), docosahexaenoic acid

(DHA) and eicosapentaenoic acid (EPA) into series of biologically active oxylipin mediators [2–4]. Cyclooxygenases largely make prostanoids (and some hydroxyeicosatetraenoic acids (HETEs)) [5,6]. Lipoxygenases make hydroperoxy-eicostetraeoic acid HpETE, HETEs, hydroxyoctadecaenoic acids (HODEs), hydroxy-DHAs, and hydroxy-EPAs—some of which are the precursors for leukotrienes [2]. The PUFA-utilizing CYP450s metabolize fatty acids into series of oxylipin mediators through a combination of either epoxidation or lipoxygenase-like or ω- and ω-1-hydroxylation [2,3]. Using arachidonic acid as an example, CYP2J2 can produce both epoxyeicosatrienoic acids (EETs) and 19-HETE by its epoxygenase- and hydroxylase-like activities, respectively [3,7]. The PGD_2 metabolite 15 deoxy-$D^{12,14}$-PGJ_2, PGI_2, 8-, 12-, and 15-HETE, 9- and 13-HODE [1], and 8-, 9-, 11-, 12-, and 14–15-EET [8] have all been shown to activate PPARs. Soluble (sEH) and microsomal (mEH) epoxide hydrolases (EH; encoded by the gene ephx2 and ephx1 respectively) combine to metabolize nearly all EpFAs in vivo [9]. sEH inhibitors (sEH-I) inhibit the breakdown of EpFAs to their more soluble but less biologically active dihydroxy counterparts and potentiate EET signaling [10–12].

While several oxylipins can signal through known or yet-to-be-identified G-protein-coupled receptors, transient increases in oxylipin ligands can also induce PPAR activation toward a variety of downstream signals [1]. PPAR activation induces heterodimerization with other nuclear receptors such as the retinoid X receptor (RXR), which enhances binding to a consensus sequence (direct repeats of 'AGGTCA') referred to as PPAR response elements (PPREs). PPAR ligands have diverse roles in the cardiovascular system, from repression of genes encoding pro-inflammatory cytokines to induction (e.g., *TNFα*, IL1, IL6) of monocytes/macrophages toward foam cell morphology [13].

The roles of oxylipins are of long-standing interest in vascular biology [11,13–20]. COX products have both cardioprotective (prostacyclin; PGI_2) and pro-thrombotic (e.g., thromboxane; TXA_2) activity [20,21]. CYP450-derived EpFAs are anti-atherosclerotic, vasodilatory and anti-inflammatory [11,22–30], with the notable exception of LA-derived dihydroxyoctadecamonoenoic acids (DHOMEs), which regulate cardiac function [31], vascular development [32], and thermal hyperalgesia [33] at low levels, but are toxic at higher levels [34]. CYP450-derived EETs, in particular, were originally described in porcine coronary artery as an endothelium-derived hyperpolarizing factor produced in response to stimulation and stretch sEH-I-treated or sEH-knockout mice show protection to injury induced vascular neointima formation [25], atherosclerosis and aneurysm formation [26], hypertension [35,36], type 2 diabetes [37], and inflammatory cell recruitment [23,30]. Interestingly, in the pulmonary circulation, although sEH inhibitors have been shown to augment hypoxia-induced vasoconstriction, sEH inhibition or overexpression of EpFA-producing enzymes such as CYP2J2 is protective in various acute lung injury models [38–40].

We previously showed that PPARs can be activated by CYP2J2 and its products in vitro and in vivo [8]. A number of protective effects of CYP2J2 or EETs have now been shown to be mediated by PPARs, including the protective effects of laminar flow on endothelial cells [24], mediating coronary reactive hyperemia [41–44] and vascular response in soluble epoxide hydrolase-null mice [45], cytoprotection of cardiomyocytes [46], inhibition of angiotensin II cardiac remodeling [47] and abdominal aortic aneurysm formation [48], inhibition of renal interstitial fibrosis and inflammation [49], improved vascular function and decreased renal injury in hypertensive obese rats [50], and promoting angiogenesis and migration in human endothelial progenitor cells from acute myocardial infarction patients [51].

Pigs have a similar heart and cardiovascular system to humans and undergo spontaneous and diet-induced atherogenesis [52]. Here, we used a lipidomic approach to study endogenous oxylipin PPAR ligand production by the large vessels of the pig: the thoracic aorta compared to the coronary and pulmonary arteries. The vessel releasates were also compared to those of aortic perivascular adipose tissue ex vivo. Perivascular adipose was investigated as it has been shown to release various cytokines that act in an endocrine and paracrine manner to regulate vascular signaling and inflammation which have been implicated in the development of atherosclerosis, hypertension, neointimal formation, aneurysm, arterial formation and vasculitis [53,54].

Using a targeted lipidomic approach, we found coronary artery releases significantly more oxylipins of almost all classes than aorta and pulmonary artery. Perivascular adipose was a particularly rich source of COX-derived PGE_2. Coronary artery was the highest source of CYP450-derived EpFAs PPAR ligands. The use of a sEH inhibitor TPPU on pig primary coronary artery vascular smooth muscle cells in culture showed strong anti-inflammatory activity consistent with PPAR activation.

2. Materials and Methods

2.1. Materials

Authentic oxylipins (EETs, DHEQ, and HDPA) were from Cayman Chemical Company (Cambridge Bioscience, Cambridge, UK). SYBR green was from Takara. TPPU (*N*-[1-(1-oxopropyl)-4-piperidinyl]-N′-[4-(trifluoromethoxy)phenyl]-urea) was synthesized as previously described [55]. Unless stated, all other reagents were from Sigma-Aldrich (Poole, Dorset, UK).

2.2. Vessel Organ Culture

Abattoir pig vessels largely from white X female pigs aged 8–10 weeks old were obtained from the Royal Veterinary College. Fresh tissue was collected and used within 4 h. The 50–500 mg segments of vessel or perivascular aortic adventitia were cultured in serum-free DMEM supplemented with antibiotic/antimycotic mix (Sigma-Aldrich, St. Louis, MO., USA) at 37 °C, 5% CO2 and 95% air, as previously described for rat and human vessels [56,57]. Serum-free media was used, as most sera contain large amounts of oxylipins (unpublished observations). Organ culture was performed for just the first 24 h after explant in order to minimize cell differentiation. In some experiments, lipopolysaccharide (LPS; *E. coli*, 1 µg/mL) was given to induce an inflammatory response.

2.3. Cell and Tissue Culture

Primary coronary artery smooth muscle cells (pCASMCs) were obtained by explant and grown as previously described for human vascular smooth muscle cells [58]. Briefly, extraneous tissue was removed, coronary arteries were opened along the midline, gently denuded, and chopped into small explants. SMCs were grown in DMEM supplemented with antibiotic/antimycotic mix, and 20% FBS, at 37 °C, 5% CO2 and 95% air. SMCs were identified by a characteristic morphological "hill-and-valley" growth pattern and by smooth muscle α-actin immunostaining. Since FBS interferes with lipid substrate composition and the release and detection of eicosanoids, all experiments were performed with DMEM supplemented with antibiotic/antimycotic mix and without FBS.

2.4. Real-Time qRT-PCR

Pig CYP2J34, sEH, *TNFα*, *VCAM-1*, *ICAM-1*, *MCP-1* and *CD40* were measured using the SYBR Green ddCT method (see Table S1 for primer pairs). Targets were normalized to 18S expression. RNA was extracted using the ThermoScientific RNA extraction kit and 1 µg of total RNA was used to generate cDNA using Superscript II (Invitrogen) according to the manufacturer's instructions. SYBR green qPCR was performed using Premix Ex Taq II mastermix (Takara) using a Chromo-4 machine and Opticon software. Genomic sequences were obtained from the UCSC Genome Browser website (http://genome.ucsc.edu/cgi-bin/hgGateway) and primers (see Table S1) were designed from NCBI's Primer Blast website (http://www.ncbi.nlm.nih.gov/tools/primer-blast/index.cgi?LINK_LOC=BlastHome).

2.5. Oxylipin Measurements

Explants were incubated in serum-free DMEM for 24 h, which allows for detection of both the highly abundant prostaglandins and HETEs and less-abundant CYP-derived oxylipins. LC–MS/MS analysis of oxylipin products in culture supernatants was as previously described [23,59]. LC–MS/MS analytes in samples were quantified against oxylipin standard curves (Cayman Chemical) using TraceFinder v4.1 (Thermo Scientific, Waltham, MA, USA) software.

2.6. Statistical Analyses

Graphical representations, heat maps and statistical analyses between groups (*t*-tests and paired *t*-tests) were performed using GraphPad Prism v8.1. When comparing multiple groups, ANOVA was followed by Holm–Sidak correction for multiple comparisons. All distributions appeared and were assumed to be normal.

3. Results

3.1. Oxylipin Lipidomic Profiling of the Large Vessels of the Pig

Young female pigs were selected to be devoid of atherosclerosis and represent non-diseased vascular tissues. Fresh tissue explants were divided into various treatment groups and cultured in serum-free media for 24 h. Serum-free media was used, as most sera contain large amounts of oxylipins (unpublished observations). CYP-, LOX- and COX-derived oxylipins were detectable in organs culture for 24 h after explant. The most abundant oxylipin species represented in 24 h organ culture in all tissues were prostanoids derived from COX (Figure 1a). PGI_2 was the major product from aorta (190 pg/mg) and pulmonary artery (640 pg/mg), whereas PGE_2 was the major product from coronary artery (1135 pg/mg) and perivascular adipose (1390 pg/mg; Figure 1). The coronary artery generated by far the largest total amounts of measurable oxylipins followed by pulmonary artery, with the aorta producing approximately 1/8 of the prostanoids per unit weight as the coronary artery (Figure 1a). The coronary artery produced significantly more EpFA and hydroxy fatty acids than the aorta or perivascular adipose (Figure 1b; Figure S1), with the pulmonary artery production again intermediate between the aorta and coronary artery (Figures 1 and 2). The perivascular adipose produced similar amounts of PGE_2 as the coronary artery, with much lower relative levels of lipoxygenase or CYP450 products formed than any of the vessels (data not shown).

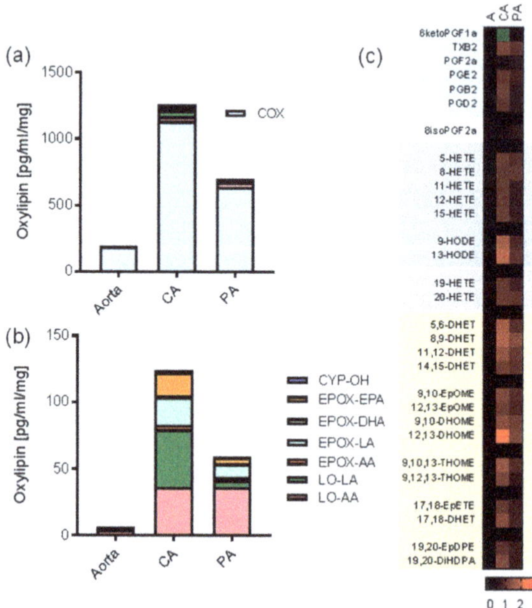

Figure 1. Characterization of oxylipin production from aorta, coronary artery, and pulmonary artery.

(**a,b**) Comparative and relative contribution of cyclooxygenase (COX), lipoxygenase-arachidonic acid (LO-AA), lipoxygenase-linoleic acid (LO-LA), CYP-epoxygenase-arachidonic acid (EPOX-AA), CYP-epoxygenase-linoleic acid (EPOX-LA), CYP-epoxygenase-DHA (EPOX-DHA), CYP-epoxygenase-EPA (EPOX-EPA), and CYP-ω-hydroxylase (CYP-OH) products to the oxylipin releasate of aorta, coronary artery (CA), and pulmonary artery (PA) in 24 h organ culture. (**a**) shows all pathways, whereas (**b**) shows all pathways minus COX. Bars are based upon the single most oxylipin abundant oxylipin product detected in each pathway which is used as a representative index of oxylipin class. (**c**) Heatmap showing Log10 fold differences in the mean amount of each oxylipin detected from coronary artery (CA) and pulmonary artery (PA) compared to aorta. The actual fold range in the coronary artery was 0.5-fold for 6-keto $PGF_{1\alpha}$ to 823-fold for 12,13-DHOME. Data represents organ culture from $n = 3$–4 separate animals.

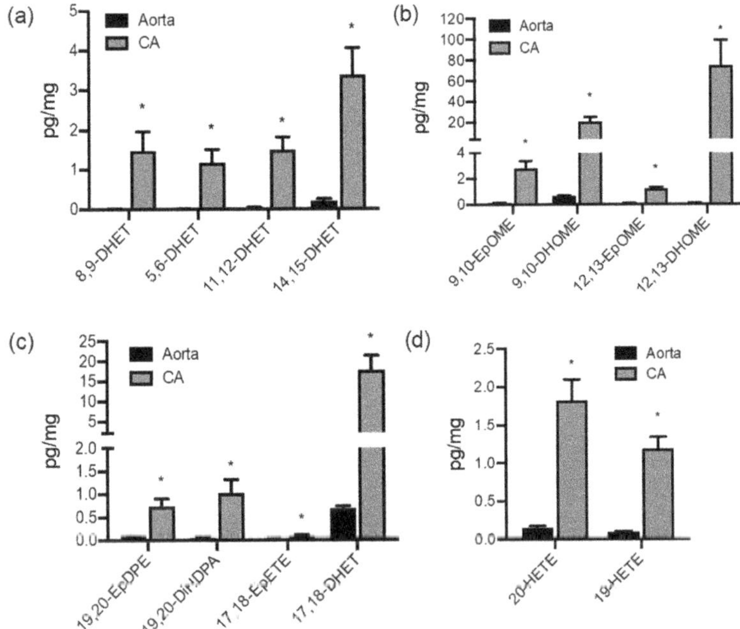

Figure 2. Coronary arteries produce high levels of CYP-derived oxylipins. Figures show detectable CYP epoxygenase (**a**) EPOX-AA, (**b**) EPOX-LA, (**c**) EPOX-DHA/EPA and (**d**) CYP-OH products released by pig aorta (black bars) and coronary artery (grey bars). Oxylipins accumulated in 24 h serum-free organ culture were measured by LC–MS/MS and expressed as pg/mg of wet tissue weight. Data represents organ culture from $n = 3$–4 separate animals. Data represents organ culture from $n = 3$–4 separate animals. * indicates $p < 0.05$ between Aorta and CA.

Interestingly, the aorta and coronary artery produced similar levels of COX products, with the notable exceptions of PGI_2, which was significantly higher from aorta compared to coronary artery, and PGE_2, which was higher in coronary artery compared to aorta ($p < 0.05$ unpaired t-test; Figures 1 and 2). Lipoxygenase-derived HETEs and HODEs were also produced in significantly higher amounts by coronary artery than the aorta (Figure 3). In particular, LA-derived oxylipin epoxygenase and lipoxygenase products were produced at considerably higher levels (up to 90-fold) by coronary artery than aorta (Figures 1–3).

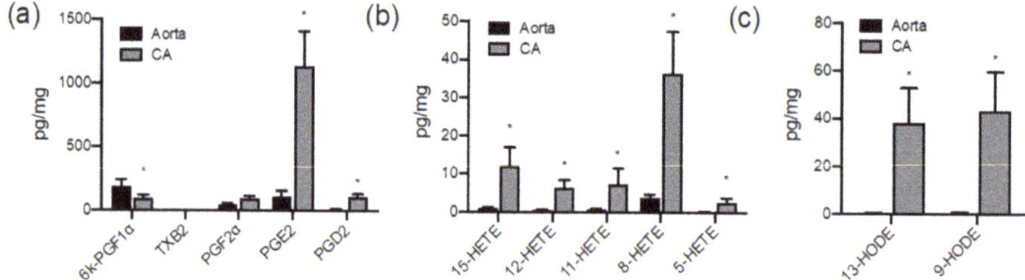

Figure 3. Comparison of aortic and coronary artery production of cyclooxygenase and 'lipoxygenase' oxylipin products. Figures show (**a**) cyclooxygenase, (**b**) 'lipoxygenase' products of arachidonic acid (HETEs) and (**c**) linoleic acid (HODEs) products released by pig aorta (black bars) and coronary artery (grey bars). Oxylipins accumulated in 24 h serum-free organ culture were measured by LC–MS/MS and expressed as pg/mg of wet tissue weight. Data represents organ culture from n = 3–4 separate animals. * indicates $p < 0.05$ between Aorta and CA.

3.2. Regulation of Oxylipin Generation in the Large Vessels of the Pig by Inflammatory Stimuli: LPS/TLR4 Activation

Consistent with the well-established sensitivity of COX-2 induction, LPS elevated prostanoids in aorta, coronary artery, and pulmonary artery. Interestingly, LPS did not induce prostanoids in aortic perivascular adipose tissue (Figure 4a). In particular, the major vascular prostanoids PGI$_2$ and PGE$_2$ were significantly induced by LPS in vascular tissue (Figure 4). The 11-HETE, 15-HETE, 9-HODE and 13-HODE were significantly increased in the aorta and pulmonary artery, but not the coronary artery. With some exceptions, notably 19,20-EpDPE in pulmonary artery and 19-HETE in aorta (Figure 4a; Figure S2), LPS did not consistently alter lipoxygenase or CYP450 product levels (Figure 4a).

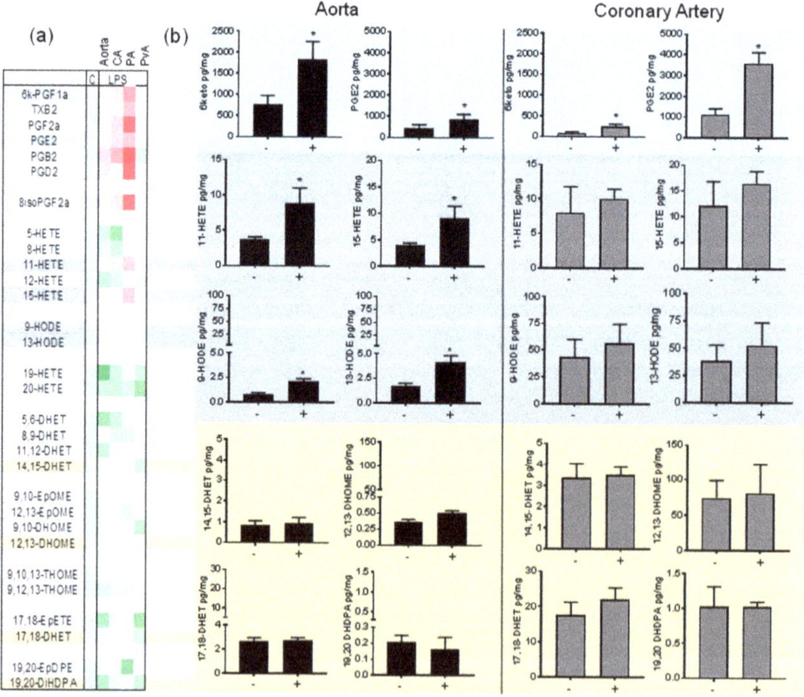

Figure 4. Regulation of oxylipin production in large vessels by LPS/TLR4 activation. (**a**) Heatmap showing

summary of fold differences in the mean oxylipin generation in aorta, coronary artery (CA), pulmonary artery (PA) and perivascular adipose (PvA) untreated tissue (C) compared to tissue treated with LPS (1 µg/mL) ex vivo. The range of fold differences was from 0.5- (19-HETE; Aorta) to 9-fold (PGB$_2$; PVA). (**b**) Comparison of major oxylipin production: 6-ketoPGF$_{1\alpha}$, PGE$_2$, 11-HETE, 15-HETE, 9-HODE, 13-HODE, 14,15-DHET, 12,13-DHOME, 17,18-DHET and 19,20-DHDPA in aorta and coronary artery treated in the absence (-) or presence regulation by LPS (1 µg/mL; +). * indicates $p < 0.05$ by unpaired *t*-test between tissue treated in the presence of absence of LPS. Data represents organ culture from n = 3–4 separate animals.

3.3. The sEH Inhibitor TPPU Reduces TLR-4 Induced Inflammation in pCASMCs

LPS did not induce the pig CYP2J homologue CYP2J34 in organ culture tissue (pulmonary artery and coronary artery) at 24 h or in primary pCASMCs (Figure 5a) at 4 h. By contrast, LPS strongly induced *TNFα* mRNA in both organ culture tissue and pCASMCs (Figure 5a). Although not induced by LPS, the endogenously produced EpFAs were anti-inflammatory in pCASMCs, as co-treatment of pCASMCs with the sEH-I TPPU (1 uM) significantly reduced LPS-induced *TNFα, ICAM-1, VCAM-1, MCP-1 (CCL2)*, and *CD40* mRNA (Figure 5b).

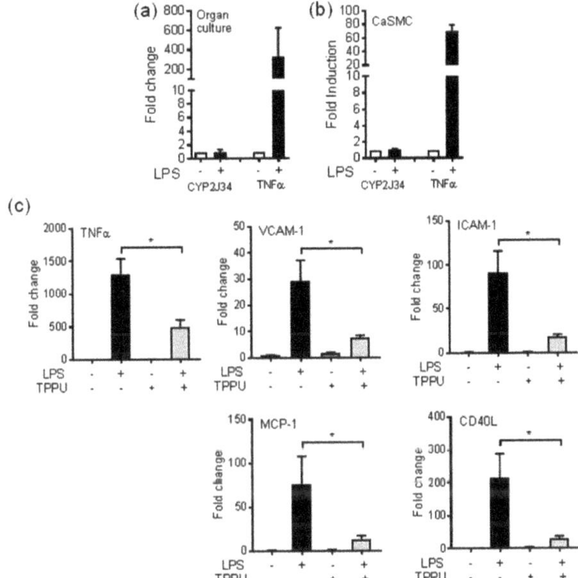

Figure 5. The sEH inhibitor TPPU is anti-inflammatory in coronary artery vascular smooth muscle cells. Expression of TNFα and CYP2J34 mRNA in (**a**) combined pig coronary and pulmonary artery vessels in organ culture at 24 h (n = 4), and (**b**) pig primary coronary artery cells at 4 h (CaSMCs) in the presence or absence of LPS (1 µg/mL). mRNA was measured by qRT-PCR and fold levels normalized to 18S. (**c**) Inflammatory target gene expression of TNFα, VCAM-1, ICAM-1, MCP-1 (*CCL2*) and *CD40* in cultures of pCASMCs in the presence or absence LPS (1 µg/mL; 4 h), and/or sEH inhibitor TPPU (1 µM; given as a 1 h pretreatment before addition of LPS). * indicates $p < 0.05$ by paired *t*-test between cells treated with TPPU in the presence of absence of LPS. Data represents mean ± SE from n = 4 cultures from two separate animals.

4. Discussion

We used a targeted lipidomic approach to identify the profile of oxylipins produced by pig coronary artery, aorta, pulmonary artery and aortic perivascular adipose tissue. In particular, the coronary artery was a major source of epoxygenase-derived oxylipins. Bovine and porcine coronary artery was one of the original sites for the discovery of vasoactive CYP450-derived EETs [60]. Here, we show that the coronary artery also produces CYP450 EpFAs from linoleic acid, EPA and DHA in significantly greater amounts than other large pig vessels. Outside of primates, the pig cardiovascular system is considered the most relevant to human biology. The pig heart is a similar size to human heart, and pigs can spontaneously undergo coronary artery disease [52]. The increased size of the pig compared to rodent models also means that it is also relatively easy to examine specific vascular responses in arteries such as the coronary artery, which would be very difficult in common rodent models.

Using fresh tissue in organ culture comes with certain caveats. Although directly comparative, these results are based upon 24 h accumulation of products. Our previous studies with organ culture indicate that fresh vessels are put under a mild inflammatory stress, which is associated with a low but significant level of COX-2 induction [56,57]. The results here are consistent with these previous studies [56,57]. Additionally, we know relatively little about the long-term stability of a number of oxylipins in media or biological fluids, but clearly there are differences. EETs for example are rapidly metabolized or reincorporated into membranes [3,61], so this 24 h accumulation analysis is likely to underestimate total EET production When we have examined acute oxylipin release (30 min) from rat aorta, prostaglandins and in particular PGI_2 are still the most abundant species detected (DBB unpublished observation). Another caveat to this analysis is whether tissue-specific oxylipin metabolism is present. For example, coronary artery endothelial cells are known to metabolize EETs to chain-shortened epoxy-hexadecadienoic acids [62]; additionally, the presence of CYP4A3 may metabolize EETs into 20-OH derivative PPAR ligands [63], which were not included in our analysis.

LPS induces COX-derived prostanoids in rat and human vessels in organ culture in vitro [56,57]. All the pig vessels tested similarly produced prostanoids in response to LPS. The responsiveness of other oxylipin pathways to LPS is less well understood. Since activation of cPLA2 appears to be common to all three pathways, we hypothesized that lipoxygenase and CYP450 pathways would also be activated. Interestingly, the COX-derived eicosanoids were the only species commonly induced by LPS in all vessels. In aorta, but not coronary artery, 11-HETE and 15-HETE were similarly induced. The 11-HETE and 15-HETE are also potentially COX products [5], so it is intriguing why they are induced in the aorta and not coronary artery. Similar findings were previously found in human whole blood treated with LPS for 18 h [63]. Interestingly, HODEs were also induced by LPS in aorta, pulmonary artery and perivascular adipose tissue, which shows a selective induction of linoleic acid- and arachidonic acid-lipoxygenase pathways [64,65]. Unlike HETE induction, this HODE induction was not previously observed in human whole blood treated with LPS [66] but has been observed in the circulation of mice treated with LPS [67]. These lipidomic results clearly show a high compartmentalization between substrate generation and delivery to individual COX, lipoxygenase and CYP450 pathways. This data provides further impetus to look at the actions of these other oxylipin species. EPA and DHA are considered key components of the purported cardiovascular health benefits of oily fish. Supplementation of human or rodent diets with DHA and EPA increases DHA and EPA EpFAs [68,69]. Coronary artery metabolism of EPA and DHA into EpFAs could therefore contribute to these dietary lifestyle modifications in cardiovascular disease. Additionally, further investigations are required to understand the role of PPAR signaling in any effects.

The coronary artery is of particular interest for vascular research, since coronary artery disease and occlusion is the major cause of heart attacks in humans. There has been considerable interest in both the potential cardioprotective effects of PPAR ligands and testing sEH inhibitors in cardiovascular disease [25,26]. As recently reviewed PPAR ligands in experimental animal models have all shown to reduce aortic atherosclerosis [70,71]. There has been considerable interest in whether these findings translate into humans [70,71]. Both PPARα and PPARγ agonists have shown some mild clinical efficacy in reducing cardiovascular event [71]. However, the clinical efficacy of the PPARγ ligand rosiglitazone has been questioned as it appeared to increase cardiovascular events in an early trial [71]. Nonetheless, there has been considerably recent interest in developing selective modulators, and dual-

and pan-PPAR agonists, that have increased efficacy and reduced side effects [70,71]. We hypothesize that the potential endogenous oxylipin PPAR ligands are more likely to act as pan/dual or selective modulator-type agonists. sEH inhibitors are anti-hypertensive, anti-diabetic, anti-obesity and reduce the development of aortic atherosclerosis in mouse models [25,26,59,72,73]. Importantly, atherosclerosis is rarely investigated in the coronary circulation of mice. More often, aortic atherosclerotic plaque formation is used as a surrogate for coronary artery disease. Our oxylipin profiling suggests that aorta CYP activity underestimates that found in coronary arteries and likely underpredicts the role of CYP-derived oxylipins in coronary artery atherosclerosis. Preservation of the coronary circulation therefore may be so critical that it has evolved this higher EpFA system to maintain flow and limit inflammation. Originally, these positive benefits of sEH inhibitors were attributed to lipid-lowering actions [26]. Human coronary artery disease, in particular obstructive coronary artery disease, is associated with decreased circulating EETs [74,75]. In the heart and coronary circulation, CYP2J2 or EETs mediate coronary reactive hyperemia [41–44], cytoprotection of cardiomyocytes [46], and inhibition of angiotensin II cardiac remodeling [47] in part by the activation of PPARs. sEH inhibition acts to maintain higher levels of EETs/EpFAs or may shunt to alternative PPAR ligands such as the 20-OH CYP4A derivatives. Although, DHETs are also PPAR activators [76], the concentrations required are 10–100-fold higher than published for EETs [8,64]; thus, sEH inhibition will act to promote the PPAR agonist activity of CYP-derived epoxides.

The coronary artery was the largest source of epoxygenase products of all the major vessels we tested, suggesting that multiple oxylipin species may have particularly important roles at this site. The pig homologue of human CYP2J2 is CYP2J34 [77]. In human monocytes and endothelial cells, we found that LPS induced CYP2J2 [28,29,78]. By contrast, LPS did not induce CYP2J34 in pig vessels or monocytes (Figure 5a, DBB unpublished observations), indicating at least one difference between human and pig CYP2J enzymes. We previously reported differences in intimal and medial SMC phenotypes isolated from the rat [79,80]. Medial SMCs but not intimal SMCs were sensitive to the anti-inflammatory actions of sEH inhibitors [81]. The sEH inhibitor TPPU inhibited inflammatory mediators induced by TLR-4 activation in primary pCASMCs. This is the first time anti-inflammatory actions have been described in coronary tissue for sEH inhibitors, and this further supports an anti-inflammatory/pro-resolution role for the sEH pathway in mediating cardioprotective actions. The pCASMCs we cultured represent a classical medial SMC phenotype, with a classical spindle shape and hill-and-valley morphology. No epithelial cell types were observed in these primary cultures. We have yet to determine whether distinct porcine 'intimal' SMC phenotypes can be identified that share these different properties.

The aorta produced the lowest levels of oxylipins, with the pulmonary artery in between the aorta and coronary artery. The lack of activity in the aorta may just reflect the aorta's main role as a conduit vessel and one not particularly responsive to vasoactive mediators or a major site for human vascular disease initiation. The pulmonary artery produced the highest levels of basal and LPS-inducible PGI_2, consistent with the importance of this eicosanoid in maintaining pulmonary health [79], which may be in part mediated by activation of PPARβ/δ (or PPARα) [1,82]. CYP450-derived eicosanoids contribute to hypoxia-induced pulmonary hypertension [83] and are protective in models of inflammation [38–40,84]. The production of these oxylipin mediators from the pulmonary artery suggest that the pig may also be a useful translational model to study oxylipins and PPARs on pulmonary health and disease. Here, we show perivascular adipose is also a large potential source of oxylipins, in particular PGE_2, and further suggest a role for oxylipins from alternative cellular sources as potential mediators of vascular health and disease. Similarly, the relative contribution of vascular cell types—endothelial, smooth muscle phenotypes, adventitial fibroblasts and adipose—require further investigation.

5. Conclusions

We have performed a lipidomic analysis on large vessels and perivascular adipose from the pig. Although prostanoids were the dominant detectable species from all tissue, the coronary artery produced considerably more oxylipins in terms of species and amounts when compared to the aorta and pulmonary artery, in particular those from CYP450 pathways. Using porcine pCASMCs, we showed using the sEH inhibitor TPPU that endogenous CYP-derived epoxy-oxylipin PPAR ligands were strongly anti-inflammatory. The CYP450 pathway in the coronary artery not only provides vasodilator tone, but here we propose an anti-inflammatory tone that helps to prevent processes of vascular disease progression. These results also further highlight the potential for sEH inhibitors as therapies for cardiovascular and inflammatory diseases.

Supplementary Materials: The following are available online at http://www.mdpi.com/2073-4409/9/5/1096/s1, Table S1. Primer pairs, Figure S1. Aorta produces low but detectable levels of CYP-derived oxylipins, and Figure S2. Regulation of oxylipin production in large vessels by LPS/TLR4 activation.

Author Contributions: Conceptualization, D.B.-B.; methodology, D.B.-B., M.L.E., F.B.L., and S.T.; formal analysis, D.B.-B., M.L.E., F.B.L., and S.T.; investigation, D.B.-B., M.L.E., and S.T.; resources, D.B.-B., B.D.H., and D.C.Z.; writing—original draft preparation, D.B.-B.; writing—review and editing, D.B.-B., M.L.E., B.D.H., and D.C.Z.; project administration, D.B.-B., B.D.H., and D.C.Z.; funding acquisition, D.B.-B., B.D.H., and D.C.Z. All authors have read and agreed to the published version of the manuscript.

Funding: This work was supported by funding in part from the British Heart Foundation to D.B-B. (PG/11/39/28890), and from the Intramural Research Program of the National Institutes of Health, National Institute of Environmental Health Sciences Z01 ES025034 to D.C.Z. and in part by NIEHS R35 ES030443-01 RIVER Award, NIEHS P42 ES004699 Superfund Program, and NIDDK R01 DK103616 grant NIH – NIEHS (RIVER Award) to B.D.H.

Conflicts of Interest: The authors declare no conflict of interest. The funders had no role in the design of the study; in the collection, analyses, or interpretation of data; in the writing of the manuscript, or in the decision to publish the results.

Abbreviations

AA	arachidonic acid
CA	coronary artery
COX	cyclooxygenase
CYP	cytochrome P450
DHA	docosahexaenoic acid
DHDPA	dihydroxydocosapentaenoic acid
DHET	dihydroxyeicosatrienoic acid
DHOME	dihydroxyoctadecamonoenoic acid
EET	epoxyeicosatrienoic acid
EPA	eicosapentaenoic acid
EpDPE	epoxydocosapentaenoic acid
EpETE	epoxyeicosatetraenoic acid
EpFA	epoxy fatty acid
EpOME	epoxyoctadecamonoenoic acid
HETE	hydroxyoctadecaenoic acid
HODE	hydroxyoctadecaenoic acid
LA	linoleic acid
LOX	lipoxygenase
PA	pulmonary artery
pCASMC	primary coronary artery smooth muscle cell
PPAR	peroxisome proliferator-activated receptor
PPRE	PPAR response element
PvA	perivascular adipose
sEH	soluble epoxide hydrolase
TPPU	(*N*-[1-(1-oxopropyl)-4-piperidinyl]-N′-[4-(trifluoromethoxy)phenyl]-urea)

References

1. Bishop-Bailey, D.; Wray, J. Peroxisome proliferator-activated receptors: A critical review on endogenous pathways for ligand generation. *Prostaglandins Other Lipid Mediat.* **2003**, *71*, 1–22. [CrossRef]

2. Smilowitz, J.T.; Zivkovic, A.M.; Wan, Y.J.; Watkins, S.M.; Nording, M.L.; Hammock, B.D.; German, J.B. Nutritional lipidomics: Molecular metabolism, analytics, and diagnostics. *Mol. Nutr. Food Res.* **2013**, *57*, 1319–1335. [CrossRef]

3. Zeldin, D.C. Epoxygenase pathways of arachidonic acid metabolism. *J. Biol. Chem.* **2001**, *276*, 36059–36062. [CrossRef]

4. Bishop-Bailey, D.; Thomson, S.; Askari, A.; Faulkner, A.; Wheeler-Jones, C. Lipid-metabolizing CYPs in the regulation and dysregulation of metabolism. *Annu. Rev. Nutr.* **2014**, *34*, 261–279. [CrossRef]

5. Bailey, J.M.; Bryant, R.W.; Whiting, J.; Salata, K. Characterization of 11-HETE and 15-HETE, together with prostacyclin, as major products of the cyclooxygenase pathway in cultured rat aorta smooth muscle cells. *J. Lipid Res.* **1983**, *24*, 1419–1428.

6. Thuresson, E.D.; Lakkides, K.M.; Smith, W.L. Different catalytically competent arrangements of arachidonic acid within the cyclooxygenase active site of prostaglandin endoperoxide H synthase-1 lead to the formation of different oxygenated products. *J. Biol. Chem.* **2000**, *275*, 8501–8507. [CrossRef]

7. Wu, S.; Moomaw, C.R.; Tomer, K.B.; Falck, J.R.; Zeldin, D.C. Molecular cloning and expression of CYP2J2, a human cytochrome P450 arachidonic acid epoxygenase highly expressed in heart. *J. Biol. Chem.* **1996**, *271*, 3460–3468. [CrossRef]

8. Wray, J.A.; Sugden, M.C.; Zeldin, D.C.; Greenwood, G.K.; Samsuddin, S.; Miller-Degraff, L.; Bradbury, J.A.; Holness, M.J.; Warner, T.D.; Bishop-Bailey, D. The epoxygenases CYP2J2 activates the nuclear receptor PPARalpha in vitro and in vivo. *PLoS ONE* **2009**, *4*, e7421. [CrossRef]

9. Edin, M.L.; Hamedani, B.G.; Gruzdev, A.; Graves, J.P.; Lih, F.B.; Arbes, S.J., 3rd; Singh, R.; Orjuela Leon, A.C.; Bradbury, J.A.; DeGraff, L.M.; et al. Epoxide hydrolase 1 (EPHX1) hydrolyzes epoxyeicosanoids and impairs cardiac recovery after ischemia. *J. Biol. Chem.* **2018**, *293*, 3281–3292. [CrossRef]

10. Hwang, S.H.; Wecksler, A.T.; Wagner, K.; Hammock, B.D. Rationally designed multitarget agents against inflammation and pain. *Curr. Med. Chem.* **2013**, *20*, 1783–1799. [CrossRef]

11. Imig, J.D.; Hammock, B.D. Soluble epoxide hydrolase as a therapeutic target for cardiovascular diseases. *Nat. Rev. Drug Discov.* **2009**, *8*, 794–805. [CrossRef]

12. Morisseau, C.; Hammock, B.D. Impact of soluble epoxide hydrolase and epoxyeicosanoids on human health. *Annu. Rev. Pharmacol. Toxicol.* **2013**, *53*, 37–58. [CrossRef]

13. Wray, J.; Bishop-Bailey, D. Epoxygenases and peroxisome proliferator-activated receptors in mammalian vascular biology. *Exp. Physiol.* **2008**, *93*, 148–154. [CrossRef]

14. Bishop-Bailey, D.; Mitchell, J.A.; Warner, T.D. COX-2 in cardiovascular disease. *Arter. Thromb. Vasc. Biol.* **2006**, *26*, 956–958. [CrossRef]

15. Baum, S.J.; Kris-Etherton, P.M.; Willett, W.C.; Lichtenstein, A.H.; Rudel, L.L.; Maki, K.C.; Whelan, J.; Ramsden, C.E.; Block, R.C. Fatty acids in cardiovascular health and disease: A comprehensive update. *J. Clin. Lipidol.* **2012**, *6*, 216–234. [CrossRef]

16. Spiecker, M.; Liao, J.K. Vascular protective effects of cytochrome p450 epoxygenase-derived eicosanoids. *Arch Biochem. Biophys.* **2005**, *433*, 413–420. [CrossRef]

17. Capra, V.; Back, M.; Barbieri, S.S.; Camera, M.; Tremoli, E.; Rovati, G.E. Eicosanoids and their drugs in cardiovascular diseases: Focus on atherosclerosis and stroke. *Med. Res. Rev.* **2013**, *33*, 364–438.

18. Hersberger, M. Potential role of the lipoxygenase derived lipid mediators in atherosclerosis: Leukotrienes, lipoxins and resolvins. *Clin. Chem. Lab. Med.* **2010**, *48*, 1063–1073. [CrossRef]

19. Fleming, I. The factor in EDHF: Cytochrome P450 derived lipid mediators and vascular signaling. *Vasc. Pharm.* **2016**, *86*, 31–40. [CrossRef]

20. Mitchell, J.A.; Kirkby, N.S. Eicosanoids, prostacyclin and cyclooxygenase in the cardiovascular system. *Br. J Pharm.* **2019**, *176*, 1038–1050. [CrossRef]

21. Mitchell, J.A.; Warner, T.D. COX isoforms in the cardiovascular system: Understanding the activities of non-steroidal anti-inflammatory drugs. *Nat. Rev. Drug Discov.* **2006**, *5*, 75–86. [CrossRef]

22. Chaudhary, K.R.; Zordoky, B.N.; Edin, M.L.; Alsaleh, N.; El-Kadi, A.O.; Zeldin, D.C.; Seubert, J.M. Differential effects of soluble epoxide hydrolase inhibition and CYP2J2 overexpression on postischemic cardiac function in aged mice. *Prostaglandins Other Lipid Mediat.* **2013**, *104–105*, 8–17. [CrossRef]

23. Deng, Y.; Edin, M.L.; Theken, K.N.; Schuck, R.N.; Flake, G.P.; Kannon, M.A.; Degraff, L.M.; Lih, F.B.; Foley, J.; Bradbury, J.A.; et al. Endothelial CYP epoxygenase overexpression and soluble epoxide hydrolase disruption attenuate acute vascular inflammatory responses in mice. *Faseb. J.* **2011**, *25*, 703–713. [CrossRef]

24. Liu, Y.; Zhang, Y.; Schmelzer, K.; Lee, T.S.; Fang, X.; Zhu, Y.; Spector, A.A.; Gill, S.; Morisseau, C.; Hammock, B.D.; et al. The antiinflammatory effect of laminar flow: The role of PPARgamma, epoxyeicosatrienoic acids, and soluble epoxide hydrolase. *Proc. Natl. Acad. Sci. USA* **2005**, *102*, 16747–16752. [CrossRef]

25. Revermann, M.; Schloss, M.; Barbosa-Sicard, E.; Mieth, A.; Liebner, S.; Morisseau, C.; Geisslinger, G.; Schermuly, R.T.; Fleming, I.; Hammock, B.D.; et al. Soluble epoxide hydrolase deficiency attenuates neointima formation in the femoral cuff model of hyperlipidemic mice. *Arterioscler. Thromb. Vasc. Biol.* **2010**, *30*, 909–914. [CrossRef]

26. Zhang, L.N.; Vincelette, J.; Cheng, Y.; Mehra, U.; Chen, D.; Anandan, S.K.; Gless, R.; Webb, H.K.; Wang, Y.X. Inhibition of soluble epoxide hydrolase attenuated atherosclerosis, abdominal aortic aneurysm formation, and dyslipidemia. *Arter. Thromb. Vasc. Biol.* **2009**, *29*, 1265–1270. [CrossRef]

27. Askari, A.; Thomson, S.J.; Edin, M.L.; Zeldin, D.C.; Bishop-Bailey, D. Roles of the epoxygenase CYP2J2 in the endothelium. *Prostaglandins Other Lipid Mediat* **2013**, *107*, 56–63. [CrossRef]

28. Bystrom, J.; Thomson, S.J.; Johansson, J.; Edin, M.L.; Zeldin, D.C.; Gilroy, D.W.; Smith, A.M.; Bishop-Bailey, D. Inducible CYP2J2 and its product 11,12-EET promotes bacterial phagocytosis: A role for CYP2J2 deficiency in the pathogenesis of Crohn's disease? *PLoS ONE* **2013**, *8*, e75107. [CrossRef]

29. Bystrom, J.; Wray, J.A.; Sugden, M.C.; Holness, M.J.; Swales, K.E.; Warner, T.D.; Edin, M.L.; Zeldin, D.C.; Gilroy, D.W.; Bishop-Bailey, D. Endogenous epoxygenases are modulators of monocyte/macrophage activity. *PLoS ONE* **2011**, *6*, e26591. [CrossRef]

30. Gilroy, D.W.; Edin, M.L.; De Maeyer, R.P.; Bystrom, J.; Newson, J.; Lih, F.B.; Stables, M.; Zeldin, D.C.; Bishop-Bailey, D. CYP450-derived oxylipins mediate inflammatory resolution. *Proc. Natl. Acad. Sci. USA* **2016**, *113*, E3240–E3249. [CrossRef]

31. Edin, M.L.; Wang, Z.; Bradbury, J.A.; Graves, J.P.; Lih, F.B.; DeGraff, L.M.; Foley, J.F.; Torphy, R.; Ronnekleiv, O.K.; Tomer, K.B.; et al. Endothelial expression of human cytochrome P450 epoxygenase CYP2C8 increases susceptibility to ischemia-reperfusion injury in isolated mouse heart. *Faseb. J.* **2011**, *25*, 3436–3447. [CrossRef]

32. Fromel, T.; Jungblut, B.; Hu, J.; Trouvain, C.; Barbosa-Sicard, E.; Popp, R.; Liebner, S.; Dimmeler, S.; Hammock, B.D.; Fleming, I. Soluble epoxide hydrolase regulates hematopoietic progenitor cell function via generation of fatty acid diols. *Proc. Natl. Acad. Sci. USA* **2012**, *109*, 9995–10000. [CrossRef]

33. Zimmer, B.; Angioni, C.; Osthues, T.; Toewe, A.; Thomas, D.; Pierre, S.C.; Geisslinger, G.; Scholich, K.; Sisignano, M. The oxidized linoleic acid metabolite 12,13-DiHOME mediates thermal hyperalgesia during inflammatory pain. *Biochim. Et. Biophys. Acta* **2018**, *1863*, 669–678. [CrossRef]

34. Moghaddam, M.F.; Grant, D.F.; Cheek, J.M.; Greene, J.F.; Williamson, K.C.; Hammock, B.D. Bioactivation of leukotoxins to their toxic diols by epoxide hydrolase. *Nat. Med.* **1997**, *3*, 562–566. [CrossRef]

35. Sinal, C.J.; Miyata, M.; Tohkin, M.; Nagata, K.; Bend, J.R.; Gonzalez, F.J. Targeted disruption of soluble epoxide hydrolase reveals a role in blood pressure regulation. *J. Biol. Chem.* **2000**, *275*, 40504–40510. [CrossRef]

36. Imig, J.D.; Zhao, X.; Capdevila, J.H.; Morisseau, C.; Hammock, B.D. Soluble epoxide hydrolase inhibition lowers arterial blood pressure in angiotensin II hypertension. *Hypertension* **2002**, *39*, 690–694. [CrossRef]

37. Luria, A.; Bettaieb, A.; Xi, Y.; Shieh, G.J.; Liu, H.C.; Inoue, H.; Tsai, H.J.; Imig, J.D.; Haj, F.G.; Hammock, B.D. Soluble epoxide hydrolase deficiency alters pancreatic islet size and improves glucose homeostasis in a model of insulin resistance. *Proc. Natl. Acad. Sci. USA* **2011**, *108*, 9038–9043. [CrossRef]

38. Revermann, M.; Barbosa-Sicard, E.; Dony, E.; Schermuly, R.T.; Morisseau, C.; Geisslinger, G.; Fleming, I.; Hammock, B.D.; Brandes, R.P. Inhibition of the soluble epoxide hydrolase attenuates monocrotaline-induced pulmonary hypertension in rats. *J. Hypertens.* **2009**, *27*, 322–331. [CrossRef]

39. Wang, L.; Yang, J.; Guo, L.; Uyeminami, D.; Dong, H.; Hammock, B.D.; Pinkerton, K.E. Use of a soluble epoxide hydrolase inhibitor in smoke-induced chronic obstructive pulmonary disease. *Am. J. Respir. Cell Mol. Biol.* **2012**, *46*, 614–622. [CrossRef]

40. Podolin, P.L.; Bolognese, B.J.; Foley, J.F.; Long, E., 3rd; Peck, B.; Umbrecht, S.; Zhang, X.; Zhu, P.; Schwartz, B.; Xie, W.; et al. In vitro and in vivo characterization of a novel soluble epoxide hydrolase inhibitor. *Prostaglandins Other Lipid Mediat.* **2013**, *104–105*, 25–31. [CrossRef]

41. Hanif, A.; Edin, M.L.; Zeldin, D.C.; Morisseau, C.; Falck, J.R.; Nayeem, M.A. Vascular endothelial overexpression of human CYP2J2 (Tie2-CYP2J2 Tr) modulates cardiac oxylipin profiles and enhances coronary reactive hyperemia in mice. *PLoS ONE* **2017**, *12*, e0174137. [CrossRef]

42. Hanif, A.; Edin, M.L.; Zeldin, D.C.; Morisseau, C.; Falck, J.R.; Nayeem, M.A. Vascular Endothelial Over-Expression of Human Soluble Epoxide Hydrolase (Tie2-sEH Tr) Attenuates Coronary Reactive Hyperemia in Mice: Role of Oxylipins and omega-Hydroxylases. *PLoS ONE* **2017**, *12*, e0169584. [CrossRef]

43. Hanif, A.; Edin, M.L.; Zeldin, D.C.; Morisseau, C.; Nayeem, M.A. Effect of Soluble Epoxide Hydrolase on the Modulation of Coronary Reactive Hyperemia: Role of Oxylipins and PPARgamma. *PLoS ONE* **2016**, *11*, e0162147. [CrossRef]

44. Hanif, A.; Edin, M.L.; Zeldin, D.C.; Morisseau, C.; Nayeem, M.A. Deletion of soluble epoxide hydrolase enhances coronary reactive hyperemia in isolated mouse heart: Role of oxylipins and PPARgamma. *Am. J. Physiol. Regul. Integr. Comp. Physiol.* **2016**, *311*, R676–R688. [CrossRef]

45. Nayeem, M.A.; Pradhan, I.; Mustafa, S.J.; Morisseau, C.; Falck, J.R.; Zeldin, D.C. Adenosine A2A receptor modulates vascular response in soluble epoxide hydrolase-null mice through CYP-epoxygenases and PPARgamma. *Am. J. Physiol. Regul. Integr. Comp. Physiol.* **2012**, *304*, R23–R32. [CrossRef]

46. Samokhvalov, V.; Vriend, J.; Jamieson, K.L.; Akhnokh, M.K.; Manne, R.; Falck, J.R.; Seubert, J.M. PPAR gamma signaling is required for mediating EETs protective effects in neonatal cardiomyocytes exposed to LPS. *Front Pharmacol.* **2014**, *5*, 242. [CrossRef]

47. He, Z.; Zhang, X.; Chen, C.; Wen, Z.; Hoopes, S.L.; Zeldin, D.C.; Wang, D.W. Cardiomyocyte-specific expression of CYP2J2 prevents development of cardiac remodelling induced by angiotensin II. *Cardiovasc. Res.* **2015**, *105*, 304–317. [CrossRef]

48. Cai, Z.; Zhao, G.; Yan, J.; Liu, W.; Feng, W.; Ma, B.; Yang, L.; Wang, J.A.; Tu, L.; Wang, D.W. CYP2J2 overexpression increases EETs and protects against angiotensin II-induced abdominal aortic aneurysm in mice. *J. Lipid Res.* **2013**, *54*, 1448–1456. [CrossRef]

49. Kim, J.; Imig, J.D.; Yang, J.; Hammock, B.D.; Padanilam, B.J. Inhibition of soluble epoxide hydrolase prevents renal interstitial fibrosis and inflammation. *Am. J. Physiol. Renal Physiol.* **2014**, *307*, F971–F980. [CrossRef]

50. Imig, J.D.; Walsh, K.A.; Hye Khan, M.A.; Nagasawa, T.; Cherian-Shaw, M.; Shaw, S.M.; Hammock, B.D. Soluble epoxide hydrolase inhibition and peroxisome proliferator activated receptor gamma agonist improve vascular function and decrease renal injury in hypertensive obese rats. *Exp. Biol. Med. (Maywood)* **2012**, *237*, 1402–1412. [CrossRef]

51. Xu, D.Y.; Davis, B.B.; Wang, Z.H.; Zhao, S.P.; Wasti, B.; Liu, Z.L.; Li, N.; Morisseau, C.; Chiamvimonvat, N.; Hammock, B.D. A potent soluble epoxide hydrolase inhibitor, t-AUCB, acts through PPARgamma to modulate the function of endothelial progenitor cells from patients with acute myocardial infarction. *Int. J. Cardiol.* **2012**, *167*, 1298–1304. [CrossRef]

52. Shim, J.; Al-Mashhadi, R.H.; Sorensen, C.B.; Bentzon, J.F. Large animal models of atherosclerosis—new tools for persistent problems in cardiovascular medicine. *J. Pathol.* **2016**, *238*, 257–266. [CrossRef]

53. Tanaka, K.; Sata, M. Roles of Perivascular Adipose Tissue in the Pathogenesis of Atherosclerosis. *Front. Physiol.* **2018**, *9*, 3. [CrossRef]

54. Huang Cao, Z.F.; Stoffel, E.; Cohen, P. Role of Perivascular Adipose Tissue in Vascular Physiology and Pathology. *Hypertension* **2017**, *69*, 770–777. [CrossRef]

55. Liu, J.Y.; Lin, Y.P.; Qiu, H.; Morisseau, C.; Rose, T.E.; Hwang, S.H.; Chiamvimonvat, N.; Hammock, B.D. Substituted phenyl groups improve the pharmacokinetic profile and anti-inflammatory effect of urea-based soluble epoxide hydrolase inhibitors in murine models. *Eur. J. Pharm. Sci.* **2013**, *48*, 619–627. [CrossRef]

56. Bishop-Bailey, D.; Larkin, S.W.; Warner, T.D.; Chen, G.; Mitchell, J.A. Characterization of the induction of nitric oxide synthase and cyclo-oxygenase in rat aorta in organ culture. *Br. J. Pharm.* **1997**, *121*, 125–133. [CrossRef]

57. Bishop-Bailey, D.; Pepper, J.R.; Haddad, E.B.; Newton, R.; Larkin, S.W.; Mitchell, J.A. Induction of cyclooxygenase-2 in human saphenous vein and internal mammary artery. *Arter. Thromb. Vasc. Biol.* **1997**, *17*, 1644–1648. [CrossRef]

58. Bishop-Bailey, D.; Pepper, J.R.; Larkin, S.W.; Mitchell, J.A. Differential induction of cyclooxygenase-2 in human arterial and venous smooth muscle: Role of endogenous prostanoids. *Arter. Thromb. Vasc. Biol.* **1998**, *18*, 1655–1661. [CrossRef]

59. Lee, C.R.; Imig, J.D.; Edin, M.L.; Foley, J.; DeGraff, L.M.; Bradbury, J.A.; Graves, J.P.; Lih, F.B.; Clark, J.; Myers, P.; et al. Endothelial expression of human cytochrome P450 epoxygenases lowers blood pressure and attenuates hypertension-induced renal injury in mice. *Faseb. J.* **2010**, *24*, 3770–3781. [CrossRef]

60. Hecker, M.; Bara, A.T.; Bauersachs, J.; Busse, R. Characterization of endothelium-derived hyperpolarizing factor as a cytochrome P450-derived arachidonic acid metabolite in mammals. *J. Physiol.* **1994**, *481*, 407–414. [CrossRef]

61. Fang, X.; Kaduce, T.L.; Weintraub, N.L.; Harmon, S.; Teesch, L.M.; Morisseau, C.; Thompson, D.A.; Hammock, B.D.; Spector, A.A. Pathways of epoxyeicosatrienoic acid metabolism in endothelial cells. Implications for the vascular effects of soluble epoxide hydrolase inhibition. *J. Biol. Chem.* **2001**, *276*, 14867–14874. [CrossRef]

62. Fang, X.; Weintraub, N.L.; Oltman, C.L.; Stoll, L.L.; Kaduce, T.L.; Harmon, S.; Dellsperger, K.C.; Morisseau, C.; Hammock, B.D.; Spector, A.A. Human coronary endothelial cells convert 14,15-EET to a biologically active chain-shortened epoxide. *Am. J. Physiol.* **2002**, *283*, H2306–H2314. [CrossRef]

63. Cowart, L.A.; Wei, S.; Hsu, M.H.; Johnson, E.F.; Krishna, M.U.; Falck, J.R.; Capdevila, J.H. The CYP4A isoforms hydroxylate epoxyeicosatrienoic acids to form high affinity peroxisome proliferator-activated receptor ligands. *J. Biol. Chem.* **2002**, *277*, 35105–35112.

64. Zarbock, A.; Distasi, M.R.; Smith, E.; Sanders, J.M.; Kronke, G.; Harry, B.L.; Von Vietinghoff, S.; Buscher, K.; Nadler, J.L.; Ley, K. Improved survival and reduced vascular permeability by eliminating or blocking 12/15-lipoxygenase in mouse models of acute lung injury (ALI). *J. Immunol.* **2009**, *183*, 4715–4722.

65. Serio, K.J.; Reddy, K.V.; Bigby, T.D. Lipopolysaccharide induces 5-lipoxygenase-activating protein gene expression in THP-1 cells via a NF-kappaB and C/EBP-mediated mechanism. *Am. J. Physiol. Cell Physiol.* **2005**, *288*, C1125–C1133. [CrossRef]

66. Kirkby, N.S.; Reed, D.M.; Edin, M.L.; Rauzi, F.; Mataragka, S.; Vojnovic, I.; Bishop-Bailey, D.; Milne, G.L.; Longhurst, H.; Zeldin, D.C.; et al. Inherited human group IVA cytosolic phospholipase A2 deficiency abolishes platelet, endothelial, and leucocyte eicosanoid generation. *Faseb. J.* **2015**, *29*, 4568–4578. [CrossRef]

67. Apaya, M.K.; Lin, C.Y.; Chiou, C.Y.; Yang, C.C.; Ting, C.Y.; Shyur, L.F. Simvastatin and a plant galactolipid protect animals from septic shock by regulating oxylipin mediator dynamics through the MAPK-cPLA2 signaling pathway. *Mol. Med.* **2015**, *21*, 988–1001.

68. Arnold, C.; Markovic, M.; Blossey, K.; Wallukat, G.; Fischer, R.; Dechend, R.; Konkel, A.; Von Schacky, C.; Luft, F.C.; Muller, D.N.; et al. Arachidonic acid-metabolizing cytochrome P450 enzymes are targets of 1-3 fatty acids. *J. Biol. Chem.* **2010**, *285*, 32720–32733.

69. Fischer, R.; Konkel, A.; Mehling, H.; Blossey, K.; Gapelyuk, A.; Wessel, N.; Von Schacky, C.; Dechend, R.; Muller, D.N.; Rothe, M.; et al. Dietary omega-3 fatty acids modulate the eicosanoid profile in man primarily via the CYP-epoxygenase pathway. *J. Lipid Res.* **2014**, *55*, 1150–1164.

70. Han, L.; Shen, W.J.; Bittner, S.; Kraemer, F.B.; Azhar, S. PPARs: Regulators of metabolism and as therapeutic targets in cardiovascular disease. Part I: PPAR-α. *Future Cardiol.* **2017**, *13*, 259–278.

71. Han, L.; Shen, W.J.; Bittner, S.; Kraemer, F.B.; Azhar, S. PPARs: Regulators of metabolism and as therapeutic targets in cardiovascular disease. Part II: PPAR-β/δ and PPAR-γ. *Future Cardiol.* **2017**, *13*, 279–296.

72. Luo, P.; Chang, H.H.; Zhou, Y.; Zhang, S.; Hwang, S.H.; Morisseau, C.; Wang, C.Y.; Inscho, E.W.; Hammock, B.D.; Wang, M.H. Inhibition or deletion of soluble epoxide hydrolase prevents hyperglycemia, promotes insulin secretion, and reduces islet apoptosis. *J. Phar. Exp. Ther.* **2010**, *334*, 430–438.

73. Xiao, B.; Li, X.; Yan, J.; Yu, X.; Yang, G.; Xiao, X.; Voltz, J.W.; Zeldin, D.C.; Wang, D.W. Overexpression of cytochrome P450 epoxygenases prevents development of hypertension in spontaneously hypertensive rats by enhancing atrial natriuretic peptide. *J. Phar. Exp. Ther.* **2010**, *334*, 784–794.

74. Oni-Orisan, A.; Edin, M.L.; Lee, J.A.; Wells, M.A.; Christensen, E.S.; Vendrov, K.C.; Lih, F.B.; Tomer, K.B.; Bai, X.; Taylor, J.M.; et al. Cytochrome P450-derived epoxyeicosatrienoic acids and coronary artery disease in humans: A targeted metabolomics study. *J. Lipid Res.* **2016**, *57*, 109–119.

75. Schuck, R.N.; Theken, K.N.; Edin, M.L.; Caughey, M.; Bass, A.; Ellis, K.; Tran, B.; Steele, S.; Simmons, B.P.; Lih, F.B.; et al. Cytochrome P450-derived eicosanoids and vascular dysfunction in coronary artery disease patients. *Atherosclerosis* **2013**, *227*, 442–448.

76. Ng, V.Y.; Huang, Y.; Reddy, L.M.; Falck, J.R.; Lin, E.T.; Kroetz, D.L. Cytochrome P450 eicosanoids are activators of peroxisome proliferator-activated receptor alpha. *Drug Metab. Dispos.* **2007**, *35*, 1126–1134.

77. Messina, A.; Siniscalco, A.; Puccinelli, E.; Gervasi, P.G.; Longo, V. Cloning and tissues expression of the pig CYP1B1 and CYP2J34. *Res. Vet. Sci.* **2012**, *92*, 438–443.

78. Askari, A.A.; Thomson, S.; Edin, M.L.; Lih, F.B.; Zeldin, D.C.; Bishop-Bailey, D. Basal and inducible anti-inflammatory epoxygenase activity in endothelial cells. *Biochem. Biophys. Res. Commun.* **2014**, *446*, 633–637.

79. Bishop-Bailey, D.; Hla, T.; Warner, T.D. Intimal smooth muscle cells as a target for peroxisome proliferator-activated receptor-gamma ligand therapy. *Circ. Res.* **2002**, *91*, 210–217.

80. Thomson, S.; Edin, M.L.; Lih, F.B.; Davies, M.; Yaqoob, M.M.; Hammock, B.D.; Gilroy, D.; Zeldin, D.C.; Bishop-Bailey, D. Intimal smooth muscle cells are a source but not a sensor of anti-inflammatory CYP450 derived oxylipins. *Biochem. Biophys. Res. Commun.* **2015**, *463*, 774–780.

81. Del Pozo, R.; Hernandez Gonzalez, I.; Escribano-Subias, P. The prostacyclin pathway in pulmonary arterial hypertension: A clinical review. *Expert. Rev. Respir. Med.* **2017**, *11*, 491–503. [CrossRef] [PubMed]

82. Ali, F.Y.; Egan, K.; FitzGerald, G.A.; Desvergne, B.; Wahli, W.; Bishop-Bailey, D.; Warner, T.D.; Mitchell, J.A. Role of prostacyclin versus peroxisome proliferator-activated receptor beta receptors in prostacyclin sensing by lung fibroblasts. *Am. J. Respir. Cell Mol. Biol.* **2006**, *34*, 242–246. [CrossRef] [PubMed]

83. Pokreisz, P.; Fleming, I.; Kiss, L.; Barbosa-Sicard, E.; Fisslthaler, B.; Falck, J.R.; Hammock, B.D.; Kim, I.H.; Szelid, Z.; Vermeersch, P.; et al. Cytochrome P450 epoxygenase gene function in hypoxic pulmonary vasoconstriction and pulmonary vascular remodeling. *Hypertension* **2006**, *47*, 762–770. [CrossRef] [PubMed]

84. Zheng, C.; Wang, L.; Li, R.; Ma, B.; Tu, L.; Xu, X.; Dackor, R.T.; Zeldin, D.C.; Wang, D.W. Gene delivery of cytochrome p450 epoxygenase ameliorates monocrotaline-induced pulmonary artery hypertension in rats. *Am. J. Respir. Cell Mol. Biol.* **2010**, *43*, 740–749. [CrossRef] [PubMed]

Article

Vascular PPARβ/δ Promotes Tumor Angiogenesis and Progression

Kay-Dietrich Wagner [1,†], Siyue Du [1,†], Luc Martin [1], Nathalie Leccia [2], Jean-François Michiels [2] and Nicole Wagner [1,*]

[1] Université Côte d'Azur, CNRS, INSERM, iBV, 06107 Nice, France; kwagner@unice.fr (K.-D.W.); Siyue.DU@univ-cotedazur.fr (S.D.); Luc.MARTIN@univ-cotedazur.fr (L.M.)
[2] Department of Pathology, CHU Nice, 06107 Nice, France; leccia.n@chu-nice.fr (N.L.); michiels.jf@chu-nice.fr (J.-F.M.)
* Correspondence: nwagner@unice.fr; Tel.: +33-493-377665
† These authors contributed equally to this work.

Received: 29 October 2019; Accepted: 11 December 2019; Published: 12 December 2019

Abstract: Peroxisome proliferator-activated receptors (PPARs) are nuclear receptors, which function as transcription factors. Among them, PPARβ/δ is highly expressed in endothelial cells. Pharmacological activation with PPARβ/δ agonists had been shown to increase their angiogenic properties. PPARβ/δ has been suggested to be involved in the regulation of the angiogenic switch in tumor progression. However, until now, it is not clear to what extent the expression of PPARβ/δ in tumor endothelium influences tumor progression and metastasis formation. We addressed this question using transgenic mice with an inducible conditional vascular-specific overexpression of PPARβ/δ. Following specific over-expression of PPARβ/δ in endothelial cells, we induced syngenic tumors. We observed an enhanced tumor growth, a higher vessel density, and enhanced metastasis formation in the tumors of animals with vessel-specific overexpression of PPARβ/δ. In order to identify molecular downstream targets of PPARβ/δ in the tumor endothelium, we sorted endothelial cells from the tumors and performed RNA sequencing. We identified platelet-derived growth factor receptor beta (Pdgfrb), platelet-derived growth factor subunit B (Pdgfb), and the tyrosinkinase KIT (c-Kit) as new PPARβ/δ -dependent molecules. We show here that PPARβ/δ activation, regardless of its action on different cancer cell types, leads to a higher tumor vascularization which favors tumor growth and metastasis formation.

Keywords: peroxisome-proliferator activated receptors; tumor angiogenesis; tumor progression; metastasis formation; endothelial cells; RNA sequencing

1. Introduction

Peroxisome proliferator-activated receptors (PPARs) are nuclear receptors. They function as ligand activated transcription factors. They exist in three isoforms, PPARα, PPARδ (formerly PPARβ), and PPARγ. For all PPARs, lipids are endogenous ligands, linking them directly to metabolism. PPARs are considered as important transcriptional regulators of genes involved in lipid metabolism and cardiac energy production [1]. PPARs form heterodimers with retinoic X receptors, and, upon ligand binding, modulate gene expression of downstream target genes dependent on the presence of co-repressors or co-activators. This results in cell-type specific complex regulations of proliferation, differentiation, and cell survival. Specific synthetic agonists for all PPARs are available. PPARα and PPARγ agonists are already in clinical use for the treatment of hyperlipidemia and type 2 diabetes, respectively. More recently, PPARβ/δ activation came into focus as an interesting novel approach for the treatment of metabolic syndrome and associated cardiovascular diseases.

Agonists of PPARβ/δ improve insulin sensitivity in both, murine models and in humans. Thus, they are considered a potential target in the treatment of obesity and obesity associated disorders [2]. Metabolic syndrome is regarded as a high-risk state for cancer. Targeting of PPARβ/δ was suggested for the treatment of metabolic syndrome (reviewed in [3]), but the resulting consequences for cancer risk are less clear. The function of PPARβ/δ in different types of cancer is highly controversial at present. This might result from the different experimental models used and also from the varying contribution of PPARβ/δ to endothelial cell proliferation, inflammation, and tumor cell proliferation, differentiation, or apoptosis (reviewed in [4]). As all these processes are critically involved in cancer growth; different approaches could give rise to opposing results. We recently reported that PPARβ/δ inhibits melanoma cell proliferation through the transcriptional repression of the Wilms' tumor suppressor WT1 [5]. In contrast, we found that PPARβ/δ increases liposarcoma cell proliferation through the direct repression of the adipose tissue secretory factor leptin [6].

PPARβ/δ, unlike PPARα or PPARγ, appears to be a predominantly pro-angiogenic signaling molecule. PPARβ/δ is expressed in endothelial cells and pharmacological activation of endothelial cells with PPARβ/δ agonists had been shown to increase the angiogenic properties of these cells [7]. PPARβ/δ is also involved in physiological angiogenesis. As we and others showed, treatment with the PPARβ/δ agonists GW0742 and GW501516 induced an exercise-like phenotype in the heart. Both agonists induced a surprisingly rapid (after 24 h) remodeling of mouse hearts [8] and skeletal muscle [9] by increasing micro-vessel densities. However, until now it was not clear whether either the increase of the cardiac vasculature drives the myocardial hypertrophy or the enhanced cardiac angiogenesis might be a potential indirect effect of cardiomyocyte-specific PPARβ/δ activation. In a recent work, we addressed this question through the generation of transgenic mice with an inducible conditional vascular-specific overexpression of PPARβ/δ and analyzed the normal cardiac phenotype and function as well as morphology and function under chronic ischemic heart disease conditions. We showed that inducible vessel-specific overexpression of PPARβ/δ results in a rapid induction of angiogenesis, cardiac hypertrophy, and impairment of cardiac function. Additionally, we demonstrated that after myocardial infarction, despite the higher collateral vessel formation, the animals with vascular- specific PPARβ/δ overexpression display bigger infarct lesions, higher cardiac fibrosis, and further reduced cardiac function. This points to a more careful view about the potential benefits of PPARβ/δ agonists in the treatment of cardiovascular diseases as the proper balance between cardiomyocytic and vascular PPARβ/δ seems to be crucial for cardiac health, especially under ischemic conditions [10]. Although the function of PPARβ/δ in lipid and glucose metabolism, and the remodeling of skeletal and cardiac muscle is well established, its role in tumor angiogenesis and cancer progression is unclear or at least partially controversial. It is of great importance to clarify the impact of PPARβ/δ activation in the vasculature particularly in cancer, as caution may be required when testing PPARβ/δ activation where more angiogenesis may play important pathological roles. The question of the safety of a potential use of PPAR β/δ modulation in clinical studies therefore needs to be answered urgently.

In human pancreatic tumors, PPARβ/δ expression strongly correlated with advanced tumor stage and increased risk of tumor recurrence and distant metastasis [11]. PPARβ/δ has therefore been suggested to be involved in the regulation of the angiogenic switch in tumor progression. However, until now it is not clear to what extent the expression of PPARβ/δ in the tumor endothelium influences tumor progression and metastasis formation. We determine in this study the in vivo relevance of PPARβ/δ for tumor vessel formation and cancer growth using modern mouse genetic approaches. We first tested the effects of the PPARβ/δ agonist GW0742 on tumor cell proliferation in vitro and in vivo. Although the PPARβ/δ agonist inhibited cancer cell proliferation in vitro, we could observe enhanced tumor growth upon treatment in vivo. Tumors of animals treated with the PPARβ/δ agonist displayed higher vessel densities and formed more metastases compared to controls. We next generated an inducible conditional endothelial cell-specific over-expression mouse model for PPARβ/δ to analyze the effects on tumor angiogenesis and progression. We show here that tumor angiogenesis and cancer growth as well as metastasis formation are increased upon endothelial cell-specific upregulation of PPARβ/δ. Furthermore, we identify platelet-derived growth factor receptor beta (Pdgfrb), platelet-derived growth factor subunit B (Pdgfb), and the tyrosinkinase KIT (c-Kit) as new PPARβ/δ -dependent molecules involved in tumor vessel formation based on RNA sequencing and subsequent verification applying a variety of molecular biology approaches.

2. Materials and Methods

2.1. Animals

All animal work was conducted according to national and international guidelines and was approved by the local ethics committee (Comité Institutionnel d'Éthique Pour l'Animal de Laboratoire Azur (Ciepal), agreement number: *PEA-NCE/2013/334*). PPARβ/δ-flox$^{+/-}$ [12] and Tie2-CreERT2 [13] animals were crossed to generate Tie2-CreERT2;PPARβ/δ-flox$^{+/-}$ mice, further referred to as Tie2-CreERT2;PPARβ/δ. The Tie2-CreERT2-line was back-crossed four times onto C57BL/6J. Age- and sex-matched Tie2-CreERT2;PPARβ/δ animals were injected for one week intraperitoneally either with sunflower oil (vehicle) or Tamoxifen dissolved in sunflower oil in a dose of 33 mg/kg per day [10,14,15]. Tie2-CreERT2 animals injected with Tamoxifen served as additional controls. One week after the last Tamoxifen or vehicle treatment, 1×10^6 LLC1 tumor cells were injected subcutaneously. Tumors and organs were collected after three weeks. For treatment with the PPARβ/δ agonist, ten-week-old male C57BL/6J (Janvier, France) mice were subcutaneously injected with 1×10^6 LLC1 tumor cells. GW0742 (Selleckchem, Houston, TX, USA) dissolved in DMSO was then subcutaneously injected at 1 mg/kg once every second day (100 μL). Controls received 100 μL DMSO injections [8].

2.2. Cell Culture

Human umbilical vein endothelial cells (HUVEC) were purchased from PromoCell (Heidelberg, Germany) and grown in endothelial cell growth medium (PromoCell) supplemented with gentamycin (50 μg mL^{-1}) and amphotericin B (50 ng mL^{-1}). For all experiments, we used HUVECs pooled from up to four donors, which did not exceed passage 4. Human embryonic kidney (HEK) 293 cells (ATCC CRL-1573) were grown in DMEM medium (Invitrogen, Cergy Pontoise, France) supplemented with 10% fetal calf serum (FCS), 100 IU mL^{-1} penicillin, and 100 μg mL^{-1} streptomycin (Invitrogen, Cergy Pontoise, France). C166 mouse endothelial cells (accession number CRL-2581) and LLC1 mouse lung cancer cells (accession number CRL-1642) were grown in DMEM medium (Invitrogen, Cergy Pontoise, France). Media were supplemented with 10% fetal calf serum (FCS), 100 IU mL^{-1} penicillin and 100 μg mL^{-1} streptomycin. As positive control for apoptosis assays, LLC1 mouse lung cancer cells were treated with 100 nmol/L Staurosporine (Sigma, St. Louis, MO, USA) overnight. For RNA isolation and quantitative RT-PCR experiments, HUVEC and LLC1 cells were maintained for 48 h (HUVEC) or 24 h (LLC1) in medium in the presence of GW0742 (Selleckchem, Houston, TX, USA) or GSK3787 (Selleckchem) dissolved in dimethyl sulfoxide (DMSO) at concentrations of 1 μmol/L. Controls were treated with vehicle (0.1% DMSO) only [6,16].

2.3. Detection of Cell Proliferation

After incubation for 24 h (LLC1 cells) or 48 h (HUVECs) with DMSO, GW0742, or GSK3787, bromodeoxyuridine was added and the cells incubated for 3 h. Afterwards, BrdU incorporation was measured spectrophotometrically according to manufacturer's instructions (Millipore, Molsheim, France). Alternatively, cells were labeled with a mouse monoclonal proliferating cell nuclear antigen (PCNA) antibody (PC-10, Santa Cruz Biotechnology, Heidelberg, Germany) and 4′,6-diamidino-2-phenylindole (DAPI) counterstain (Vector Laboratories, Burlingame, CA, USA). PCNA-positive cells in five random optical fields from six independent experiments each were counted at 400× magnification.

2.4. Apoptosis Assays

Apoptotic cells were detected by Terminal deoxynucleotidyl transferase dUTP nick end labeling (TUNEL) staining of HUVECs, 48 h after treatment with DMSO, GW0742, or GSK3787 using the In Situ Cell Death Detection Kit (Roche Molecular Biochemicals, Meylan, France) according to the manufacturer's instructions. LLC1 cells were incubated with APC-conjugated annexin V (Roche, Meylan, France) and counterstained with propidium iodide to distinguish necrotic from apoptotic cell death. LLC1 cells treated with 100 nmol/L Staurosporine (Sigma, St. Louis, MO, USA) overnight served as positive controls.

2.5. Immunofluorescence Assays

Cells were fixed for 10 min on ice with 4% paraformaldehyde in phosphate-buffered saline (PBS). After PBS washes, cells were incubated for 1 h at room temperature in blocking solution (1% Triton X-100, 1%BSA, 5% donkey serum in PBS). Cells were then immuno-stained overnight at 4 °C in blocking solution containing the following primary antibodies: rabbit polyclonal anti PPARβ/δ (ThermoFisher Scientific, Nimes, France, 1:200) and mouse monoclonal PDGFRB (ThermoFisher Scientific, 1:300), or goat polyclonal PDGFB antibody (Abcam, Cambridge, UK, 1:50), or mouse monoclonal anti c-Kit (Abcam, 1:500). After three washes with PBS/0.1% Triton X-100, slides were incubated for 1 h 30 min at room temperature with Dylight 488 donkey anti-mouse or Dylight 488 donkey anti-goat and Dylight 594 donkey anti-rabbit secondary antibodies in PBS containing 0.5% Triton X-100, 1%BSA, 2.5% donkey serum. Slides were mounted with Vectashield with DAPI (Vector Laboratories, Burlingame, USA). Slides were viewed under an epifluorescence microscope (DMLB, Leica, Germany) connected to a digital camera (Spot RT Slider, Diagnostic Instruments, Scotland).

2.6. Endothelial Cell Isolation

Mouse tumor endothelial cells (EC) were isolated from Tie2-CreERT2;PPARβ/δ mice treated with Tamoxifen or vehicle as previously described [10,17,18]. Tie2-CreERT2 animals injected with Tamoxifen served as an additional control. Briefly, tumor tissues were cut into small fragments and digested with 1 mg/mL collagenase A and 100 IU/mL type I DNase (Roche Diagnostics, Meylan, France) for 45 min at 37 °C. ECs were then purified from the cell suspension using CD31 MicroBeads followed by magnetic separation in LS columns (Miltenyi Biotec SAS, Paris, France).

2.7. RT-PCR and Quantitative RT-PCR

Total RNA was isolated using the Trizol reagent (Invitrogen). First-strand cDNA synthesis was performed with 0.5 µg of total RNA using the Thermo Scientific Maxima First Strand cDNA synthesis kit (Thermo Scientific). The reaction product was diluted to 200 µL and 1 µL of the diluted reaction product was taken for real time RT-PCR amplification (StepOne plus, Applied Biosystems) using the SYBR® Select Master Mix (Applied Biosystems). Expression of each gene was normalized to the respective arithmetic means of *Gapdh*, *Actnb*, and *Rplp0* expression.

Primer sequences for mouse genes are given in Table 1 and for human genes in Table 2.

Table 1. Primer sequences for mouse genes.

Name	Accession Number	Primer Sequences	Amplicon Size (bp) [1]
PPARβ/δ	NM_011145.3	F: ATGGGGGACCAGAACACAC R: GGAGGAATTCTGGGAGAGGT	62
Angptl4	NM_020581.2	F: CACCCACTTACACAGGCCG R: GAAGTCCACAGAGCCGTTCA	178
Acaca	NM_133360.2	F: GCCTTTCACATGAGATCCAGC R: CTGCAATACCATTGTTGGCGA	175
Fabp4	NM_024406.3	F: TGAAATCACCGCAGACGACA R: ACACATTCCACCACCAGCTT	141

Table 1. *Cont.*

Name	Accession Number	Primer Sequences	Amplicon Size (bp) [1]
Foxo1	NM_019739.3	F: CAAGGCCATCGAGAGCTCAG R: AATTGAATTCTTCCAGCCCGCC	130
Cna	NM_008913.5	F: AAAGCGCTACTGTTGAGGCT R: ATTCGGTCTAAGCCCTTGGC	103
Pdk4	NM_013743.2	F: TTCCAGGCCAACCAATCCAC R: TGGCCCTCATGGCATTCTTG	87
Vegf	NM_001025250.3	F: CTCACCAAAGCCAGCACATA R: AATGCTTTCTCCGCTCTGAA	198
Vegfr1	NM_010228.4	F: TACCTCACCGTGCAAGGAAC R: AAGGAGCCAAAAGAGGGTCG	93
Vegfr2	NM_010612.3	F: AGTGGTACTGGCAGCTAGAAG R: ACAAGCATACGGGCTTGTTT	65
Ets1	NM_011808.3	F: CTGACCTCAACAAGGACAAGC R: AGAAACTGCCACAGCTGGAT	88
Mmp1	NM_008607.2	F: GGCCAGAACTTCCCAACCAT R: AGCCCAGAATTTTCTCCCTCT	89
Mmp8	NM_008611.4	F: CCTGCAGGACTCCTTCTTCCT R: CCTCATAGGGTGCGTGCAA	156
Mmp9	NM_013599.4	F: CCATGCACTGGGCTTAGATCA R: GGCCTTGGGTCAGGCTTAGA	147
Wt1	NM_144783.2	F: CCAGCTCAGTGAAATGGACA R: CTGTACTGGGCACCACAGAG	97
Cd31	NM_008816.3	F: CGGTGTTCAGCGAGATCC R: CGACAGGATGGAAATCACAA	71
Vwf	NM_011708.4	F: TGTGACACATGTGAGGAGCC R: CTTTGCTGGCACACTTTCCC	127
Vcam1	NM_011693.3	F: TATGTCAACGTTGCCCCCAA R: CAGGACTGCCCTCCTCTAGT	73
Il-1	NM_008361.4	F: GCCACCTTTTGACAGTGATGAG R: AGCTTCTCCACAGCCACAAT	186
Il-6	NM_031168.2	F: CACTTCACAAGTCGGAGGCT R: TGCCATTGCACAACTCTTTTCT	86
Il-18	NM_008360.2	F: CAAAGTGCCAGTGAACCCCA R: TTCACAGAGAGGGTCACAGC	89
Pdgfb	NM_011057.4	F: GGAGTCGGCATGAATCGCT R: GCCCCATCTTCATCTACGGA	182
Pdgfra	NM_001083316.2	F: ATGAGAGTGAGATCGAAGGCA R: CGGCAAGGTATGATGGCAGAG	130
Pdgfrb	NM_001146268.1	F: CCAGCACCTTTGTTCTGACCT R: TGCCGTCCTGATTCATGGC	99
Sirt1	NM_019812.3	F: GCCGCGGATAGGTCCATA R: AACAATCTGCCACAGCGTCA	136
Sox18	NM_009236.2	F: ACTGGCGCAACAAAATCC R: CTTCTCCGCCGTGTTCAG	88
Thbs1	NM_011580.4	F: CCTGCCAGGGAAGCAACAA R: ACAGTCTATGTAGAGTTGAGCCC	115
Cd36	NM_001159555.1	F: GTGTGGAGCAACTGGTGGAT R: ACGTGGCCCGGTTCTAATTC	147
Tnfα	NM_013693.3	F: GTAGCCCACGTCGTAGCAAA R: ACAAGGTACAACCCATCGGC	137
c-kit	NM_001122733.1	F: GCCTGACGTGCATTGATCC R: AGTGGCCTCGGCTTTTTCC	110

Table 1. *Cont.*

Name	Accession Number	Primer Sequences	Amplicon Size (bp) [1]
Ccl2	NM_011333.3	F: AGCTGTAGTTTTTGTCACCAAGC R: GTGCTGAAGACCTTAGGGCA	155
Ccl5	NM_013653.3	F: TGCAGTCGTGTTTGTCACTC R: AGAGCAAGCAATGACAGGGA	152
Actnb	NM_007393.5	F: CTTCCTCCCTGGAGAAGAGC R: ATGCCACAGGATTCCATACC	124
Gapdh	NM_001289726.1	F: AGGTCGGTGTGAACGGATTTG R: TGTAGACCATGTAGTTGAGGTCA	123
Rplp0	NM_007475.5	F: CACTGGTCTAGGACCCGAGAAG R: GGTGCCTCTGGAGATTTTCG	73

[1] bp: base pairs; F: forward primer sequence; R: reverse primer sequence.

Table 2. Primer sequences for human genes.

Name	Accession Number	Primer Sequences	Amplicon Size (bp) [1]
PPARβ/δ	NM_006238.5	F: TCAGAAGAAGAACCGCAACAAGTG R: CCTGCCACCAGCTTCCTCTT	126
ANGPTL4	NM_001039667.3	F: ATCCAGCAACTCTTCCACAAGGT R: TTGAAGTCCACTGAGCCATCGT	254
FABP4	NM_001442.3	F: AAGTCAAGAGCACCATAACCTTAGATG R: TGACGCATTCCACCACCAGTT	120
CNA	NM_000944.5	F: AAAGCGCTACTGTTGAGGCT R: ATTCGGTCTAAGCCCTTGGC	103
PDK4	NM_002612.4	F: CCACATTGGAAGCATTGATCCTAACT R: TCACAGAGCATCCTTGAACACTCA	81
VEGFA	NM_001025366.3	F: GAGGAGTCCAACATCACCATGC R: CTTGCAACGCGAGTCTGTGTT	351
VEGFR1	NM_002019.4	F: GCCCGGGATATTTATAAGAAC R: CCATCCATTTTAGGGGAAGTC	70
VEGFR2	NM_002253.3	F: CAGAGTGAGGAAGGAGGACGAAGG R: GATGATGACAAGAAGTAGCCAGAAGAACA	181
CD31	NM_000442.5	F: GCCCGAAGGCAGAACTAAC R: AACAGAGCAGAAGGGTCAG	111
VWF	NM_000552.4	F: TGTGACACATGTGAGGAGCC R: CTTTGCTGGCACACTTTCCC	127
PDGFB	NM_002608.4	F: TCCGCTCCTTTGATGATCTCCAA R: GGTCATGTTCAGGTCCAACTCG	83
c-KIT	NM_000222.2	F: GCTCTGCTTCTGTACTGCC R: TAGGCAGAAGTCTTGCCCAC	160
SOX18	NM_018419.3	F: ATGGTGTGGGCAAAGGAC R: GCGTTCAGCTCCTTCCAC	107
ACTNB	NM_001101.5	F: CTCCTTAATGTCACGCACGAT R: CATGTACGTTGCTATCCAGGC	250
GAPDH	NM_002046.7	F: AGCCACATCGCTCAGACAC R: GCCCAATACGACCAAATCC	66
RPLP0	NM_001002.4	F: CAGATTGGCTACCCAACTGTT R: GGCCAGGACTCGTTTGTACC	69

[1] bp: base pairs; F: forward primer sequence; R: reverse primer sequence.

2.8. mRNA Sequencing

For sequencing, RNAs from tumor sorted endothelial cells from Tie2-CreERT2;PPARβ/δ mice treated with Tamoxifen or vehicle were used (*n* = 4 each). RNA sequencing and data analysis was performed by Novogene, Beijing, China. Briefly, RNA quality was monitored on 1% agarose gels. RNA purity was checked using a NanoPhotometer. RNA concentration was measured using the Qubit® RNA Assay Kit in a Qubit® 2.0 Flurometer (Life Technologies, CA, USA). RNA integrity was assessed using the RNA Nano 6000 Assay Kit of the Bioanalyzer 2100 system (Agilent Technologies, CA, USA). A total amount of 1 µg RNA per sample was used as input material for the RNA sample preparations. Sequencing libraries were generated using NEBNext® UltraTM RNA Library Prep Kit for Illumina® (NEB, USA) following manufacturer's instructions. Library quality was assessed on the Agilent Bioanalyzer 2100 system. Library preparations were sequenced on an Illumina Hiseq platform and 125 bp/150 bp paired-end reads were generated. Clean reads were obtained by removing reads containing adapter, reads containing ploy-N and low quality reads from raw data. HTSeq v0.6.1 was used to count the reads numbers mapped to each gene. Fragments per kilobase of transcript sequence per millions base pairs sequenced (FPKM) of each gene was calculated based on the length of the gene and reads count mapped to this gene. Differential expression analysis of the two groups was performed using the DESeq R package (1.18.0). Resulting *p*-values were adjusted using the Benjamini and Hochberg's approach. Genes with an adjusted *p*-value < 0.05 were assigned as differentially expressed. The RNA sequencing data have been deposited in NCBI's Gene Expression Omnibus [19,20] and are accessible through GEO Series accession number GSE140513 (https://www.ncbi.nlm.nih.gov/geo/query/acc.cgi?acc=GSE140513).

2.9. Bioinformatics

The maximum scoring subnetwork was calculated with the runFastHeinz function from the R BioNet package [21]. The *p*-values obtained from the differential expression were assigned to each node (gene) of the networks. The following networks were analyzed: FULL-Mouse (18 January, 2019) from signor database (http://signor.uniroma2.it/), HumanCyc metabolic pathways (http://humancyc.org/), NCI PID, Complete Interactions (http://www.cancer.gov), Biogrid (https://thebiogrid.org/), HCOP Mouse PCNet [22], ConsensusPathDB (http://cpdb.molgen.mpg.de/). All networks were downloaded from http://www.ndexbio.org [23]. Sub-network visualizations and analyses were done with Cytoscape [24] and pathway cluster analysis at http://impala.molgen.mpg.de/ [25]. Prediction of PPAR, responsive elements in differentially expressed genes was done using the oPOSSUM3 software at http://opossum.cisreg.ca/oPOSSUM3/ [26].

2.10. Tissue Samples and Immunohistology

The study adheres to the principles of the Declaration of Helsinki and to Title 45 of the U.S. Code of Federal Regulations (Part 46, Protection of Human Subjects). Paraffin-embedded samples, cut at 3µm, were used for immunohistochemical detection of PPARβ/δ, CD31, and PCNA. In total, 35 paraffin-embedded human tumor samples (7 liver carcinomas, 7 melanomas, 7 pancreas carcinomas, 7 ovary carcinomas, and 7 prostate cancers) were used for this study. For immunofluorescence double-labeling of human tumor samples, anti-CD31 mouse monoclonal antibody from Dako (Trappes, France, clone JC70A) was combined with the anti-PPARβ/δ antibody (ThermoFisher Scientific), using Dylight 488 donkey anti rabbit and Dylight 594 donkey anti mouse secondary antibodies. Negative controls were obtained by omission of first antibodies. Images were taken using a confocal ZEISS LSM Exciter microscope (Zeiss, Jena, Germany).

Mouse tissue sections were in routine stained with hematoxylin-eosin. After heat-mediated antigen retrieval and quenching of endogenous peroxidase activity, PPARβ/δ (ThermoFisher Scientific) or CD31 (rabbit polyclonal, 1:50, Abcam) were detected using the EnVisionTM Peroxidase/DAB Detection System from Dako. Negative controls were obtained by incubation of samples with a rabbit IgG Control (Abcam). Sections were counterstained with hematoxylin (Dako) and analyzed by two independent investigators, one of them an experienced pathologist. Slides were viewed under an epifluorescence microscope (DMLB, Leica, Germany) connected to a digital camera (Spot RT Slider, Diagnostic Instruments, Scotland).

2.11. Cloning and Transient Transfection Experiments

The PDGFRβ promoter was amplified from HUVEC genomic DNA using the following primers: 5′-TAGGTACCAAAGACTTAGCGGCGCAGAG-3′ (forward, position -1766), 5′-GTGAGATCTCTGCCCTCTCCCAGTTATCAG-3′ (backward, position +374) and cloned into the KpnI/BglII restriction sites of pGl3 basic [17]. The Pdgfb promoter was amplified from mouse genomic DNA using the following primers: 5′-CGGGGTACCATCAGTACCACCTCATCCA-3′ (forward, position -1199), 5′-CCCAAGCTTCTCGGGTCAGTCTGTCTA-3′ (backward, position +98) and cloned into the KpnI/HindIII restriction sites of pGl3 basic. The c-Kit promoter was a kind gift of C. Nishiyama [27]. As vector backbone, pGl3 basic (Promega) was used for all constructs. The putative PPARβ/δ responsive elements sites were deleted from the PDGFRB promoter construct using the Quik Change II site directed mutagenesis kit (Stratagene, Agilent Technologies, Massy, France) with the following oligonucleotides: PPRE1: 5′-ATATCCAATCTGTGCTGGAATCACATTCCCTCTCTGTG-3′; antisense: reverse complement; PPRE2: 5′-TCATGTGTCTCATGAGACCTAGTTCTGCCATTGCTGC-3′; antisense: reverse complement; PPRE3: 5′-ATATCCAATCTGTGCTGGAATCACATTCCCTCTCTGTG-3′; antisense: reverse complement. The putative PPARβ/δ responsive elements sites were deleted from the Pdgfb promoter construct using the Quik Change II site directed mutagenesis kit (Stratagene) with the following oligonucleotides: PPRE1: 5′-GTGGGTGGGTAGCGAACTGGGTGGGG-3′; antisense: reverse complement; PPRE2: 5′-GACAAGCAAGGAGAGGTGTAGCTGAAGGGTTC-3′; antisense: reverse complement; PPRE3: 5′-GAAGGAAAGTGACGTGCCCAAGATTTAATTAGACTCAATGGAATC-3′; antisense: reverse complement. For deletion of putative PPARβ/δ responsive element sites in the c-kit promoter we used oligonucleotides PPRE1: 5′-TACCAACAGGAACAGAAATAAATGTTCCTAATCCCTTCGCC-3′; antisense: reverse complement; PPRE2: 5′-TGGGCTCGGTCTTTTACGGGTGCCACGATC-3′; antisense: reverse complement.

HEK-293 and C166 cells at approximately 60% confluence were transfected using Fugene 6 reagent (Roche, Meylan, France). Reporter constructs (full promoter sequences and sequences with deletions of the PPARβ/δ responsive elements sites) were co-transfected with a cytomegalovirus (CMV)-driven galactosidase plasmid, and the PPARβ/δ expression construct, and assayed for luciferase- and galactosidase activity ($n = 12$ each).

2.12. Chromatin Immunoprecipitation Assay

Chromatin immunoprecipitation (ChIP) assay was performed on HEK293 (human PPREs) or C166 (mouse PPREs) cells using manufacturer's instructions (Millipore) as described [15,17,18]. One microgram of the following antibodies each were used: PPARβ/δ rabbit polyclonal, (H-74, Santa Cruz Biotechnology), PPARβ/δ goat polyclonal (K-20, Santa Cruz Biotechnology), Acetyl-Histone H3 (06-599, Upstate). Omission of primary antibodies served as a negative control for the PPARβ/δ antibodies and dilutions of the input sample as positive control. PCR products were electrophoresed on 4% agarose gels. Alternatively, samples were used in quantitative PCRs ($n = 3$ each). Fold enrichment was calculated from CT values relative to the input signal of each experiment set to 100%. The following oligonucleotides were used for PCR amplification of the CHIP products: PDGFRβPPRE1: 5′-GGTAAGCCCACTCTATATGCCCTTCTAA-3′ (forward); 5′-CCAGTTACAGACTCCTAGCCCTC AG-3′ (reverse); PDGFRβPPRE2: 5′-GGTCAGATGACTTGTGTCTCTTCCA-3′ (forward); 5′-CTTAC GCAGCAATGGCAGAGC-3′ (reverse); PDGFRβPPRE3: 5′-GGGCTTTGAGACGTGAAAAGGA-3′ (forward); 5′-ATTGGCACAGAGAGGGAATGTG-3′ (reverse); PDGFRβUTR: 5′-CAGGTCCAGGTG AGTCAT-3′ (forward); 5′-CCTCTTCCTCTTCCTCTTCT-3′ (reverse); PdgfbPPRE1: 5′-AGGTGTTAA CTGTGAGAGTG-3′ (forward); 5′-TGTTACTACCCCTCTCTGC-3′ (reverse); PdgfbPPRE2: 5′-TCAACAGACTCAAATTCAGC-3′ (forward); 5′-CTCTAAACCCACAGCCAG-3′ (reverse); Pdgfb PPRE3: 5′-ATCACAGAAGGAAAGTGACG-3′ (forward); 5′-AGAACCAGACATCTGCAAC-3′ (reverse); PdgfbUTR: 5′-GCTGGAGATAACCTTGGCTAAG -3′ (forward); 5′-GTTGGGACTCAGGA

TAGACTCA-3′ (reverse); c-kitPPRE1: 5′-TGGAGAAACTGAGCATGAAA-3′ (forward); 5′-TTCTGTTCCTGTTGGTAGAG-3′ (reverse); c-kitPPRE2: 5′-CTCTACCAACAGGAACAGAA-3′ (forward); 5′-CTTATGGTGGAGGTGTTACTA-3′ (reverse); c-kitUTR: 5′-CGATCTCATGTGG TCCAA-3′ (forward); 5′-CGCCTTGTTCATTACTACTG-3′ (reverse).

2.13. Statistical Analysis

Data are expressed as mean ± SEM. Student's t-tests (Instat, GraphPad) were performed to determine statistical significance. A *p*-value of less than 0.05 was considered significant.

3. Results and Discussion

3.1. The PPARβ/δ Agonist GW0742 Decreases LLC1 Lewis Lung Cancer Cell Proliferation In Vitro

As a prerequisite before in vivo treatment of tumor-bearing mice with a PPARβ/δ agonist, we investigated the in vitro effects on syngenic tumor cells, which we aimed to inject into animals. LLC1 cells were treated for 24 h with the PPARβ/δ agonist GW0742 or the PPARβ/δ antagonist GSK3787 at a concentration of 1 µmol/L. Thereafter, we performed bromodeoxyuridine (BrdU) incorporation assays and immunostaining for proliferating cell nuclear antigen (PCNA), with subsequent quantification of PCNA-positive cells relative to the total cell number. GW0742 decreased BrdU incorporation and the fraction of PCNA-positive cells significantly, GSK3787 had the opposite effects (Figure 1a–c). These findings are in line with the effects of PPARβ/δ modulation on the proliferation of melanoma cells reported by our group [5] and recently confirmed in a study where a PPARβ/δ antagonist enhanced melanoma progression [28]. To determine whether apoptosis might contribute to the observed effects on cell growth upon GW0742 treatment, we used Annexin V/propidium iodide (PI) labeling to detect apoptotic events followed by fluorescence-activated cell sorting (FACS) analysis. Apoptosis of LLC1 cells was neither influenced by the PPARβ/δ agonist GW0742, nor the antagonist GSK3787 (Figure 1d).

As mentioned before, the role of PPARβ/δ in cancer is highly controversial. Concerning lung cancer, two studies reported an increased expression of PPARβ/δ in human non-small cell lung cancer (NSCLC) compared to normal lung and observed in NSCLC cell lines that PPARβ/δ activation promoted proliferation [29,30]. However, an earlier report observed an inhibition of proliferation in a NSCLC cell line upon PPARβ/δ activation [31] and finally, neither growth promoting nor inhibiting effects were observed in another study [32]. Thus, it is difficult to draw a definitive conclusion regarding the biological effect of PPARβ/δ in human lung cancer. There is also a high variability in the kind and concentration of PPARβ/δ agonists used, which might further contribute to these conflicting results. Nevertheless, the aim of our study was not to determine the effects of PPARβ/δ modulation in one cancer cell line, but to clarify the in vivo relevance of PPARβ/δ for tumor vessel formation and cancer growth, which necessitated to distinguish between direct effects on tumor cells and effects on the tumor microenvironment, especially tumor vessels.

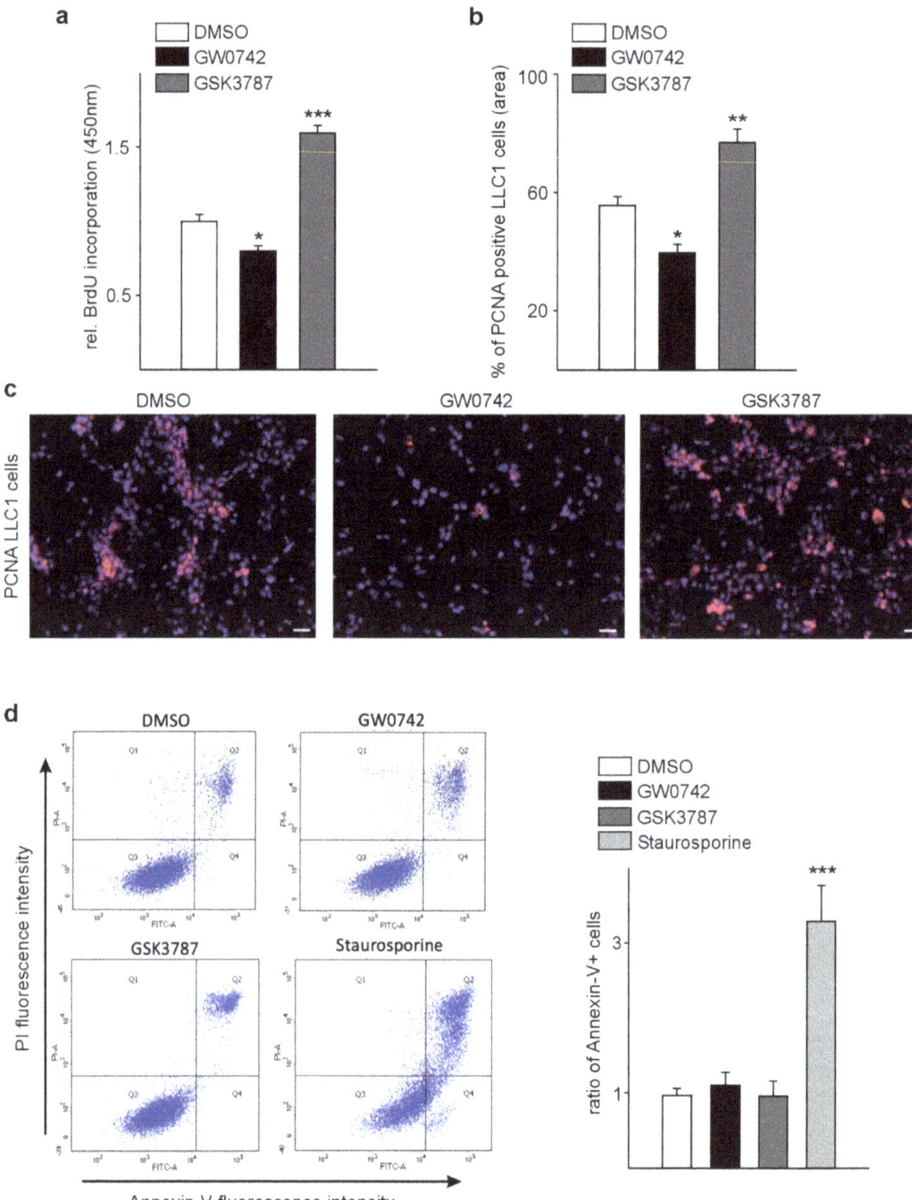

Figure 1. The PPARβ/δ agonist GW0742 decreases and the PPARβ/δ antagonist GSK3787 increases LLC1 proliferation. (**a**) Quantification of bromodeoxyuridine (BrdU) incorporation assays (*n* = 8) (**b**) and the proportion of proliferating cell nuclear antigen (PCNA)-positive cells (**c**) as measures of proliferation (*n* = 4). (**d**) fluorescence-activated cell sorting (FACS) analysis of Annexin V/propidiumiodide labeled LLC1 cells after modulation of PPARβ/δ to measure apoptosis. Staurosporin-treated cells were used as positive control (*n* = 4). Scale bars indicate 50μm (c). Data are mean ± SEM. * *p* < 0.05, ** *p* < 0.01, *** *p* < 0.001.

3.2. The PPARβ/δ Agonist GW0742 Increases LLC1 Lewis Lung Cancer Progression In Vivo

We investigated the effect of the PPARβ/δ agonist GW0742 on tumor growth by subcutaneous implantation of LLC1 tumor cells and subsequent injections of GW0742. Control animals received DMSO injections in the same frequency. Tumor growth was comparable between the two groups in the first days and started to increase in the animals receiving the PPARβ/δ agonist in the second week (Figure 2a). After three weeks, we determined tumor-to-body weight ratios in both groups and observed significantly higher tumor weights in GW0742 treated animals (Figure 2b,c). Next, we analyzed whether treatment with a PPARβ/δ agonist might modify metastasis occurrence. We investigated serial lung and liver sections of both groups for metastases formation. All animals, which had received the PPARβ/δ agonist displayed lung metastases formation, whereas only 30% of the animals in the control group had metastases occurrence in the lung. Lung metastases were not only more frequent in the treated group, but also of a bigger size than in controls. Liver metastases could be observed in 16% of the GW0742 treated mice, whereas none of the control animals had liver metastases (Figure 2d). Vessel density in tumors was determined by immunostaining for Cd31. Strikingly, in tumors of GW0742 treated mice, vessel densities were more than doubled compared to controls (Figure 2e). Double-staining for Cd31 and PCNA suggested a higher number of PCNA-positive endothelial cells in the tumors of PPARβ/δ agonist treated animals (Figure 2f).

Although the PPARβ/δ agonist GW0742 inhibited tumor cell proliferation in vitro, we could observe a strikingly different situation in vivo, where tumor growth and metastasis formation were enhanced. Tumor vascularization was strongly increased, which is in line with previous studies reporting a rapid boost of vascularization after pharmacological PPARβ/δ activation [8–10]. The increased tumor vascularization upon PPARβ/δ activation seems therefore to be sufficient to dominate over the anti-proliferative effect of the PPARβ/δ agonist on tumor cells. Increasing neovascularization is not only required for further expansion of the tumor-cell population, but also correlates with a rising rate of metastasis. In agreement with our findings, a recent study observed a marked inhibition of tumor angiogenesis and growth in PPARβ/δ$^{-/-}$ mouse models of subcutaneous Lewis lung carcinoma and B16 melanoma, with occurrence of immature, leaky microvascular structures [33]. However, another group demonstrated enhanced extravasation of B16 melanoma cells in PPARβ/δ$^{-/-}$ mice using an experimental model of metastases formation by injecting the tumor cells in the tail vein of animals, resulting in enhanced pulmonary metastases as compared to wildtype animals [28]. In contrast to the patho-physiologically occurring spontaneous metastasis formation, this assay does not require the formation of functional blood vessels necessary for the propagation of the primary tumor cells to other organs. The enhanced pulmonary metastases formation observed in the PPARβ/δ$^{-/-}$ animals could therefore simply be due to the presence of non-functional leaky microvessels [33], facilitating the anchorage of the tumor cells in the lung.

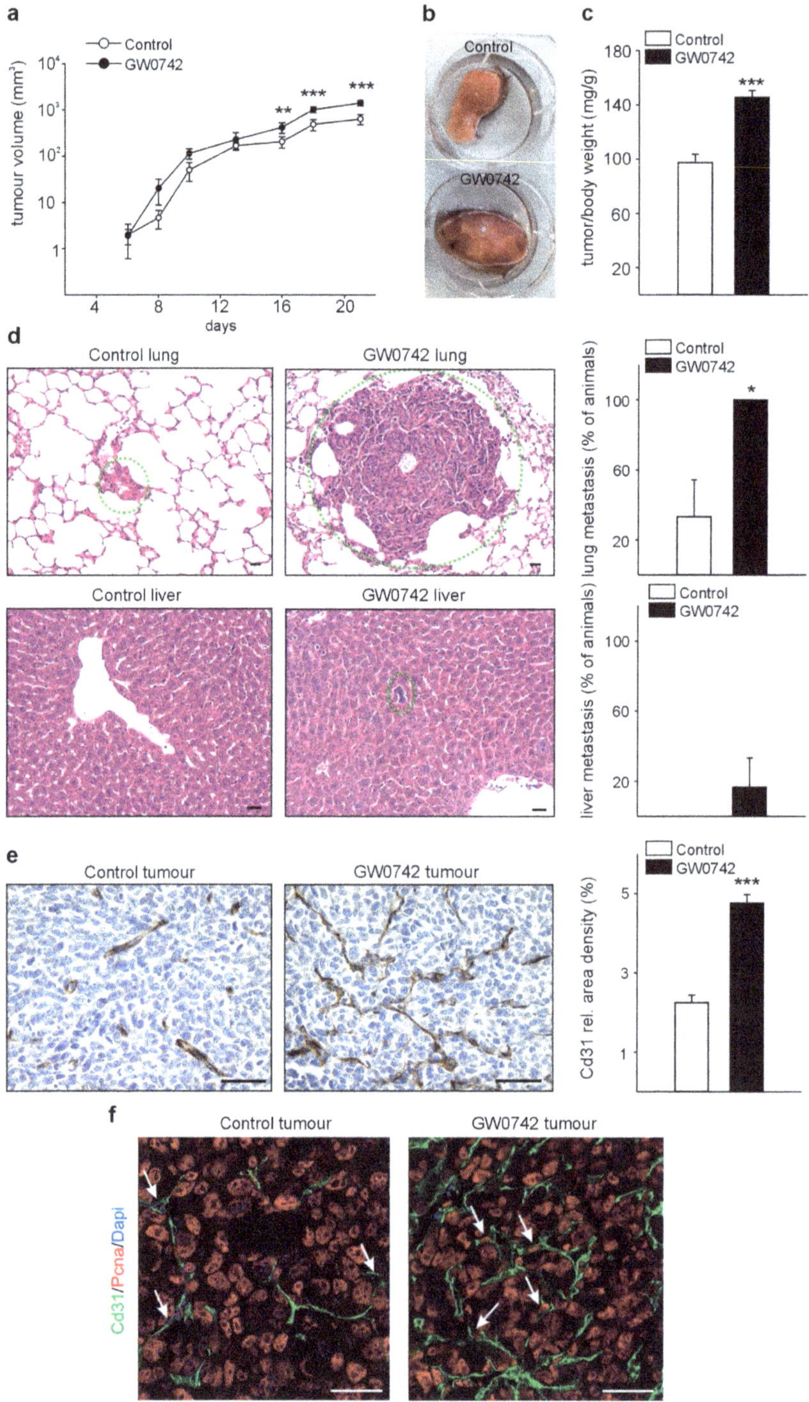

Figure 2. Treatment with the PPARβ/δ agonist GW0742 increases LLC1 growth, angiogenesis, and metastases formation in vivo. (**a**) LLC1 tumor growth curves in animals treated with GW0742 and respective controls (DMSO) (*n* = 6). Tumor volume data were calculated from Caliper measurements as

described [15]. (**b**) LLC1 tumor in mice treated with GW0742 and respective control (DMSO). (**c**) Quantification of tumor/body weights in GW0742 and DMSO treated mice. (**d**) Representative photomicrographs of lung (upper panel) and liver (lower panel) metastases in the two groups of mice. Metastases are indicated with a dotted green circle. Note that in the animals with vehicle (DMSO), no liver metastasis was detectable. Graphs on the right show quantification of the percentage of animals with lung (upper graph) or liver (lower graph) metastasis from LLC1 tumors. (**e**) Cd31 immunostaining in LLC1 tumors and quantification of Cd31 signal area densities. (**f**) PCNA/Cd31 double-labeling of LLC1 tumors. Scale bars indicate 50 μm. Data are mean ± SEM. * $p < 0.05$, ** $p < 0.01$, *** $p < 0.001$.

3.3. PPARβ/δ Is Expressed in the Vasculature of Human Tumors and PPARβ/δ Modulation Impacts Proliferation and Expression of Pro-Angiogenic Factors in Human Endothelial Cells

To confirm a relevance for human pathophysiology, we investigated human tumor samples (liver, pancreas, ovary, and prostate carcinoma, as well as melanoma) for expression of PPARβ/δ in the tumor vasculature. Confocal microscopy of tumor samples labeled for PPARβ/δ and CD31 confirmed co-localization of PPARβ/δ in endothelial cells (Figure 3). Furthermore, PPARβ/δ could be detected in tumor cells at various degrees in different tumor types.

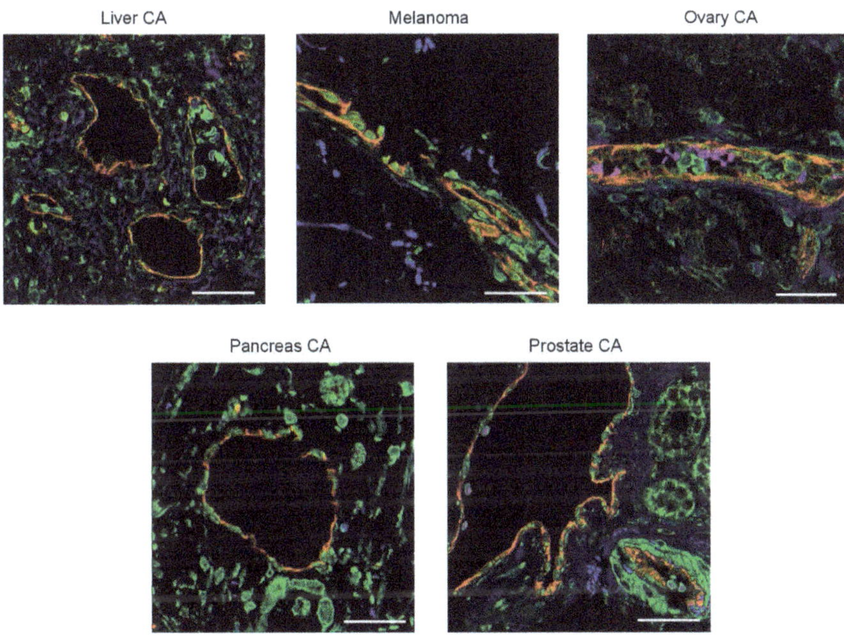

Figure 3. PPARβ/δ is expressed in human tumor vessels. Double labeling of PPARβ/δ (green) and cell adhesion molecule-1 (CD31) (red) in human tumor samples. Nuclei were counterstained with DAPI. Scale bars indicate 50 μm.

Treatment of human umbilical vein endothelial cells (HUVECs) with GW0742 increased proliferation, while the PPARβ/δ antagonist GSK3787 tended to reduce HUVEC proliferation (Figure 4a). This is in agreement with previous works reporting enhanced endothelial cell proliferation [34] and activation of angiogenesis upon PPARβ/δ activation [7] and the general view of PPARβ/δ as a pro-angiogenic factor [4]. As an initial study suggested that PPARβ/δ is involved in endothelial cell apoptosis [35], we tested for differences in apoptosis upon modulation of PPARβ/δ. Rarely,

TUNEL-positive endothelial cells were observed under any of the culture conditions (Figure 4b), suggesting that PPARβ/δ modulation does not influence apoptosis in vascular cells.

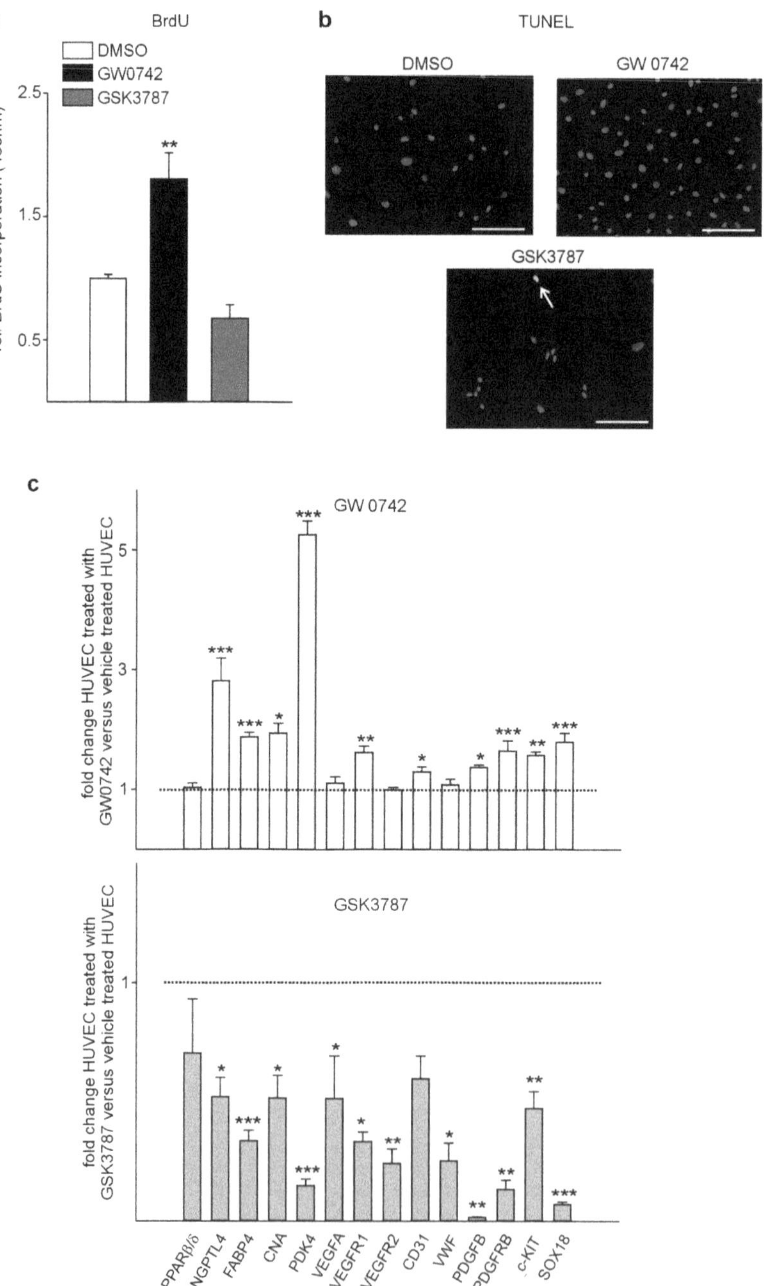

Figure 4. PPARβ/δ agonist GW0742 increases human umbilical vein endothelial cells (HUVEC) proliferation and upregulates expression of angiogenic genes. (**a**) Quantification of BrdU incorporation assays (*n* = 7). (**b**) TUNEL-labeling as a marker for apoptosis (*n* = 3). Note that nearly no TUNEL labeling

could be observed independent of the condition. The white arrow points to one of the rare TUNEL positive cells. (**c**) Quantitative RT-PCRs of HUVECs treated with the PPARβ/δ agonist GW0742 and the antagonist GSK3787 (n = 6). Scale bars indicate 50 μm. Data are mean ± SEM. * $p < 0.05$, ** $p < 0.01$, *** $p < 0.001$.

To identify molecular events in the observed pro-angiogenic effect of the PPAR agonist GW0742 on HUVECs, we performed quantitative RT-PCRs. First, we confirmed up-regulation of established PPARβ/δ target genes as Angiopoietin-like 4 (ANGPTL4) [36], Fatty acid binding protein 4 (FABP4) [37], Calcineurin (CNA) [8], and Pyruvate dehydrogenase kinase isoform 4 (PDK4) [38]. Next, we evaluated the expression of genes known to be involved in angiogenesis and found many of them to be up-regulated upon agonist treatment: Vascular endothelial growth factor receptor 1 (Vegfr1), which is critical for endothelial cell survival [39], and platelet/endothelial cell adhesion molecule-1 (CD31), regulating the endothelial cell vascular permeability barrier [40], Platelet-derived growth factor subunit B (PDGFB) and PDGF beta-receptor (PDGFRb), enhancing angiogenesis [17,41], Tyrosinekinase Kit (c-KIT), expressed on endothelial cells [42], implicated in tumor angiogenesis [43], and increasing migration and tube formation of endothelial cells [15,44], and also SRY-related HMG-box transcription factor 18 (SOX18), mediating physiologic and pathologic angiogenesis [45]. Although it has been suggested that Vascular endothelial growth factor (VEGF) is regulated by PPARβ/δ [7,46], we did not observe a statistically significant increase in expression, and no changes in Vascular endothelial growth factor receptor 2 (VEGFR2) expression and Von Willebrand factor (VWF), controlling blood vessel formation [47] (Figure 4c, upper panel). However, in HUVECs treated with the PPARβ/δ antagonist GSK3787, all of the mentioned genes were downregulated (Figure 4c, lower panel) supporting a pro-angiogenic function of PPARβ/δ.

3.4. Inducible Vascular-Specific Overexpression of PPARβ/δ Promotes Tumor Angiogenesis, Growth, and Spontaneous Metastases Formation In Vivo

To determine further the functional relevance of PPARβ/δ for tumor vessel formation and tumorigenesis, we used PPARβ/δ-flox$^{+/-}$ mice [12] crossed with Tamoxifen-inducible Tie2-CreERT2 animals. This Cre becomes activated in endothelial cells upon Tamoxifen induction [13]. Using this strategy, we obtained Tie2-CreERT2;PPARβ/δ animals with an inducible conditional vascular-specific overexpression of PPARβ/δ [10]. In adult Tie2-CreERT2;PPARβ/δ, Cre was activated by tamoxifen injection. Tie2-CreERT2;PPARβ/δ injected with vehicle and Tie2-CreERT2 mice injected with tamoxifen served as controls. We investigated the effects of vascular specific PPARβ/δ overexpression on tumor growth by overexpressing PPARβ/δ in Tie2$^+$ cells, followed by subcutaneous implantation of LLC1 tumor cells. Tumor growth curves revealed increased tumor growth rates in mice with vessel- specific PPARβ/δ overexpression (Figure 5a). Three weeks after tumor cell injection, we determined tumor/body weight ratios. Body weights were comparable in all groups of mice (*CreERT2*;PPARβ/δ + vehicle 36.57 g ± 0.44 g, Tie2-CreERT2 + Tamoxifen 38.75 g ± 0.12 g, Tie2-CreERT2;PPARβ/δ + Tamoxifen 36.77 g ± 0.27 g). Of note, tumor weights were nearly doubled in the mice with vascular PPARβ/δ overexpression (Figure 5b,c). Next, we analyzed whether conditional overexpression of PPARβ/δ in vessels might modify metastasis occurrence. In this respect, the LLC1 model is very useful as it forms spontaneous metastases [15]. We investigated serial lung and liver sections of Tie2-CreERT2;PPARβ/δ + Tamoxifen mice and the respective control groups for metastases. All mice with vascular PPARβ/δ overexpression had lung metastases compared to 30% in the control groups (Figure 5d), and 50% displayed liver metastases whereas in the controls no liver metastases formation could be observed (Figure 5e). These observations strengthen the hypothesis that vascular PPARβ/δ enhances tumor growth and spontaneous metastatic spreading; and they are in perfect agreement with tumor growth inhibition and prolonged survival observed in PPARβ/δ$^{-/-}$ mice [33].

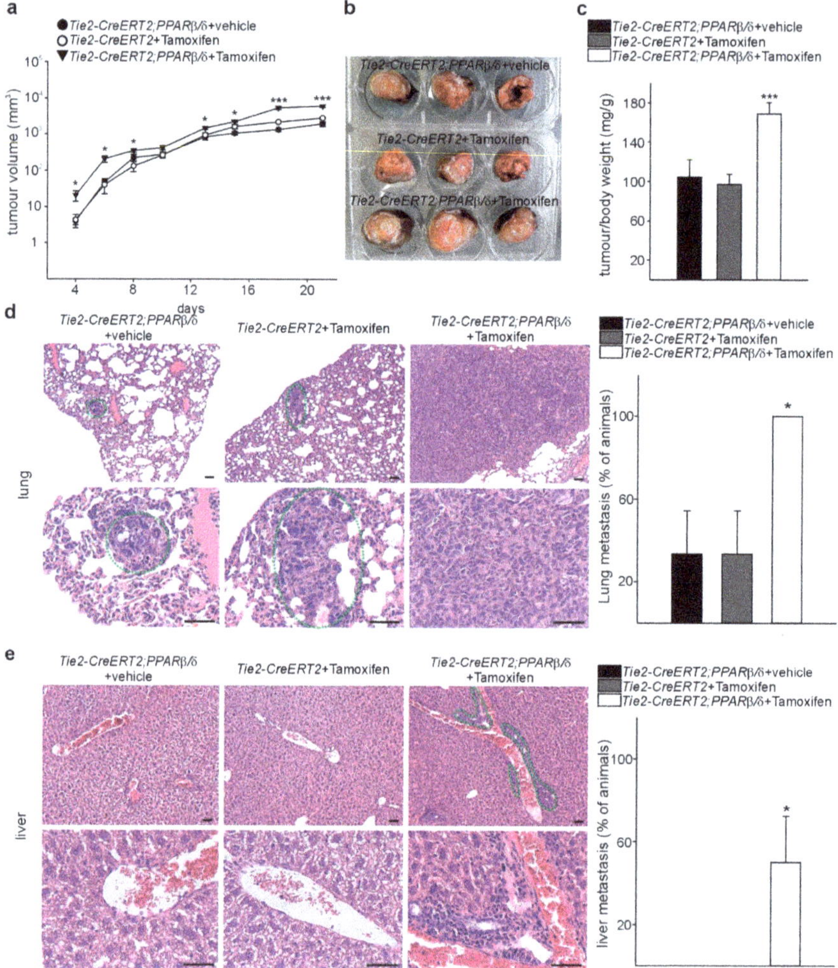

Figure 5. Vascular- specific PPARβ/δ overexpression increases LLC1 tumor growth and spontaneous metastases formation in vivo. (**a**) LLC1 tumor growth curves in Tie2-CreERT2;PPARβ/δ + Tamoxifen animals ($n = 6$) and respective controls (Tie2-CreERT2;PPARβ/δ + vehicle ($n = 6$) and Tie2-CreERT2 + Tamoxifen ($n = 6$). (**b**) LLC1 tumors in Tie2-CreERT2;PPARβ/δ + Tamoxifen mice and control groups. (**c**) Quantification of tumor/body weights. (**d**) Representative photomicrographs of lung metastases in the three groups of mice. Metastases are indicated with a dotted green circle. Note that in Tie2-CreERT2;PPARβ/δ + Tamoxifen animals, metastases were so considerable in size that the whole photomicrograph shows a metastasis only. The plot on the right shows quantification of the percentage of animals with lung metastases. (**e**) Representative photomicrographs of liver sections from the three groups of animals. Metastases are indicated with a dotted green circle. In the two control groups, no liver metastases were detectable. Quantification of the percentage of animals with liver metastases is shown on the right. Scale bars indicate 50 μm. Data are mean ± SEM. $^{*} p < 0.05$, $^{***} p < 0.001$.

We analyzed next PPARβ/δ and Cd31 expression in the tumors by immunostaining. Tie2-CreERT2;PPARβ/δ + Tamoxifen tumors displayed much higher PPARβ/δ and Cd31 positive vessel-and area-densities than both control groups (Figure 6a). To confirm the endothelial PPARβ/δ overexpression on the RNA level, we sorted endothelial cells from the tumors and performed quantitative RT-PCRs. PPARβ/δ expression was nearly doubled in the tumor-derived endothelial cells from Tie2-CreERT2;PPARβ/δ + Tamoxifen mice compared to both controls (Figure 6b). Co-immunostaining for PCNA and Cd31 demonstrated more PCNA-positive endothelial cells in the tumors from animals with vessel-specific overexpression of PPARβ/δ than in the controls, indicating higher proliferation rates of endothelial cells, like it had been the case in the tumors of GW0742 agonist treated mice (Figure 6c).

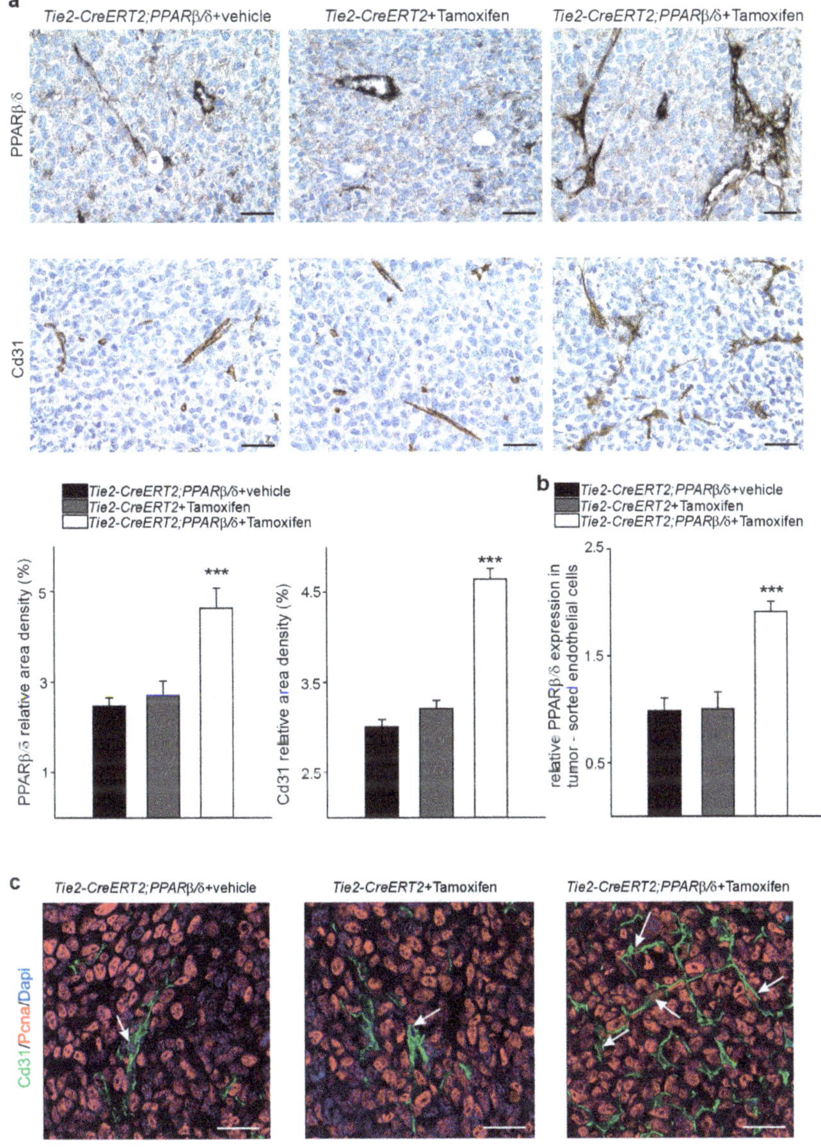

Figure 6. PPARβ/δ and Cd31 are increased in LLC1 tumors from mice with vascular PPARβ/δ overexpression. (**a**) PPARβ/δ (upper panel) and Cd31 (lower panel) immunostainings in LLC1 tumors and quantification of PPARβ/δ and Cd31 signal area densities. (**b**) Quantitative RT-PCRs for PPARβ/δ from tumor sorted endothelial cells of Tie2-CreERT2;PPARβ/δ + Tamoxifen animals and respective

controls (*n* = 6 each). (**c**) PCNA/Cd31 double-labeling of LLC1 tumors from Tie2-CreERT2;PPARβ/δ + Tamoxifen animals and respective controls. Scale bars indicate 50 μm. Data are mean ± SEM. *** $p <$ 0.001.

3.5. PPARβ/δ Overexpression Upregulates Angiogenic and Metastases Promoting Molecules in Endothelial Cells In Vivo

To confirm the differential expression of genes we had observed upon pharmacological PPARβ/δ modulation in human endothelial cells and to elucidate further the molecular mechanisms upon PPARβ/δ overexpression in vascular cells in vivo, we performed quantitative RT-PCRs from tumor-sorted endothelial cells (Figure 7). Again, established PPARβ/δ target genes were upregulated except Pdk4, as likewise genes with established angiogenic functions: Vegf, Vegfr1 and 2, Cd31, vwf, Pdgfb, Pdgfrα, and β, Sox18, and c-kit. Furthermore, we found an increase of tumor-angiogenesis related genes such as E26 transformation specific factor 1 (Ets-1) [48,49], and Wt1 (Wilms' tumor suppressor 1) [15,48]. Interestingly, we could demonstrate that PPARβ/δ represses WT1 in human melanoma cells leading to decreased melanoma cell proliferation [5]. The upregulation of Wt1 upon overexpression of PPARβ/δ in endothelial cells observed here, could be due to cell type-specific expression of co-activators/co-repressors, which requires further studies. We additionally analyzed expression of matrix metalloproteinases 1, 8, and 9. It has been observed that pharmacological activation of PPARβ/δ in vascular cells increased MMP9 mRNA levels [7]. However, we could only find an increase for MMP8, which is known to positively regulate angiogenesis [50] in our model. Interleukin 1 and 6 (Il 1 and Il 6) expression was increased, pointing to an inflammatory state of endothelial cells; yet, Interleukin 18 (Il 18) and Tumor necrosis factor alpha (TNFα) expression were unchanged. Sirt1 and Vcam-1 were not differentially expressed. The mRNA expression of anti-angiogenic Thrombospondin -1 and its receptor Cd36 [51] was unaffected by PPARβ/δ overexpression. It seems that vascular overexpression of PPARβ/δ promotes exclusively a hyper-angiogenic phenotype rather than repressing anti-angiogenic pathways. This might contribute to the observed relatively weak phenotype of PPARβ/δ knockout animals, which are viable, develop normally, are fertile, and display only light growth, skin, adipose tissue, and myelinization alterations [52]. Finally, we observed upregulation of the metastases promoting [53,54] CC-chemokine ligands 2 and 5 (Ccl 2 and 5) in tumor- derived PPARβ/δ overexpressing endothelial cells (Figure 7).

In summary of these findings, the use of PPARβ/δ agonists, initially developed to treat hyperlipidemia or cardiovascular diseases, as anti-cancer drugs, even in the setting of an antiproliferative effect on the tumor cells, seems irresponsible. Although the PPARβ/δ agonist 501,516 entered clinical trials for the treatment of metabolic syndrome and diabetes in the beginning of 2000, these trials were stopped in 2007 due to multiple appearance of cancers in mice and rats (cited in [55]. Unfortunately, upon publication of the beneficial effects of PPARβ/δ activation against obesity and on exercise endurance in mice [12,56,57], the PPARβ/δ agonist GW501516 became very popular in the athletes community.

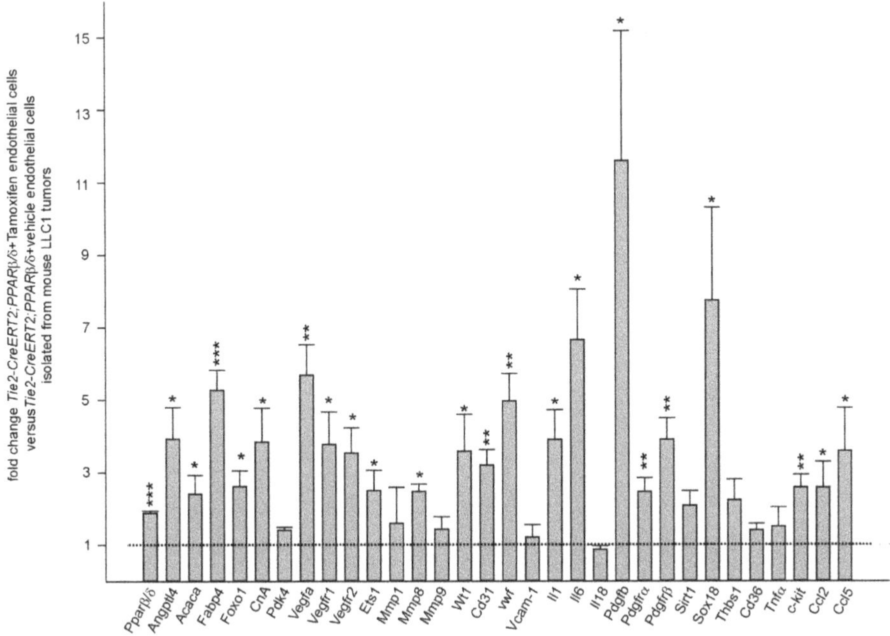

Figure 7. Quantitative RT-PCR corroborates angiogenic functions of PPARβ/δ in endothelial cells. Quantitative RT-PCR of known PPARβ/δ target genes, inflammation, and angiogenesis markers in tumor- sorted endothelial cells from LLC1 bearing Tie2-CreERT2;PPARβ/δ + Tamoxifen animals and Tie2-CreERT2;PPARβ/δ +vehicle mice as controls (*n* = 6 each). Data are mean ± SEM. *P < 0.05, **P < 0.01, ***P < 0.001.

3.6. RNA Sequencing Further Certifies the Acquisition of a Highly angiogenic Endothelial Phenotype Upon PPARβ/δ Overexpression and Identifies Pro-Tumor-Angiogenic Signaling Networks

To analyze the transcriptome of gene expression patterns in PPARβ/δ overexpressing tumor-derived endothelial cells, we sorted endothelial cells from the tumors of Tie2-CreERT2;PPARβ/δ injected with Tamoxifen, or vehicle as controls. mRNAs were after purity and quality controls submitted to sequencing. 283 genes were found to be differentially expressed in vascular cells overexpressing PPARβ/δ, most of them were up-regulated (Figure 8a and Supplementary Table S1). Cluster analysis of differentially expressed genes confirms mostly up-regulation of genes upon vascular overexpression of PPARβ/δ (Figure 8b). Again, this is in line with our observation in the quantitative PCRs, which indicate an angiogenesis boosting effect rather than a repression of anti-angiogenic molecules to enhance angiogenesis. Interestingly, gene ontology term analysis identified angiogenesis as the most upregulated biological process upon PPARβ/δ overexpression, followed by cell adhesion (Figure 8c). Top ten network analysis combined with a search for PPAR responsive elements (PPREs) identified in total six genes with potential PPREs: the three Vegf receptors 1 (Flt1), 2 (Kdr), and 3 (Flt4), all of them known to be implicated in the promotion of tumor angiogenesis [58,59], and Pdgfrβ, Pdgfb, and c-kit, which we had already found to be upregulated upon PPARβ/δ overexpression (Figure 7). Initial clinical efforts to inhibit tumor angiogenesis mainly focused on inhibition of the VEGF/VEGFR signaling, but often further disease progression could be observed. Angiogenesis involves multiple pathways, and it became clear that inhibition of only one pathway could be compensated by the others resulting in tumor progression. One of these compensatory pathways is platelet-derived growth factor PDGF and PDGFR signaling which led to the development of new anti-angiogenic therapies in the treatment of cancer [60]. Additionally, c-Kit belongs to the group of tumor angiogenesis-promoting molecules and new tyrosine kinase receptor inhibitors have been developed which target also c-Kit [61].

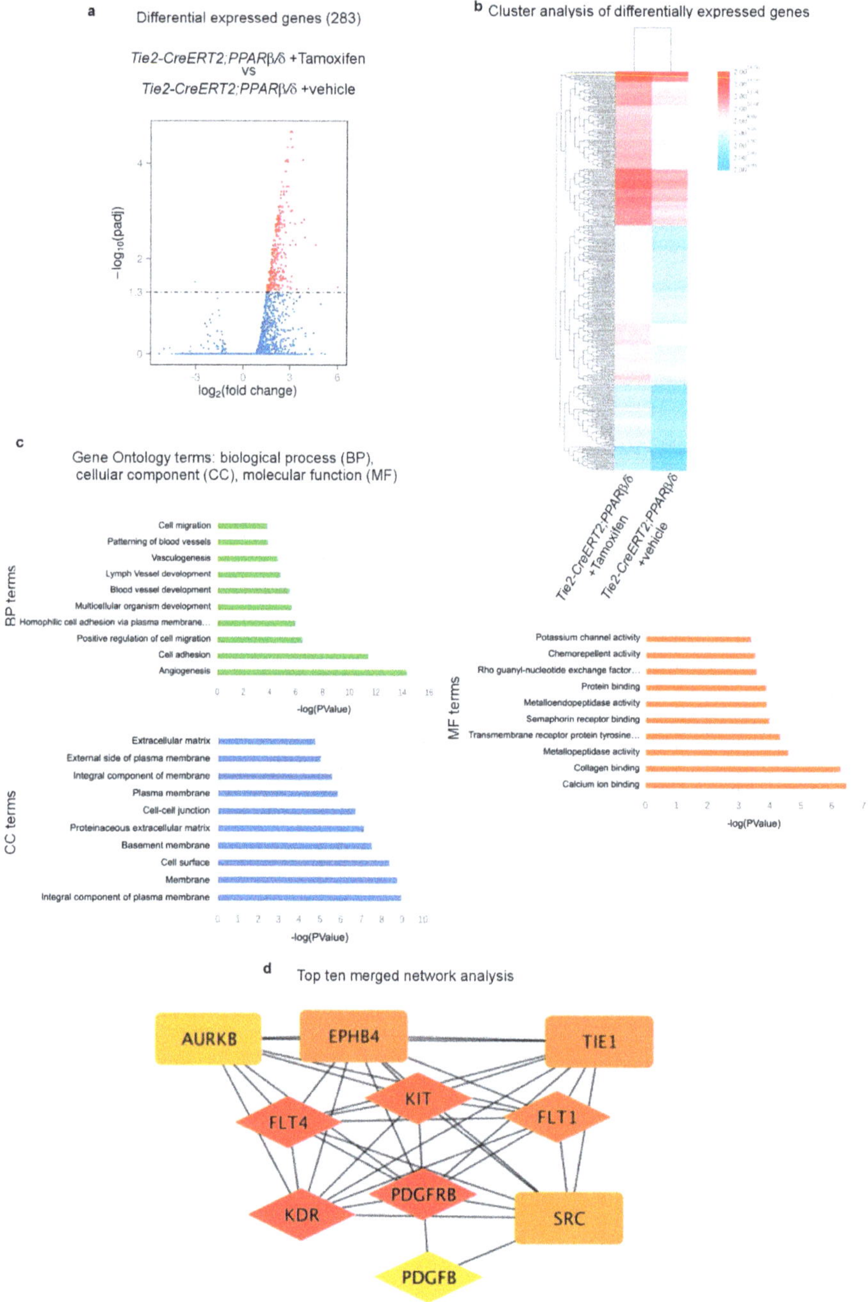

Figure 8. RNA sequencing of tumor- sorted endothelial cells confirms the tumor angiogenesis promoting effect of PPARβ/δ. (**a**) Volcano plot analysis of differentially expressed genes in tumor endothelial cells

sorted from LLC1 tumors of Tie2-CreERT2;PPARβ/δ + Tamoxifen animals and Tie2-CreERT2;PPARβ/δ + vehicle mice as controls. (*n* = 5 each). (**b**) Cluster analysis of differentially expressed genes. (**c**) Most significantly changed gene ontology terms (BP: biological process, MF: molecular function, CC: cellular component). (**d**) Top ten network analysis merged with the prediction of PPAR-responsive elements (PPRE): diamonds and rectangles represent the top ten network; but only diamonds exhibited PPRE prediction. As the Vascular endothelial growth factor (VEGF) pathway has been suggested already to be regulated by PPARβ/δ [7,46], we decided to investigate further a potential regulation of Pdgfrβ, Pdgfb, and c-kit.

3.7. PPARβ/δ Directly Activates PDGFRβ, PDGFB, and c-Kit

As a prerequisite for transcriptional regulation, we first demonstrated co-localization of PPARβ/δ and PDGFRβ (Figure 9a), PDGFB (Figure 9d), and c-Kit (Figure 9g) in endothelial cells. Transient co-transfection with a PPARβ/δ expression construct increased the activities of PDGFRβ (Figure 9b), PDGFB (Figure 9e), and c-Kit (Figure 9h) promoter reporter constructs. Using in silico analysis, PPAR responsive elements (PPREs) were identified in the respective promoter sequences (Figure 10a, Figure 11a, Figure 12a). Deletion of each of the three identified PPREs in the PDGFRβ promoter was sufficient to abolish transactivation by PPARβ/δ (Figure 9c). Regarding the PDGFB promoter, only the combined deletion of the three identified PPREs abrogated activation upon co-transfection of the PPARβ/δ construct, indicating a collaborative interaction between these PPREs (Figure 9f). Finally, in the c-Kit promoter, each deletion for the two identified binding regions was sufficient to cancel activation by PPARβ/δ (Figure 9i).

Binding of PPARβ/δ to the predicted PPREs was confirmed by chromatin immunoprecipitation (ChIP) assays. An antibody against acetylated histone H3 was used to check for nucleosome integrity. Specificity of the interaction of PPARβ/δ with the PPAR responsive elements in the PDGFRβ promoter is indicated by the lack of a PCR product when the same samples were amplified with primers specific for the 3′ UTR. We confirmed binding to the identified PPREs in the PDGFRβ (Figure 10b,c), PDGFB (Figure 11b,c), and the c-Kit promoter (Figure 12b,c).

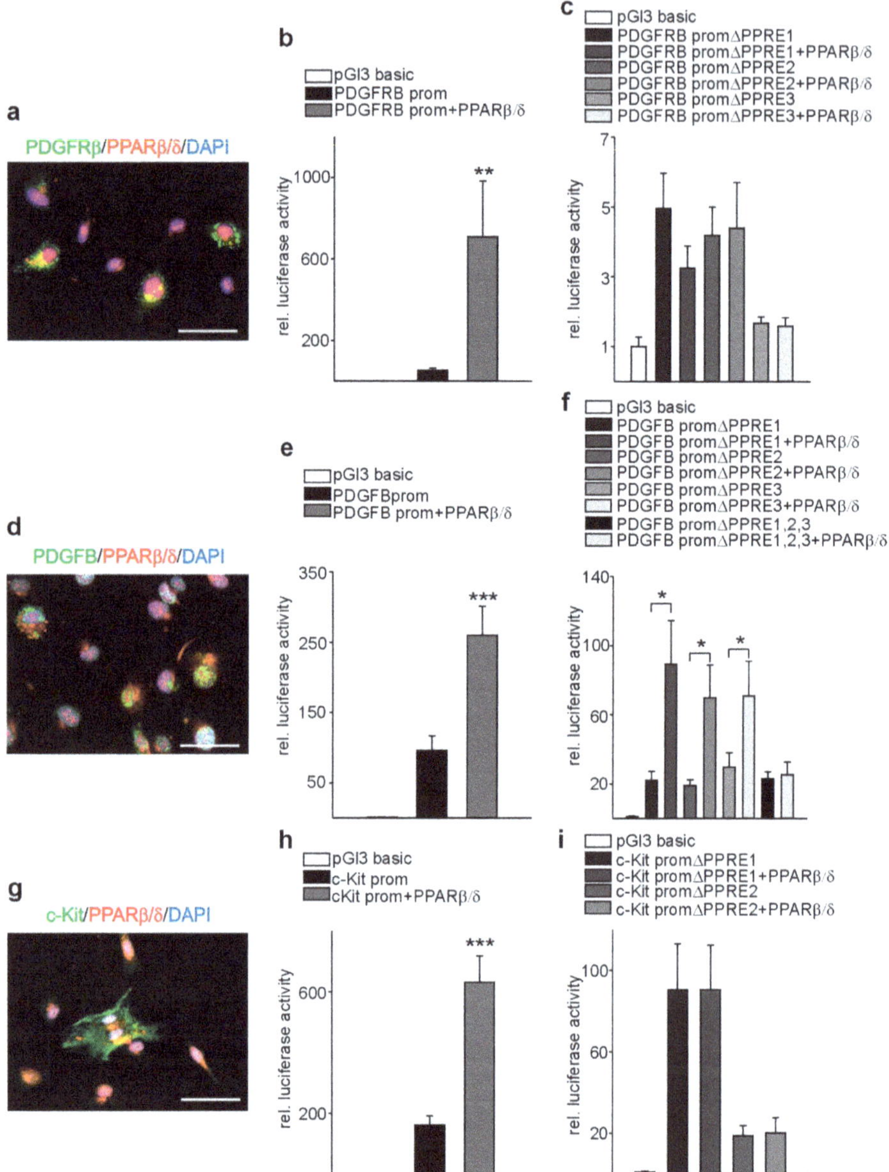

Figure 9. PPARβ/δ transactivates PDGFRb, PDGFB, and c-Kit promoters. (**a**) Co-labeling for PPARβ/δ (red) and PDGFRβ (green) in endothelial cells. (**b**) Luciferase activity of reporter constructs carrying the PDGFRβ promoter in the presence of the PPARβ/δ expression construct (*n* = 12 each). (**c**) Transient transfections of luciferase constructs carrying the PDGFRβ promoter with deletion of the identified PPREs (ΔPPRE) in the presence of the PPARβ/δ expression construct (*n* = 12 each). (**d**) Co-immunostaining for PPARβ/δ (red) and PDGFB (green) in endothelial cells. (**e**) Luciferase activity of reporter constructs carrying the PDGFB promoter in the presence of the PPARβ/δ expression construct (*n* = 12 each). (**f**) Transient transfections of luciferase constructs carrying the PDGFB promoter with deletion of the identified PPREs (ΔPPRE) in the presence of the PPARβ/δ expression construct (*n* =16

each). (**g**) Co-labeling for PPARβ/δ (red) and c-Kit (green) in endothelial cells. (**h**) Transient transfections of luciferase reporter constructs carrying the c-Kit promoter in the presence of the PPARβ/δ expression construct (*n* = 10 each). (**i**) Luciferase activity of constructs carrying the c-Kit promoter with deletion of the identified PPREs (ΔPPRE) in the presence of the PPARβ/δ expression construct (*n* = 6 each). The promoter-less luciferase expression construct (pGl3basic) served as a negative control in all transfection experiments. Luciferase activities were normalized for the activity of co-transfected β-galactosidase. Scale bars indicate 50 μm. Data are mean ± SEM. *$P < 0.05$, **$P < 0.01$, ***$P < 0.001$.

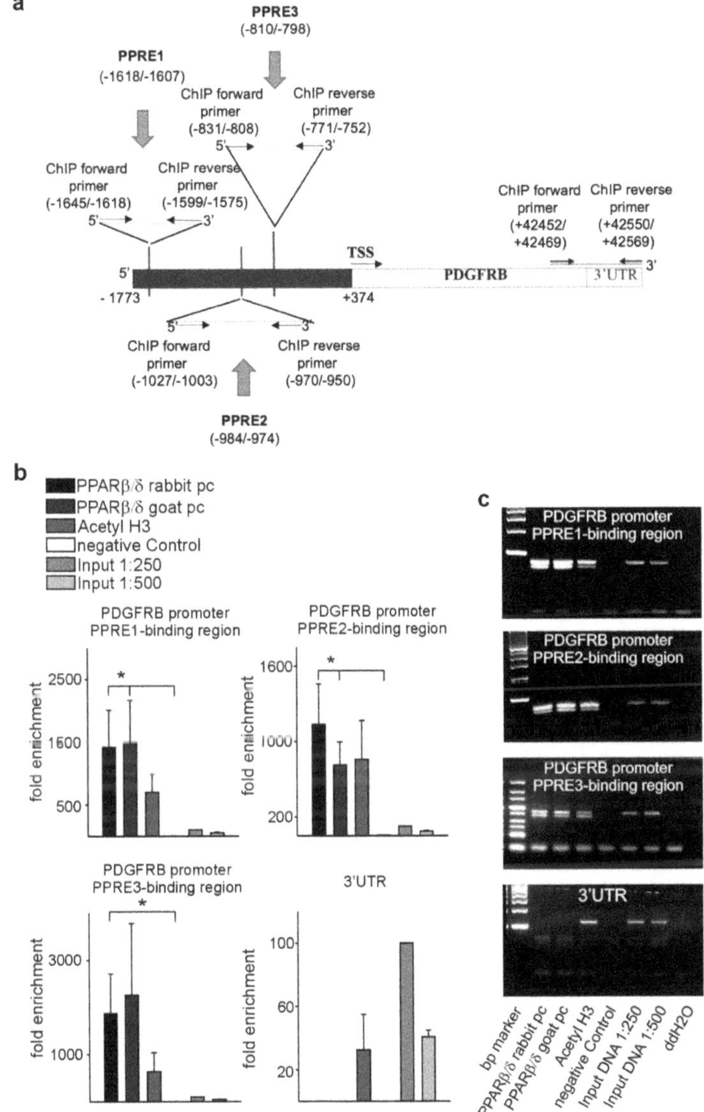

Figure 10. PPARβ/δ binds to the PDGFRβ promoter. (**a**) Schematic representation of the putative PPREs in the PDGFRβ promoter. Positions of the cloned promoters relative to the transcription start site, positions and sequences of the putative PPREs, and positions of the oligonucleotides used for

CHIP analyses are indicated. For promoter-deletion constructs, the indicated PPREs were removed from the promoter reporter constructs. Chromatin immunoprecipitation (ChIP, $n = 4$) was performed using polyclonal antibodies against PPARβ/δ or anti-acetyl-histone H3 antibody. Input DNA was used as a positive control for quantitative PCRs (**b**) or semiquantitative PCRs ((**c**), representative agarose gels) for the PDGFRβ promoter and respective 3′UTR sequences. Data are mean ± SEM. *$p < 0.05$.

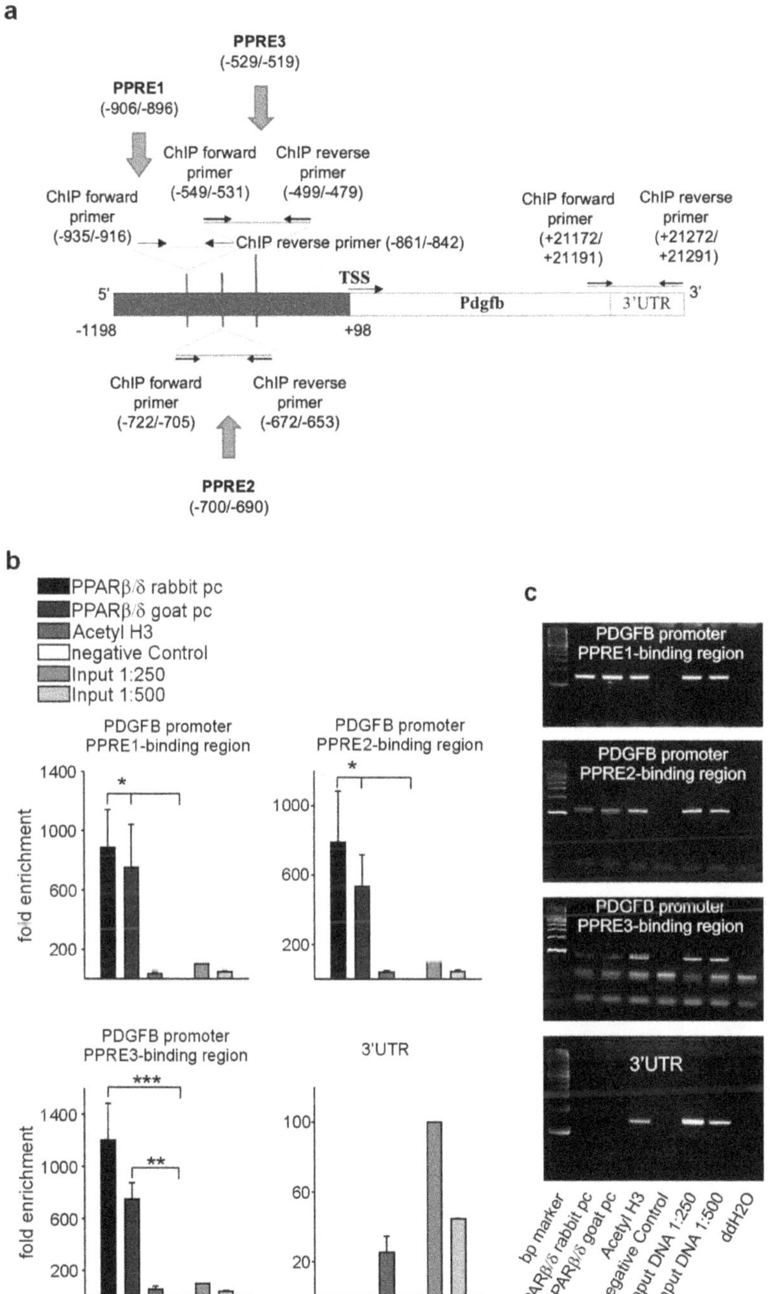

Figure 11. PPARβ/δ binds to the PDGFB promoter. (**a**) Schematic representation of the putative PPREs in the PDGFB promoter. Positions of the cloned promoters relative to the transcription start site, positions and sequences of the putative PPREs, and positions of the oligonucleotides used for CHIP analyses are indicated. For promoter-deletion constructs, the indicated PPREs were removed from the

promoter reporter constructs. Chromatin immunoprecipitation (ChIP, $n = 3$) was performed using polyclonal antibodies against PPARβ/δ or anti-acetyl-histone H3 antibody. Input DNA was used as a positive control for quantitative PCRs (**b**) or semiquantitative PCRs ((**c**), representative agarose gels) for the PDGFRβ promoter and respective 3′UTR sequences. Data are mean ± SEM. $^*p < 0.05$, $^{**}p < 0.01$.

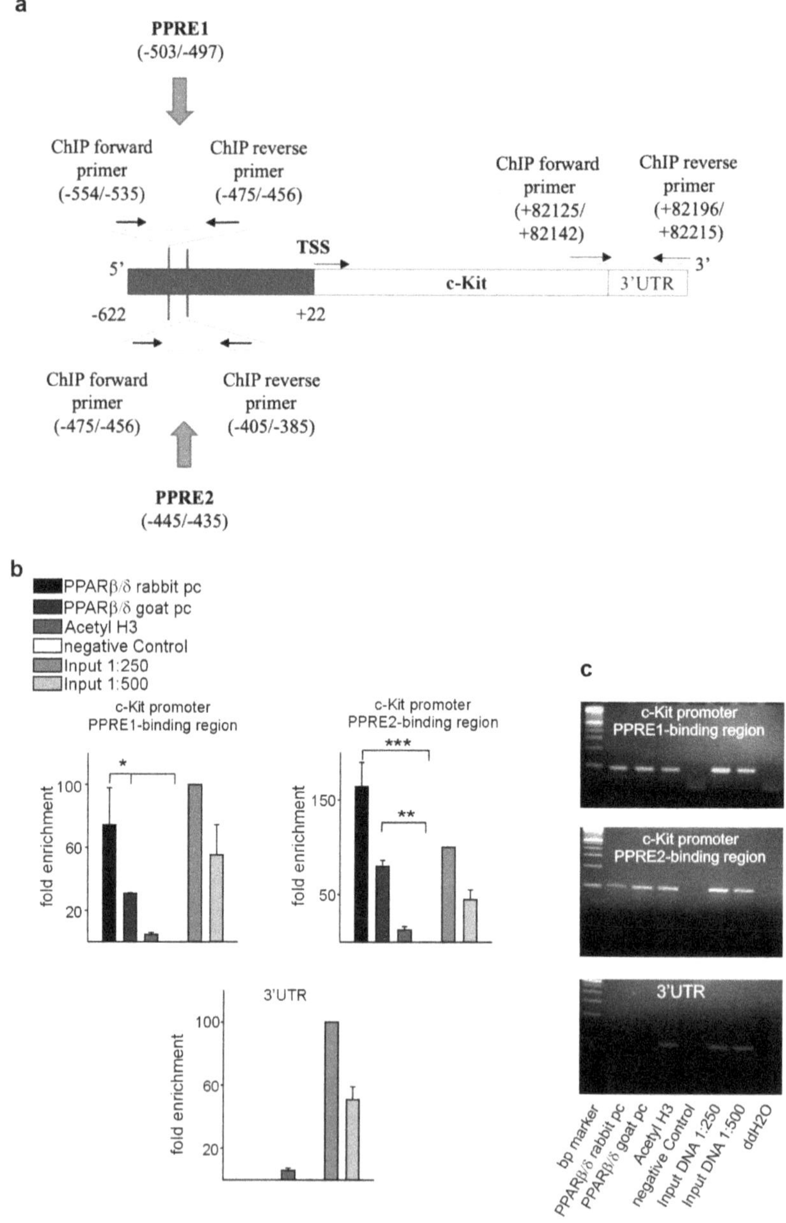

Figure 12. PPARβ/δ binds to the c-Kit promoter. (**a**) Schematic representation of the putative PPREs in the c-Kit promoter. Positions of the cloned promoters relative to the transcription start site, positions and sequences of the putative PPREs, and positions of the oligonucleotides used for CHIP analyses are indicated. For promoter-deletion constructs, the indicated PPREs were excluded from the promoter

reporter constructs. Chromatin immunoprecipitation (ChIP, $n = 3$) was performed using polyclonal antibodies against PPARβ/δ or anti-acetyl-histone H3 antibody. Input DNA was used as a positive control for quantitative PCRs (**b**) or semiquantitative PCRs ((**c**), representative agarose gels) for the PDGFRβ promoter and respective 3′UTR sequences. Data are mean ± SEM. $^*p < 0.05$, $^{**}p < 0.01$, $^{***}p < 0.001$.

4. Conclusions

We show here that pharmacological PPARβ/δ activation increases endothelial cell proliferation in vitro and intensifies tumor angiogenesis in vivo, enhancing tumor progression and metastases formation. Conditional vascular specific overexpression of PPARβ/δ also resulted in enhanced tumor angiogenesis, growth, and metastases formation, suggesting an endothelial cell- specific mechanism for PPARβ/δ function in tumor progression, independent from its contrasting effects on specific tumor cell types.

Tumor-angiogenesis promoting effects of PPARβ/δ are mediated via activation of the PDGF/PDGFR pathway, c-Kit, and probably the VEGF/VEGFR pathway. The pharmacological use of PPARβ/δ agonists should be considered as dangerous regarding its consistent consequences on tumor angiogenesis. Consideration of PPARβ/δ antagonists for inhibition of tumor angiogenesis and cancer growth is unlikely to represent a novel direction due to their controversial effects on the proliferation of different tumor cell types.

Supplementary Materials: The following is available online at http://www.mdpi.com/2073-4409/8/12/1623/s1, Table S1.

Author Contributions: Conceptualization, K.-D.W., J.-F.M., and N.W.; methodology, K.-D.W., N.W.; formal analysis, K.-D.W., N.W., L.M., S.D., N.L., J.-F.M.; investigation, K.-D.W., N.W., S.D.; writing—original draft preparation, K.-D.W., N.W.; writing—review and editing, K.-D.W., N.W.; visualization, N.W.; supervision, K.-D.W., N.W.; project administration, K.-D.W., N.W.; funding acquisition, N.W., K.-D.W.

Funding: This research was funded by "Fondation ARC pour la recherche sur le cancer", grant number n°PJA 20161204650 (N.W.), Gemluc (N.W.), and Plan Cancer INSERM, Fondation pour la Recherche Médicale (K.-D.W.). S.D. was supported by a PhD fellowship from the China Scholarship Council.

Acknowledgments: The authors thank S. Destree, A. Borderie, A. Biancardini, A. Martres A. Loubat, and M. Cutajar-Bossert for technical assistance.

Conflicts of Interest: The authors declare no conflict of interest.

References

1. Wagner, K.D.; Wagner, N. Peroxisome proliferator-activated receptor beta/delta (PPARbeta/delta) acts as regulator of metabolism linked to multiple cellular functions. *Pharmacol. Ther.* **2010**, *125*, 423–435. [CrossRef] [PubMed]

2. Grygiel-Górniak, B. Peroxisome proliferator-activated receptors and their ligands: Nutritional and clinical implications—A review. *Nutr. J.* **2014**, *13*, 17. [CrossRef] [PubMed]

3. Grimaldi, P.A. Regulatory functions of PPARbeta in metabolism: Implications for the treatment of metabolic syndrome. *Biochim. Biophys. Acta* **2007**, *1771*, 983–990. [CrossRef] [PubMed]

4. Bishop-Bailey, D.; Swales, K.E. The Role of PPARs in the Endothelium: Implications for Cancer Therapy. *PPAR Res.* **2008**, *2008*, 904251. [CrossRef]

5. Michiels, J.F.; Perrin, C.; Leccia, N.; Massi, D.; Grimaldi, P.; Wagner, N. PPARbeta activation inhibits melanoma cell proliferation involving repression of the Wilms' tumour suppressor WT1. *Pflügers Arch.* **2010**, *459*, 689–703. [CrossRef]

6. Wagner, K.D.; Benchetrit, M.; Bianchini, L.; Michiels, J.F.; Wagner, N. Peroxisome proliferator-activated receptor β/δ (PPARβ/δ) is highly expressed in liposarcoma and promotes migration and proliferation. *J. Pathol.* **2011**, *224*, 575–588. [CrossRef]

7. Piqueras, L.; Reynolds, A.R.; Hodivala-Dilke, K.M.; Alfranca, A.; Redondo, J.M.; Hatae, T.; Tanabe, T.; Warner, T.D.; Bishop-Bailey, D. Activation of PPARbeta/delta induces endothelial cell proliferation and angiogenesis. *Arter. Thromb. Vasc. Biol.* **2007**, *27*, 63–69. [CrossRef]

8. Wagner, N.; Jehl-Piétri, C.; Lopez, P.; Murdaca, J.; Giordano, C.; Schwartz, C.; Gounon, P.; Hatem, S.N.; Grimaldi, P.; Wagner, K.D. Peroxisome proliferator-activated receptor beta stimulation induces rapid cardiac growth and angiogenesis via direct activation of calcineurin. *Cardiovasc. Res.* **2009**, *83*, 61–71. [CrossRef]

9. Gaudel, C.; Schwartz, C.; Giordano, C.; Abumrad, N.A.; Grimaldi, P.A. Pharmacological activation of PPARbeta promotes rapid and calcineurin-dependent fiber remodeling and angiogenesis in mouse skeletal muscle. *Am. J. Physiol. Endocrinol. Metab.* **2008**, *295*, E297–E304. [CrossRef]

10. Wagner, K.D.; Vukolic, A.; Baudouy, D.; Michiels, J.F.; Wagner, N. Inducible Conditional Vascular-Specific Overexpression of Peroxisome Proliferator-Activated Receptor Beta/Delta Leads to Rapid Cardiac Hypertrophy. *PPAR Res.* **2016**, *2016*, 7631085. [CrossRef]

11. Abdollahi, A.; Schwager, C.; Kleeff, J.; Esposito, I.; Domhan, S.; Peschke, P.; Hauser, K.; Hahnfeldt, P.; Hlatky, L.; Debus, J.; et al. Transcriptional network governing the angiogenic switch in human pancreatic cancer. *Proc. Natl. Acad. Sci. USA* **2007**, *104*, 12890–12895. [CrossRef] [PubMed]

12. Luquet, S.; Lopez-Soriano, J.; Holst, D.; Fredenrich, A.; Melki, J.; Rassoulzadegan, M.; Grimaldi, P.A. Peroxisome proliferator-activated receptor delta controls muscle development and oxidative capability. *FASEB J.* **2003**, *17*, 2299–2301. [CrossRef] [PubMed]

13. Forde, A.; Constien, R.; Gröne, H.J.; Hämmerling, G.; Arnold, B. Temporal Cre-mediated recombination exclusively in endothelial cells using Tie2 regulatory elements. *Genesis* **2002**, *33*, 191–197. [CrossRef] [PubMed]

14. Wagner, K.D.; Wagner, N.; Guo, J.K.; Elger, M.; Dallman, M.J.; Bugeon, L.; Schedl, A. An inducible mouse model for PAX2-dependent glomerular disease: Insights into a complex pathogenesis. *Curr. Biol.* **2006**, *16*, 793–800. [CrossRef] [PubMed]

15. Wagner, K.D.; Cherfils-Vicini, J.; Hosen, N.; Hohenstein, P.; Gilson, E.; Hastie, N.D.; Michiels, J.F.; Wagner, N. The Wilms' tumour suppressor Wt1 is a major regulator of tumour angiogenesis and progression. *Nat. Commun.* **2014**, *5*, 5852. [CrossRef] [PubMed]

16. Wagner, N.; Panelos, J.; Massi, D.; Wagner, K.D. The Wilms' tumor suppressor WT1 is associated with melanoma proliferation. *Pflügers Arch.* **2008**, *455*, 839–847. [CrossRef] [PubMed]

17. El Maï, M.; Wagner, K.D.; Michiels, J.F.; Ambrosetti, D.; Borderie, A.; Destree, S.; Renault, V.; Djerbi, N.; Giraud-Panis, M.J.; Gilson, E.; et al. The Telomeric Protein TRF2 Regulates Angiogenesis by Binding and Activating the PDGFRβ Promoter. *Cell Rep.* **2014**, *9*, 1047–1060. [CrossRef]

18. Wagner, K.D.; El Maï, M.; Ladomery, M.; Belali, T.; Leccia, N.; Michiels, J.F.; Wagner, N. Altered VEGF Splicing Isoform Balance in Tumor Endothelium Involves Activation of Splicing Factors Srpk1 and Srsf1 by the Wilms' Tumor Suppressor Wt1. *Cells* **2019**, *8*, 41. [CrossRef]

19. Edgar, R.; Domrachev, M.; Lash, A.E. Gene Expression Omnibus: NCBI gene expression and hybridization array data repository. *Nucleic Acids Res.* **2002**, *30*, 207–210. [CrossRef]

20. Barrett, T.; Wilhite, S.E.; Ledoux, P.; Evangelista, C.; Kim, I.F.; Tomashevsky, M.; Marshall, K.A.; Phillippy, K.H.; Sherman, P.M.; Holko, M.; et al. NCBI GEO: Archive for functional genomics data sets—Update. *Nucleic Acids Res.* **2013**, *41*, D991–D995. [CrossRef]

21. Beisser, D.; Klau, G.W.; Dandekar, T.; Müller, T.; Dittrich, M.T. BioNet: An R-Package for the functional analysis of biological networks. *Bioinformatics* **2010**, *26*, 1129–1130. [CrossRef] [PubMed]

22. Huang, J.K.; Carlin, D.E.; Yu, M.K.; Zhang, W.; Kreisberg, J.F.; Tamayo, P.; Ideker, T. Systematic Evaluation of Molecular Networks for Discovery of Disease Genes. *Cell Syst.* **2018**, *6*, 484–495. [CrossRef] [PubMed]

23. Pillich, R.T.; Chen, J.; Rynkov, V.; Welker, D.; Pratt, D. NDEx: A Community Resource for Sharing and Publishing of Biological Networks. *Methods Mol. Biol.* **2017**, *1558*, 271–301. [CrossRef] [PubMed]

24. Shannon, P.T.; Grimes, M.; Kutlu, B.; Bot, J.J.; Galas, D.J. RCytoscape: Tools for exploratory network analysis. *BMC Bioinform.* **2013**, *14*, 217. [CrossRef]

25. Kamburov, A.; Cavill, R.; Ebbels, T.M.; Herwig, R.; Keun, H.C. Integrated pathway-level analysis of transcriptomics and metabolomics data with IMPaLA. *Bioinformatics* **2011**, *27*, 2917–2918. [CrossRef]

26. Kwon, A.T.; Arenillas, D.J.; Worsley Hunt, R.; Wasserman, W.W. oPOSSUM-3: Advanced analysis of regulatory motif over-representation across genes or ChIP-Seq datasets. *Genes Genomes Genet.* **2012**, *2*, 987–1002. [CrossRef]

27. Maeda, K.; Nishiyama, C.; Ogawa, H.; Okumura, K. GATA2 and Sp1 positively regulate the c-kit promoter in mast cells. *J. Immunol.* **2010**, *185*, 4252–4260. [CrossRef]
28. Lim, J.C.W.; Kwan, Y.P.; Tan, M.S.; Teo, M.H.Y.; Chiba, S.; Wahli, W.; Wang, X. The Role of PPARβ/δ in Melanoma Metastasis. *Int. J. Mol. Sci.* **2018**, *19*, 2860. [CrossRef]
29. Pedchenko, T.V.; Gonzalez, A.L.; Wang, D.; DuBois, R.N.; Massion, P.P. Peroxisome proliferator-activated receptor beta/delta expression and activation in lung cancer. *Am. J. Respir. Cell Mol. Biol.* **2008**, *39*, 689–696. [CrossRef]
30. Genini, D.; Garcia-Escudero, R.; Carbone, G.M.; Catapano, C.V. Transcriptional and Non-Transcriptional Functions of PPARβ/δ in Non-Small Cell Lung Cancer. *PLoS ONE* **2012**, *7*, e46009. [CrossRef]
31. Fukumoto, K.; Yano, Y.; Virgona, N.; Hagiwara, H.; Sato, H.; Senba, H.; Suzuki, K.; Asano, R.; Yamada, K.; Yano, T. Peroxisome proliferator-activated receptor delta as a molecular target to regulate lung cancer cell growth. *FEBS Lett.* **2005**, *579*, 3829–3836. [CrossRef] [PubMed]
32. He, P.; Borland, M.G.; Zhu, B.; Sharma, A.K.; Amin, S.; El-Bayoumy, K.; Gonzalez, F.J.; Peters, J.M. Effect of ligand activation of peroxisome proliferator-activated receptor-beta/delta (PPARbeta/delta) in human lung cancer cell lines. *Toxicology* **2008**, *254*, 112–117. [CrossRef] [PubMed]
33. Müller-Brüsselbach, S.; Kömhoff, M.; Rieck, M.; Meissner, W.; Kaddatz, K.; Adamkiewicz, J.; Keil, B.; Klose, K.J.; Moll, R.; Burdick, A.D.; et al. Deregulation of tumor angiogenesis and blockade of tumor growth in PPARbeta-deficient mice. *EMBO J.* **2007**, *26*, 3686–3698. [CrossRef] [PubMed]
34. Stephen, R.L.; Gustafsson, M.C.; Jarvis, M.; Tatoud, R.; Marshall, B.R.; Knight, D.; Ehrenborg, E.; Harris, A.L.; Wolf, C.R.; Palmer, C.N. Activation of peroxisome proliferator-activated receptor delta stimulates the proliferation of human breast and prostate cancer cell lines. *Cancer Res.* **2004**, *64*, 3162–3170. [CrossRef]
35. Hatae, T.; Wada, M.; Yokoyama, C.; Shimonishi, M.; Tanabe, T. Prostacyclin-dependent apoptosis mediated by PPAR delta. *J. Biol. Chem.* **2001**, *276*, 46260–46267. [CrossRef]
36. Staiger, H.; Haas, C.; Machann, J.; Werner, R.; Weisser, M.; Schick, F.; Machicao, F.; Stefan, N.; Fritsche, A.; Häring, H.U. Muscle-derived angiopoietin-like protein 4 is induced by fatty acids via peroxisome proliferator-activated receptor (PPAR)-delta and is of metabolic relevance in humans. *Diabetes* **2009**, *58*, 579–589. [CrossRef]
37. Shin, J.; Li, B.; Davis, M.E.; Suh, Y.; Lee, K. Comparative analysis of fatty acid-binding protein 4 promoters: Conservation of peroxisome proliferator-activated receptor binding sites. *J. Anim. Sci.* **2009**, *87*, 3923–3934. [CrossRef]
38. Degenhardt, T.; Saramäki, A.; Malinen, M.; Rieck, M.; Väisänen, S.; Huotari, A.; Herzig, K.H.; Müller, R.; Carlberg, C. Three members of the human pyruvate dehydrogenase kinase gene family are direct targets of the peroxisome proliferator-activated receptor beta/delta. *J. Mol. Biol.* **2007**, *372*, 341–355. [CrossRef]
39. Zhang, Z.; Neiva, K.G.; Lingen, M.W.; Ellis, L.M.; Nör, J.E. VEGF-dependent tumor angiogenesis requires inverse and reciprocal regulation of VEGFR1 and VEGFR2. *Cell Death Differ.* **2010**, *17*, 499–512. [CrossRef]
40. Lertkiatmongkol, P.; Liao, D.; Mei, H.; Hu, Y.; Newman, P.J. Endothelial functions of platelet/endothelial cell adhesion molecule-1 (CD31). *Curr. Opin. Hematol.* **2016**, *23*, 253–259. [CrossRef]
41. Battegay, E.J.; Rupp, J.; Iruela-Arispe, L.; Sage, E.H.; Pech, M. PDGF-BB modulates endothelial proliferation and angiogenesis in vitro via PDGF beta-receptors. *J. Cell Biol.* **1994**, *125*, 917–928. [CrossRef] [PubMed]
42. Broudy, V.C.; Kovach, N.L.; Bennett, L.G.; Lin, N.; Jacobsen, F.W.; Kidd, P.G. Human umbilical vein endothelial cells display high-affinity c-kit receptors and produce a soluble form of the c-kit receptor. *Blood* **1994**, *83*, 2145–2152. [CrossRef]
43. Sihto, H.; Tynninen, O.; Bützow, R.; Saarialho-Kere, U.; Joensuu, H. Endothelial cell KIT expression in human tumours. *J. Pathol.* **2007**, *211*, 481–488. [CrossRef]
44. Matsui, J.; Wakabayashi, T.; Asada, M.; Yoshimatsu, K.; Okada, M. Stem cell factor/c-kit signaling promotes the survival, migration, and capillary tube formation of human umbilical vein endothelial cells. *J. Biol. Chem.* **2004**, *279*, 18600–18607. [CrossRef]
45. Young, N.; Hahn, C.N.; Poh, A.; Dong, C.; Wilhelm, D.; Olsson, J.; Muscat, G.E.; Parsons, P.; Gamble, J.R.; Koopman, P. Effect of disrupted SOX18 transcription factor function on tumor growth, vascularization, and endothelial development. *J. Natl. Cancer Inst.* **2006**, *98*, 1060–1067. [CrossRef]
46. Wang, D.; Wang, H.; Guo, Y.; Ning, W.; Katkuri, S.; Wahli, W.; Desvergne, B.; Dey, S.K.; DuBois, R.N. Crosstalk between peroxisome proliferator-activated receptor delta and VEGF stimulates cancer progression. *Proc. Natl. Acad. Sci. USA* **2006**, *103*, 19069–19074. [CrossRef]

47. Randi, A.M.; Laffan, M.A. Von Willebrand factor and angiogenesis: Basic and applied issues. *J. Thromb. Haemost.* **2017**, *15*, 13–20. [CrossRef]

48. Wagner, N.; Michiels, J.F.; Schedl, A.; Wagner, K.D. The Wilms' tumour suppressor WT1 is involved in endothelial cell proliferation and migration: Expression in tumour vessels in vivo. *Oncogene* **2008**, *27*, 3662–3672. [CrossRef]

49. Oettgen, P. The role of ets factors in tumor angiogenesis. *J. Oncol.* **2010**, *2010*, 767384. [CrossRef]

50. Fang, C.; Wen, G.; Zhang, L.; Lin, L.; Moore, A.; Wu, S.; Ye, S.; Xiao, Q. An important role of matrix metalloproteinase-8 in angiogenesis in vitro and in vivo. *Cardiovasc. Res.* **2013**, *99*, 146–155. [CrossRef]

51. Chu, L.Y.; Ramakrishnan, D.P.; Silverstein, R.L. Thrombospondin-1 modulates VEGF signaling via CD36 by recruiting SHP-1 to VEGFR2 complex in microvascular endothelial cells. *Blood* **2013**, *122*, 1822–1832. [CrossRef] [PubMed]

52. Peters, J.M.; Lee, S.S.; Li, W.; Ward, J.M.; Gavrilova, O.; Everett, C.; Reitman, M.L.; Hudson, L.D.; Gonzalez, F.J. Growth, adipose, brain, and skin alterations resulting from targeted disruption of the mouse peroxisome proliferator-activated receptor beta(delta). *Mol. Cell Biol.* **2000**, *20*, 5119–5128. [CrossRef] [PubMed]

53. Roblek, M.; Protsyuk, D.; Becker, P.F.; Stefanescu, C.; Gorzelanny, C.; Glaus Garzon, J.F.; Knopfova, L.; Heikenwalder, M.; Luckow, B.; Schneider, S.W.; et al. CCL2 Is a Vascular Permeability Factor Inducing CCR2-Dependent Endothelial Retraction during Lung Metastasis. *Mol. Cancer Res.* **2019**, *17*, 783–793. [CrossRef] [PubMed]

54. Läubli, H.; Spanaus, K.S.; Borsig, L. Selectin-mediated activation of endothelial cells induces expression of CCL5 and promotes metastasis through recruitment of monocytes. *Blood* **2009**, *114*, 4583–4591. [CrossRef] [PubMed]

55. Mitchell, J.A.; Bishop-Bailey, D. PPARβ/δ a potential target in pulmonary hypertension blighted by cancer risk. *Pulm. Circ.* **2019**, *9*. [CrossRef]

56. Wang, Y.X.; Lee, C.H.; Tiep, S.; Yu, R.T.; Ham, J.; Kang, H.; Evans, R.M. Peroxisome-proliferator-activated receptor delta activates fat metabolism to prevent obesity. *Cell* **2003**, *113*, 159–170. [CrossRef]

57. Fan, W.; Waizenegger, W.; Lin, C.S.; Sorrentino, V.; He, M.X.; Wall, C.E.; Li, H.; Liddle, C.; Yu, R.T.; Atkins, A.R.; et al. PPARδ Promotes Running Endurance by Preserving Glucose. *Cell Metab.* **2017**, *25*, 1186–1193. [CrossRef]

58. Sadremomtaz, A.; Mansouri, K.; Alemzadeh, G.; Safa, M.; Rastaghi, A.E.; Asghari, S.M. Dual blockade of VEGFR1 and VEGFR2 by a novel peptide abrogates VEGF-driven angiogenesis, tumor growth, and metastasis through PI3K/AKT and MAPK/ERK1/2 pathway. *Biochim. Biophys. Acta* **2018**, *1862*, 2688–2700. [CrossRef]

59. Tammela, T.; Zarkada, G.; Wallgard, E.; Murtomäki, A.; Suchting, S.; Wirzenius, M.; Waltari, M.; Hellström, M.; Schomber, T.; Peltonen, R.; et al. Blocking VEGFR-3 suppresses angiogenic sprouting and vascular network formation. *Nature* **2008**, *454*, 656–660. [CrossRef]

60. Zhao, Y.; Adjei, A.A. Targeting Angiogenesis in Cancer Therapy: Moving Beyond Vascular Endothelial Growth Factor. *Oncologist* **2015**, *20*, 660–673. [CrossRef]

61. Qin, S.; Li, A.; Yi, M.; Yu, S.; Zhang, M.; Wu, K. Recent advances on anti-angiogenesis receptor tyrosine kinase inhibitors in cancer therapy. *J. Hematol. Oncol.* **2019**, *12*, 27. [CrossRef] [PubMed]

MDPI
St. Alban-Anlage 66
4052 Basel
Switzerland
Tel. +41 61 683 77 34
Fax +41 61 302 89 18
www.mdpi.com

Cells Editorial Office
E-mail: cells@mdpi.com
www.mdpi.com/journal/cells